# Springer Series in Computational Neuroscience

**Computational Neuroscience** gathers monographs and edited volumes on all aspects of computational neuroscience, including theoretical and mathematical neuroscience, biophysics of the brain, models of neurons and neural networks, and methods of data analysis (e.g. information theory). The editors welcome suggestions and projects for inclusion in the series.

**About the Editors:** Alain Destexhe is Research Director at the Centre National de la Recherche Scientifique (CNRS), France, and Romain Brette is Research Director at the Institut National de la Santé et de la Recherche Médicale (INSERM), France.

More information about this series at http://www.springer.com/series/8164

Boris I. Prilutsky • Donald H. Edwards
Editors

# Neuromechanical Modeling of Posture and Locomotion

Springer

*Editors*
Boris I. Prilutsky
Georgia Institute of Technology
Atlanta
Georgia
USA

Donald H. Edwards
Georgia State University
Atlanta
Georgia
USA

ISSN 2197-1900 ISSN 2197-1919 (electronic)
Springer Series in Computational Neuroscience
ISBN 978-1-4939-3266-5 ISBN 978-1-4939-3267-2 (eBook)
DOI 10.1007/978-1-4939-3267-2

Library of Congress Control Number: 2015948177

Springer New York Heidelberg Dordrecht London


Printed on acid-free paper

Springer Science+Business Media LLC New York is part of Springer Science+Business Media (www.springer.com)

# Preface

Neuromechanics is a new, quickly growing field of neuroscience research that merges neurophysiology, biomechanics and motor control and aims at understanding motor behavior of living systems through interactions between their neural and mechanical dynamic properties. Although research in Neuromechanics is not limited by computational approaches, neuromechanical modeling is a powerful tool that allows for integration of massive knowledge gained in the past several decades of organization of motion related brain and spinal cord activity, various body sensors and reflex pathways, muscle mechanical and physiological properties and detailed quantitative morphology of musculoskeletal systems. Recent work in neuromechanical modeling, featured in this book, has demonstrated advantages of such an integrative approach and led to discoveries of new emergent properties of neuromechanical systems.

This book is largely based on the presentations at the workshop Neuromechanical Modeling of Posture and Locomotion, which was a part of the Computational Neuroscience Symposium in Atlanta and Decatur, Georgia, USA in July 2012. The goals of this workshop were to bring together neuromechanics researchers, discuss new developments in the field of neuromechanical modeling and inform computational neuroscience community of this new and exciting area of research. The goals of this book are similar. This book is the first to present a comprehensive and diverse collection of neuromechanical modeling studies of posture and locomotion.

It seems natural that neuromechanical modeling sprang from studies of neural control of posture and locomotion. Over the last 100 years, increasingly more sophisticated experimental paradigms, animal preparations and recording methods have been developed to obtain quantitative information about the mechanical, morphological and physiological properties of the musculoskeletal system, on the one hand, and about the biophysics and physiology of neurons, neuronal networks, reflex pathways and neural pattern generating circuits that control the musculoskeletal system, on the other hand. Neuromechanical modeling appears to be a very useful tool for integrating this massive body of information in models that recreate complex motor behaviors and reveal the mechanisms by which these behaviors emerge.

This book has several unique features. It is the first book on the topic of neuromechanics in general and on neuromechanical modeling of posture and locomotion specifically. The majority of research groups working in the area of neuromechanical modeling contributed chapters to the book. The book covers a wide range of topics from theoretical studies linking the organization of reflex pathways and central pattern generating circuits with morphology and mechanics of the musculoskeletal system (Nichols et al., Chap. 3; Burkholder, Chap. 4 and Shevtsova et al., Chap. 5) to detailed neuromechanical models of postural and locomotor control (Bunderson and Bingham, Chap. 1; Markin et al., Chap. 2 and Aoi, Chap. 8). Furthermore, diverse modeling approaches are presented in the book including theoretical analyses of muscle non-linear transformations of neural signals (Burkholder, Chap. 4; Hooper et al., Chap. 6; Peterka, Chap. 9), detailed neuromechanical modelling incorporating multi-joint musculoskeletal models with afferent feedback signals and central pattern generating networks (Bunderson and Bingham, Chap. 1 and Markin et al., Chap. 2), theoretical and computational dynamic analyses of activity regime transitions in a multistable half-center oscillator (Bondy et al., Chap. 12), and others.

We hope this book will be a useful contribution to the field, inform the computational neuroscience community of this relatively new area of research and help attracting new talented students and researchers.

# Contents

**Part I Innovative Modeling Approaches in Neuromechanical Research**

1 **Better Science Through Predictive Modeling: Numerical Tools for Understanding Neuromechanical Interactions** ........ 3
Nathan E. Bunderson and Jeffrey Bingham

2 **A Neuromechanical Model of Spinal Control of Locomotion** ........ 21
Sergey N. Markin, Alexander N. Klishko, Natalia A. Shevtsova, Michel A. Lemay, Boris I. Prilutsky and Ilya A. Rybak

**Part II Organization of Afferent Signals, Central Neural Circuits and the Musculoskeletal System: Insights from Neuromechanical Modeling**

3 **Neural Regulation of Limb Mechanics: Insights from the Organization of Proprioceptive Circuits** ........ 69
T. Richard Nichols, Nathan E. Bunderson and Mark A. Lyle

4 **Model-Based Approaches to Understanding Musculoskeletal Filtering of Neural Signals** ........ 103
Thomas J. Burkholder

5 **Modeling the Organization of Spinal Cord Neural Circuits Controlling Two-Joint Muscles** ........ 121
Natalia A. Shevtsova, Khaldoun Hamade, Samit Chakrabarty, Sergey N. Markin, Boris I. Prilutsky and Ilya A. Rybak

6 **Muscles: Non-linear Transformers of Motor Neuron Activity** ........ 163
Scott L. Hooper, Christoph Guschlbauer, Marcus Blümel, Arndt von Twickel, Kevin H. Hobbs, Jeffrey B. Thuma and Ansgar Büschges

**Part III Neuromechanical Modeling of Posture and Postural Control**

7 **Why Is Neuromechanical Modeling of Balance and Locomotion So Hard?** ........ 197
Jessica L. Allen and Lena H. Ting

8 **Neuromusculoskeletal Modeling for the Adaptive Control of Posture During Locomotion** ........ 225
Shinya Aoi

9 **Model-Based Interpretations of Experimental Data Related to the Control of Balance During Stance and Gait in Humans** ........ 245
Robert J. Peterka

**Part IV Neuromechanical Modeling of Locomotion**

10 **Computing Motion Dependent Afferent Activity During Cat Locomotion Using a Forward Dynamics Musculoskeletal Model** ........ 273
Boris I. Prilutsky, Alexander N. Klishko, Douglas J. Weber and Michel A. Lemay

11 **Modeling and Optimality Analysis of Pectoral Fin Locomotion** ........ 309
Xinmin Liu, Frank Fish, R. Scott Russo, Silvia S. Blemker and Tetsuya Iwasaki

12 **Control of Cat Walking and Paw-Shake by a Multifunctional Central Pattern Generator** ........ 333
Brian Bondy, Alexander N. Klishko, Donald H. Edwards, Boris I. Prilutsky and Gennady Cymbalyuk

**Index** ........ 361

# Contributors

**Jessica L. Allen** Coulter Department of Biomedical Engineering at Georgia Institute of Technology and Emory University, Atlanta, GA, USA

**Shinya Aoi** Department of Aeronautics and Astronautics, Graduate School of Engineering, Kyoto University, Kyoto, Japan

**Ansgar Büschges** Zoologisches Institut, Universität zu Köln, Cologne, Germany

**Jeffrey Bingham** School of Mechanical Engineering, Georgia Institute of Technology, Atlanta, GA, USA

**Marcus Blümel** Institut für Biologie und ihre Didaktik, Freiburg, Germany

**Silvia S. Blemker** Biomedical Engineering, University of Virginia, Charlottesville, VA, USA

**Brian Bondy** Neuroscience Institute, Georgia State University, Atlanta, GA, USA

**Nathan E. Bunderson** School of Applied Physiology, Georgia Institute of Technology, Atlanta, GA, USA

Autonomous Solutions, Mendon, UT, USA

**Thomas J. Burkholder** School of Applied Physiology and Interdisciplinary Bioengineering Program, Georgia Institute of Technology, Atlanta, GA, USA

**Samit Chakrabarty** School of Biomedical Sciences, Faculty of Biology, University of Leeds, Leeds, West Yorkshire, UK

**Gennady Cymbalyuk** Neuroscience Institute, Georgia State University, Atlanta, GA, USA

**Donald H. Edwards** Neuroscience Institute, Georgia State University, Atlanta, GA, USA

**Frank Fish** Department of Biology, Liquid Life Laboratory, West Chester University, West Chester, PA, USA

**Christoph Guschlbauer** Zoologisches Institut, Universität zu Köln, Cologne, Germany

**Khaldoun Hamade** Department of Neurobiology and Anatomy, Drexel University College of Medicine, Philadelphia, PA, USA

**Kevin H. Hobbs** Department of Biological Sciences, Irvine Hall, Athens, OH, USA

**Scott L. Hooper** Department of Biological Sciences, Irvine Hall, Athens, OH, USA

**Tetsuya Iwasaki** Mechanical and Aerospace Engineering, University of California, Los Angeles, CA, USA

**Alexander N. Klishko** School of Applied Physiology, Center for Human Movement Studies, Georgia Institute of Technology, Atlanta, GA, USA

**Michel A. Lemay** Department of Bioengineering, Temple University, Philadelphia, PA, USA

**Xinmin Liu** Mechanical and Aerospace Engineering, University of California, Los Angeles, CA, USA

**Mark A. Lyle** School of Applied Physiology, Georgia Institute of Technology, Atlanta, GA, USA

**Sergey N. Markin** Department of Neurobiology and Anatomy, Drexel University College of Medicine, Philadelphia, PA, USA

**T. Richard Nichols** School of Applied Physiology, Georgia Institute of Technology, Atlanta, GA, USA

**Robert J. Peterka** Department of Biomedical Engineering, Oregon Health & Science University, Portland, OR, USA

**Boris I. Prilutsky** School of Applied Physiology, Center for Human Movement Studies, Georgia Institute of Technology, Atlanta, GA, USA

**R. Scott Russo** Malvern, PA, USA

**Ilya A. Rybak** Department of Neurobiology and Anatomy, Drexel University College of Medicine, Philadelphia, PA, USA

**Natalia A. Shevtsova** Department of Neurobiology and Anatomy, Drexel University College of Medicine, Philadelphia, PA, USA

**Jeffrey B. Thuma** Department of Biological Sciences, Irvine Hall, Athens, OH, USA

**Lena H. Ting** Coulter Department of Biomedical Engineering at Georgia Institute of Technology and Emory University, Atlanta, GA, USA

**Arndt von Twickel** Zoologisches Institut, Universität zu Köln, Cologne, Germany

**Douglas J. Weber** Department of Physical Medicine and Rehabilitation, Department of Bioengineering, University of Pittsburgh, Pittsburgh, PA, USA

# About the editors

**Boris I. Prilutsky PhD** received his BS degrees in physical education and applied mathematics/mechanics from Central Institute of Physical Culture in Moscow, Russia and Moscow Institute of Electronic Engineering, respectively, and a PhD in animal movement biomechanics and physiology from Latvian Research Institute of Traumatology and Orthopedics. He is currently an associate professor in the School of Applied Physiology and director of Biomechanics and Motor Control laboratory at the Georgia Institute of Technology and an adjunct associate professor in the Division of Physical Therapy at Emory University School of Medicine. His research interests are biomechanics and neural control of normal and pathological movement.

**Donald H. Edwards PhD** received a B.S. degree in electrical engineering from the Massachusetts Institute of Technology and a Ph.D. in neurobiology from Yale University. He studied sensori-motor integration as a postdoctoral research associate with Donald Kennedy at Stanford University and with Brian Mulloney at University of California, Davis. He joined the faculty at Georgia State University as Assistant Professor of Biology and is currently Regents' Professor of Neuroscience. His research interests include sensori-motor integration, neuromechanics, and the neural control of behavior.

# Part I
# Innovative Modeling Approaches in Neuromechanical Research

# Chapter 1
# Better Science Through Predictive Modeling: Numerical Tools for Understanding Neuromechanical Interactions

**Nathan E. Bunderson and Jeffrey Bingham**

**Abstract** The animal kingdom is filled with amazing examples of coordinated locomotor and balance behavior. The intricate interaction of the neuromechanics of the combined skeletal, muscular, and neural systems that underlie these behaviors only adds to their impressiveness. To wit, the neuromechanics must deal with fantastically nonlinear dynamics, delayed and noisy sensory input, and multiple stability regimes in unpredictable environments. Because of these underlying complex interactions, an integrative systems approach is required to understand the performance of the locomotor and balance behavior that emerges. In this chapter, we propose the use of predictive modeling to facilitate the investigation of neuromechanics using our software platform, Neuromechanic. With this technique the dynamics of constituent neuromechanical systems are modeled and the resulting emergent behaviors studied; holistic behaviors are an output rather than an input for simulation. We describe three ways in which software can aid in a predictive approach to neuromechanical modeling: first, use of tools that emphasize control and optimization for predictive modeling; second, visualization and organization to aid in careful parameterization necessary to account for the variation found in biological specimens; third, building confidence in modeling results through the use of sensitivity analysis. We offer examples of these techniques using Neuromechanic, which is designed to simplify the prototyping of neural control strategies, formulate optimization criteria, visualize key parameters that effect model performance, and succinctly perform sensitivity analysis.

**Keywords** Optimization · Control · Neuromechanics · Stability · Simulation · Biomechanics · Sensitivity

---

N. E. Bunderson (✉)
School of Applied Physiology, Georgia Institute of Technology,
555 14th St NW, Atlanta, GA 30332-0356, USA
e-mail: nbunderson@gatech.edu

J. Bingham
School of Mechanical Engineering, Georgia Institute of Technology,
313 Ferst Drive, Atlanta, GA 30332, USA
e-mail: bingjeff@gmail.com



B. I. Prilutsky, D. H. Edwards (eds.), *Neuromechanical Modeling of Posture and Locomotion*, Springer Series in Computational Neuroscience,
DOI 10.1007/978-1-4939-3267-2_1

## 1.1 The Need for and Value of Predictive Neuromechanical Models of Posture and Locomotion

The pirouette of a dancer, the leap of a receiver catching a football or a toddler clumsily taking her first steps are examples of the sophisticated interaction of the neural and musculoskeletal systems. We call these interactions neuromechanics and they are the bases for movement and balance. Our neuromechanics may evolve as we grow into adulthood, train for a sport, or suffer from injury or disease. Understanding human neuromechanics gives insight into how we are able to achieve grace and efficiency in our movements and offers a way to improve human health. Applying this knowledge to the fields of rehabilitation, robotics, and prosthetics will undoubtedly lead to better fitness training, new methods of injury prevention, improved treatment of neuromusculoskeletal disorders, and better engineered robotic systems.

A tremendous amount of knowledge about the neural, skeletal and muscular systems has come through reductionism and observation (Sherrington 1910; Liddell and Sherrington 1924; Fitts 1954; Huxley 1957; Gordon et al. 1966). However, significantly less advancement has been made in understanding the interplay of these systems during functional behavior. This is in part due to the complexity of each system, but also because the systems are strongly interdependent. Computational modeling is becoming an increasingly powerful tool to analyze these interdependencies and is commonly used to describe the behavior of constituent systems. Models have an advantage over experimentation because they allow complete control over the level of complexity of each system and over how the systems are combined. Also, models can be used to generate a considerably richer set of data for analysis that would likely be prohibitive with physical experimentation. However, *descriptive* modeling is not sufficient to advance science and does not take advantage of the opportunities of a virtual environment.

Compared with descriptive modeling, predictive modeling is less interested in deconstructing a particular behavior than in providing a prediction for *how* the behavior emerges from the interdependent neuromechanical systems. A predictive model can be used to provide constructive arguments, offering an additional logic tool for exploring a particular hypothesis. The constructive nature of predictive modeling often gives important insight about the nature of the problems faced and solved by neuromechanical systems even if the results do not directly explain how they are solved. Properly utilized a predictive model can be used to develop theories and inform the design of specific physical experimentation.

Making predictions about how postural and locomotor behaviors emerge requires greater emphasis on developing neural control theories. Excellent control strategies and optimization techniques have been developed for joint torque-based biomimetic robotic simulations (Brock and Khatib 2002; Jain et al. 2009; Coros et al. 2010; Erez et al. 2013). While these strategies have achieved a remarkable diversity of behaviors in diverse environments and contexts (as do neuromechanical systems) they make little or no attempt to provide an implementation framework for a neuromechanical system. They are nevertheless important to neuromechanical predictive modeling since they directly address the question of how robust biomimetic behaviors are generated. Others have attempted to generate control strategies which incorporate muscle models and a variety of neural control structures including reflexive mechanisms (Welch and Ting 2008; Geyer and Herr 2010; Bingham et al. 2011; Geijtenbeek et al. 2013) 10/8/2015, central pattern generators (Ijspeert 2008; Markin et al. 2010), basal ganglia (Tomita and Yano 2007; Sarvestani et al. 2013), and the cerebellum (Jo and Massaquoi 2004, 2007). While these are generally not able to achieve the robust behavior of the more abstract biomimetic robotics simulations and are usually implemented with dramatically simplified mechanics they do generate emergent behaviors based on underlying neuromechanical principles.

## 1.2 What Tools Facilitate Predictive Neuromechanical Modeling?

Neuromechanical models have been implemented using a variety of software tools including Matlab (Mathworks, Natick Ma), OpenSim (Delp et al. 2007), MuJoCo (Todorov et al. 2012), AnimatLab (Cofer et al. 2010), DART (Bai et al. 2012), MSMS (Davoodi and Loeb 2012), and Neuromechanic (Bunderson et al. 2012). The goal of this paper is not to discuss the relative merits of each of these packages. Our purpose is to describe how Neuromechanic is being developed specifically to facilitate predictive neuromechanical modeling. There are three primary considerations in predictive neuromechanical modeling that we are incorporating into the design of Neuromechanic.

First, as discussed previously, predictive neuromechanical modeling requires an emphasis on control and optimization. Neuromechanic is designed to simplify the prototyping of neural control strategies and optimization. In the first section we will describe how an integrated control-oriented approach, which allows for emergent rather than pre-determined behavior, has greater potential for taking neuromechanical modeling from description to prediction.

Second, model parameterization for neuromechanical systems involves greater care and effort than for engineered systems due to the natural variance of parameters in biological systems and the difficulty of accurately measuring many of these parameters. Neuromechanic has visualization tools for several key parameters which we have determined have the greatest effect on model performance. In the second section we describe the parameter visualization tools we are implementing in Neuromechanic that allow the modeler to quickly compare, and preliminarily validate, neuromechanical models.

The third aspect of the predictive approach is the ability to assess the reliability and robustness of modeling results by performing sensitivity analyses. Neuromechanical systems are extremely variable and the behavior is amazingly robust. Any neuromechanical control paradigm that is highly sensitive to parameter variation should be viewed with a great deal of suspicion. Further, conclusions based on the results of a single model, not to mention a single simulation, do not characterize robustness and are missing the great opportunity mathematical modeling provides to explore the parameter space. In the third section we discuss some sensitivity analysis tools which have been implemented in Neuromechanic.

## 1.3 Developing Control Strategies in Neuromechanic

In the descriptive framework, the model is generally used to answer "What …" questions, as in, "What is the force contribution of muscle X to the vertical ground reaction force during vertical leaping?". These types of questions can and often do generate predictions but they rely heavily on predetermined behaviors. On the other hand in the predictive approach behaviors are an output rather than input and research questions are framed as "Can …" or "How …", as in, "How high can my model jump?". Where the descriptive questions tend to be reductive, predictive questions tend to be holistic. There is no doubt that asking the reductive questions has been and will continue to be a very productive line of questioning with neuromechanical models. However, there is a strong case to be made for asking holistic questions of neuromechanical models that predict instead of describe behavior. Furthermore, even if a model is being used to answer reductive questions it is still useful and informative to know how it performs holistically (How high can it jump? How fast can it run? How stable is it against external perturbations?) before asking the important detailed reductive questions. The implication for a neuromechanical modeling package is that tools that assist in answering the holistic questions can be extremely useful for predictive modeling.

So what are the tools we need to be able to answer holistic questions such as "How high can a model jump?" Since the predictive approach requires the generation of new behaviors it requires an emphasis on optimization tools and control strategies to generate those behaviors. In the example of maximal vertical leaping we need a control strategy that produces leaping and a process to optimize the parameters of that control strategy to obtain maximum performance. This process may be enhanced by kinematic and kinetic data from experimentally measured behaviors but does not rely on it. We present one example of this approach in another section. Neuromechanic has several tools and control-oriented design features to facilitate predictive modeling.

### *1.3.1 Python Application Programmer's Interface and Tree-Like Data Structure*

With its focus on control and optimization, predictive simulation requires that the neuromechanical modeling package have a scripting environment that allows the modeler to quickly implement and test control and optimization strategies.

The scripting environment should be extensible, have a mathematical toolset, and give the modeler access to all the relevant parameters and dynamic variables of the model. Neuromechanic comes with Python as the scripting environment and includes the scientific computing package numpy. All parameters of the model are available from the Python scripting environment through a model tree that mirrors the input format for intuitive access (Fig. 1.1).

The scripting tools were used to construct a virtual prosthesis environment and prototype a new multi-functional prosthetic controller (Bunderson 2014). Able-bodied subjects and a shoulder disarticulation subject successfully used the system to complete a virtual box and blocks task. An early experimental setup, virtual environment, and the control structure for the virtual prosthesis are shown in Fig. 1.2. A command signal, generated from real-time processing of electromyograms (EMG)

***Input file in XML format (e.g. 2392.nmcb)***

```
<NeuromechanicFile>
 ...
 <Bodies>
  <RigidBody Name="pelvis">
   <FrameLocation Parent="ground">.0 .0 .0</FrameLocation>
   <Mass>11.7</Mass>
   <Inertia>.1028 .0871 .0579 0. 0. 0.</Inertia>
   <CenterOfMass>-.0707 .0 .0</CenterOfMass>
  </RigidBody>
  <RigidBody Name="left_femur">
   ...
  </RigidBody>
 ...
 </Bodies>
<NeuromechanicFile>
```

***Accessing data with python API:***

```
>>>import nmextension as nme
>>>nmtree = nme.Tree('model')
>>>print(nmtree.NeuromechanicFile.Bodies.RigidBody['Name=pelvis'].Mass.x_)
11.7
```

**Fig. 1.1** Example input file and access commands to model parameters through the Python Application Programmer's Interface (API). Model parameters and dynamic variables can be accessed from the API with a tree class that matches the input XML file format

**Fig. 1.2** A virtual prosthesis was developed in Neuromechanic and four able-bodied subjects and one shoulder disarticulation subject were able to complete a virtual box and blocks task using real-time EMG control signals

in a separate process, was retrieved in Neuromechanic through shared memory resources. The controller converts the command signal into the joint torques necessary to drive the virtual prosthesis in Neuromechanic in real time. The joint torque calculation relies on dynamic variables (such as frictional contact forces and rigid body dynamics) accessed from the model tree. The virtual prosthesis and box and blocks task implemented in Neuromechanic are extensible and modifiable and can serve as a platform and baseline to quickly and cheaply prototype prosthetic arms and control strategies.

### *1.3.2 Linearization and Eigenanalysis Tools*

Stability is an essential feature of control and Neuromechanic has built in tools for assessing the local stability of a neuromechanical system. A model can be linearized about a state and neuromuscular parameters calculated to equilibrate dynamic variables in that state. In a previous study (Bunderson et al. 2008), we found that the classical muscle redundancy problem was reduced when a stability criterion (in addition to the endpoint force criteria) was used to choose patterns of muscle activation. We used the linearization and Eigenanalysis tools in Neuromechanic and found that the instabilities most likely to occur in the cat hindlimb are in the sagittal plane (Fig. 1.3). By comparing the local (joint-level) stiffness of muscles to the stability of the limb modes we were able to identify destabilizing muscles and construct an optimization cost function to enhance limb stability. This optimization cost function can be used in constructing controllers for locomotion and posture to predict behavior.

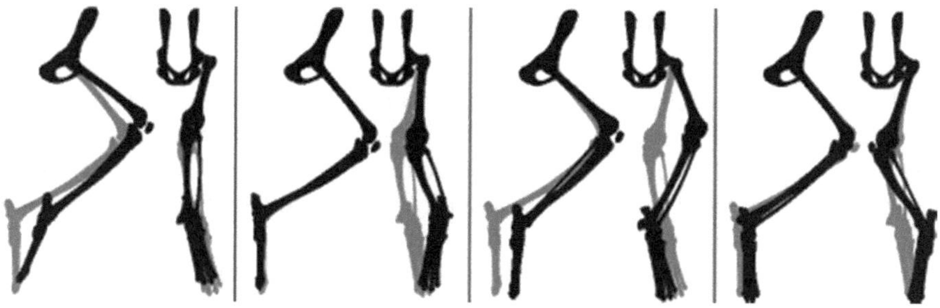

**Fig. 1.3** Four dominant modes were found from the hindlimb stability analysis ranging from most unstable (*left*) to most stable (*right*)

## 1.4 Visualization Tools for Developing Model Parameter Intuition

The parameterization of neuromechanical models poses several difficulties. To begin with, the higher the fidelity at which the model represents the neuromechanical system the more parameters are introduced and the more difficult it is for the modeler to track parameter values. Moreover, because of the inherent variability in biological systems there may not be a single "correct value" for a particular parameter. In Neuromechanic we are developing tools to more completely visualize and "gain a feel for" the parameterization of the animal.

Most neuromechanical modeling packages have visualization tools for a limited set of the parameters of the system. For example most packages provide a visual representation of the global location of important points, vectors, and surfaces such as joint axes, muscle attachment points, and muscle wrapping surfaces. Most packages also render the muscles in colors that reflect the activation level of the muscle which can give a fast and intuitive impression of muscle coordination. Often polygon meshes representing bone surfaces or skin are also included (however, these usually do not represent any of the actual parameters that describe the model and are included for reference only). Neuromechanic also provides these visualizations but has additional tools to give the modeler an intuitive feel for other parameters which, in our experience, have the greatest impact on simulations. This includes the parameters directly impacting (1) the force generating capacity of the muscles, (2) the moment arm of the muscles and (3) the inertia of the body segments. To demonstrate these tools we have included figures generated from a Neuromechanic implementation of the 2392 human lower limb model that is included in the OpenSim (Delp et al. 2007) distribution. The name for the model derives from the fact that it models 23 kinematic degrees of freedom and 92 muscles from the lower body (legs and lumbar region). The head, arms, and trunk (HAT) are modeled as a single rigid body attached to the pelvis with a spherical (3 rotational degrees of freedom) joint.

### 1.4.1 Visualizing the Force Generating Capacity of Muscle

Hill-type muscle models, where the force produced by a muscle is a function of its length, velocity and activation and is scaled by a maximum isometric force ($F_{max}$), are commonly used in neuromechanical modeling. Zajac expanded a basic Hill-type model to incorporate first order activation dynamics and the effects of pennation (Zajac 1989). His model is the basis for the two primary models incorporated in Neuromechanic. In both of these models the maximum isometric force that can be generated by a muscle is given by $F_{max}a_{max}\theta$ where $a_{max}$ represents the maximum activation level of the muscle (almost always 1) and $\theta$ represents the pennation angle. Since any neural control strategy depends critically on the ability of muscles to generate force, Neuromechanic provides a visual representation of this maximum force. Muscles are rendered in a fusiform shape (Fig. 1.4) with a maximum cross-sectional area ($A_{max}$) at the midpoint of the muscle lengthwise which is defined by

$$A_{max} = \frac{F_{max}\, a_{max} cos(\theta)}{\sigma}$$

where the proportionality constant $\sigma$ is a global parameter (i.e. the same value for all muscles in Neuromechanic) which represents the specific tension of muscle. This can be set by the user and is 22.5N/cm$^2$ by default.

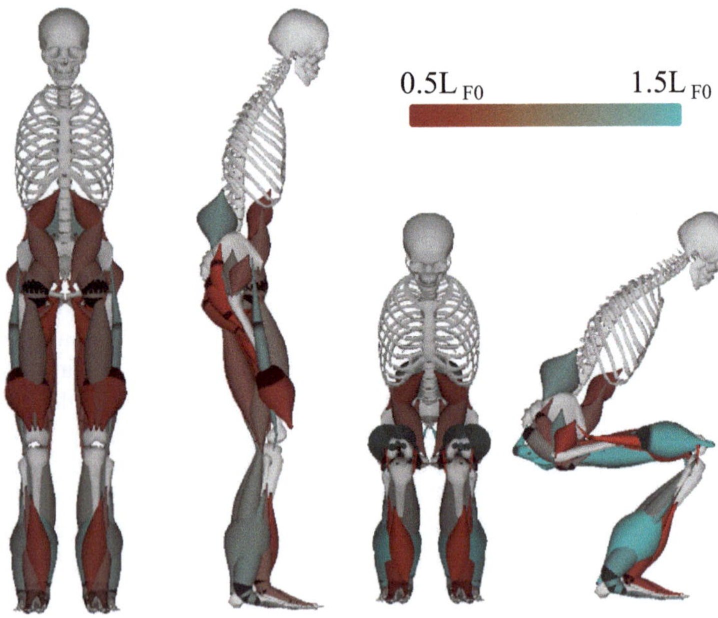


**Fig. 1.4** The force generating capacity of muscle can be visualized in Neuromechanic by rendering the muscles in a fusiform shape. The maximum cross-sectional area is proportional to the maximum force that can be produced by the muscle. The color of the muscle can be chosen as a function of the operating length upon which the force generated by muscle depends critically

The force generated by a muscle also depends strongly upon the operating length of a muscle due primarily to variations in actin-myosin overlap of the sarcomeres. The operating length of muscle is not a parameter but a variable that depends on several other time dependent variables (the configuration or joint positions of the animal and fiber lengths), several parameters (the location of the muscle attachment and via points, muscle wrapping surfaces, the number of sarcomeres, pennation angle, and tendon slack lengths), and the particular muscle model in use. In Neuromechanic muscles can be rendered with colors that reflect the current normalized operating length of the muscles giving the modeler immediate insight into whether the muscles will be able to generate appropriate levels of force. In Fig. 1.4 the color of the muscles indicate their respective normalized operating lengths in an inelastic tendon model. The lengths vary from 50 % shorter (red) to 50 % longer (cyan) than the optimal operating length ($L_{F0}$). At the nominal upright posture on the left of Fig. 1.4 Vastus lateralis of the 2392 model operates at 52 % of optimal fiber length while at the squatting position shown on the right it is operating at 132 % of optimal fiber length.

### 1.4.2 Visualizing the Moment Arms of Muscle

Muscle moment arms also strongly impact the performance of neuromechanical models since muscle forces affect the dynamics of the body only after being scaled by the moment arms of the muscles at each joint. Like operating length, moment arms are state dependent variables rather than parameters. They depend on the muscle attachment point parameters, muscle wrapping surface parameters, and joint definition parameters. In particular when attachment points or wrapping surfaces are in close proximity to joint axes small changes in their location can have a large impact on the moment arm. We have implemented a moment arm visualization tool which scales and renders the moment arms at the joints as points along the joint axis. The moment arm of the Rectus femoris muscle in a standing and squatting posture is shown in Fig. 1.5a. The cyan line goes through the hip flexion/extension axis and the extent of the line to the right and left correspond to a 10x scaling of the maximum flexion and extension moment arm of *all* muscles at that joint. The cyan circle represents a 10x scaling of the Rectus femoris moment arm at that joint. The tool indicates that while Rectus femoris is a primary flexor of the hip at the upright posture its moment arm is relatively decreased in the squatting posture. The red line and circle reflect the scaled moment arm of Rectus femoris at the knee flexion/extension joint. Other packages such as OpenSim have excellent moment arm plotting tools which we are also incorporating into Neuromechanic to complement this visualization tool.

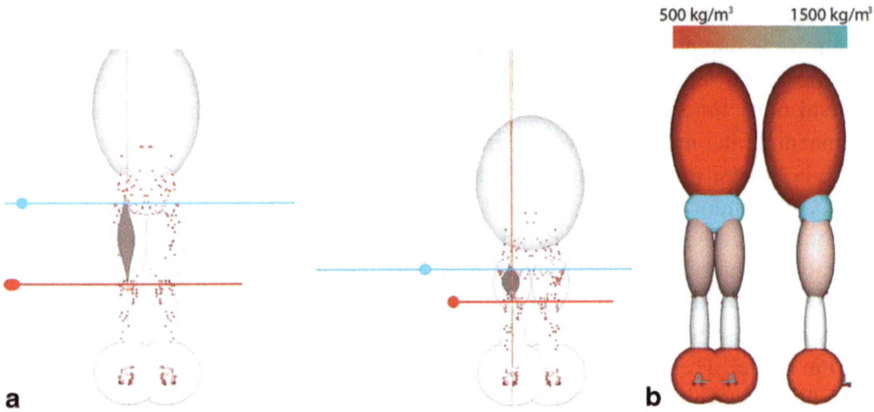


**Fig. 1.5** **a** The moment arms of the Rectus femoris muscle of the 2392 OpenSim model (Delp et al. 2007) are represented visually as points along the joint axes. The extent of the cyan line represents the maximum moment arms from all muscles for hip flexion and extension in the shown posture. The cyan dot represents the moment arm of the selected muscle (Rectus femoris) in that posture. The *red* is for knee flexion/extension. **b** The seven inertial parameters of the rigid body segments in the 2392 OpenSim model are represented in Neuromechanic with shaded ellipsoids

### *1.4.3 Visualizing the Inertia of Body Segments*

The dynamics of movement also depend fundamentally on the inertia of the body. Neuromechanic, like most neuromechanical modeling packages uses rigid body dynamics where the inertia of each segment is completely described by seven parameters including mass (translational inertia) and six components of rotational inertia. These seven parameters can be uniquely transformed into an equivalent inertial ellipsoid representation which can be easily visualized. In the equivalent inertial ellipsoid representation the seven parameters can be visualized as the size (3 parameters), orientation (3 parameters), and density (1 parameter) of an ellipsoid. If the average density of the animal is known then the modeler can quickly get a feel for the appropriateness of the inertial parameters by rendering the ellipsoid with color corresponding to density (Fig. 1.5b). In Fig. 1.5 the bones and most muscles of the 2392 model are not rendered so as not to obfuscate the information conveyed by the rendered inertial ellipsoids. Most segments of the 2392 model are between 50 % below and 50 % above the average density of human body (1062 kg/m^3) with the HAT (head, arms, trunk) and pelvis lying at the extremes of this range. Two segments, the talus and toes, saturate the scale with densities of less than 10kg/m^3 for the talus (large spherical red ellipsoid near the ankle) and 6700 kg/m^3 for the toes (small very thin cyan ellipsoid).

We emphasize that none of these tools are meant to give final validation to the choice of parameters. We believe however that they are very useful for a quick appraisal of general appropriateness of the parameters that will have the greatest effect on performance of the model. In the case of the 2392 model this visualization suggests that some inertial parameters need revisiting.

## 1.5 Strategies and Tools for Determining Model Sensitivity

Ultimately the modeler must assess the reliability of model results. Modeling results are used to answer questions like: Is it safe to use a particular physical design of a prosthesis or orthosis? Is it safe to promote a gait modification to alleviate joint stress? Is it safe to perform a tendon transfer surgery? Due to the variety of anatomical structure and variation in observed behavior it is necessary to quantify how model parameters affect simulated behavior to gain confidence in model predictions. In short, the model prediction is only as good as the model itself, and the model is dependent on the accuracy of its parameters and the verisimilitude of its dynamics. Therefore, to vette a particular model often requires determining how accurate a parameter must be measured or predicted to achieve a reasonable variation in simulated behavior. Furthermore, it is also necessary to discern whether variations in observed behavior are the result of differences in parameters or actual differences in the behavior itself. Quantifying the sensitivity of behavior to changes in parameters also helps to identify parameters that are most important for a particular behavior and may give insight into the underlying neuromechanical interactions.

Quantifying sensitivity of neuromechanical systems is challenging due to their non-linear nature and multitude of parameters. However, using dynamical systems theory the stability of the neuromechanical system for a particular behavior and parameter set can be used to identify sensitivity of behavior to parameter changes. Therefore, Neuromechanic implements methods to determine equilibrium behavior, linearization of the full neuromechanical system and eigenvalue analysis. These tools have been discussed here in a previous section.

In Neuromechanic we have also implemented the stability radius technique as a tool for identifying different sets of parameter values that produce similar behavior, or quantifying the sensitivity of a behavior to parameter changes. Briefly, stability radius is a technique from robust control theory that produces a single number that measures the smallest change to any system parameter that would result in unstable behavior. This technique has advantages over other methods in that it is possible to compare dynamical behavior as parameters change for one system against itself, as well as against entirely different systems.

As an example, it is possible to use stability radius to compare the stability of different equilibrium postures during a station-keeping task. Here we use a model of a cat hind-limb having seven degrees-of-freedom and 31 Hill-type muscles (Burkholder and Nichols 2004) starting in a nominal configuration. The pelvis was held fixed and the leg was allowed to hang against gravity and minimum muscle activation was solved to hold the limb in the desired configuration. The hip angle was varied $-30$ to $+10°$ from the nominal posture.

Stability radius was computed for each posture by linearizing this system about the equilibrium state, $\tilde{x}$, to form the linearized system with a system matrix, **A**:

$$\dot{\tilde{x}} = \boldsymbol{A}\tilde{x}$$

The resulting characteristic equation of the system can be written as:

$$C.E. = z\boldsymbol{I} - \boldsymbol{A}$$

Finally, the stability radius, $r$, was calculated using the following relation where, $z \in \mathbb{C}, \Re(z) = 0$ implies $z$ is purely imaginary, and $\sigma_{min}$ was the operator that returned the smallest singular value:

$$r = \min_{\Re(z)=0} \sigma_{min}(z\boldsymbol{I} - \boldsymbol{A})$$

Stability radius was calculated for different biomechanical postures across hip angles. Across the range of postures the maximum stability radius ($r = 0.79$) was at a hip angle of $-4°$ (Fig. 1.6b). This posture was least sensitive to changes in system parameters and implies that this posture produced the most stable behavior in the presence of modeling errors, state estimate error, and external disturbances. A similar level of stability could also be found between two postures one more flexed and the other more extended. This similar level of stability can be used to predict similar dynamic responses across different postures (Bingham and Ting 2013).

## 1.6 Predictive Modeling in Neuromechanic: Vertical Leap Example

Finally, we provide a brief example of the predictive approach with an analysis of maximal vertical leaping. The analysis is not intended to be comprehensive but provides an approximation of the maximal leap behavior and predicts major

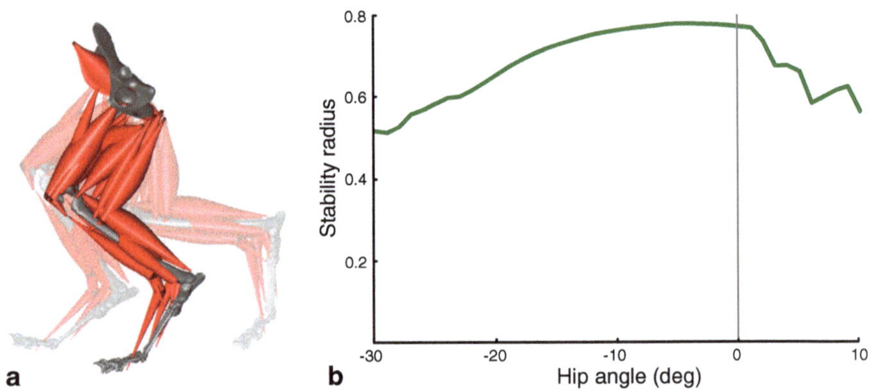


**Fig. 1.6** **a** Cat hind-limb in neutral posture and *greyed* out postures showing limits of hip angle. **b** Stability radius across different hip angles with maximum stability occurring near the nominal posture

influencing factors by relying on optimization techniques instead of experimental data. The optimization is implemented in a Python module and the necessary files to reproduce the data are available at www.neuromechanic.com.

The leaping behavior is generated using the 2392 model with immobilized lumbar flexion/extension and metatarsophalangeal joints. A sequence of postures spanning a vertical center of mass displacement of 60 cm were generated with each posture satisfying the condition that no horizontal plane displacements of the center of mass took place and no angular momentum is required to transition between postures. A single normalized velocity $\dot{v}_i$ was calculated at each posture as the average displacement between the current posture and the preceding and succeeding postures. Each normalized velocity was then replicated and scaled so that the vertical component of the center of mass velocity ranged between 0 and 4 m/s in intervals of 0.2 m/s. For each of the 2400 posture ($q_i$) and velocity ($\dot{q}_i$) combinations the maximum joint accelerations ($a_i$) that the model can produce which transitions the state to the "next" posture ($q_i+1$) was approximated by maximizing the vertical ground reaction force subject to:

$$\left[ diag\left(-2\left(q_{i+1}-q_i\right)\right) \quad \dot{v}_{i+1}^2 \right] \begin{bmatrix} a_i \\ \alpha^2 \end{bmatrix} = \dot{q}_i^2$$

where $\alpha$ is a free parameter that scales the dimensionless velocity $\left(\dot{v}_{i+1}\right)$ at the $q_{i+1}$ posture. The optimization was further constrained by the equations of motion (which relate muscle forces, endpoint forces, and joint accelerations) and the maximum and minimum forces that can be generated by each muscle. A Hill-type muscle model (Zajac 1989) was employed with an infinitely stiff series elastic element. The upper and lower muscle force bounds ultimately determine the maximum acceleration that can be generated and depend both upon instantaneous muscle kinematics defined by the particular posture/velocity combination and upon parameters of the muscle model including the active force-length curve.

To determine the sensitivity of maximal leap performance to the force/length relationship of muscle we calculated the maximal accelerations for each posture/velocity combination using three normalized force-length curve conditions called "constant", "Thelen", and "splined". In the "constant" condition the active force-length curve is a constant value of 1 so that the force generated by a muscle does not depend on its length. The "Thelen" condition uses the Gaussian shaped Thelen-muscle model (Thelen 2003) implemented in OpenSim which allows muscles to produce active force even when operating outside physiological ranges. The "splined" condition uses a splined fit to a dimensionless adaptation of the original isometric force-length relationship (Gordon et al. 1966).

The maximum vertical center of mass acceleration determined for each posture/velocity combination under each of the three force length curve conditions are shown as contour plots in Fig. 1.7. White space in the contour plots indicates a state space location for which no feasible set of muscle forces can bring to pass the desired vertical transition. The relationship between fiber length (normalized to

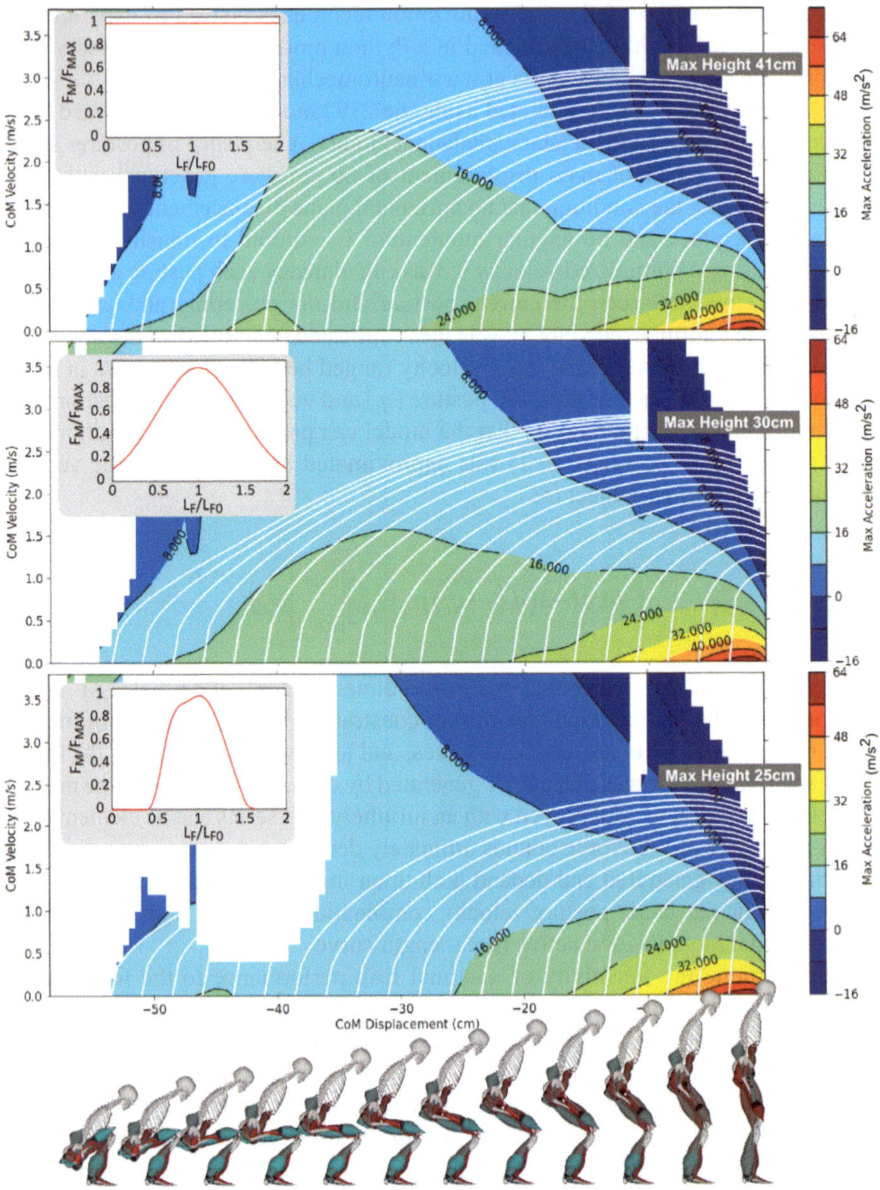


**Fig. 1.7** Maximum acceleration contour plots for the 2392 OpenSim model during a vertical displacement of the center of mass under three force-length relationships: constant (*top*), Gaussian (*middle*) as in Thelen et al. 2003, and a splined fit of the force-length relationship reported in Gordon et al. 1966 (*bottom*). *White lines* through the contour plots indicate a trajectory which begins at zero joint velocity and moves vertically upward through the trajectories. The maximum final velocity achieved near the upright posture would result in a maximum vertical displacement of 41, 30, and 25 cm for the constant, Thelen, and splined relationships respectively

optimal fiber length) and isometric force (normalized to maximum force) for each condition are plotted and inset into the corresponding contour plot. Trajectories that follow the maximum accelerations and begin from a variety of initial conditions (various initial postures and zero joint velocities) are shown as white lines over the contour plots. Dynamic counter movements (where the center of mass moves down before moving up) provide no benefit in this model since the muscles do not have series elasticity for energy storage.

The contours of the maximum accelerations (and correspondingly the maximum leap trajectory) are smaller for the splined force-length curve (25 cm) than for the Thelen force-length curve (30 cm) and the constant force-length curve (41 cm). These values are not necessarily the true global maximum for each condition since some constraints placed on the system are overly stringent. For example allowing small net moments and horizontal plane ground reaction forces might increase the leap height as may slightly different joint trajectories, unlocking the metatarsophalangeal joint, and, of course, modeling the energy storage capacity of the musculotendon unit. However since the maximal leap behavior was not dictated *a priori* by experimental data but, instead, discovered through optimization techniques each of these possible constraints may be relaxed providing new behaviors and testable predictions in each case.

While knowing the performance characteristics (e.g. how high can my model jump?) of the neuromechanical model is useful it is more useful still for the modeler to have an understanding of the sensitivity of those performance characteristics to the particular parameters used. In this case the variation in contour plots between the three force-length relationship conditions provides an understanding of how the choice of muscle model affects real performance. In particular we can see the range of limb postures where the Thelen-model allows state transitions while a more traditional force-length curve prohibits them. A more complete analysis would assess sensitivity to the parameters discussed previously.

## 1.7 Conclusion

To assist in predictive modeling we have developed a platform, Neuromechanic, that emphasizes an integrated, performance and control-oriented approach to neuromusculoskeletal modeling. We believe that the predict and test paradigm is an important step towards simulations that are capable of aiding in understanding performance and safety of workers, athletes and soldiers. These tools also help to form testable hypotheses for understanding the fundamental interactions between neural control and biomechanics as well as for designing better rehabilitation and preventative strategies for a wide range of injuries and neurological and musculoskeletal disorders.

## References

Bai YF, Siu K, Liu CKR (2012) Synthesis of concurrent object manipulation task. Acm Trans Graph 31

Bingham JT, Ting LH (2013) Stability radius as a method for comparing the dynamics of neuromechanical systems. IEEE Trans Neural Syst Rehabil Eng 21:840–848

Bingham JT, Choi JT, Ting LH (2011) Stability in a frontal plane model of balance requires coupled changes to postural configuration and neural feedback control. J Neurophysiol 106:437–448

Brock O, Khatib O (2002) Elastic strips: a framework for motion generation in human environments. Int J Robot Res 21:1031–1052

Bunderson NE (2014) Real-time control of an interactive impulsive virtual prosthesis. IEEE Trans Neural Syst Rehabil Eng 22:363–370

Bunderson NE, Burkholder TJ, Ting LH (2008) Reduction of neuromuscular redundancy for postural force generation using an intrinsic stability criterion. J Biomech 41:1537–1544

Bunderson NE, Bingham JT, Sohn MH, Ting LH, Burkholder TJ (2012) Neuromechanic: a computational platform for simulation and analysis of the neural control of movement. Int J Numer Meth BioMed Eng 28:1015–1027

Burkholder TJ, Nichols TR (2004) Three-dimensional model of the feline hindlimb. J Morphol 261:118–129

Cofer D, Cymbalyuk G, Reid J, Zhu Y, Heitler WJ, Edwards DH (2010) AnimatLab: a 3D graphics environment for neuromechanical simulations. J Neurosci Meth 187:280–288

Coros S, Beaudoin P, van de Panne M (2010) Generalized biped walking control. Acm T Graphic 29

Davoodi R, Loeb GE (2012) Real-Time animation software for customized training to use motor prosthetic systems. IEEE Trans Neural Syst Rehabil Eng 20:134–142

Delp SL, Anderson FC, Arnold AS, Loan P, Habib A, John CT et al. (2007) OpenSim: open-source software to create and analyze dynamic simulations of movement. IEEE Trans Biomed Eng 54:1940–1950

Erez T, Lowrey K, Tassa Y, Kumar V, Kolev S, Todorov E (2013) An integrated system for real-time model predictive control of humanoid robots. IEEE/RAS International Conference on Humanoid Robot. (ed.), Atlanta GA

Fitts PM (1954) The information capacity of the human motor system in controlling the amplitude of movement. J Exp Psychol 47:381–391

Geijtenbeek T, van de Panne M, van der Stappen AF (2013) Flexible muscle-based locomotion for bipedal creatures. Acm T Graphic 32

Geyer H, Herr H (2010) A muscle-reflex model that encodes principles of legged mechanics produces human walking dynamics and muscle activities. IEEE Trans Neural Syst Rehabil Eng 18:263–273

Gordon AM, Huxley AF, Julian FJ (1966) The variation in isometric tension with sarcomere length in vertebrate muscle fibres. J Physiol 184:170–192

Huxley AF (1957) Muscle structure and theories of contraction. Prog Biophys Biophys Chem 7:255–318

Ijspeert AJ (2008) Central pattern generators for locomotion control in animals and robots: a review. Neural Netw 21:642–653

Jain S, Ye YT, Liu CK (2009) Optimization-Based interactive motion synthesis. Acm Trans Graphic 28

Jo S, Massaquoi SG (2004) A model of cerebellum stabilized and scheduled hybrid long-loop control of upright balance. Biol Cybern 91:188–202

Jo S, Massaquoi SG (2007) A model of cerebrocerebello-spinomuscular interaction in the sagittal control of human walking. Biol Cybern 96:279–307

Liddell EGT, Sherrington C (1924) Reflexes in response to stretch (myotatic reflexes). Proc R Soc Lond Ser B Biol Sci 96:212–242

Markin SN, Klishko AN, Shevtsova NA, Lemay MA, Prilutsky BI, Rybak IA (2010) Afferent control of locomotor CPG: insights from a simple neuromechanical model. Ann N Y Acad Sci 1198:21–34

Sarvestani IK, Kozlov A, Harischandra N, Grillner S, Ekeberg O (2013) A computational model of visually guided locomotion in lamprey. Biol Cybern 107:497–512

Sherrington CS (1910) Flexion-reflex of the limb, crossed extension-reflex, and reflex stepping and standing. J Physiol 40:28–121

Thelen DG (2003) Adjustment of muscle mechanics model parameters to simulate dynamic contractions in older adults. J Biomech Eng 125:70–77

Todorov E, Erez T, Tassa Y (2012) MuJoCo: a physics engine for model-based control. 2012 Ieee/Rsj international conference on intelligent robots and systems (Iros), pp 5026–5033

Tomita N, Yano M (2007) Bipedal robot controlled by the basal ganglia and brainstem systems adjusting to indefinite environment. 2007 Ieee/Icme international conference on complex medical engineering, vol 1–4, pp 116–121

Welch TDJ, Ting LH (2008) A feedback model reproduces muscle activity during human postural responses to support-surface translations. J Neurophysiol 99:1032–1038

Zajac FE (1989) Muscle and tendon—properties, models, scaling, and application to biomechanics and motor control. Crit Rev Biomed Eng 17:359–411

# Chapter 2
# A Neuromechanical Model of Spinal Control of Locomotion

**Sergey N. Markin, Alexander N. Klishko, Natalia A. Shevtsova, Michel A. Lemay, Boris I. Prilutsky and Ilya A. Rybak**

**Abstract** We have developed a neuromechanical computational model of cat hindlimb locomotion controlled by spinal central pattern generators (CPGs, one per hindlimb) and motion-dependent afferent feedback. Each CPG represents an extension of previously developed two-level model (Rybak et al. J Physiol 577:617–639, 2006a, J Physiol 577:641–658, 2006b) and includes a half-center rhythm generator (RG), generating the locomotor rhythm, and a pattern formation (PF) network operating under control of RG and managing the synergetic activity of different hindlimb motoneuronal pools. The basic two-level CPG model was extended by incorporating additional neural circuits allowing the CPG to generate the complex activity patterns of motoneurons controlling proximal two-joint muscles (Shevtsova et al., Chap. 5, Neuromechanical modeling of posture and locomotion, Springer, New York, 2015). The spinal cord circuitry in the model includes reflex circuits mediating reciprocal inhibition between flexor and extensor motoneurons and disynaptic excitation of extensor motoneurons by load-sensitive afferents. The hindlimbs and trunk were modeled as a 2D system of rigid segments driven by Hill-type muscle actuators with force-length-velocity dependent properties. The musculoskeletal model has been tuned to reproduce the mechanics of locomotion; as a result, the computed motion-dependent activity of muscle group Ia, Ib, and II afferents and the paw-pad cutaneous afferents matched well the cat in vivo afferent recordings reported in the literature (Prilutsky et al., Chap. 10, Neuromechanical modeling of posture and locomotion, Springer, New York, 2015). In the neuromechanical model, the CPG operation is adjusted by afferent feedback from the moving hindlimbs. The

---

S. N. Markin (✉) · N. A. Shevtsova · I. A. Rybak
Department of Neurobiology and Anatomy, Drexel University College of Medicine, 2900 W. Queen Lane, Philadelphia, PA 19129, USA
e-mail: smarkin@drexelmed.edu

A. N. Klishko · B. I. Prilutsky
School of Applied Physiology, Center for Human Movement Studies, Georgia Institute of Technology, 555 14th Street NW, Atlanta, GA 30332, USA
e-mail: aklishko3@gatech.edu

M. A. Lemay
Department of Bioengineering, Temple University, 1947 N 12th St, Philadelphia, PA 19122, USA



B. I. Prilutsky, D. H. Edwards (eds.), *Neuromechanical Modeling of Posture and Locomotion,* Springer Series in Computational Neuroscience,
DOI 10.1007/978-1-4939-3267-2_2

model demonstrates stable locomotion with realistic mechanical characteristics and exhibits realistic patterns of muscle activity. The model can be used as a testbed to study spinal control of locomotion in various normal and pathological conditions.

**Keywords** Neuromechanical modeling · Central pattern generator · Afferent feedback · Locomotion · Cat

## 2.1 Introduction

The mammalian spinal cord contains neural circuits that can generate a basic locomotor rhythm in the absence of rhythmic input from higher brain centers and peripheral afferent feedback (Brown 1911; Grillner 1981; Pearson 1995; Rossignol 1996; Orlovsky et al. 1999) These circuits are commonly referred to as the *central pattern generator* (CPG). During normal locomotion, however, the spinal CPG operates under the control of afferent feedback and descending signals from supraspinal centers, which both modify the locomotor pattern generated by the CPG and adjust it to the particular motor task and external environment (Conway et al. 1987; Gossard et al. 1994; Guertin et al. 1995; McCrea et al. 1995; Whelan 1996; Fouad and Pearson 1997; Pearson et al. 1998; Hiebert and Pearson 1999; Orlovsky et al. 1999; Lam and Pearson 2002; Frigon et al. 2010; Gottschall and Nichols 2011). Although the spinal reflexes continue to operate during locomotion, their pathways and relative contribution to motoneuronal activity during locomotion are modified. These modifications range from changes in reflex gain to complete reorganization of reflex pathways and emergence of new reflexes during locomotion (Pearson and Collins 1993; Guertin et al. 1995; McCrea et al. 1995; Pearson 1995; Perreault et al. 1995; Angel et al. 1996; Degtyarenko et al. 1998; Pearson et al. 1998; Burke 1999; Menard et al. 1999; Perreault et al. 1999; Gosgnach et al. 2000; Quevedo et al. 2000; Burke et al. 2001; McCrea 2001; Ross and Nichols 2009; Gottschall and Nichols 2011). An important finding has been that electrical stimulation of the group I extensor afferents enhances extensor activity if delivered during the extensor phase of locomotion and resets the rhythm to extension if delivered during the flexor phase (Conway et al. 1987; Guertin et al. 1995). In addition, the influence of the muscle force-sensitive group Ib afferents on ankle extensor activity is reversed from inhibition during non-locomotor conditions to excitation during locomotor activity (Pearson and Collins 1993; Gossard et al. 1994; McCrea et al. 1995), thus providing an additional mechanism for regulation of extensor activity depending on the load on the leg. However, the experiments in spinal cats trained to locomote on a treadmill have shown that these pathways cannot compensate for the total loss of cutaneous feedback from the paw (Bouyer and Rossignol 2003b). The length-dependent afferent feedback from the hip flexors is also important for control of stepping and is involved in the initiation of the swing phase and entrainment of locomotor activity (Andersson and Grillner 1983; Kriellaars et al. 1994; Hiebert et al. 1996; Lam and Pearson 2002).

Despite the significant amount of data on changes in locomotor activity produced by mechanical and electrical stimulations of muscles and neural circuits in the mammalian spinal cord, the structure and operation of spinal locomotor CPG(s) remain unknown (Grillner et al. 2008; McCrea and Rybak 2008; Gossard et al. 2011; Kiehn 2011; Yakovenko 2011; Guertin 2012). Computational models of the mammalian spinal circuitry and musculoskeletal system can complement experimental studies and propose explanations for the complex mechanisms of locomotor pattern generation. Several models of locomotor CPG have been developed based on data from so-called fictive locomotion generated within the spinal cord without afferent feedback from moving limbs (Cohen et al. 1982; Collins and Richmond 1994; Beer et al. 1999; Rybak et al. 2006a). However, as discussed above, afferent signals from moving limbs can reset the locomotor rhythm, advance or delay the phase transitions and modulate the duration of flexor and extensor phases. To understand the contribution of afferent feedback in locomotion, the computational models of spinal circuitries should include afferent feedback from the moving musculoskeletal system. Several such models have been developed and the possible mechanisms for sensory control of the CPG suggested (Taga 1995a, b; Wadden and Ekeberg 1998; Rybak et al. 2002; Ivashko et al. 2003; Ekeberg and Pearson 2005; Maufroy et al. 2008; Aoi et al. 2013; Toth et al. 2013; Nassour et al. 2014); see also Chap. 8 in this book (Aoi 2015). Those models, however, have not attempted to reproduce the locomotor patterns of motoneuronal and afferent activity. Neither have they accurately reproduced the exact kinematics and kinetics of walking.

Our study has focused on the development of a comprehensive neuromechanical model of cat spinal locomotion. The neural subsystem of this model includes a locomotor CPG. The model of this CPG is based on the previously developed two-level model (Rybak et al. 2006a, b). The basic two-level CPG model included separate *rhythm generation* (RG) and *pattern formation* (PF) networks. This basic model has been extended to accommodate and reproduce the realistic activity patterns of motoneurons controlling complex two-joint muscles (see Chap. 5 by Shevtsova et al. 2015). In this study we connected the extended CPG model with the comprehensive hindlimb musculoskeletal model simulating biomechanics of cat walking and providing motion-dependent afferent feedback to the CPG (Prilutsky et al., Chap. 10, in this volume). The combined neuromechanical model demonstrates the ability to generate stable locomotion showing realistic patterns of muscle activity and mechanical characteristics of walking. The model has been used for the investigation of the role of particular afferent pathways for stable walking.

## 2.2 Musculoskeletal Model of Hindlimbs

The musculoskeletal model of the cat hindlimbs is described in details in Chap. 10 of this book (Prilutsky et al. 2015) and only its brief description is provided here. The two cat's hindlimbs, pelvis and trunk are modeled as a 2D, 10 degrees-of-freedom (DOF) system of rigid segments interconnected by frictionless revolute joints (Fig. 2.1a and b). Interactions of hindlimbs with the ground and the trunk with the forelimbs,

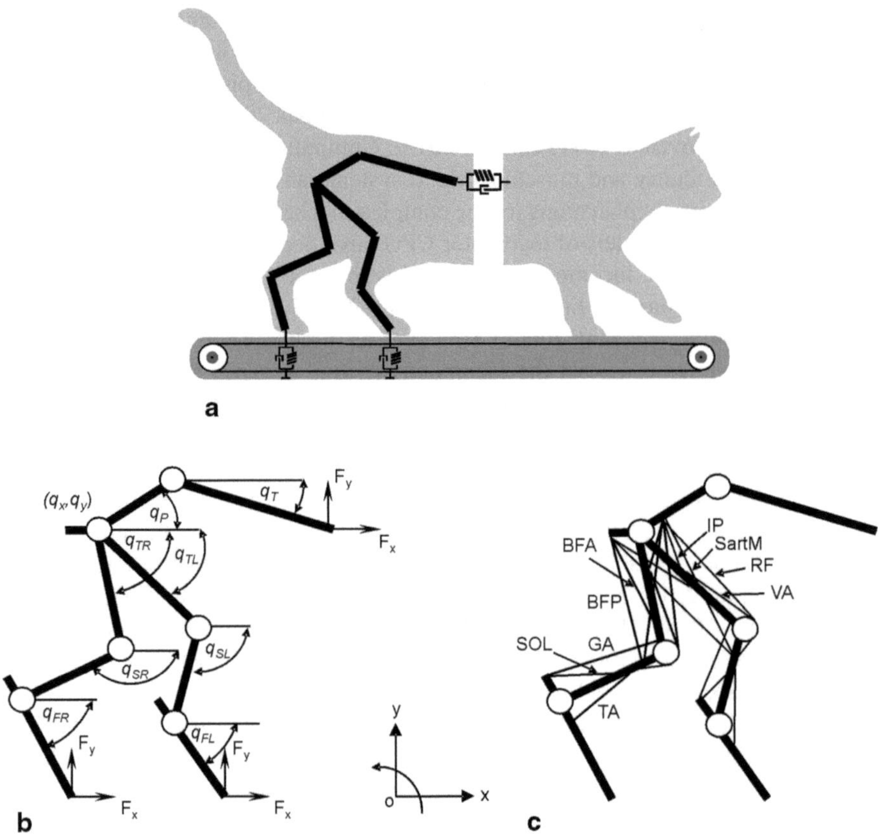


**Fig. 2.1** Schematic representation of the musculoskeletal model of the cat hindlimbs and trunk. **a** The hindlimbs and posterior trunk interact with the ground and the anterior trunk and forelimbs. These interactions are modeled as viscoelastic forces. **b** A 10-DOF planar system of rigid segments with frictionless revolute joints representing two hindlimbs, pelvis and posterior trunk. **c** Schematic representation of muscles of the model: *IP* iliopsoas, *BFA* biceps femoris anterior, *RF* rectus femoris, *BFP* biceps femoris posterior, *SartM* sartorius medial, *VA* vastii, *GA* gastrocnemii, *TA* tibialis anterior, and *SOL* soleus. (Adopted from Prilutsky et al., Chap. 10, in this volume)

neck and head are modeled as linear springs with dampers. The inertial parameters of hindlimb segments are computed from the measured mass of the cat and length of each hindlimb segment using the regression equations (Hoy and Zernicke 1985). The equations of motion are derived from the Lagrange equations. The generalized coordinates of the model include the horizontal and vertical positions of the hip and the segment angles (Fig. 2.1b). The equations of hindlimbs dynamics include the vectors of segmental generalized velocities and accelerations, Coriolis and centrifugal forces, gravitational forces, ground and trunk reaction forces, muscle forces, and viscoelastic forces at the joints (for details see Prilutsky et al. 2015, Chap. 10, in this volume).

Each hindlimb in the model is actuated by 9 muscles described by Hill-type models (Fig. 2.1c) with realistic tendon force-length properties, contractile element force-length-velocity properties, muscle mass and angle of pennation as well as a parallel elastic component (Prilutsky et al. 2015; Chap. 10, in this volume). The description of contractile and activation dynamics of the muscle-tendon actuator can be found in Chap. 10 of this volume (Prilutsky et al. 2015).

Parameters of the musculoskeletal model (constants for the viscoelastic elements producing reaction forces, the tendon slack length, tendon force-length relationship parameters, maximal muscle activation, activation and deactivation time constants for each muscle-tendon unit, etc.; see Prilutsky et al. 2015, Chap. 10, in this volume), were identified by minimizing the mismatch between the simulated and experimentally obtained cat locomotion variables—muscle fascicle lengths/velocities, joint angles, joint moments and ground reaction forces—using a parallel simulated annealing optimization algorithm (Corana et al. 1987). The simulated walking mechanics were obtained by integrating the equations of the limb and muscle dynamics, using the recorded activity of 9 muscles as input and the recorded position and velocity of each generalized hindlimb coordinate at the walking cycle onset as the initial conditions (Prilutsky et al. 2015, Chap. 10, in this volume). The obtained parameters of the musculoskeletal model were within physiological ranges reported in the literature (e.g., Spector et al. 1980; Sacks and Roy 1982; Baratta et al. 1993, 1995; Brown et al. 1996) and allowed for a close match (typically within one standard deviation) between the simulated and recorded joint angles and moments as well as ground reaction forces during walking.

The firing rates of spindle length-sensitive group Ia and II afferents and force-sensitive Golgi tendon organ group Ib afferents are closely correlated with the instantaneous muscle length/stretch velocity and tendon force, respectively, as observed during walking in the cat (Prochazka et al. 1997; Prochazka and Gorassini 1998). This fact makes it possible to estimate the firing rates of spindle and Golgi tendon organ afferents as functions of muscle fascicle length and velocity and tendon force of each muscle-tendon unit in the musculoskeletal model using equations similar to those proposed by Prochazka et al. (Prochazka and Gorassini 1998; Prochazka 1999). Another important afferent signal that indicates the stance phase of locomotion and influences the CPG operation and locomotor rhythm is activity of load-sensitive cutaneous afferents from the paw pad (McCrea 2001). The firing rate of these afferents is computed as the function of the ground reaction force and its time derivative (Prilutsky et al. 2015, Chap. 10, in this volume).

## 2.3 Model of Spinal Circuitry

### *2.3.1 Neuron Models*

The model of the spinal circuitry in this study represents a modified version of the two-level locomotor CPG model described in Chap. 5 of this volume (Shevtsova et al. 2015). The model includes a bipartite (half-center) rhythm generator, pattern formation network and other interneurons and motoneurons. The interneurons provide basic reflex circuits including reciprocal inhibition of antagonistic motoneurons, recurrent inhibition of motoneurons via Renshaw cells, disynaptic excitation of some motoneuron types, etc. The CPG model of Shevtsova et al. 2015 (see Chap. 5) was simplified so that each neuronal population was described by an activity-based (non-spiking) neuron model. Two types of neuron models were implemented: one for rhythm-generating RG and PF neurons and motoneurons and the other for all other neurons.

The membrane potentials ($V$) of principal neurons at RG, PF levels and motoneurons are described by the following equation:

$$C \cdot \frac{dV}{dt} = -I_{NaP} - I_K - I_{Leak} - I_{SynE} - I_{SynI}\,. \tag{2.1}$$

The membrane potential of all other neurons is described as:

$$C \cdot \frac{dV}{dt} = -I_{Leak} - I_{SynE} - I_{SynI}\,, \tag{2.2}$$

where $C$ is the neuronal capacitance, $I_{\text{Leak}}$ is the leakage current, $I_K$ is potassium rectifier current, $I_{NaP}$ is persistent sodium current; $I_{\text{SynE}}$ and $I_{\text{SynI}}$ are the excitatory and inhibitory synaptic currents, respectively. The ionic currents are described as follows:

$$\begin{aligned}
I_{NaP} &= \bar{g}_{NaP} \cdot m_{NaP} \cdot h_{NaP} \cdot (V - E_{Na});\\
I_K &= \bar{g}_K \cdot m_K^4 \cdot (V - E_K);\\
I_{Leak} &= \bar{g}_{Leak} \cdot (V - E_{Leak});\\
I_{SynE,i} &= \bar{g}_{SynE} \cdot (V_i - E_{SynE}) \cdot \left( \sum_j a_{ji} \cdot f(V_j) + \sum c_{mi} \cdot d_m + \sum_k w_{ki} \cdot fb_k \right);\\
I_{SynI,i} &= \bar{g}_{SynI} \cdot (V_i - E_{SynI}) \cdot \sum_j b_{ji} \cdot f(V_j),
\end{aligned} \tag{2.3}$$

where $\bar{g}_{NaP}$, $\bar{g}_{Leak}$, $\bar{g}_{SynE}$, and $\bar{g}_{SynI}$ are the maximal conductances of the corresponding ionic channels; $E_{Na}$, $E_K$, $E_L$, $E_{SynE}$, and $E_{SynI}$ are the corresponding reversal potentials; $a_{ji}$ defines the weight of the excitatory synaptic input from neuron $j$ to neuron $i$; $b_{ji}$ defines the weight of the inhibitory input from neuron $j$ to neuron $i$; $c_{mi}$ defines the weight of the excitatory drive $d_m$ to neuron $i$; $w_{ki}$ defines the synaptic weight of afferent feedback $fb_k$ ($k$=Ia, Ib, II, cutaneous) to neuron $i$; (see Tables 2.1, 2.2, 2.3, 2.4, 2.5, 2.6, 2.7, 2.8, and 2.9 in Appendix). Activation of the potassium delayed rectifier and persistent sodium currents is considered instantaneous. Voltage dependent activation and inactivation variables and time constant for the potassium delayed rectifier and persistent sodium channels are described as follows:

$$\begin{aligned}
m_K &= 1/(1 + \exp(-(V + 44.5)/5)),\\
m_{NaP} &= 1/(1 + \exp(-(V + 47.1)/3.1)),\\
\tau_{hNaP} \cdot \frac{d}{dt} h_{NaP} &= h_{\infty NaP} - h_{NaP},\\
h_{\infty NaP} &= 1/(1 + \exp((V + 51)/4)),\\
\tau_{hNaP} &= \tau_{hNaP\max} / \cosh((V + 51)/8).
\end{aligned} \tag{2.4}$$

The neuron output activity is defined by a nonlinear function *f(V)*:

$$f(V) = \begin{cases} 1/(1+\exp\ (-(V - V_{1/2})/k)), & if\ \ V \geq V_{tr}; \\ 0, \ if \quad V < V_{tr}, \end{cases} \tag{2.5}$$

where $V_{1/2}$ is the half-activation voltage, *k* defines the slope of the output function and $V_{tr}$ is the threshold.

The following values of neuronal parameters were used: $C = 20$ pF; $E_{Na} = 55$ mV, $E_K = -80$ mV, $E_{SynE} = -10$ mV, $E_{SynI} = -70$ mV, $E_{Leak} = -64$ mV for RG, PF neurons and motoneurons and $-60$ mV for all other neurons; $\overline{g}_K = 4.5$ nS, $\overline{g}_{Leak} = 1.60$ nS, $\overline{g}_{SynE} = \overline{g}_{SynI} = 10.0$ nS, $\overline{g}_{NaP} = 3.5$ nS for RG neurons, 0.5 nS for PF neurons, and 0.3 nS for motoneurons; $\tau_{hNaP} = 600$ ms. Parameters of $f(V)$ function were $V_{1/2} = -30$ mV, $V_{tr} = -50$ mV, $k = 3$ mV for motoneurons and 8 mV for other neurons.

The model of locomotor center (Fig. 2.2) incorporates the model of CPG and basic reflex circuits mediating the reciprocal inhibition of antagonistic motoneurons via Ia inhibitory interneurons, recurrent inhibition of motoneurons via Renshaw cells (RC), non-reciprocal motoneuron inhibition (Ib cells) and disynaptic excitation of extensor motoneurons (Ia and Ib cells) (Fig. 2.3).

The conceptual architecture of the CPG model is based on the idea of a two-level locomotor CPG (Rybak et al. 2006a, b; McCrea and Rybak 2007, 2008). According to this hypothesis, the locomotor CPG consists of a half-center rhythm generator and multiple pattern formation circuits controlling different synergist and antagonist motoneuron pools (see Figs. 2.2 and 2.3). Depending on the input from the RG and the interactions within the PF network, each PF neuron is active within the particular phase(s) of the locomotor cycle and produces a phase-specific activity pattern. The specific principal PF elements control the corresponding group of synergistic motoneurons that are active synchronously. Organization of multiple neural circuits that control the activation of synergistic motoneuron groups is mainly unknown. Previous analysis based on the onset and offset times in motoneuron/muscle activity allowed to identify several synergistic groups of motoneurons operating during locomotion (Markin et al. 2012). The identified groups (see Fig. 2.4) include hip flexors (IP, SartM), hip extensor (BFA), knee extensor (VA), ankle flexor (TA), ankle extensors (GA, SOL), and two two-joint muscles BFP and RF, which demonstrate activity in both swing and stance phases. Figure 2.5 shows the proposed organization of rhythm generator and pattern formation circuits in the CPG controlling one hindlimb. All PF circuits receive excitatory and inhibitory inputs from rhythm generator and control flexor and extensor motoneurons operating at hip, knee and ankle as well as motoneurons controlling two-joint muscles (BFP, RF). Each joint-related PF circuitry is a half-center network consisting of PF-F and PF-E neurons reciprocally inhibiting each other via Inpf-F and Inpf-E inhibitory interneurons, respectively. The detailing description of PF organization can be found in Chap. 5 of this volume (Shevtsova et al. 2015).

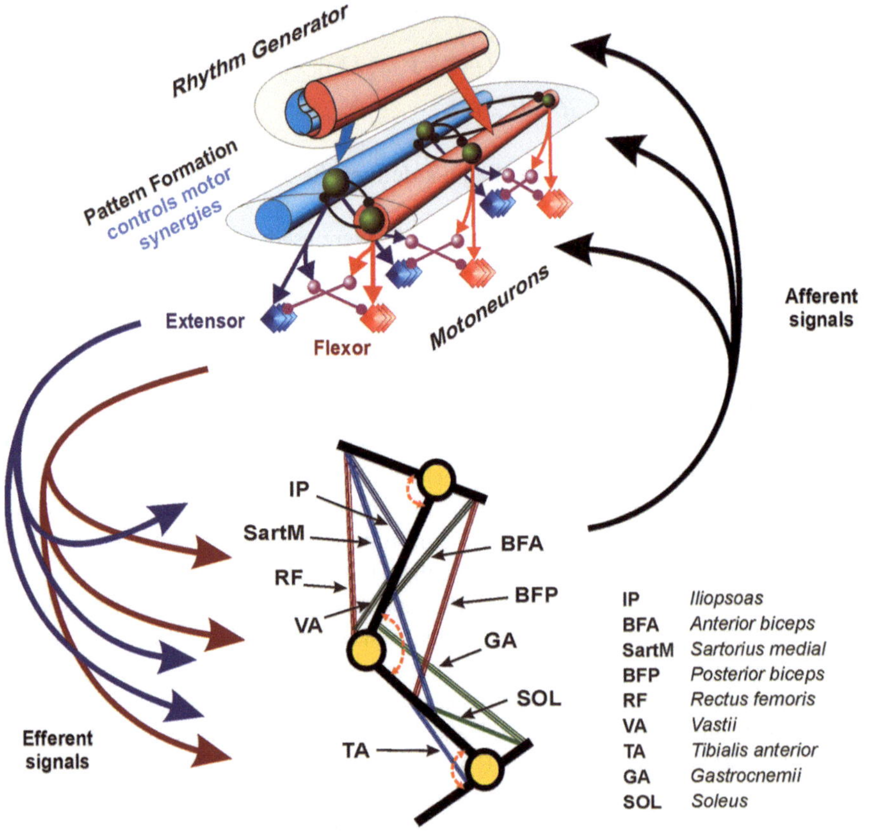


**Fig. 2.2** General schematics of the neuromechanical model of the cat hindlimb locomotor control. The neural part of the model consists of the two-level locomotor CPG that controls the hindlimb musculoskeletal model. The activity of corresponding motoneuron pools controls major hindlimb muscles that drive the 10-DOF cat hindlimb model. The generated somatosensory feedback signals from the moving musculoskeletal hindlimb model (i.e., firing rates of group I and II muscle and paw pad cutaneous afferents) project onto both levels of the CPG (*RG* rhythm generator and *PF* pattern formation) and motoneuron level as well

## 2.4 Control of CPG by Afferent Feedback

Although the locomotor CPG can generate rhythm in the absence of sensory feedback signals, the sensory feedback plays a critical role in regulating phase transitions, stabilizing locomotor movements, contributing to weight support during the stance phase, and adjusting the locomotor pattern to the constantly changing external environment. A possible organization of afferent pathways to the CPG for a simple 1-DOF musculoskeletal system has been recently proposed (Markin et al. 2010). According to this organization, the stance-swing (extensor-flexor) phase transition was controlled by both the reduction of force-dependent afferent activity from the

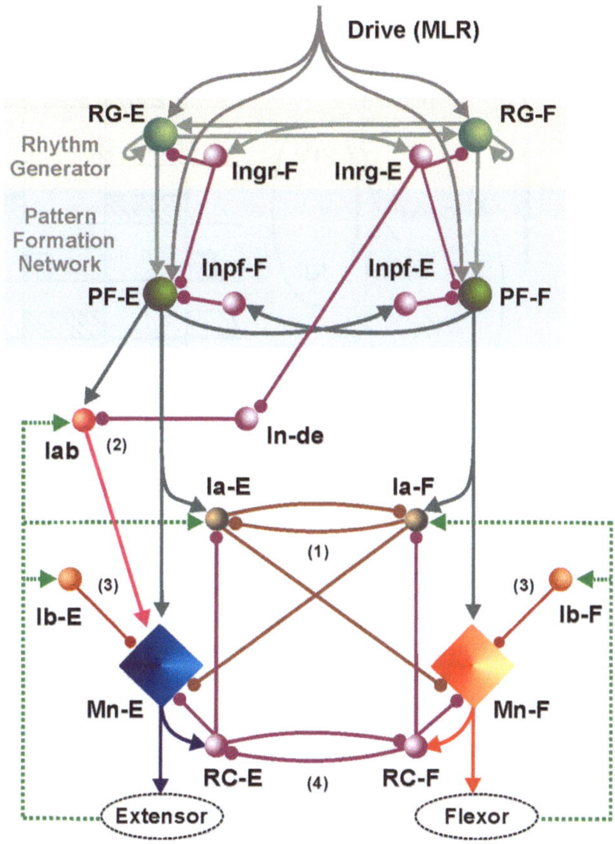


**Fig. 2.3** Basic model of the two-level locomotor CPG model by (Rybak et al. 2006a, b) including reflex circuits: *1* reciprocal inhibition of antagonistic motoneurons via Ia inhibitory interneurons (*Ia-E* and *Ia-F*, correspondingly), *2* disynaptic excitation of extensor muscle motoneurons via Iab interneurons, *3* non-reciprocal inhibition of motoneurons via Ib inhibitory interneurons, and *4* recurrent inhibition via Renshaw cells (*RC-E* and *RC-F*, respectively)

extensor muscles and the increase in length-dependent afferent activity from the flexor muscles. Because of this organization, the duration of the stance phase depended on the locomotor speed. In contrast, the timing of the swing-stance (flexor-extensor) phase transition was mainly controlled by the length/velocity-dependent afferent activity from the hip extensor muscles; this feedback signal adjusted the duration of the flexor phase to limb kinematics during the swing phase, keeping the swing duration relatively constant.

In the present, more realistic neuromechanical model we have adopted the organization of sensory pathways between muscle afferents and RG-interneurons from the previous work (Markin et al. 2010). The type and organization of afferent pathways from the moving musculoskeletal system to the CPG have been chosen based

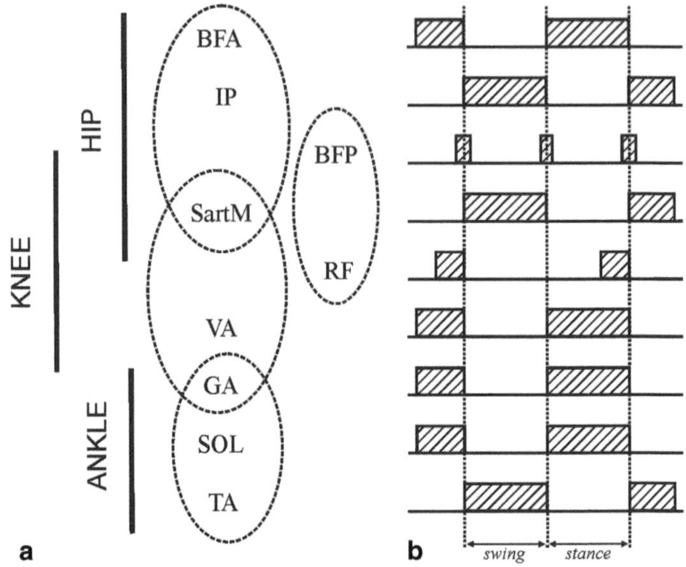


**Fig. 2.4** Phases of activity of cat hindlimb muscles during real locomotion. **a** The possible organization of synergist motoneuron groups at the pattern formation level. Three circuits at the pattern formation level (hip-PF, knee-PF and ankle-PF) are introduced to control motoneuron groups innervating joint specific muscles: (1) BFA, IP, SartM as hip muscles; (2) VA, SartM, GA as knee muscles; and (3) GA, SOL, TA as ankle muscles. Note that two two-joint muscles (SartM and GA) receive control signal from hip-PF/knee-PF and knee-PF/ankle-PF sub-networks, respectively. The fourth circuit at pattern formation level specifically controls BFP and RF muscles that are partially active during both flexor and extensor phases. **b** Schematic representation of periods of EMG activity during level walking in the cat. While most of the hindlimb muscles are active during most of swing (flexor) or stance (extensor) phases, the two-joint BFP and RF muscles are only active at the swing-stance or stance-swing phase transition (BFP) or in the later part of the stance phase (RF)

on the following experimental data: (1) muscle length-sensitive spindle afferents of hip extensor and flexor muscles influence the flexor-extensor and extensor-flexor phase transitions (Perreault et al. 1995; Lam and Pearson 2002; McVea et al. 2005); and (2) activation of both group I and II afferents of ankle flexors can terminate flexor and initiate extensor phases during fictive locomotion (Perreault et al. 1995; Stecina et al. 2005); (3) group Ib afferents from the Golgi tendons organs of ankle extensors are responsible for prolongation of the stance phase (Duysens and Pearson 1980; Pearson 2008); (4) stimulation of cutaneous afferents innervating the paw pad can prolong the stance phase and is responsible for terminating the ongoing swing and initiating the stance phase (McCrea 2001; Rossignol et al. 2006). A possible organization of afferent signals at the RG-level for the CPG model is presented in Fig. 2.6. Two additional interneurons (Frg-F and Frg-E) are incorporated into the CPG model. These neurons receive the multi-modal afferent input signals from the afferents listed above and project their excitatory activity onto the corresponding neurons at the RG level.

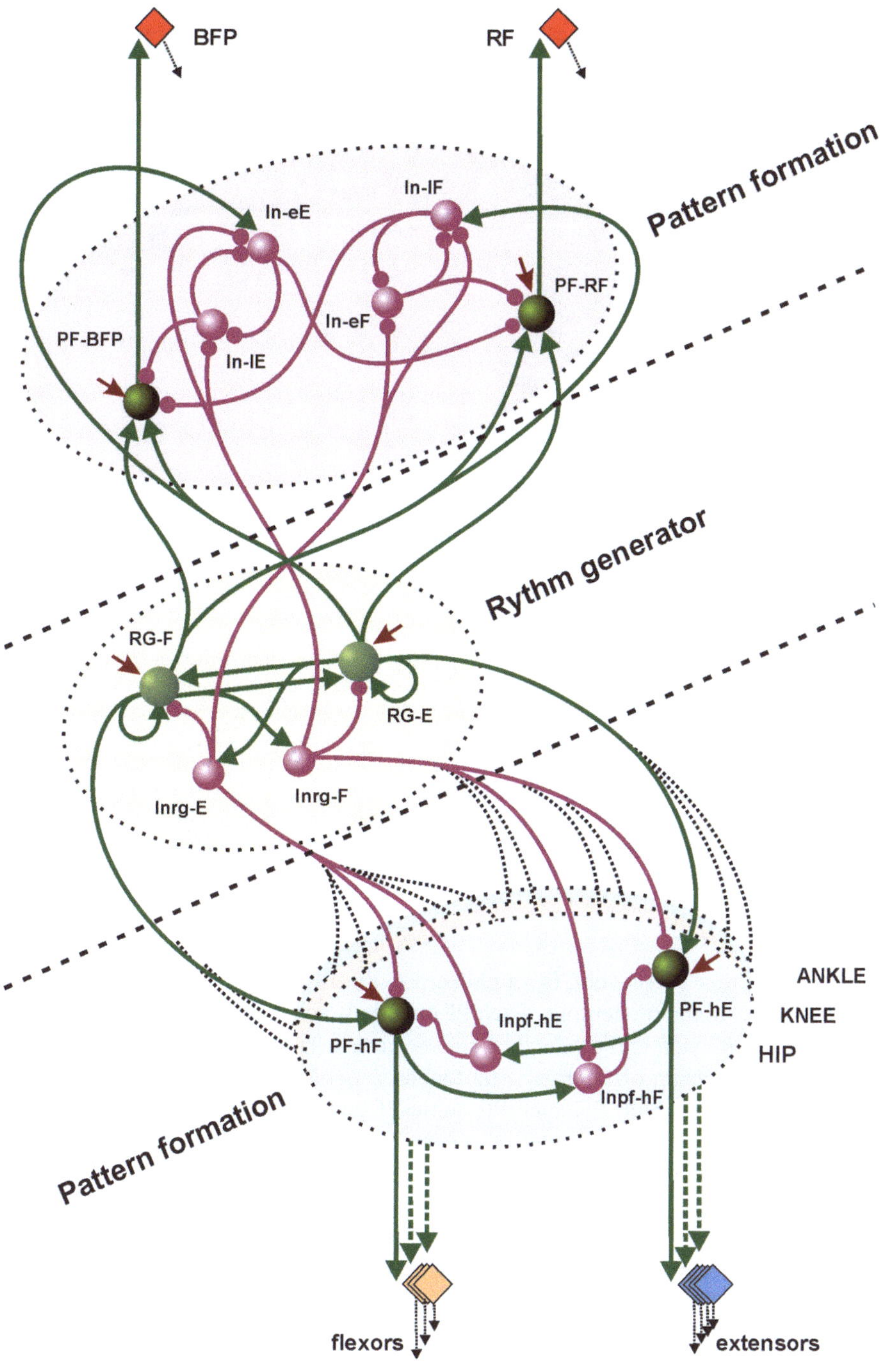


**Fig. 2.5** A general diagram of the comprehensive model of locomotor circuitry. The rhythm generator (in the *center* of the figure) mediates rhythmic excitatory and inhibitory drive to the three pattern formation networks that control distinct synergistic motoneuron groups innervating mus-

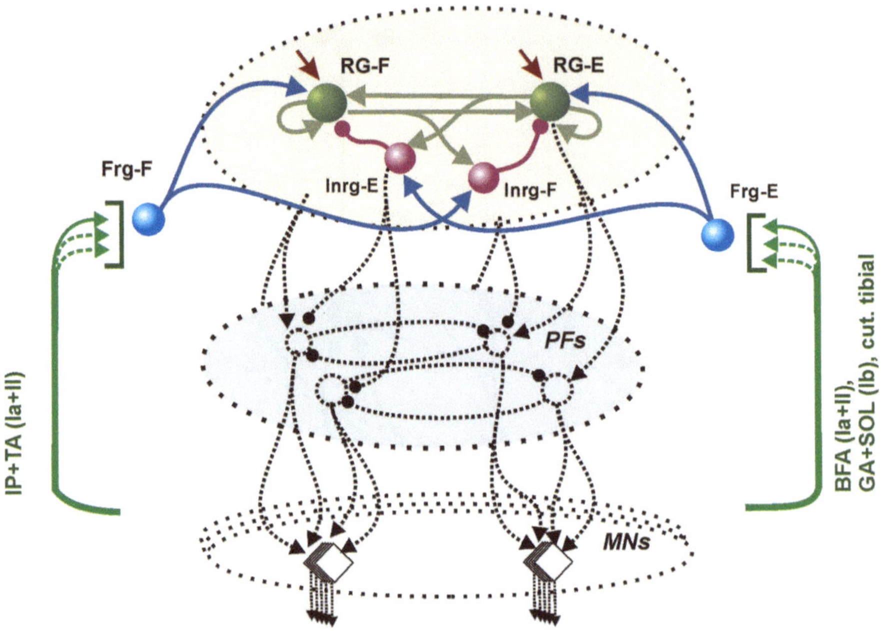


**Fig. 2.6** Organization of afferent feedback projections onto the rhythm generator *(RG)*. Two additional interneurons (Frg-F and Frg-E) receive proprioceptive feedback signals from hip and ankle muscles and cutaneous paw pad afferents and distribute them between flexor and extensor parts of the rhythm generator. Specifically, Frg-F neuron receives excitatory inputs from IP-Ia, IP-II, TA-Ia, TA-II afferents and mediates the excitation to the flexor part of the rhythm generator (RG-F and Inrg-F, respectively); afferent feedback signals from the BFA-Ia, BFA-II, GA-Ib, SOL-Ib and cutaneous paw pad afferents converge to and excite the extensor half-center of the rhythm generator (RG-E and Inrg-E, respectively) via Frg-E. Weights of synaptic connections between afferent feedback signals and the corresponding interneurons at the RG level are given in Tables 2.1 and 2.9 in Appendix

In the current two-level CPG architecture, the PF networks are interposed between the RG and motoneurons. Each PF-neuron is active during a particular phase of the walking cycle and produces a phase-specific synchronized activation of the corresponding group of synergetic motoneuron pools. We hypothesized that each principal PF neuron (hip-PF, knee-PF, ankle-PF and BFP-RF-PF) receives proprioceptive feedback from the group of synergetic muscles that are regulated by this neuron. The schematic of afferent pathways based on this approach is presented in

cles of individual joints (*lower* part of the figure) and a pattern formation network that controls the two separate motoneuron groups innervating two-joint *BFP* and *RF* muscles (*upper* part of the figure). Here and further in the figures, the inhibitory and excitatory connections are shown as *solid circles* and *arrows*, respectively; *red arrows* indicate supra-spinal tonic drive. Weights of synaptic connections within locomotor CPG circuitry are given in Tables 2.1–2.5 in Appendix

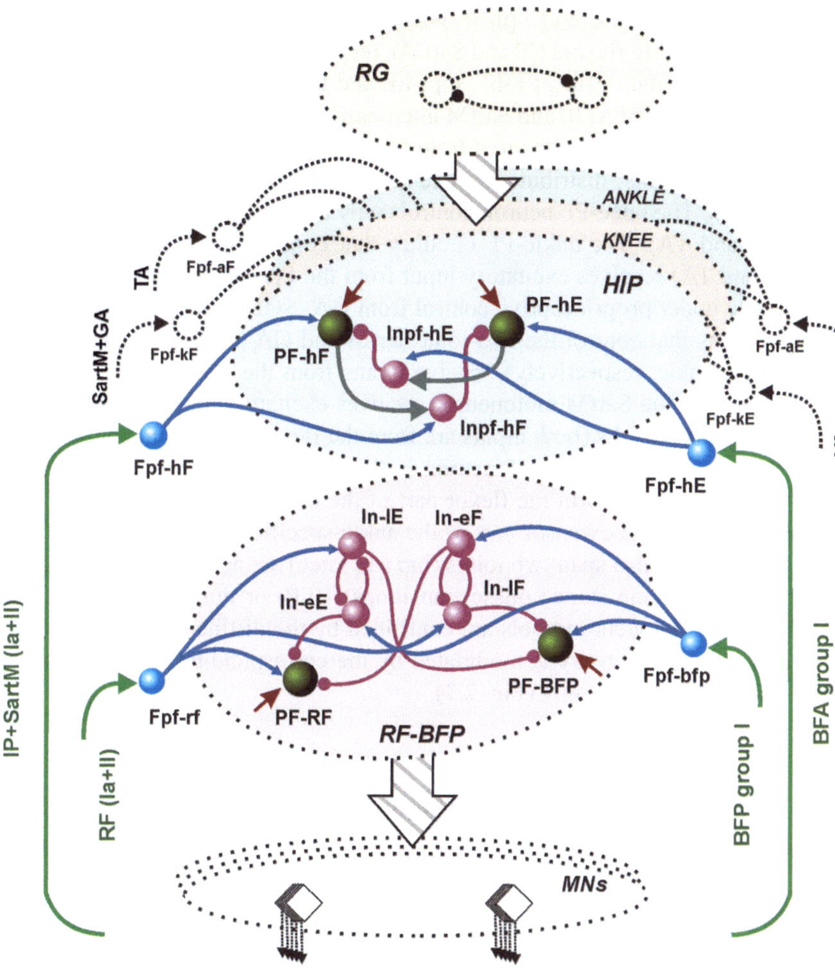


**Fig. 2.7** Organization of afferent feedback signals projected onto the pattern formation *(PF)* networks. Eight interneurons (Fpf-hE, Fpf-hF, Fpf-kE, Fpf-kF, Fpf-aE, Fpf-aF, Fpf-rf and Fpf-bfp) receive afferent signals from the musculoskeletal system and excite the corresponding interneurons at the pattern formation level. Tables 2.2–2.5 and 2.9 in Appendix provide additional information about weights of synaptic connections between afferent signals and correspondent interneurons of locomotor CPG at the PF level

Fig. 2.7. Eight additional neurons at the PF-level (Fpf-hE, Fpf-hF, Fpf-kE, Fpf-kF, Fpf-aE, Fpf-aF, Fpf-rf and Fpf-bfp) receive proprioceptive feedback from the corresponding muscles and distribute the excitatory signal among interneurons within the particular PF neurons that control muscles belonging to the same muscle synergy. The flexor part of each joint-specific PF sub-network receives input from groups Ia and II spindle afferents of the corresponding muscles, whereas the extensor half-centers receive afferent input from group I afferents (Ia and Ib) of the corresponding muscles. Thus, the Fpf-hE and Fpf-hF neurons receive afferent signals from hip extensor (BFA) and hip flexors (IP and SartM), respectively. These neurons mediate excitatory inputs to the PF-hE, PF-hF, Inpf-hE and Inpf-hF neurons, which in turn control the activity of BFA, IP and SartM interneurons. The Fpf-kE and Fpf-kF neurons, that receive proprioceptive input from knee flexors (SartM and GA) and knee extensor (VA) afferents, distribute the excitatory signals within the knee-specific PF sub-network. The knee-PF neuron controls only knee related motoneuron group (SartM, GA and VA). The ankle-PF circuitry that controls ankle related muscles (GA, SOL and TA) receives excitatory input from the Fpf-aE and Fpf-aF neurons, which operate under proprioceptive control from GA, SOL and TA muscles. Note, the motoneurons that control the two-joint SartM and GA muscles (spanning hip/knee and knee/ankle, respectively), receive inputs from the corresponding PF-circuits. Specifically, the SartM motoneuron receives excitatory signals from the hip and knee PF sub-networks (both inputs are from the flexor parts of the corresponding PF circuitry), and the GA motoneuron is controlled by the knee and ankle PF sub-networks (one input form the flexor part of the knee-specific PF circuitry and the other input from the extensor part of the ankle-specific PF circuitry). The BFP and RF muscles that also span two joints (hip and knee) are active either during the stance-swing and swing-stance phase transitions (BFP) or during the late stance (RF). These two motoneuron pools are controlled by the distinct PF network (RF-BFP circuitry) whose activity is modulated by the corresponding feedback inputs from RF and BPF muscles only (Fig. 2.7).

## 2.5 Organization of Spinal Autogenic and Heterogenic Reflex Pathways

Besides the CPG circuits the neural sub-system in the model contains reflex circuits and pathways that allow the motoneurons to receive afferent feedback input from the muscles they innervate (autogenic pathways) and from other muscles (heterogenic pathways) or skin afferents (see also Chap. 3 in this volume by Nichols et al. 2015). These pathways and circuits provide: (1) disynaptic excitation of extensors by group I extensor afferents during extensor phase of locomotion (Angel et al. 1996; Rybak et al. 2006b); (2) excitation of extensors via load-sensitive paw pad cutaneous afferents (McCrea 2001; Bouyer and Rossignol 2003b); (3) recurrent inhibition of motoneurons via Renshaw cells (McCrea et al. 1980; Nishimaru et al. 2006); (4) Ia-evoked monosynaptic excitation of homonymous and synergist motoneuron pools (Eccles et al. 1957a; Eccles and Lundberg 1958; Nichols et al. 1999, 2015); (5) the reciprocal disynaptic inhibition of antagonistic motoneuron pools via the Ia inhibitory interneurons (Feldman and Orlovsky 1975; Pratt and Jordan 1987); and (6) non-reciprocal Ib inhibition of synergist motoneuron pools (Eccles et al. 1957b; Nichols et al. 1999, 2015).

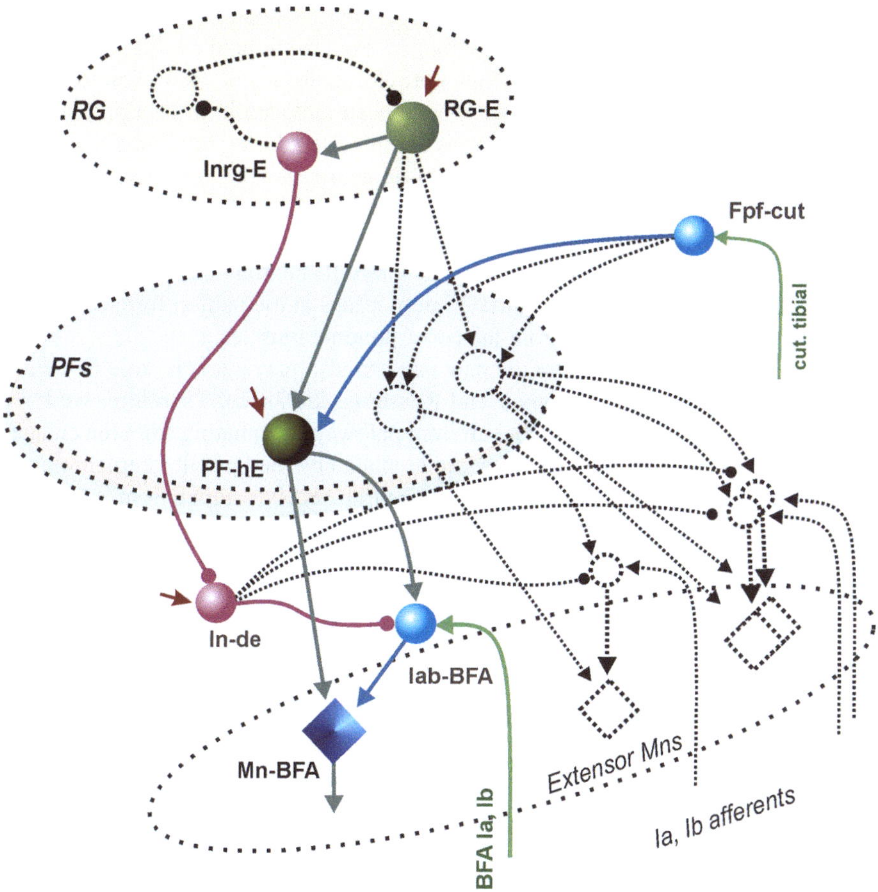


**Fig. 2.8** Organization of the disynaptic excitation of extensor motoneurons during the extensor phase of locomotion. The Iab-BFA interneuron provides phase-dependent disynaptic excitation of extensor motoneuron (Mn-BFA) by extensor group I afferents. During the stance phase of locomotion, Iab-BFA interneuron is released from inhibition by the In-de interneuron and receives additional excitatory signal from the PF-hE at the PF level. The hypothetical Fpf-cut interneuron receives signals from cutaneous paw pad afferents and mediates additional excitation on extensor parts of the joint-specific PF circuitry. (see Table 2.6 in Appendix)

## *2.5.1 Disynaptic Excitation of Extensor Motoneurons by Group I Extensor and Cutaneous Paw Pad Afferents*

An important function of proprioceptive feedback during locomotion is to reinforce ongoing motor activity during stance (Pearson 1995). One of the mechanisms of reinforcement of ongoing extensors activity is additional disynaptic excitatory input from group I afferents of extensor muscles. Disynaptic excitation of extensor motoneurons was demonstrated in fictive locomotion when during the extensor phase extensor afferents were electrically stimulated at group I strength (McCrea et al. 1995; Angel et al. 1996; Degtyarenko et al. 1998). Figure 2.8 shows the organization of additional circuitry that provides the disynaptic reflex excitation to extensor motoneurons (Rybak et al. 2006a, b). Four additional excitatory neurons (Iab-BFA, Iab-VA, Iab-GA and Iab-SOL) providing disynaptic excitation to extensor motoneurons (BFA, VA, GA, SOL) have been included in the neural circuitry of the model (see Fig. 2.8). These neurons are inhibited during the swing phase due to inhibitory input from the hypothetical In-de neuron that in turn receives inhibitory input from the Inrg-E interneuron. During the stance phase of locomotion (i.e., when the Inrg-E inhibitory interneuron is active), the In-de neuron is inhibited and thereby the Iab-BFA, Iab-VA, Iab-GA and Iab-SOL neurons can be activated by both excitatory inputs from appropriate interneurons at the pattern formation level and group I afferent feedback from the corresponding muscles.

The cutaneous afferents from the paw pad can also play a crucial role in weight support during locomotion (Bouyer and Rossignol 2003a, b). Therefore, we have incorporated in the model additional afferent pathways originating from the cutaneous paw pad afferents (Fig. 2.8). They are mediated by the Fpf-cut interneuron and contribute to weight support. The Fpf-cut neuron receives input from cutaneous paw pad afferents and excites the principal neurons at the pattern formation level (namely, PF-hE, PF-kE and PF-aE), which activate BFA (hip extensors), VA (knee extensors), GA and SOL (ankle extensors) muscles.

### *2.5.2 Recurrent Inhibition via Renshaw Cells*

The activity of motoneurons is also regulated by Renshaw cells. These interneurons fire rhythmically during locomotion and produce feedback inhibition to homonymous and synergetic motoneurons (McCrea et al. 1980; Pratt and Jordan 1987). They also regulate the activity of interneurons which provide Ia reciprocal inhibition (Jankowska 1992; Alvarez and Fyffe 2007). Interestingly, Renshaw cells receive rhythmic inhibitory inputs in both the active and silent phases during fictive locomotion (Nishimaru et al. 2006). It has been shown that Renshaw cells can receive inhibition from Renshaw cells excited by antagonist motoneuron pools (Ryall 1970; Windhorst 1996), which may explain the inhibition of Renshaw cells in the non-active phase but not in the active phase. One of the possible sources of this inhibition could be ipsilateral populations of interneurons of locomotor circuitry (Nishimaru et al. 2006). The implementation of a possible connectivity for Renshaw cells in the model is presented in Fig. 2.9.

### *2.5.3 Other Reflexes*

In the cat, activation of group Ia muscle spindle afferents evokes the stretch reflex, a monosynaptic excitation of synergist motoneurons with disynaptic inhibition of antagonist motoneurons mediated by Ia inhibitory interneurons (Eccles et al. 1956;

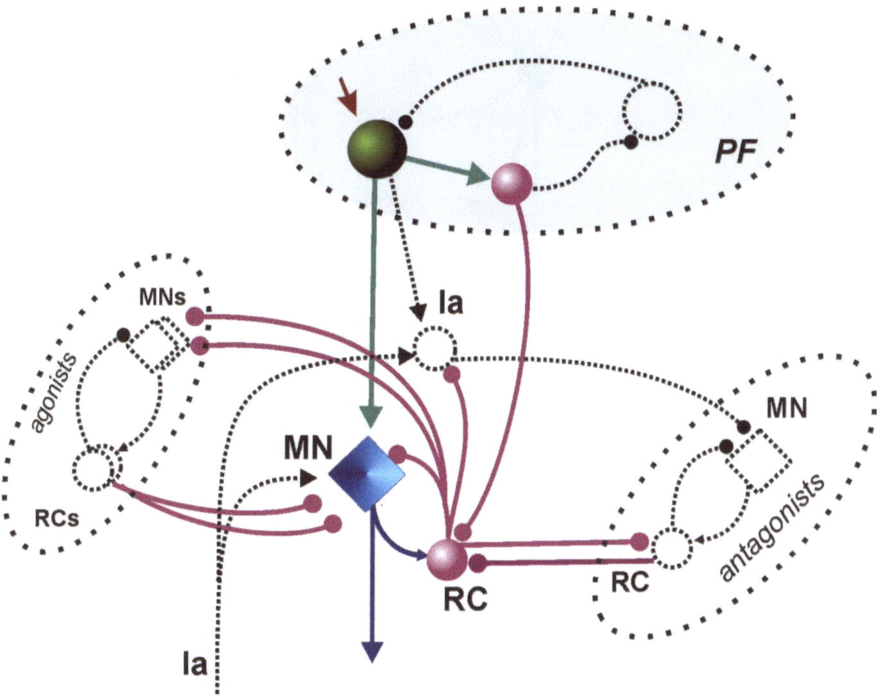


**Fig. 2.9** Possible organization of interconnections between Renshaw cells *(RC)*, motoneurons *(MN)* and other interneurons in the locomotor CPG. Renshaw cells receive excitatory input from particular motoneurons and inhibit the same and the synergist motoneurons. Renshaw cells also provide inhibitory signals to Ia interneurons and receive the ipsilateral inhibition from CPG circuitry and Renshaw cell of the antagonist motoneurons. Here and further the unlabeled *green circle* represents a principal interneuron at PF level that controls a corresponding motoneuron; the unlabeled *purple circle* inside the PF level represents inhibitory interneuron at the PF level that receives excitatory input from corresponding principal interneuron. (see Tables 2.2–2.5, and 2.7 in Appendix for details)

Baldissera et al. 1981; Jankowska 1992). A simplified scheme of the monosynaptic stretch reflex with disynaptic Ia reciprocal inhibition circuitry is shown in Fig. 2.10a. The additional Ia interneurons which receive Ia afferent input and mediate inhibition to antagonist motoneurons are introduced for the following pairs of antagonist muscles: IP/BFA, SartM/RF, SartM/VA, RF/BFP, BFP/VA, GA/RF, GA/VA, GA/TA, SOL/TA (Lloyd 1946; Eccles and Lundberg 1958). These interneurons also receive mutual inhibition from the antagonist Ia interneurons and the inhibitory input from the Renshaw cells (RC) coupled with the corresponding synergist motoneurons (Jankowska 1992).

*Monosynaptic excitation* via Ia afferent feedback has been implemented for all motoneurons of the locomotor spinal circuitry (Fig. 2.10a). To simulate the tonic presynaptic depression of Ia monosynaptic excitation of motoneurons occurring during locomotion (Gosgnach et al. 2000), the weights of monosynaptic excitatory inputs of Ia afferents to the corresponding motoneurons have been set to be relatively small values.

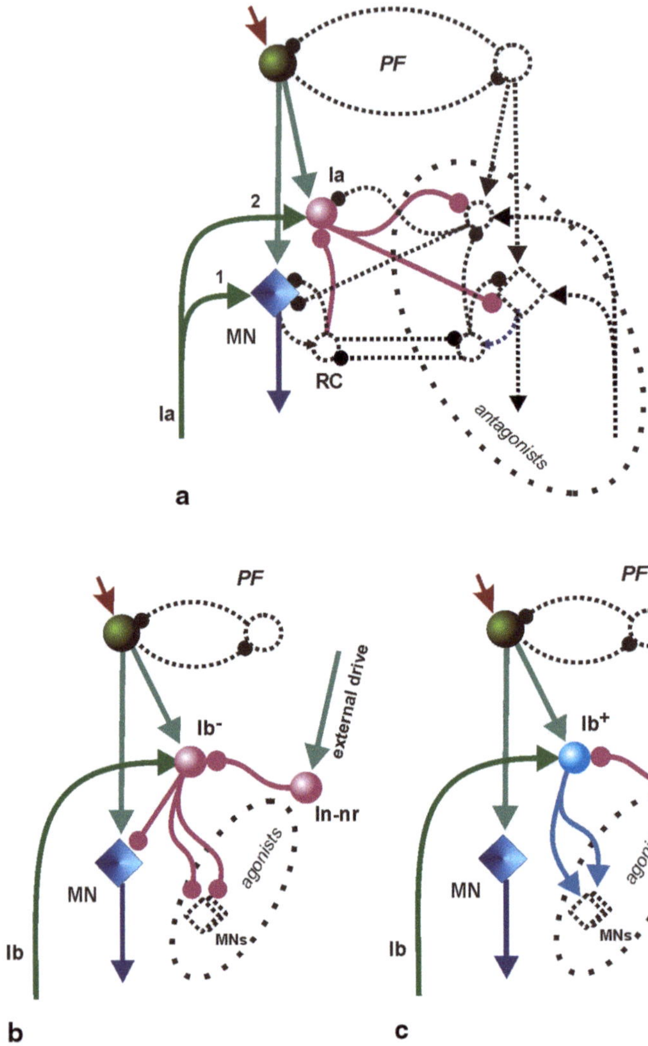


**Fig. 2.10** General schematic of basic reflexes implemented in the model of the locomotor circuitry (for further details see Table 2.7 in Appendix). **a** Monosynaptic stretch reflex *(1)* and Ia-reciprocal inhibition (*2*) pathways. The Ia interneuron is excited by the primary muscle spindle afferents from the corresponding muscle, and mediates Ia reciprocal inhibition of antagonist motoneurons (see (Jankowska 1992)). **b** Non-reciprocal inhibition circuitry. The $Ib^-$ inhibitory interneuron receives the excitatory input from Ib afferents of a muscle and inhibits the motoneurons innervating the corresponding muscle and its synergists (Jami 1992). **c** Force-dependent excitation between motoneurons innervating a two-joint muscle and its one-joint antagonist (e.g., from GA to TA) has been postulated based on a biomechanical analysis of muscle coordination (Prilutsky 2000). Such force-dependent excitation has been observed during muscle stretch evoked responses in the decerebrate cat. (Nichols et al. 2015)

Activation of Golgi tendon organ Ib afferents evokes a short latency inhibition of synergist motoneurons (Jami 1992; Jankowska 1992; Nichols et al. 1999). The principal circuitry that *provides non-reciprocal inhibition* of motoneurons is shown in Fig. 2.10b. This schematic is based on the circuitry described previously in (Rybak et al. 2006a, b). We introduced the additional $Ib^-$ interneuron that receives an excitatory Ib feedback from the corresponding muscle and an excitatory signal from interneurons at the pattern formation level. Because Ib inhibition is enhanced in decerebrate animals compared to intact animals (Whelan and Pearson 1997), we suggested an additional interneuron (In-nr) that might receive excitatory supra-spinal signals and mediate inhibitory signal on $Ib^-$ interneuron. The In-nr interneuron can be used to modulate activity of the corresponding motoneuron pool during execution of specific locomotor tasks. The $Ib^-$ interneuron relays the inhibitory signal to the same motoneuron and motoneurons innervating the synergists: BFP→(BFA, SartM), RF→(IP, VA), GA→(BFP, SartM, SOL).

A *force-dependent excitation* (Fig. 2.10c) is somewhat opposite to non-reciprocal inhibition. It is mediated by Ib afferents from two-joint muscles to their one-joint antagonists (e.g., from GA to TA). The force-dependent excitation has been postulated based on a biomechanical analysis of muscle coordination (Prilutsky 2000). Such force-dependent excitation has been observed during muscle stretch evoked responses in the decerebrate cat (Nichols et al. 2015).

### 2.5.4 Left-Right Hindlimb Coordination

The neuromechanical computational model of the locomotor circuitry has been constructed using the hypothesis that each hindlimb is controlled by a single CPG (Orlovsky et al. 1999), and the two hindlimb CPGs are connected via a coordinating neural network. The interaction between the left and right parts of the locomotor circuitry controlling the two hindlimbs is implemented via several excitatory and inhibitory populations of so-called *commissural interneurons* (CINs) relaying coordinating signals between the left and right CPGs. These interneuron populations project their connections onto the contralateral side of the spinal cord and coordinate operation of the contralateral CPG (Butt and Kiehn 2003; Lanuza et al. 2004; Jankowska et al. 2009; Zhong et al. 2012; Rybak et al. 2013). The inhibitory CINs mediate mutual inhibition between the ipsi- and contra-lateral CPGs and provide the alternation in activation of left and right homonymous CPG half-centers during normal walking. The excitatory CINs may provide direct excitation of the contralateral CPG, promoting synchronization between the homonymous CPG half-centers (as in the case of hopping or galloping), or contribute to mutual inhibition and alternation by acting via inhibitory interneurons located on each side. Because this chapter is focusing on normal walking we consider the possible organization of circuitry between CINs and contra-lateral CPG which provides alternating activity between left and right hindlimbs. In addition, according to recent experimental data obtained by (Jankowska et al. 2009), commissural interneurons (both excitatory and inhibi-

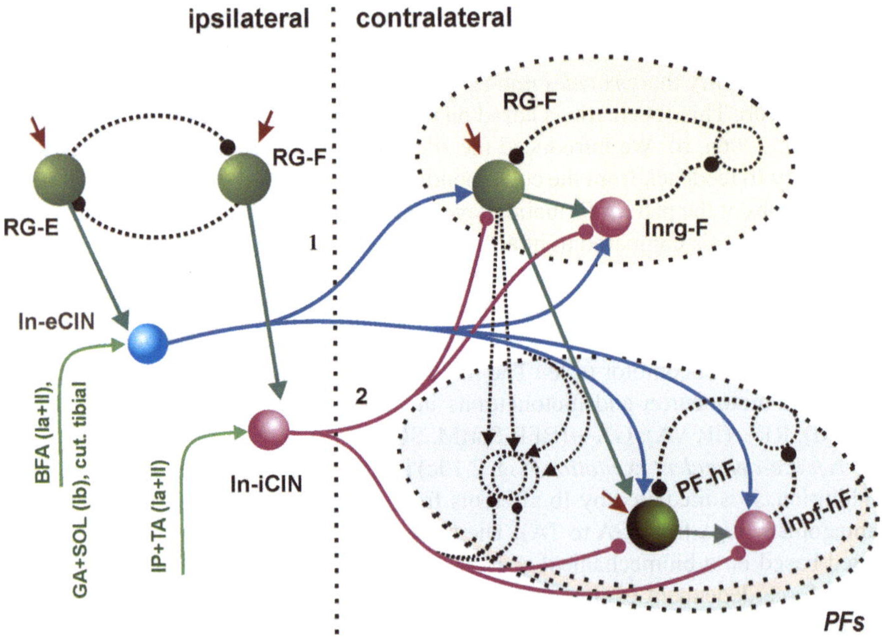


**Fig. 2.11** Possible connectivity between commissural interneurons and neurons of the contralateral locomotor circuitry. Two pathways providing alternation between *left* and *right* limbs during normal walking are presented in the figure: **1** excitatory pathways from the excitatory neuron *In-eCIN* that project excitation to the interneurons of the flexor part of the contra-lateral CPG (*blue* pathways); **2** inhibitory pathways from the inhibitory neuron *In-iCIN* to the interneurons at the *RG* and *PF* levels of the contra-lateral CPG (*purple* pathways). Tables 2.8 and 2.9 in Appendix provide additional information about weights of synaptic connections within *left-right* coordination circuitry

tory) receive afferent feedback from group I and/or group II afferents from a number of muscles including *quadriceps, sartorius medial, biceps femoris posterior* and *semitendinosus, biceps femoris anterior* and *semimembranosus, gastrocnemius* and *soleus, plantaris, flexor digitorum* and *hallucis longus*, and *deep peroneal* muscles.

Based on the above information, we propose the general schematic of circuitry providing left-right coordination between the corresponding hindlimbs (Fig. 2.11). Two types of CIN neurons (excitatory and inhibitory; In-eCIN and In-iCIN, respectively) were introduced to the locomotor circuitry model. The excitatory signals from the ipsilateral interneurons at the rhythm generator level (RG-E and RG-F) project their action to the flexor part of the contralateral CPG (see Fig. 2.11). They also receive monosynaptic multi-modal afferent input from muscles such as: IP and TA (Ia+II) to In-iCIN; BFA (Ia+II) and GA+SOL (Ib) to In-eCIN (the same afferent feedback signals are being used to control phase transition at the RG level of the ipsilateral CPG). These interneurons mediate inhibitory and excitatory actions to the flexor parts of the contralateral locomotor CPG at the RG and PFs levels, namely to the following interneurons: RG-F, Inrg-F; PF-hF, Inpf-hF; PF-kF, Inpf-kF; PF-aF, Inpf-aF; PF-BFP, In-eF, In-lF (see Fig. 2.7).

Previously, it has been suggested that cutaneous afferents are also involved in the coordination between left and right sides of locomotor circuitry by providing polysynaptic actions onto the flexor part of the contralateral CPG (Bouyer and Rossignol 2003a, b). Therefore, we hypothesized that feedback from cutaneous paw pad afferents also provides excitatory action onto the flexor part of the contralateral locomotor CPG at the RG and PFs levels via the appropriate commissural interneurons (In-eCIN).

## 2.6 Neuromechanical Simulations of Cat Level Walking

The neuromechanical hindlimb model and the CPG model were interconnected through feedforward (from motoneurons to muscles) and feedback (from muscles to neural circuits) signals. Weights of feedback signals were adjusted manually to obtain stable locomotion and to match the simulated motoneuronal activity with the muscle electromyographic (EMG) activity recorded during walking. The weights of the feedback signals were tuned using muscle EMG activity and locomotor mechanical characteristics recorded in five cats during overground and treadmill walking. The experimental procedures were approved by the Institutional Animal Care and Use Committees of Drexel University and Georgia Institute of Technology and described, along with the experimental results, in details elsewhere (Gregor et al. 2006; Ollivier-Lanvin et al. 2011; Prilutsky et al. 2011; Markin et al. 2012; Prilutsky et al. 2015). Briefly, the animals were trained for 3–4 weeks to walk on a Plexiglas enclosed walkway with 3 embedded 6-component force plates ($16 \times 11$ cm$^2$, 360 Hz sampling rate, Bertec, USA) with self-selected speed or on a treadmill at constant speed between 0.3 and 0.6 m/s. Kinematics of the two hindlimbs were recorded using a 6-camera motion capture system Vicon (120 Hz, UK) and reflective markers attached to the joints. After locomotor training, EMG electrodes were implanted in 9 muscles under aseptic conditions and isoflurane anesthesia, and walking mechanics and EMG activity (sampling rates 120 and 3000 Hz) were recorded and analyzed.

### *2.6.1 Model Performance*

After adjusting weights of afferent feedback signals, the neuromechanical model demonstrated stable locomotion with realistic walking mechanics and muscle activity. An increase in the supra-spinal drive to the CPG in the model (Fig. 2.12) increased the speed of locomotion, so that the walking cycle duration decreased due to the shortening of the stance phase with a relatively constant swing phase. The computed joint angles and ground reaction forces during walking with speeds of 0.4 and 0.55 m/s (see Fig. 2.12a and b, respectively) showed peak values and patterns similar to those recorded in walking cats (Gregor et al. 2006; Prilutsky et al. 2011).

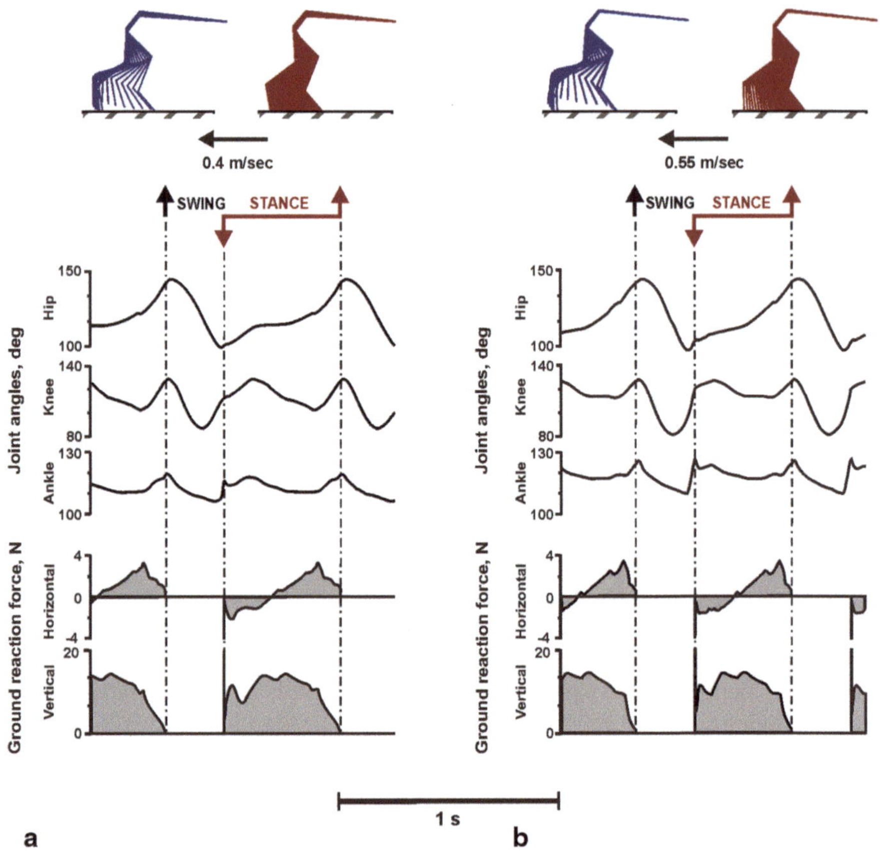


**Fig. 2.12** Model performance at different speeds of treadmill locomotion: 0.4 m/s (**a**) and 0.55 m/s (**b**). The figure panels show (from *top* to *bottom*) stick figures, *joint angles*, and *horizontal* and *vertical* ground reaction forces. The simulated mechanics of walking closely match the experiment. (Gregor et al. 2006; Prilutsky et al. 2015)

Simulated patterns of muscle activity showed the correct phases of bursts and patterns similar to the recorded EMG (Markin et al. 2012) (Fig. 2.13).

### *2.6.2 Afferent Feedback Signals Controlling Phase Transitions*

The neuromechanical model has been used for initial investigations of the effects of somatosensory afferent feedback on the control of locomotion. All results are obtained for a walking speed of 0.4 m/sec.

Our simulations have shown that the afferent feedback organization in the model and selected weights allow the model to maintain stable locomotion. Specifically, the transition between extensor and flexor phases in the CPG is mainly triggered

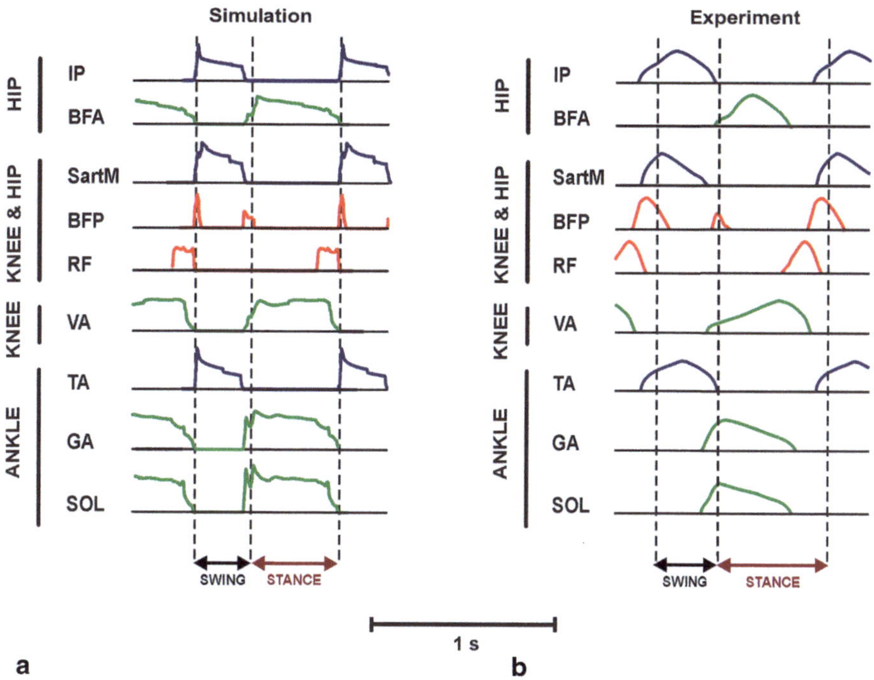


**Fig. 2.13** Comparison of simulated (**a**) and recorded (**b**) patterns of normalized muscle activity obtained for walking on a treadmill (panel **b** reproduced with permission from (Markin et al. 2012)). Simulations were conducted for 0.4 m/s walking speed. All muscles of the model demonstrated the correct phases of activity and patterns similar to the recorded EMGs

by the length- and velocity-dependent afferent activity of IP (see Fig. 2.14). The force-dependent afferent feedback from ankle extensors (SOL and GA) and *cutaneous* paw pad feedback signal also contribute to the control of the transition between extensor and flexor phases by prolonging the duration of the extensor phase. In contrast, the timing of the flexor-extensor transition in the CPG is controlled by the length- and velocity-dependent hip extensor afferent activity (BFA) which adjusts the duration of flexor phase to limb kinematics during swing phase.

Further analysis of the possible role of afferent feedback signals in controlling phase transitions during normal locomotion has shown that some of them are critical for providing stable locomotion during level walking. In our computational experiments we selectively blocked the specific length- and load-sensitive afferent pathways at the RG level of CPG during locomotion by setting the corresponding synaptic weights to zero. The results of selective blocking the afferent feedback signals are presented in Fig. 2.15. The kinematics of the model is shown in the left panels of the figure by stick diagrams. The blue stick figures represent the model performance under normal conditions. After 40 s of steady-state locomotion, the specific afferent feedback signals are blocked at the RG levels of both left and right hindlimbs' CPGs. The behavior of the cat model after blocking the specific afferent signal is shown by red stick figures. The right panels of Fig. 2.15 represent the

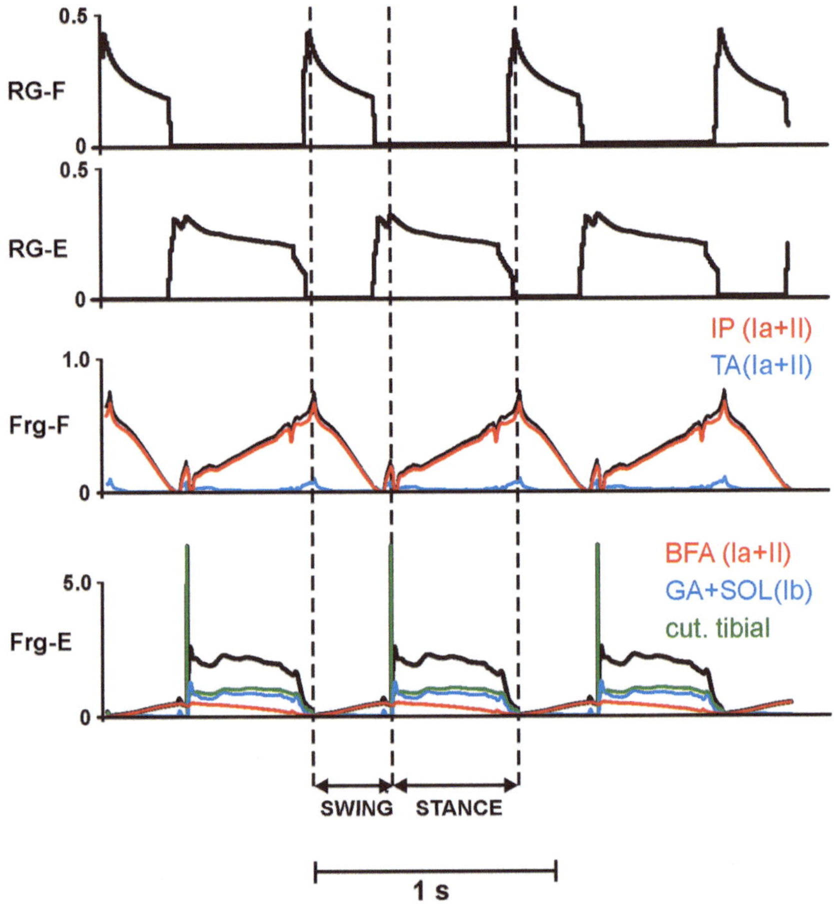


**Fig. 2.14** Rhythmic activity of RG neurons (RG-F and RG-E) and normalized feedback signals to corresponding interneurons (Frg-F and Frg-E) during locomotion (see Fig. 2.5). The activity of RG-F and RG-E represent changes of the output neural variable of each neuron *f*(V). The summarized afferent feedback inputs to both Frg-F and Frg-E interneurons are represented by *black lines*. The afferent feedback signals from different sources are shown by *red*, IP (Ia + II) or BFA (Ia + II); blue, TA (Ia + II) or GA + SOL (Ib); and *green* (cutaneous) *lines*. The *vertical dashed lines* separate the swing and stance phases

swing and stance phase durations of each cycle before and after blocking the corresponding afferent signals.

After blocking feedback of group I and II afferents from both IP and TA muscles, the duration of stance phase increased at the afferent block onset (see right panel of Fig. 2.15a), whereas the duration of the swing phase in the first stride after the afferent block did not change. The model could produce only 2 strides without length-dependent feedback from IP and TA muscles.

The removal of length-sensitive spindle afferents of the hip extensor (see Fig. 2.15b) affects the swing-stance phase transition. Namely, it increases the swing

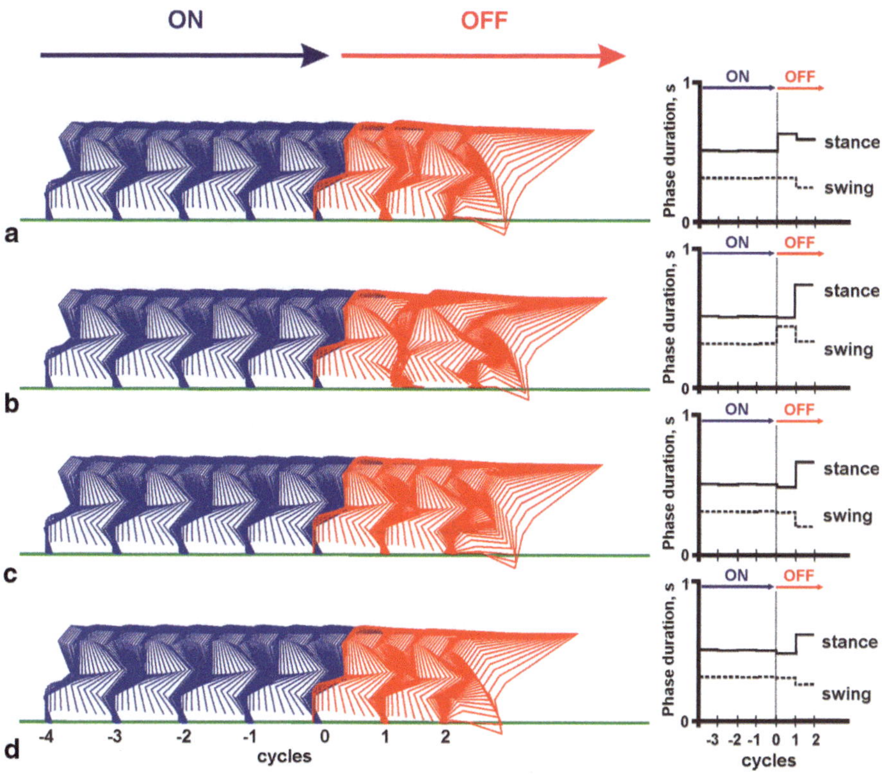


**Fig. 2.15** The role of specific afferent feedback signals in maintaining stable locomotion. The *left* panel shows walking kinematics of the model before (*blue stick* figures) and after (*red stick* figures) blocking specific afferent feedback signals at the RG level. The *right* panel shows swing (*dotted lines*) and stance (*solid lines*) phase durations of each walking cycle before and after blocking the corresponding afferent signal. **a** Both IP and TA primary and secondary spindle afferents (Ia and II) are blocked; **b** BFA primary spindle afferents (Ia) are blocked; **c** Both SOL and GA Ib afferents are blocked; **d** Cutaneous paw pad afferents are blocked. Here and further cycle 0 indicates the onset of blocking the corresponding afferent signal(s)

duration after blocking the correspondent afferents and causes the cat model to collapse after 2 strides. The changes in stance and swing phase durations shown in Fig. 2.15a and b are in agreement with physiological experiments in which hip flexion and extension phases during locomotion were perturbed to modulate the afferent input from hip flexors and extensors (Perreault et al. 1995; Lam and Pearson 2002; McVea et al. 2005; Gregor et al. 2006). Our computer simulation experiments have also shown that the contribution of afferent feedback activity to regulation of the stance-swing transition is much greater from IP than from TA (Fig. 2.14), which contradicts the experimental data obtained in fictive locomotion preparations (Perreault et al. 1995; Stecina et al. 2005).

Figures 2.15c and d demonstrate model behavior after block of group Ib afferent pathways from ankle extensors (SOL/GA) and from *cutaneous paw pad* afferents,

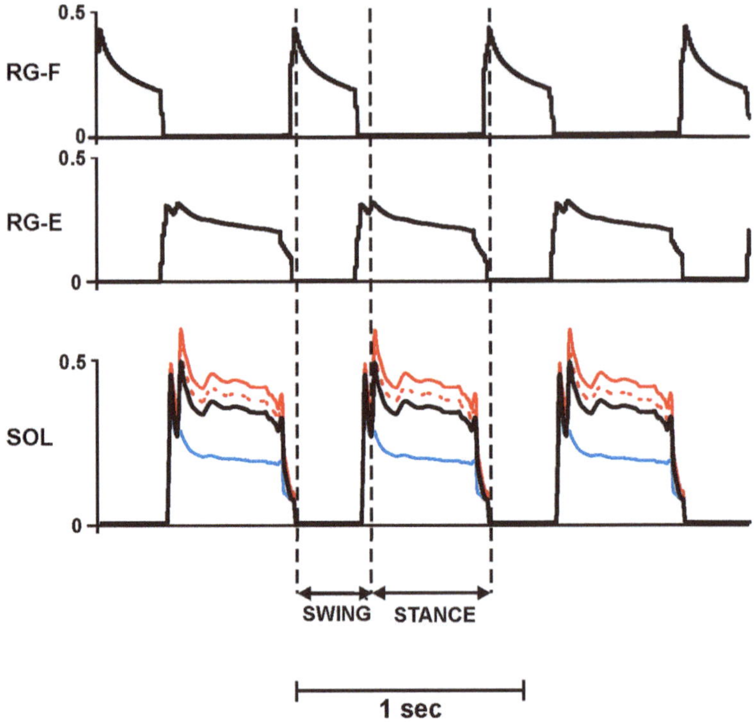


**Fig. 2.16** Regulation of activity of soleus motoneuron (*SOL*) by disynaptic excitation and Renshaw cells. The *upper* and *middle* traces represent activity of the output neural variable *f*(V) of the principal neurons at RG level. The *lower* traces represent normalized activity of *SOL* motoneuron with and without the disynaptic excitation or recurrent inhibition from Renshaw cells: *black* and *blue lines* indicate *SOL* motoneuron activity with and without the effects of disynaptic excitation, respectively; *red* continuous and *dashed lines* show *SOL* motoneuron activity when all Renshaw cells are completely inhibited and when only Renshaw cells of agonistic motoneurons are blocked, respectively. The *vertical dashed lines* separate the *swing* and *stance* phases

correspondingly. Both feedback signals play similar roles in regulating the phase transition and are responsible for prolongation of the stance phase. The stance phase duration (see right panel of Fig. 2.15c and d) is slightly reduced in the first stride after blocking either Ib afferents from SOL and GA or *cutaneous paw pad* afferents, whereas the duration of the swing phase is relatively stable. The block of either feedback signal also causes the cat model to collapse after the second walking stride.

### *2.6.3 Regulation of Motoneuron Activity by Disynaptic Excitation and Renshaw Cells*

Computer simulation experiments with the neuromechanical model have also shown the importance of afferent feedback for maintaining the ongoing motor activity during the stance phase of locomotion. Figure 2.16 demonstrates the activity

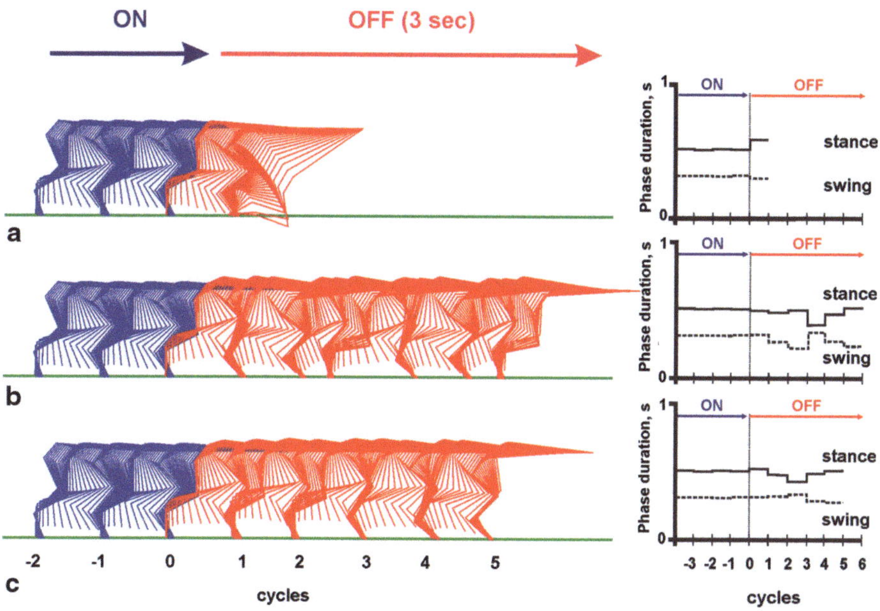


**Fig. 2.17** Regulation of motoneuronal activity by the disynaptic excitation and recurrent inhibition during locomotion. The *left* panel represents simulated kinematics of the model before (*blue stick figures*) and after (*red stick figures*) blocking the disynaptic excitation or recurrent inhibition. The *right* panel represents *swing* (*dotted lines*) and *stance* (*solid lines*) phase durations of each simulated walking cycle before and after the blocking intervention. **a** Blocking the disynaptic excitatory feedback from group Ia and Ib afferents from all extensor muscles that normally reinforces motor activity during the extensor phase. **b** All Renshaw cells are completely inhibited. **c** Renshaw cells of agonistic motoneurons are blocked

of SOL motoneuron with and without inputs from group Ib (or Ia and Ib) afferents from SOL muscle that provide disynaptic excitation of hindlimb extensor motoneurons during the stance phase. The obtained contribution of disynaptic excitation to the amplitude of SOL motoneuronal activity is about 30 %, which is similar to the afferent contributions estimated in other experimental and computational studies (Stein et al. 2000; Yakovenko et al. 2004).

Our simulations show that Renshaw cells may play an important role during locomotion. According to results of computational experiments (Fig. 2.16), Renshaw cells reduce the activity of synergistic motoneurons by recurrent inhibition, whereas inhibition of Renshaw cells by antagonistic motoneuron pools reduces Renshaw cell inhibitory influences on the synergist motoneurons.

Other reflex pathways included in the model—monosynaptic stretch reflex, non-reciprocal Ib-inhibition, have negligible contribution to regulation of the amplitude of motoneuronal activity.

In order to investigate how locomotor behavior of the model depends on the disynaptic excitation during the extensor phase and on Renshaw cell activity, we selectively switch off these pathways to all motoneurons for 3 s after 40 s of intact walking simulations (see Fig. 2.17). The results of selective blocking load-dependent disynaptic excitation of extensor motoneurons demonstrate the importance of these feedback signals in providing weight support during the stance phase of locomotion. After blocking the extensor group I disynaptic pathway, the model was able to perform only one stride and then collapsed (Fig. 2.17a). Interestingly, the stance phase duration in these conditions increases, whereas the swing phase duration practically does not change (see Fig. 2.17a, right panel). In addition, the simulations have shown that *cutaneous* input from the paw pad also plays a crucial role in weight support during locomotion. Blocking the *cutaneous paw pad* pathway (see Fig. 2.8), which reinforces the ongoing extensor motor activity during stance phase, also results in collapse of the system. The model performance in this case is very similar to blocking the disynaptic excitation (see Fig. 2.17a).

Renshaw cells appear to be mostly responsible for regulating stability of locomotion. In our computational experiments, turning off the Renshaw cell pathways, which either inhibit the synergist motoneurons or inhibit the antagonist Renshaw cells (Fig. 2.9), affects the model walking kinematics (see Fig. 2.17b and c, left panels). In both cases, the flexor and extensor motoneurons are overexcited, which distorts walking kinematics. It is interesting to note that turning off all Renshaw cells (Fig. 2.17b) brings the system into a less stable locomotion regime compared to the situation when only the inhibitory pathways to the agonistic Renshaw cells are blocked (Fig. 2.17c). For example, the comparison of swing and stance phase durations in both cases shows that they become more variable than in the control condition (see Fig. 2.17b and c, right panels). Depending on the onset time of the Renshaw pathway block and its duration, the model may collapse after several walking cycles.

### 2.6.4 Role of Length-Dependent Ia and Force-Dependent Ib Triceps Surae Afferents in Controlling Locomotor Activity

The relative contribution of length-dependent (Ia) and force-dependent (Ib) muscle afferents in the control of locomotor activity is difficult to establish experimentally due to similar activation thresholds of these afferents and simultaneous increases in muscle length and force during muscle stretch. In order to illustrate how the role of Ia and Ib afferents of triceps surae muscle (SOL+GA) in controlling locomotion can be studied using the developed neuromechanical model, the following computational experiments were performed. After 40 s of steady-state locomotion, the pathways from triceps surae group I afferents were selectively blocked for 6 s. Specifically, we selectively blocked Ib feedback signals from GA and SOL muscles which projected onto the circuitry of (1) disynaptic excitation and (2) pattern formation network. Also, (3) all pathways of Ia feedback signals from triceps surae were blocked. Model performance is shown as stick figures in Fig. (2.18a) and the stance and swing phase

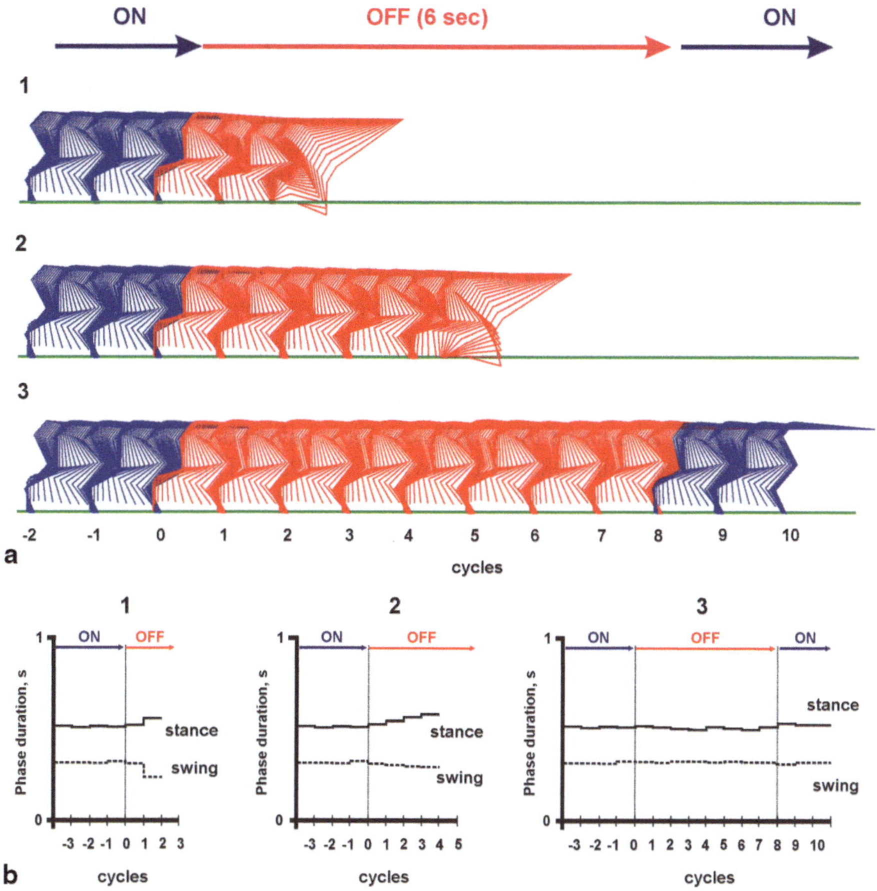


**Fig. 2.18** Stick-figure diagrams (**a**) and cycle phase durations (**b**) of simulated walking before (*blue stick figures*) and after (*red stick figures*) selective removal of Ia and Ib feedback signals from the triceps surae muscle. *(1)* The disynaptic excitation circuitry does not receive triceps Ib afferent input; *(2)* the pattern formation circuitry does not receive triceps Ib afferent input; *(3)* all levels of the locomotor CPG do not receive triceps Ia afferent input

durations (Fig. 2.18b) for the above three cases of selective removal of afferent pathways.

Based on these computer experiments we concluded that selective removal of Ia feedback signals from triceps surae muscle should not substantially change the activity patterns of this and other muscles and the model maintains stable locomotion (Fig. 2.18a3 and b3). On the other hand, when the Ib afferent feedback signals from triceps surae were reduced to zero, the model was unable to produce stable locomotion and collapsed after several strides. Moreover, according to the model, different Ib pathways play different roles in providing stable locomotion. For example, Ib feedback signals from GA and SOL muscles projecting onto the autogenic circuitry of disynaptic excitation were critical for reinforcement of ongoing GA-SOL motoneuronal activity (Fig. 2.18a1 and b1). The projection of Ib feedback signals from triceps surae onto the pattern formation network played an important role in regulation of the stance phase duration during level locomotion. The results of selective removal of this pathway (see Fig. 2.18b2) showed that the duration of stance phase was gradually increasing in each cycle after removal of this feedback eventually leading to collapse of the system after four walking cycles. Some of simulation results with selective removal of Ia afferent signals from SOL and GA (see Fig. 2.18) can be compared with experiments in which the monosynaptic action of Ia afferents is removed by self-reinnervation of these muscles in the cat (Abelew et al. 2000; Maas et al. 2007). In these studies, the muscle nerves were transected and surgically repaired and the proximal nerve stumps were allowed to regenerate for at least 3 months, an amount of time sufficient for reinnervation of the denervated muscle (Gordon and Stein 1982; O'Donovan et al. 1985). This self-reinnervation procedure removes the monosynaptic stretch reflex from self-reinnervated muscles (Cope and Clark 1993; Cope et al. 1994) due to permanent loss of Ia synapses on the cell bodies of motoneurons (Alvarez et al. 2011). Self-reinnervation of the entire triceps surae (Abelew et al. 2000), two heads of GA, medial gastrocnemius and lateral gastrocnemius, (Maas et al. 2007) or SOL and lateral gastrocnemius (Gregor et al. 2014) does not cause noticeable changes in hindlimb kinematics during level walking (Fig. 2.19), however an excessive ankle yield in stance during downslope walking has been reported (Abelew et al. 2000; Maas et al. 2007). Hindlimb joint angles during simulated walking with (blue lines) and without Ia input (red lines) from SOL and GA (Fig. 2.19a) have similar patterns and magnitudes and closely resemble joint angles recorded during walking of a representative cat before (blue lines) and after self-reinnervation of SOL and lateral gastrocnemius muscles (red lines, Fig. 2.19b).

## 2.7 Conclusion

This chapter describes a comprehensive neuromechanical model of spinal neural control of locomotion. The neuromechanical model includes a model of spinal circuits with the locomotor central pattern generator (Rybak et al. 2006a, b) controlled by afferent signals from 18 hindlimb Hill-type muscle actuators (Prilutsky et al. 2015). The spinal network incorporates basic circuits involved in the monosynaptic stretch reflex, Ia reciprocal inhibition, recurrent inhibition via Renshaw cells, and disynaptic excitation of extensor muscles via Ia and Ib afferents.

A special care was taken to tune model parameters since many of them cannot be determined experimentally. Three groups of model parameters, representing parameters of three subsystems of the model, were tuned independently. Parameters

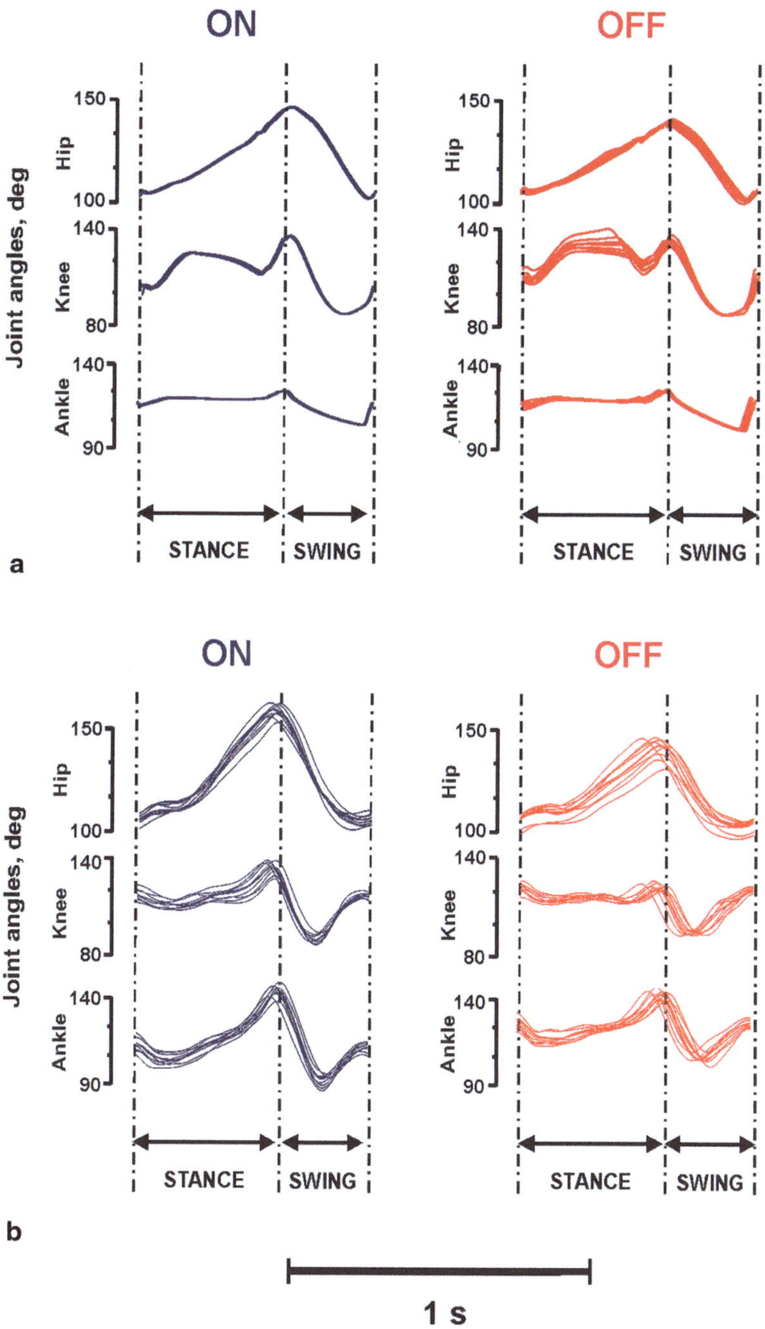


**Fig. 2.19** Comparison of simulated (**a**) and recorded (**b**) hindlimb joint angles during a cycle of level walking before (*blue lines*) and after (*red lines*) removal of Ia afferent input from GA and SOL muscles. Simulated joint angles are shown for 10 consecutive cycles of walking at 0.4 m/s.

of the CPG model were manually selected to reproduce the effects of deletions and afferent stimulations observed during fictive locomotion in the cat (Rybak et al. 2006a, b). Parameters of the musculoskeletal model of the hindlimbs were optimized so that the difference between computed and recorded joint angles, muscle fascicle length changes, joint moments, and ground reaction forces during walking in the cat is minimized (Prilutsky et al. 2015). Finally, weight coefficients of motion-dependent sensory feedback signals to the spinal circuits were manually tuned without changing the parameters of the CPG and musculoskeletal models to reproduce realistic locomotor patterns of muscle activity and mechanical characteristics of locomotion.

The model provides important insights into the neural mechanisms for control of locomotion and can be used as a testbed to study control of locomotion in normal and pathological conditions.

**Acknowledgements** This work was supported by the Biomedical Engineering Partnership Program of the National Institutes of Health (NIH) and the National Institute of Neurological Disorders and Stroke (Grant R01 NS-048844) and the National Institute of Biomedical Imaging and Bioengineering (Grant R01 EB-012855). B. I. Prilutsky was also supported by NIH Grant P01 HD-032571 and by the Center for Human Movement Studies, Georgia Tech.

---

In experiments, Ia afferent input was removed by self-reinnervation of SOL and lateral gastrocnemius muscles in the cat (Cope et al. 1994; Maas et al. 2007; Alvarez et al. 2011). The experimental data are unpublished results from Prilutsky's laboratory. The experimental procedures have been described in details elsewhere (Maas et al. 2007; Prilutsky et al. 2011). Experimental joint angles are shown for 10 cycles of level walking with self-selected speed recorded before (mean ± SD speed: 0.571 ± 0.082 m/s) and at least 12 weeks after self-reinnervation of SOL and lateral gastrocnemius muscles (0.573 ± 0.026 m/s)

# APPENDIX 2 Parameters of Locomotor Circuitry

Neuronal parameters of locomotor circuitry are based on parameters of simplified version of the CPG model described in (Markin et al. 2010).

**Table 2.1** Weights of synaptic connections within rhythm generator circuitry. (see Figs. 2.5 and 2.6)

| Targets | RG-F | RG-E | Inrg-F | Inrg-E |
|---|---|---|---|---|
| Sources | *Excitatory* | | | |
| Drive | 0.12 | 0.12 | | |
| RG-F | 0.01 | | 0.4 | |
| RG-E | | 0.01 | | 0.4 |
| Frg-F | 0.16 | | 0.12 | |
| Frg-E | | 0.44 | | 0.28 |
| | *Inhibitory* | | | |
| Inrg-F | | 2.1 | | |
| Inrg-E | 2.1 | | | |

**Table 2.2** Weights of synaptic connections within pattern formation circuitry that controls hip synergists. (see Figs. 2.5 and 2.7)

| Targets | PFh-F | PFh-E | Inpf-hF | Inpf-hE |
|---|---|---|---|---|
| Sources | *Excitatory* | | | |
| Drive | 0.62 | 0.62 | | |
| RG-F | 0.5 | | | |
| RG-E | | 0.5 | | |
| PF-hF | | | 0.4 | |
| PF-hE | | | | 0.4 |
| Fpf-hF | 0.05 | | 0.15 | |
| Fpf-hE | | 0.17 | | 0.25 |
| Fpf-cut | | 0.19 | | |
| | *Inhibitory* | | | |
| Inrg-F | | 1.2 | | |
| Inrg -E | 1.2 | | | |
| Inpf-hF | | 4.5 | | |
| Inpf-hE | 4.5 | | | |

**Table 2.3** Weights of synaptic connections within pattern formation circuitry that controls knee synergists. (see Figs. 2.5 and 2.7)

| Targets | PFk-F | PFk-E | Inpf-kF | Inpf-kE |
|---|---|---|---|---|
| Sources | *Excitatory* | | | |
| Drive | 0.62 | 0.62 | | |
| RG-F | 0.3 | | | |
| RG-E | | 0.3 | | |
| PF-kF | | | 0.4 | |
| PF-kE | | | | 0.4 |
| Fpf-kF | 0.12 | | 0.25 | |
| Fpf-kE | | 0.55 | | 0.47 |
| Fpf-cut | | | | 0.22 |
| | *Inhibitory* | | | |
| Inrg-F | | 1.6 | | |
| Inrg-E | 1.6 | | | |
| Inpf-kF | | 3.4 | | |
| Inpf-kE | 3.4 | | | |

**Table 2.4** Weights of synaptic connections within pattern formation circuitry that controls ankle synergists. (see Figs. 2.5 and 2.7)

| Targets | PF-aF | PF-aE | Inpf-aF | Inpf-aE |
|---|---|---|---|---|
| Sources | *Excitatory* | | | |
| Drive | 0.6 | 0.6 | | |
| RG-F | 0.4 | | | |
| RG-E | | 0.4 | | |
| PF-aF | | | 0.4 | |
| PF-aE | | | | 0.4 |
| Fpf-aF | 0.2 | | 1.0 | |
| Fpf-aE | | 0.45 | | 0.25 |
| Fpf-cut | | | | 0.1 |
| | *Inhibitory* | | | |
| Inrg-F | | 2.6 | | |
| Inrg-E | 2.6 | | | |
| Inpf-aF | | 3.9 | | |
| Inpf-aE | 3.9 | | | |

**Table 2.5** Weights of synaptic connections within pattern formation circuitry that specifically controls two-joint BFP and RF muscles. (see Figs. 2.5 and 2.7)

| Targets | PF-BFP | PF-RF | In-eF | In-lF | In-eE | In-lE |
|---|---|---|---|---|---|---|
| Sources | *Excitatory* | | | | | |
| Drive | 0.04 | 0.09 | | | | |
| RG-F | 0.5 | 0.06 | 0.14 | 0.08 | | |
| RG-E | 0.5 | 0.9 | | | 0.4 | |
| PF-BFP | | | | | | |
| PF-RF | | | | | | |
| Fpf-rf | | 0.3 | | | 0.2 | |
| Fpf-bfp | 0.2 | | 0.2 | | | 0.1 |
| | *Inhibitory* | | | | | |
| Inrg-F | | 1.3 | | | 0.3 | 3.3 |
| Inrg-E | 1.2 | | 1.4 | 3.5 | | 1.2 |
| In-eF | | | | 1.0 | | |
| In-lF | 3.9 | 1.8 | 2.7 | | | |
| In-eE | | 3.1 | | | | 0.8 |
| In-lE | 0.4 | | | | 3.6 | |

**Table 2.6** Weights of synaptic connections within disynaptic excitation and weight support circuitry (excluding motoneurons). (see Fig. 2.8)

| Targets | PFh-E | PFk-E | PFa-E | Iab-BFA | Iab-VA | Iab-GA | Iab-SOL | In-de |
|---|---|---|---|---|---|---|---|---|
| Sources | *Excitatory* | | | | | | | |
| Drive | | | | | | | | 0.2 |
| RG-E | | | | | | | | |
| PFh-E | | | | 0.26 | | | | |
| PFk-E | | | | | 0.26 | | | |
| PFa-E | | | | | | 0.26 | 0.24 | |
| Fpf-cut | 0.19 | 0.22 | 0.12 | | | | | |
| | *Inhibitory* | | | | | | | |
| Inrg-E | | | | | | | | 0.3 |
| In-de | | | | 0.8 | 0.8 | 0.8 | 0.8 | |

**Table 2.7** Weights of synaptic connections within reflexes circuitry (including motoneurons). (see Figs. 2.9 and 2.10)

| Targets | Mn-IP | RC-IP | Ia-IP | Ib-IP- |
|---|---|---|---|---|
| Sources | *Excitatory* | | | |
| PF-hF | 1.6 | | 0.3 | 0.35 |
| Mn-IP | | 0.2 | | |
| Ib-BFP + | 0.1 | | | |
| | *Inhibitory* | | | |
| Inpf-hF | | 0.03 | | |
| RC-IP | 0.3 | | 0.2 | |
| RC-SartM | 0.15 | | | |
| RC-TA | 0.23 | | | |
| RC-BFA | | 0.1 | | |
| Ia-BFA | 0.2 | | 0.2 | |
| Ib-IP- | 0.3 | | | |
| Ib-RF- | 0.9 | | | |
| | *IP afferents* | | | |
| Ia | 0.2 | | 0.4 | |
| Ib | | | | 0.4 |

| Targets | Mn-SartM | RC-SartM | Ia-SartM | Ib-SartM- |
|---|---|---|---|---|
| Sources | *Excitatory* | | | |
| PF-hF | 0.9 | | 0.1 | 0.1 |
| PF-kF | 1.0 | | 0.2 | 0.27 |
| Mn-SartM | | 0.2 | | |
| | *Inhibitory* | | | |
| Inpf-hF | | 0.01 | | |
| Inpf-kF | | 0.02 | | |
| RC-SartM | 0.2 | | 0.2 | |
| RC-IP | 0.15 | | | |
| RC-TA | 0.1 | | | |
| RC-VA | | 0.2 | | |
| RC-BFA | | 0.1 | | |
| Ia-VA | 0.2 | | 0.15 | |
| Ia-RF | 0.2 | | 0.1 | |
| Ia-BFA | 0.1 | | 0.1 | |
| Ib-SartM- | 0.17 | | | |
| Ib-BFP- | 0.8 | | | |
| | *SartM afferents* | | | |
| Ia | 0.1 | | 0.1 | |
| Ib | | | | 0.23 |

| Targets | Mn-BFA | RC-BFA | Ia-BFA |
|---|---|---|---|
| Sources | *Excitatory* | | |
| PF-hE | 1.8 | | 0.4 |
| Mn-BFA | | 0.2 | |
| Iab-BFA | 0.5 | | |
| | *Inhibitory* | | |
| Inpf-hE | | 0.04 | |
| RC-IP | | 0.1 | |
| RC-BFA | 0.2 | | 0.2 |
| RC-VA | 0.08 | | |
| RC-GA | 0.09 | | |
| RC-SOL | 0.14 | | |
| Ia-IP | 0.25 | | 0.2 |
| Ia-SartM | 0.15 | | 0.15 |
| Ib-BFP- | 1.1 | | |
| | *BFA afferents* | | |
| Ia | 0.2 | | 0.2 |

| Targets | Mn-BFP | RC-BFP | Ia-BFP | Ib-BFP- | Ib-BFP + |
|---|---|---|---|---|---|
| Sources | *Excitatory* | | | | |
| PF-BFP | 1.4 | | 0.8 | 0.1 | 0.1 |
| Mn-BFP | | 0.2 | | | |
| | *Inhibitory* | | | | |
| RC-BFP | 0.2 | | 0.2 | | |
| RC-RF | | 0.2 | | | |
| Ia-RF | 0.2 | | 0.2 | | |
| Ia-SartM | 0.15 | | 0.05 | | |
| | *BFP afferents* | | | | |
| Ia | 0.13 | | 0.1 | | |
| Ib | | | | 0.7 | 0.7 |

| Targets | Mn-RF | RC-RF | Ia-RF | Ib-RF- |
|---|---|---|---|---|
| Sources | *Excitatory* | | | |
| PF-RF | 1.6 | | 0.85 | 0.1 |
| Mn-RF | | 0.2 | | |
| | *Inhibitory* | | | |
| RC-RF | 0.2 | | 0.2 | |
| RC-BFP | | 0.2 | | |
| Ia-GA | 0.1 | | 0.1 | |
| Ia-BFP | 0.2 | | 0.2 | |
| | *RF afferents* | | | |
| Ia | 0.15 | | 0.3 | |
| Ib | | | | 0.6 |

| Targets | Mn-VA | RC-VA | Ia-VA | Ib-VA- |
|---|---|---|---|---|
| Sources | *Excitatory* | | | |
| PF-kE | 1.75 | | 0.37 | 0.1 |
| Mn-VA | | 0.2 | | |
| Iab-VA | 0.43 | | | |
| Ib-GA + | 0.09 | | | |
| | *Inhibitory* | | | |
| Inpf-kE | | | | |
| RC-SartM | 0.2 | 0.2 | | |
| RC-VA | 0.2 | | 0.2 | |
| RC-BFA | 0.09 | | | |
| RC-GA | 0.14 | | | |
| RC-SOL | 0.06 | | | |
| Ia-GA | 0.1 | | 0.1 | |
| Ia-BFP | 0.3 | | | |
| Ia-SartM | 0.22 | | 0.02 | |
| Ib-RF- | 0.1 | | | |
| | *VA afferents* | | | |
| Ia | 0.13 | | 0.1 | |
| Ib | | | | 0.1 |

| Targets | Mn-GA | RC-GA | Ia-GA | Ib-GA- | Ib-GA + |
|---|---|---|---|---|---|
| Sources | *Excit-atory* | | | | |
| PF-kF | 0.2 | | 0.05 | 0.01 | 0.01 |
| PF-aE | 1.5 | | 0.45 | 0.1 | 0.1 |
| Mn-GA | | 0.2 | | | |
| Iab-GA | 0.33 | | | | |
| | *Inhibitory* | | | | |
| Inpf-aE | | 0.06 | | | |
| Inpf-kF | | 0.01 | | | |
| RC-TA | 0.1 | 0.1 | | | |
| RC-GA | 0.2 | | 0.2 | | |
| RC-BFA | 0.1 | | | | |
| RC-VA | 0.06 | | | | |
| RC-SOL | 0.06 | | | | |
| Ia-TA | 0.3 | | 0.2 | | |
| Ia-RF | 0.2 | | 0.2 | | |
| Ia-VA | 0.1 | | 0.2 | | |
| | *GA afferents* | | | | |
| Ia | 0.2 | | 0.2 | | |
| Ib | | | | 0.25 | 0.15 |

| Targets | Mn-TA | RC-TA | Ia-TA | Ib-TA- |
|---|---|---|---|---|
| Sources | *Excitatory* | | | |
| PF-aF | 2.1 | | 0.6 | 0.4 |
| Mn-TA | | 0.2 | | |
| Ib-GA + | 0.1 | | | |
| | *Inhibitory* | | | |
| Inpf-aF | | 0.03 | | |
| RC-TA | 0.3 | | 0.2 | |
| RC-IP | 0.2 | | | |
| RC-SartM | 0.1 | | | |
| Ia-GA | 0.2 | 0.1 | 0.2 | |
| Ia-SOL | 0.5 | 0.1 | 0.2 | |
| Ib-TA- | 0.2 | | | |
| | *TA afferents* | | | |
| Ia | 0.1 | | 0.3 | |
| Ib | | | | 0.35 |

| Targets | Mn-SOL | RC-SOL | Ia-SOL | Ib-SOL- |
|---|---|---|---|---|
| Sources | *Excitatory* | | | |
| PF-aE | 1.75 | | 0.37 | 0.1 |
| Mn-SOL | | 0.2 | | |
| Iab-SOL | 0.31 | | | |
| | *Inhibitory* | | | |
| Inpf-aE | | 0.05 | | |
| RC-TA | | 0.1 | | |
| RC-SOL | 0.2 | | 0.2 | |
| RC-BFA | 0.1 | | | |
| RC-VA | 0.04 | | | |
| RC-GA | 0.08 | | | |
| Ia-TA | 0.5 | | 0.2 | |
| Ib-GA- | 0.23 | | | |
| | *SOL afferents* | | | |
| Ia | 0.21 | | 0.3 | |
| Ib | | | | 0.1 |

**Table 2.8** Weights of synaptic connections within left-right coordination circuitry. (see Fig. 2.11)

| Trg | (c)RG-F | (c)Inrg-F | (c)PF-hF | (c)Inpf-hF | (c)PF-kF | (c) Inpf-kF | (c)PF-aF | (c) Inpf-aF |
|---|---|---|---|---|---|---|---|---|
| Src | *Excit-atory* | | | | | | | |
| In-eCIN | 0.3 | 0.02 | 0.2 | 0.04 | 0.15 | 0.03 | 0.15 | 0.03 |
| | *Inhibitory* | | | | | | | |
| In-iCIN | 2.6 | 0.05 | 3.2 | 0.07 | 2.4 | 0.04 | 2.6 | 0.05 |

**Table 2.9** Weights of synaptic connections feedbacks. (see Figs. 2.6, 2.7, 2.8, 2.10, and 2.11)

| Targets | Frg-F | Frg-E | Fpf-hF | Fpf-hE | Fpf-kF |
|---|---|---|---|---|---|
| Sources | IP afferents | | | | |
| Ia | 0.43 | | 0.15 | | |
| II | 0.2 | | 0.3 | | |
| | *SartM afferents* | | | | |
| Ia | | | | | 0.28 |
| II | | | | | 0.75 |
| | *BFA afferents* | | | | |
| Ia | | 0.06 | | 0.06 | |
| II | | 0.08 | | 0.11 | |
| Ib | | | | 0.1 | |
| | *GA afferents* | | | | |
| Ia | | | | | 0.03 |
| II | | | | | 0.1 |
| Ib | | 0.05 | | | |
| | *TA afferents* | | | | |
| Ia | 0.1 | | | | |
| II | 0.1 | | | | |
| | *SOL afferents* | | | | |
| Ia | | | | | |
| Ib | | 0.18 | | | |
| | *Cutaneous* | | | | |
| | | 0.25 | | | |

| | Fpf-kE | Fpf-aF | Fpf-aE | Fpf-bfp | Fpf-rf |
|---|---|---|---|---|---|
| | *BFP afferents* | | | | |
| Ia | | | | 0.25 | |
| II | | | | 0.15 | |
| | *RF afferents* | | | | |
| Ia | | | | | 0.5 |
| II | | | | | 0.5 |
| | *VA afferents* | | | | |
| Ia | 0.15 | | | | |
| II | 0.24 | | | | |
| Ib | 0.3 | | | | |
| | *GA afferents* | | | | |
| Ia | | | | 0.02 | |
| II | | | | 0.04 | |
| Ib | | | | 0.05 | |
| | *TA afferents* | | | | |
| Ia | | 0.3 | | | |
| II | | 0.6 | | | |
| | *SOL afferents* | | | | |
| Ia | | | 0.03 | | |
| II | | | 0.06 | | |
| Ib | | | 0.07 | | |

| | Fpf-cut | Iab-BFA | Iab-VA | Iab-GA | Iab-SOL |
|---|---|---|---|---|---|
| | *BFA afferents* | | | | |
| Ia | | 0.02 | | | |
| Ib | | 0.01 | | | |
| | *VA afferents* | | | | |
| Ia | | | 0.05 | | |
| Ib | | | 0.31 | | |
| | *GA afferents* | | | | |
| Ia | | | | 0.28 | |
| Ib | | | | 1.4 | |
| | *TA afferents* | | | | |
| Ia | | | | | 0.75 |
| Ib | | | | | 0.75 |
| | *Cutaneous* | | | | |
| | 1 | | | | |

| | In-eCIN | In-iCIN |
|---|---|---|
| | *IP afferents* | |
| Ia | | 0.43 |
| II | | 0.2 |
| | *BFA afferents* | |
| Ia | 0.06 | |
| II | 0.08 | |
| | *GA afferents* | |
| Ib | 0.05 | |
| | *TA afferents* | |
| Ia | | 0.1 |
| II | | 0.1 |
| | *SOL afferents* | |
| Ib | 0.18 | |
| | *Cutaneous* | |
| | 0.25 | |

## References


Abelew TA, Miller MD, Cope TC, Nichols TR (2000) Local loss of proprioception results in disruption of interjoint coordination during locomotion in the cat. J Neurophysiol 84:2709–2714

Alvarez FJ, Fyffe RE (2007) The continuing case for the Renshaw cell. J Physiol 584:31–45

Alvarez FJ, Titus-Mitchell HE, Bullinger KL, Kraszpulski M, Nardelli P, Cope TC (2011) Permanent central synaptic disconnection of proprioceptors after nerve injury and regeneration. I. Loss of VGLUT1/IA synapses on motoneurons. J Neurophysiol 106:2450–2470

Andersson O, Grillner S (1983) Peripheral control of the cat's step cycle. II. Entrainment of the central pattern generators for locomotion by sinusoidal hip movements during "fictive locomotion.". Acta Physiol Scand 118:229–239

Angel MJ, Guertin P, Jimenez I, McCrea DA (1996) Group I extensor afferents evoke disynaptic EPSPs in cat hindlimb extensor motorneurones during fictive locomotion. J Physiol 494(Pt 3):851–861

Aoi S (2015) Neuromusculoskeletal modeling for the adaptive control of posture during locomotion. In: Prilutsky BI, Edwards DH Jr (eds) Neuromechanical modeling of posture and locomotion. Springer, New York

Aoi S, Kondo T, Hayashi N, Yanagihara D, Aoki S, Yamaura H et al (2013) Contributions of phase resetting and interlimb coordination to the adaptive control of hindlimb obstacle avoidance during locomotion in rats: a simulation study. Biol Cybern 107:201–216

Baldissera F, Hultborn H, Illert M (1981) Integration in spinal neuronal systems. In: Brooks VB (ed) Handbook of physiology. Section 1: the nervous system. Vol. III. motor control, part 1. American Physiological Society, Washington, DC, pp 509–595

Baratta RV, Solomonow M, Best R, D'Ambrosia R (1993) Isotonic length/force models of nine different skeletal muscles. Med Biol Eng Comput 31:449–458

Baratta RV, Solomonow M, Best R, Zembo M, D'Ambrosia R (1995) Architecture-based force-velocity models of load-moving skeletal muscles. Clin Biomech (Bristol Avon) 10:149–155

Beer RD, Chiel HJ, Gallagher JC (1999) Evolution and analysis of model CPGs for walking: II. General principles and individual variability. J Comput Neurosci 7:119–147

Bouyer LJ, Rossignol S (2003a) Contribution of cutaneous inputs from the hindpaw to the control of locomotion. I. Intact cats. J Neurophysiol 90:3625–3639

Bouyer LJ, Rossignol S (2003b) Contribution of cutaneous inputs from the hindpaw to the control of locomotion. II. Spinal cats. J Neurophysiol 90:3640–3653

Brown TG (1911) The intrinsic factors in the act of progression in the mammal. Proc R Soc B 84:308–319

Brown IE, Liinamaa TL, Loeb GE (1996) Relationships between range of motion, lo, and passive force in five strap-like muscles of the feline hind limb. J Morphol 230:69–77

Burke RE (1999) The use of state-dependent modulation of spinal reflexes as a tool to investigate the organization of spinal interneurons. Exp Brain Res 128:263–277

Burke RE, Degtyarenko AM, Simon ES (2001) Patterns of locomotor drive to motoneurons and last-order interneurons: clues to the structure of the CPG. J Neurophysiol 86:447–462

Butt SJ, Kiehn O (2003) Functional identification of interneurons responsible for left-right coordination of hindlimbs in mammals. Neuron 38:953–963

Cohen AH, Holmes PJ, Rand RH (1982) The nature of the coupling between segmental oscillators of the lamprey spinal generator for locomotion: a mathematical model. J Math Biol 13:345–369

Collins JJ, Richmond SA (1994) Hard-wired central pattern generators for quadrupedal locomotion. Biol Cybern 71:375–385

Conway BA, Hultborn H, Kiehn O (1987) Proprioceptive input resets central locomotor rhythm in the spinal cat. Exp Brain Res 68:643–656

Cope TC, Clark BD (1993) Motor-unit recruitment in self-reinnervated muscle. J Neurophysiol 70:1787–1796

Cope TC, Bonasera SJ, Nichols TR (1994) Reinnervated muscles fail to produce stretch reflexes. J Neurophysiol 71:817–820

Corana A, Marchesi M, Martini C, Ridella S (1987) Minimizing multimodal functions of continuous variables with the "simulated annealing" algorithm. ACM Trans Math Softw 13:263–280
Degtyarenko AM, Simon ES, Norden-Krichmar T, Burke RE (1998) Modulation of oligosynaptic cutaneous and muscle afferent reflex pathways during fictive locomotion and scratching in the cat. J Neurophysiol 79:447–463
Duysens J, Pearson KG (1980) Inhibition of flexor burst generation by loading ankle extensor muscles in walking cats. Brain Res 187:321–332
Eccles RM, Lundberg A (1958) Integrative pattern of Ia synaptic actions on motoneurones of hip and knee muscles. J Physiol 144:271–298
Eccles JC, Fatt P, Landgren S (1956) Central pathway for direct inhibitory action of impulses in largest afferent nerve fibres to muscle. J Neurophysiol 19:75–98
Eccles JC, Eccles RM, Lundberg A (1957a) The convergence of monosynaptic excitatory afferents on to many different species of alpha motoneurones. J Physiol 137:22–50
Eccles JC, Eccles RM, Lundberg A (1957b) Synaptic actions on motoneurones caused by impulses in Golgi tendon organ afferents. J Physiol 138:227–252
Ekeberg O, Pearson K (2005) Computer simulation of stepping in the hind legs of the cat: an examination of mechanisms regulating the stance-to-swing transition. J Neurophysiol 94:4256–4268
Feldman AG, Orlovsky GN (1975) Activity of interneurons mediating reciprocal 1a inhibition during locomotion. Brain Res 84:181–194
Fouad K, Pearson KG (1997) Effects of extensor muscle afferents on the timing of locomotor activity during walking in adult rats. Brain Res 749:320–328
Frigon A, Sirois J, Gossard JP (2010) Effects of ankle and hip muscle afferent inputs on rhythm generation during fictive locomotion. J Neurophysiol 103:1591–1605
Gordon T, Stein RB (1982) Time course and extent of recovery in reinnervated motor units of cat triceps surae muscles. J Physiol 323:307–323
Gosgnach S, Quevedo J, Fedirchuk B, McCrea DA (2000) Depression of group Ia monosynaptic EPSPs in cat hindlimb motoneurones during fictive locomotion. J Physiol 526(Pt 3):639–652
Gossard JP, Brownstone RM, Barajon I, Hultborn H (1994) Transmission in a locomotor-related group Ib pathway from hindlimb extensor muscles in the cat. Exp Brain Res 98:213–228
Gossard JP, Sirois J, Noue P, Cote MP, Menard A, Leblond H et al (2011) Chapter 2-the spinal generation of phases and cycle duration. Prog Brain Res 188:15–29
Gottschall JS, Nichols TR (2011) Neuromuscular strategies for the transitions between level and hill surfaces during walking. Philos Trans R Soc Lond B Biol Sci 366:1565–1579
Gregor RJ, Smith DW, Prilutsky BI (2006) Mechanics of slope walking in the cat: quantification of muscle load, length change, and ankle extensor EMG patterns. J Neurophysiol 95:1397–1409
Gregor RJ, Bulgakova MA, Maas H, Oliver A, Prilutsky BI (2014) Locomotor activity of feline ankle extensors and kinematics during level and slope walking after removal of stretch reflex from soleus and lateral gastrocnemius by self-reinnervation. In: 2014 Neuroscience Meeting Planner, Program # 827-01. (ed), Society for Neuroscience. Online, Washington, DC
Grillner S (1981) Control of locomotion in bipeds, tetrapods, and fish. In: Brooks V (ed) Handbook of physiology. Section I. The nervous system. American Physiological Society, Bethesda, pp 1179–1236
Grillner S, Wallen P, Saitoh K, Kozlov A, Robertson B (2008) Neural bases of goal-directed locomotion in vertebrates-an overview. Brain Res Rev 57:2–12
Guertin PA (2012) Central pattern generator for locomotion: anatomical, physiological, and pathophysiological considerations. Front Neurol 3:183
Guertin P, Angel MJ, Perreault MC, McCrea DA (1995) Ankle extensor group I afferents excite extensors throughout the hindlimb during fictive locomotion in the cat. J Physiol 487(Pt 1):197–209
Hiebert GW, Pearson KG (1999) Contribution of sensory feedback to the generation of extensor activity during walking in the decerebrate cat. J Neurophysiol 81:758–770
Hiebert GW, Whelan PJ, Prochazka A, Pearson KG (1996) Contribution of hind limb flexor muscle afferents to the timing of phase transitions in the cat step cycle. J Neurophysiol 75:1126–1137

Hoy MG, Zernicke RF (1985) Modulation of limb dynamics in the swing phase of locomotion. J Biomech 18:49–60

Ivashko DG, Prilutsky BI, Markin SN, Chapin JK, Rybak IA (2003) Modeling the spinal cord neural circuitry controlling cat hindlimb movement during locomotion. Neurocomputing 52–54:621–629

Jami L (1992) Golgi tendon organs in mammalian skeletal muscle: functional properties and central actions. Physiol Rev 72:623–666

Jankowska E (1992) Interneuronal relay in spinal pathways from proprioceptors. Prog Neurobiol 38:335–378

Jankowska E, Bannatyne BA, Stecina K, Hammar I, Cabaj A, Maxwell DJ (2009) Commissural interneurons with input from group I and II muscle afferents in feline lumbar segments: neurotransmitters, projections and target cells. J Physiol 587:401–418

Kiehn O (2011) Development and functional organization of spinal locomotor circuits. Curr Opin Neurobiol 21:100–109

Kriellaars DJ, Brownstone RM, Noga BR, Jordan LM (1994) Mechanical entrainment of fictive locomotion in the decerebrate cat. J Neurophysiol 71:2074–2086

Lam T, Pearson KG (2002) Sartorius muscle afferents influence the amplitude and timing of flexor activity in walking decerebrate cats. Exp Brain Res 147:175–185

Lanuza GM, Gosgnach S, Pierani A, Jessell TM, Goulding M (2004) Genetic identification of spinal interneurons that coordinate left-right locomotor activity necessary for walking movements. Neuron 42:375–386

Lloyd DP (1946) Integrative pattern of excitation and inhibition in two-neuron reflex arcs. J Neurophysiol 9:439–444

Maas H, Prilutsky BI, Nichols TR, Gregor RJ (2007) The effects of self-reinnervation of cat medial and lateral gastrocnemius muscles on hindlimb kinematics in slope walking. Exp Brain Res 181:377–393

Markin SN, Klishko AN, Shevtsova NA, Lemay MA, Prilutsky BI, Rybak IA (2010) Afferent control of locomotor CPG: insights from a simple neuromechanical model. Ann N Y Acad Sci 1198:21–34

Markin SN, Lemay MA, Prilutsky BI, Rybak IA (2012) Motoneuronal and muscle synergies involved in cat hindlimb control during fictive and real locomotion: a comparison study. J Neurophysiol 107:2057–2071

Maufroy C, Kimura H, Takase K (2008) Towards a general neural controller for quadrupedal locomotion. Neural Netw 21:667–681

McCrea DA (2001) Spinal circuitry of sensorimotor control of locomotion. J Physiol 533:41–50

McCrea DA, Rybak IA (2007) Modeling the mammalian locomotor CPG: insights from mistakes and perturbations. Prog Brain Res 165:235–253

McCrea DA, Rybak IA (2008) Organization of mammalian locomotor rhythm and pattern generation. Brain Res Rev 57:134–146

McCrea DA, Pratt CA, Jordan LM (1980) Renshaw cell activity and recurrent effects on motoneurons during fictive locomotion. J Neurophysiol 44:475–488

McCrea DA, Shefchyk SJ, Stephens MJ, Pearson KG (1995) Disynaptic group I excitation of synergist ankle extensor motoneurones during fictive locomotion in the cat. J Physiol 487(Pt 2):527–539

McVea DA, Donelan JM, Tachibana A, Pearson KG (2005) A role for hip position in initiating the swing-to-stance transition in walking cats. J Neurophysiol 94:3497–3508

Menard A, Leblond H, Gossard JP (1999) The modulation of presynaptic inhibition in single muscle primary afferents during fictive locomotion in the cat. J Neurosci 19:391–400

Nassour J, Henaff P, Benouezdou F, Cheng G (2014) Multi-layered multi-pattern CPG for adaptive locomotion of humanoid robots. Biol Cybern 108(3):291–303

Nichols TR, Bunderson N, Lyle M (2015) Neural regulation of limb mechanics: insights from the organization of proprioceptive circuits. In: Prilutsky BI, Edwards DH Jr (eds) Neuromechanical modeling of posture and locomotion. Springer, New York

Nichols TR, Cope TC, Abelew TA (1999) Rapid spinal mechanisms of motor coordination. Exerc Sport Sci Rev 27:255–284
Nishimaru H, Restrepo CE, Kiehn O (2006) Activity of Renshaw cells during locomotor-like rhythmic activity in the isolated spinal cord of neonatal mice. J Neurosci 26:5320–5328
O'Donovan MJ, Pinter MJ, Dum RP, Burke RE (1985) Kinesiological studies of self- and cross-reinnervated FDL and soleus muscles in freely moving cats. J Neurophysiol 54:852–866
Ollivier-Lanvin K, Krupka AJ, AuYong N, Miller K, Prilutsky BI, Lemay MA (2011) Electrical stimulation of the sural cutaneous afferent nerve controls the amplitude and onset of the swing phase of locomotion in the spinal cat. J Neurophysiol 105:2297–2308
Orlovsky GN, Deliagina TG, Grillner S (1999) Neuronal control of locomotion: from mollusc to man. Oxford University Press, New York
Pearson KG (1995) Proprioceptive regulation of locomotion. Curr Opin Neurobiol 5:786–791
Pearson KG (2008) Role of sensory feedback in the control of stance duration in walking cats. Brain Res Rev 57:222–227
Pearson KG, Collins DF (1993) Reversal of the influence of group Ib afferents from plantaris on activity in medial gastrocnemius muscle during locomotor activity. J Neurophysiol 70:1009–1017
Pearson KG, Misiaszek JE, Fouad K (1998) Enhancement and resetting of locomotor activity by muscle afferents. Ann N Y Acad Sci 860:203–215
Perreault MC, Angel MJ, Guertin P, McCrea DA (1995) Effects of stimulation of hindlimb flexor group II afferents during fictive locomotion in the cat. J Physiol 487(Pt 1):211–220
Perreault MC, Shefchyk SJ, Jimenez I, McCrea DA (1999) Depression of muscle and cutaneous afferent-evoked monosynaptic field potentials during fictive locomotion in the cat. J Physiol 521(Pt 3):691–703
Pratt CA, Jordan LM (1987) Ia inhibitory interneurons and Renshaw cells as contributors to the spinal mechanisms of fictive locomotion. J Neurophysiol 57:56–71
Prilutsky BI (2000) Coordination of two- and one-joint muscles: functional consequences and implications for motor control. Motor Control 4:1–44
Prilutsky BI, Klishko AN, Weber DJ, Lemay MA (2015) Computing motion dependent afferent activity during cat locomotion using a forward dynamics musculoskeletal model. In: Prilutsky BI, Edwards DH Jr (eds) Neuromechanical modeling of posture and locomotion. Springer, New York
Prilutsky BI, Maas H, Bulgakova M, Hodson-Tole EF, Gregor RJ (2011) Short-term motor compensations to denervation of feline soleus and lateral gastrocnemius result in preservation of ankle mechanical output during locomotion. Cells Tissues Organs 193:310–324
Prochazka A (1999) Quantifying proprioception. Prog Brain Res 123:133–142
Prochazka A, Gorassini M (1998) Models of ensemble firing of muscle spindle afferents recorded during normal locomotion in cats. J Physiol 507(Pt 1):277–291
Prochazka A, Gillard D, Bennett DJ (1997) Positive force feedback control of muscles. J Neurophysiol 77:3226–3236
Quevedo J, Fedirchuk B, Gosgnach S, McCrea DA (2000) Group I disynaptic excitation of cat hindlimb flexor and bifunctional motoneurones during fictive locomotion. J Physiol 525(Pt 2):549–564
Ross KT, Nichols TR (2009) Heterogenic feedback between hindlimb extensors in the spontaneously locomoting premammillary cat. J Neurophysiol 101:184–197
Rossignol S (1996) Neural control of stereotypic limb movements. In: Rowell LB, Sheperd JT (eds) Handbook of physiology, section 12. Exercise: regulation and integration of multiple systems. Vol 12. American Physiological Society, Oxford, pp 173–216
Rossignol S, Dubuc R, Gossard JP (2006) Dynamic sensorimotor interactions in locomotion. Physiol Rev 86:89–154
Ryall RW (1970) Renshaw cell mediated inhibition of Renshaw cells: patterns of excitation and inhibition from impulses in motor axon collaterals. J Neurophysiol 33:257–270

Rybak IA, Ivashko DG, Prilutsky BI, Lewis MA, Chapin JK (2002) Modeling neural control of locomotion: Integration of reflex circuits with CPG. In: Dorronsoro J (ed) Artificial neural networks. Springer Berlin Heidelberg, pp 99–104
Rybak IA, Shevtsova NA, Lafreniere-Roula M, McCrea DA (2006a) Modelling spinal circuitry involved in locomotor pattern generation: insights from deletions during fictive locomotion. J Physiol 577:617–639
Rybak IA, Stecina K, Shevtsova NA, McCrea DA (2006b) Modelling spinal circuitry involved in locomotor pattern generation: insights from the effects of afferent stimulation. J Physiol 577:641–658
Rybak IA, Shevtsova NA, Kiehn O (2013) Modelling genetic reorganization in the mouse spinal cord affecting left-right coordination during locomotion. J Physiol 591:5491–5508
Sacks RD, Roy RR (1982) Architecture of the hind limb muscles of cats: functional significance. J Morphol 173:185–195
Shevtsova NA, Hamade K, Chakrabarty S, Markin SN, Prilutsky BI, Rybak IA (2015) Modeling the organization of central neural circuits controlling two-joint muscles. In: Prilutsky BI, Edwards DH Jr (eds) Neuromechanical modeling of posture and locomotion. Springer, New York
Spector SA, Gardiner PF, Zernicke RF, Roy RR, Edgerton VR (1980) Muscle architecture and force-velocity characteristics of cat soleus and medial gastrocnemius: implications for motor control. J Neurophysiol 44:951–960
Stecina K, Quevedo J, McCrea DA (2005) Parallel reflex pathways from flexor muscle afferents evoking resetting and flexion enhancement during fictive locomotion and scratch in the cat. J Physiol 569:275–290
Stein RB, Misiaszek JE, Pearson KG (2000) Functional role of muscle reflexes for force generation in the decerebrate walking cat. J Physiol 525(Pt 3):781–791
Taga G (1995a) A model of the neuro-musculo-skeletal system for human locomotion. I. Emergence of basic gait. Biol Cybern 73:97–111
Taga G (1995b) A model of the neuro-musculo-skeletal system for human locomotion. II Real-time adaptability under various constraints. Biol Cybern 73:113–121
Toth TI, Schmidt J, Buschges A, Daun-Gruhn S (2013) A neuro-mechanical model of a single leg joint highlighting the basic physiological role of fast and slow muscle fibres of an insect muscle system. PLoS ONE 8:e78247
Wadden T, Ekeberg O (1998) A neuro-mechanical model of legged locomotion: single leg control. Biol Cybern 79:161–173
Whelan PJ (1996) Control of locomotion in the decerebrate cat. Prog Neurobiol 49:481–515
Whelan PJ, Pearson KG (1997) Comparison of the effects of stimulating extensor group I afferents on cycle period during walking in conscious and decerebrate cats. Exp Brain Res 117:444–452
Windhorst U (1996) On the role of recurrent inhibitory feedback in motor control. Prog Neurobiol 49:517–587
Yakovenko S (2011) Chapter 10-a hierarchical perspective on rhythm generation for locomotor control. Prog Brain Res 188:151–166
Yakovenko S, Gritsenko V, Prochazka A (2004) Contribution of stretch reflexes to locomotor control: a modeling study. Biol Cybern 90:146–155
Zhong G, Shevtsova NA, Rybak IA, Harris-Warrick RM (2012) Neuronal activity in the isolated mouse spinal cord during spontaneous deletions in fictive locomotion: insights into locomotor central pattern generator organization. J Physiol 590:4735–4759

# Part II
# Organization of Afferent Signals, Central Neural Circuits and the Musculoskeletal System: Insights from Neuromechanical Modeling

# Chapter 3
# Neural Regulation of Limb Mechanics: Insights from the Organization of Proprioceptive Circuits

**T. Richard Nichols, Nathan E. Bunderson and Mark A. Lyle**

**Abstract** Sensory feedback arising from muscle spindle receptors and Golgi tendon organs are known to influence limb mechanics during postural and locomotor tasks. The purpose of this chapter is to synthesize data concerning the organization and actions of these proprioceptive pathways, and then to propose how current models can be used to promote understanding of their functional role in regulating whole limb stiffness. Following a historical introduction, the role of length feedback in transforming the mechanical properties of muscles into more spring-like actuators is reviewed. Next, we describe the organization of intermuscular length and force feedback circuits in the context of the mechanical interrelationships of the muscles involved. Finally, we provide a conceptual framework for understanding the role of proprioceptive feedback in the regulation of limb mechanics across a continuum of behaviors, and show how a developing computational model can be used to understand how these pathways are integrated to regulate limb stiffness. We conclude from a qualitative appraisal of the data that intermuscular length feedback reinforces the mechanical relationships between antagonists and between synergistic muscles that cross the same or different joints. Furthermore, inhibitory force feedback is organized to manage the distribution of stiffness across joints as well as intersegmental dynamics due to the inertial properties of the limb segments.

**Keywords** Length feedback · Force feedback · Limb stiffness · Neuromechanical integration · Limb mechanics

T. R. Nichols (✉) · M. A. Lyle
School of Applied Physiology, Georgia Institute of Technology, 555 14th St NW, Atlanta, GA 30332-0356, USA
e-mail: trn@gatech.edu

N. E. Bunderson
Autonomous Solutions, 990 North 8000 West Mendon, UT 84325
e-mail: nbunderson@gatech.edu

M. A. Lyle
e-mail: mlyle@ap.gatech.edu



B. I. Prilutsky, D. H. Edwards (eds.), *Neuromechanical Modeling of Posture and Locomotion,* Springer Series in Computational Neuroscience,
DOI 10.1007/978-1-4939-3267-2_3

## 3.1 Introduction

It is generally recognized that sensory feedback arising from muscle receptors and other proprioceptors has a strong influence on muscular activity during postural tasks as well as during locomotion. A definitive understanding of how and to what extent sensory feedback contributes to whole limb coordination has been hampered, however, by several factors, including the complexities of the organization of the neuromechanical system. This complexity arises both from the mechanics of the sensory receptors themselves and from the interactions between neural pathways and the mechanical circuits[1] of the musculoskeletal system. Moreover, systematic experimental testing of the components of the neuromuscular system is challenging. We propose that computational models can help understand these interrelationships. The purpose of this chapter is to synthesize existing experimental data concerning the organization and actions of proprioceptive feedback from muscle spindle receptors and Golgi tendon organs, and then to propose how current models can be used to promote understanding of the manner in which sensory feedback contributes to motor control.

Given the extensive information on feline motor control that is available, this chapter provides a synthesis of data primarily from the feline model concerning the role of proprioceptive feedback on limb mechanics. We first provide some fundamental concepts from the historical literature concerning the interactions between limb mechanics and sensory feedback. The role of length feedback in transforming the mechanical properties of muscles into more spring-like actuators will then be reviewed. Next, we describe the organization of length and force feedback in spinal circuits with reference to the mechanical interrelationships of the muscles involved. Finally, we will synthesize this information to provide a conceptual framework for understanding the role of proprioceptive feedback in the regulation of limb mechanics, and show how a developing computational model can be used to understand how these pathways are integrated to regulate limb stiffness.

Our organizational scheme for length and force feedback is based on two primary sources. First, we utilized data reported by Eccles et al. (1957a, b). These studies have provided an extensive mapping, in anesthetized cats, of proprioceptive pathways arising from group I receptors from most of the major hindlimb muscles by stimulating different peripheral nerves and recording intracellular potentials from motoneurons. The second source is a series of more recent experiments in which the stretch of selected muscles in decerebrate animals was used to probe the sensory network (Nichols and Koffler-Smulevitz 1991; Bonasera and Nichols 1994, 1996; Hyde et al. 1999; Nichols 1999; Wilmink and Nichols 2003; Ross and Nichols 2009). Although there is a substantial literature on connectivity of pathways from group II receptors through a number of interneuronal pathways (Jankowska and Edgley 2010), these pathways have not yet been systematically related to specific

[1] The musculoskeletal system can be represented as a mechanical network with signals corresponding to the mechanical variables of length and force and their derivatives. These signals become represented in the associated neural circuits of the central nervous system through sensory transduction.

**Table 3.1** Abbreviations

| | |
|---|---|
| ADF | Adductor femoris |
| ADL | Adductor longus |
| BFA | Biceps femoris anterior |
| BFP | Biceps femoris posterior |
| EDL | Extensor digitorum longus |
| FDL | Flexor digitorum longus |
| FHL | Flexor hallucis longus |
| GMAX | Gluteus maximus |
| GMED | Gluteus medius |
| GMIN | Gluteus minimus |
| GRAC | Gracilis |
| ILIO | Iliopsoas |
| LG | Lateral gastrocnemius |
| MG | Medial gastrocnemius |
| PB | Peroneus brevis |
| PEC | Pectineus |
| PL | Peroneus longus |
| PLAN | Plantaris |
| PT | Peroneus tertius |
| PYR | Pyriformis |
| QF | Quadratus femoris |
| RF | Rectus femoris |
| SART | Sartorius |
| SOL | Soleus |
| SM | Semimembranosus |
| ST | Semitendinosus |
| TA | Tibialis anterior |
| TP | Tibialis posterior |
| VI | Vastus intermedius |
| VL | Vastus lateralis |
| VM | Vastus medialis |

muscles. Furthermore, we will confine the discussion mainly to pathways that are thought to be located in the spinal cord, given that substantial information about the organization of these pathways is available. Abbreviations for the 31 muscles represented in the model are given in Table 3.1.

## 3.2 Background: Limb Mechanics and Sensory Feedback

We believe that proprioceptive feedback within a limb is organized to regulate the mechanical properties of the entire limb in addition to the mechanical properties of individual muscles. The limb-centric (as opposed to muscle- or joint- centric) organization of neural feedback reflects our view that limbs, rather than single muscles or joints, are the basic units for producing coordinated movement. This view acknowledges the strong mechanical coupling that exists in a limb due to dynamic properties such as inertial coupling and the multi-joint structure of many muscles. While individual muscles are the output units of the motor system, they are components of an integrated mechanical structure and are seldom activated in isolation. Even if they were activated in isolation, numerous muscles cross more than one joint and/or axes of rotation, and muscles transmit forces through fascia and can be coupled through shared tendons (Carrasco and English 1999; Lawrence and Nichols 1999a; Stahl 2010). Besides the viscoelastic coupling provided by these attachments through tendons and fascia, motions of the limb segments are linked dynamically through inertial interactions as well (Zajac 1993). Because of these physical properties, isolated muscle activation and single joint or limb segment movements are generally not observed in postural and locomotor tasks in nature. It is of course possible to control single joints individually, but this requires integrated action of muscles elsewhere in the limb due to inertial and elastic coupling. Therefore postural and locomotor tasks require organization of muscle activations throughout the entire limb rather than across single joints. Our belief that neural feedback arising from muscles within a limb is organized to regulate both the mechanical properties of muscle as well as intersegmental dynamics is supported by observations of patients with large fiber sensory neuropathy, a global loss of sensation from larger myelinated sensory axons. These patients have great difficulty controlling inertial coupling of limb segments and coordinating motions of the joints (Sainburg et al. 1993, 1995; Gordon et al. 1995).

Initial analyses of proprioceptive circuits focused on the influence of feedback on individual muscles (Liddell and Sherrington 1924; Merton 1953), apparently leaving the more global control of the mechanics of the musculoskeletal system to another level of neural integration. Proprioceptive feedback generally has a profound effect on the mechanical properties of muscles, as reviewed below, but it has become increasingly appreciated that proprioceptive circuits provide neural linkages between muscles and are highly integrated at the level of spinal segments. That is, some proprioceptive pathways project beyond the muscle of origin, indicating that motor control at the spinal level is not based on individual muscles. Given that length feedback from muscle spindle receptors projects to synergists and antagonists, length feedback would appear to provide regulation at the level of joints rather than muscles. However, it has also been shown that the projections of length feedback are not limited by the restrictive definition of the myotatic unit proposed by Lloyd (1946). According to this scheme, Ia connections cannot link muscles crossing different joints, each joint is regulated by one myotatic unit, and any given muscle cannot be a member of more than one myotatic unit. This scheme is an idealized conception and does not accurately reflect the organization of most Ia pathways (Eccles and Lundberg 1958b; Nichols et al. 1999a). For example, the flexor hallucis longus (FHL) and flexor digitorum longus (FDL) muscles in the cat are linked as synergists with respect to toe flexion, and FHL (which has a strong plantarflexion action at the ankle) is linked to the pretibial flexors by reciprocal inhibition (Nichols et al. 1999a). For muscles crossing the hip and knee, the organization of Ia excitation and inhibition is even more complex (Eccles and Lundberg 1958b). That is, a biarticular muscle (for example RF) may have connections through Ia afferents to other muscles that span either or both joints. These observations suggest that joint level control is too restrictive for even the most basic spinal reflex.

Evidence that proprioceptive circuits regulate limb mechanics is to be found in the projections of Ib afferents from Golgi tendon organs. During locomotion, excitatory force feedback is enabled (Pearson 1995; Donelan and Pearson 2004), resulting in an increase in stiffness in the muscles in which it is expressed (Ross 2006) over and above that contributed by length feedback. Excitatory force feedback has been observed mainly in biarticular muscles, so an important action of this feedback is to promote mechanical coupling between the joints that are spanned by the muscles as well as increased stiffness of the joints across which the moment arms of the muscles are greatest. However, a more comprehensive understanding of the actions of the proprioceptive network awaits the construction of suitable computational models.

The distribution of inhibitory force feedback is even wider than the distribution of length and excitatory force feedback. Projections of inhibitory force feedback across joints and axes of rotation are substantially stronger than autogenic (self) projections (Eccles et al. 1957b; Nichols 1994; Nichols et al. 1999a; Nichols and Ross 2009), indicating that the integration of length and force feedback (excitatory and inhibitory) occurs at the level of proprioceptive networks rather than at the level of single muscles. The integration of proprioceptive feedback is more than the superposition of individual ("autogenic") reflex pathways ("composite stretch reflexes" (Liddell and Sherrington 1924), but rather is instantiated by a network of neural circuits that are fully integrated with the mechanical circuits of the musculoskeletal system.

The principle that force feedback is organized to regulate limb mechanics is further illustrated by the absence of Ib projections from muscles that probably do not contribute greatly to intersegmental dynamics, such as FDL (Bonasera and Nichols 1994). This muscle controls the terminal segments of the limb (toes and claws), that have a lesser dynamic influence on the rest of the limb due to their low mass. Muscles that control more proximal joints are generally linked in a variety of combinations by length and force feedback. This observation further supports the hypothesis that proprioceptive pathways contribute to the regulation of whole limb mechanics. The actions of proprioceptive feedback are fully integrated with the intrinsic mechanical properties of muscles and presumably with the more global mechanical properties of the limb.

The organization of the proprioceptive networks combined with a knowledge of the interactions between length and force feedback provides a basis on which to understand their functions. The mechanical properties of limbs can be summarized by the property of impedance that includes components related to elasticity, viscosity and inertia. The terms corresponding to elasticity and viscosity are generally nonlinearly related to position and velocity (Houk et al. 2002), respectively, and frequently lumped together in the motor control literature as "stiffness" despite the fact that the term "stiffness" properly refers to the static mechanical properties of a system (Latash and Zatsiorsky 1993). Houk (1972b, 1979) originally proposed that the combined feedback from muscle spindle receptors and Golgi tendon organs should result in the regulation of muscular stiffness, with the magnitude of stiffness determined by the balance between length feedback and inhibitory force feedback. It was found that stiffness of the soleus muscle is indeed regulated in the decerebrate cat (Nichols and Houk 1976). It was later discovered, however, that this regulation could be attributed to length feedback alone during steady force production (Houk et al. 1981) and measurements of the strength of autogenic, inhibitory force feedback showed it to be quite weak (Rymer and Hasan 1980). Since inhibitory force feedback is mainly heterogenic and widely distributed, the integration of length and force feedback, and therefore the regulation of stiffness, occurs at the level of proprioceptive networks that regulate limb mechanics rather than the level of individual muscles. The loss of proprioceptive feedback from specific muscles results not only in a reduction of stiffness of the corresponding joints, but alterations in interjoint coordination as well (Cope et al. 1994; Abelew et al. 2000; Maas et al. 2007; Chang et al. 2009).

In addition to regulating the mechanical responses of limbs, proprioceptive feedback in conjunction with other sources of sensory information can provoke the expression of new patterns of activity when the motor task is changed. Feedback from sources extrinsic to the limb is generally concerned with influencing the patterns of muscular activity in a task-specific manner, presumably through the pattern formation network in the spinal cord (Lafreniere-Roula and McCrea 2005). For example, integrated sensory information from the otolith organs and from muscle spindle receptors in the muscles of the neck, representing a body-orientation signal (Brink et al. 1985; Marchand et al. 1987), regulates the pattern of muscular activity during ramp walking when compared to that observed during level walking (Smith and Carlson-Kuhta 1995; Gottschall and Nichols 2007; Nichols et al. 2014). The data suggest that, during downslope walking, the body orientation signal results in reduced activity in the gastrocnemius muscles, absence or greatly reduced activity of muscles of propulsion, and activation of hip flexors (Gottschall and Nichols 2007, 2011). These actions contribute to the braking action of the limbs required for controlled downslope locomotion (Nichols et al. 2014).

Besides changes in activation pattern, the body orientation signal has been shown to reduce limb stiffness through enhanced inhibitory force feedback (Nichols et al. 2014). This reduction in limb stiffness complements the braking action provided by distal anti-gravity and hip flexor muscles. The weightings of different inhibitory pathways (Bonasera and Nichols 1994; Wilmink and Nichols 2003; Ross and Nichols 2009) suggest that force feedback can be organized to determine how stiffness is distributed across the joints of the limb. The emergence of excitatory force feedback during locomotion (Pearson 1995) and modulation of inhibitory force feedback under different conditions (Nichols et al. 2014) strongly suggests that force feedback is an important variable for regulating limb stiffness in a task dependent manner. The modulation of muscular activity and proprioceptive pathways is consistent with the task specific use of the limb.

Although the modification of muscular activation patterns can be attributed mainly to sources extrinsic to the limb, including descending and propriospinal sources, it has also been suggested that feedback from within the limb can alter the pattern of activity during locomotion. For example, the semitendinosus and posterior biceps femoris muscles exhibit a double bursting pattern, one burst occurring at the transition from extension to flexion (i.e. swing initiation), and another at the transition from flexion to extension (i.e. terminal swing) (Smith et al. 1993; Krouchev et al. 2006; Markin et al. 2012). This pattern is not observed during fictive locomotion, where there is usually only a single burst at the beginning of flexion (locomotor-like patterns of muscle nerve activity in a paralyzed, decerebrate animal) (Markin et al. 2012). Since there is little or no feedback during fictive locomotion, the second burst has been attributed to feedback from within the limb. The role of intrinsic proprioceptive pathways in regulating patterns of muscular activity remains to be fully explored.

As discussed here, the functions of proprioceptive pathways can best be understood in terms of the regulation of whole limb mechanics. Until recently most studies of the proprioceptive pathways focused on measuring the output of individual muscles or individual or populations of neurons. However measuring whole limb properties such as limb stiffness can provide a metric more appropriate for the framework described here. We have recently adapted methods used to quantify endpoint stiffness in the upper extremity by Eric Perreault et al. (2004, 2008; Krutky et al. 2013) to the cat fore- and hindlimbs in an effort to more directly link proprioceptive pathways with whole limb mechanics. Moreover, a computational model can be extremely useful in connecting and synthesizing the experimental data from the previous more modular approach with experimental data from the current whole limb approach. A model incorporating the full 3 dimensional representation of muscular action of a feline hindlimb was originally developed by Thomas Burkholder (Burkholder and Nichols 2000, 2004) and later ported to the Neuromechanic software environment (Bunderson et al. 2012), see also Chap. 1 in this book (Bunderson and Bingham 2015). Using this model whole limb properties such as limb stability (Bunderson et al. 2008) and endpoint stiffness (Bunderson et al. 2010) have been obtained. We are currently utilizing this model to understand the integrated function of proprioceptive pathways.

## 3.3 Transformation of Muscular Mechanics by Autogenic Feedback

The analysis of proprioceptive feedback begins with the actions of autogenic (self) feedback on the muscle of origin. The most familiar and thoroughly studied example of autogenic feedback is the autogenic stretch reflex (Liddell and Sherrington 1924). The stretch reflex includes monosynaptic feedback from primary spindle afferents to motoneurons as well as possible contributions from group II muscle afferents (Stuart et al. 1970; Stauffer et al. 1976; Sypert et al. 1980; Munson et al. 1982). The original functional hypothesis for this reflex was the "Follow-Up Length Servo Hypothesis" (Merton 1953) based on length as the controlled variable. Subsequent work as well as more recent thinking about what variables are controlled by the nervous system (Stein 1982) justified rejection of this hypothesis (Houk 1972a). The current understanding of this and other proprioceptive pathways requires a more comprehensive understanding of the mechanics of movement and muscle physiology. Indeed, knowledge of the interactions between neural feedback and intrinsic muscle mechanics is crucial to the understanding of motor control (Nichols et al. 1999b; Dickinson et al. 2000; Huyghues-Despointes et al. 2003a, b; Sponberg et al. 2011).

Skeletal muscles have complex mechanical properties. Both length-tension and force-velocity relationships depend critically on the rate at which motor units are activated (Joyce et al. 1969; Rack and Westbury 1969). Furthermore, skeletal muscle exhibits "thixotropic" properties, such that the responses of the muscle to a given length change are altered by prior movement (Kirsch et al. 1994; Proske and Morgan 1999; Campbell and Moss 2000; Huyghues-Despointes et al. 2003b). Complementary history-dependence is also exhibited by muscle spindle receptors (Haftel et al. 2004; Nichols and Cope 2004). Finally, short-range stiffness represents the initial response of a muscle to length change, with higher stiffness than the muscle presents subsequently (Joyce et al. 1969; Rack and Westbury 1974; Malamud et al. 1996; Cui et al. 2008). Short-range stiffness provides a response to perturbations before the central nervous system can react (Dickinson et al. 2000; Nishikawa et al. 2007; Daley et al. 2009). However, short-range stiffness depends on background force, so it provides a relatively small contribution under conditions of quiet standing. Neural mechanisms are required to decouple stiffness and force.

Autogenic length feedback transforms both the steady-state and transient properties of muscle in ways that are appropriate for motor control. Steady-state properties intrinsic to the muscle are determined predominantly by muscle length, rate of motor unit activation (Rack and Westbury 1969) and level of motor unit recruitment (Boskov and Heckman 1996; Nichols et al. 1999a). The rates of activation of motor units are usually subtetanic (Grillner and Udo 1971), and force output is relatively smooth due to the asynchronous activation of motor units (Rack and Westbury 1969). At physiological rates, the force-length relationship maintains a positive slope throughout the normal range of movement (Rack and Westbury 1969), so that muscular stiffness remains positive and non-zero (Fig. 3.1). The stiffness of the muscle scales according to the level of recruitment, since motor units are arranged mechanically in parallel (Fig. 3.1). As more motor units are recruited, both force and stiffness increase.

An important function of proprioceptive length feedback is to decouple force and stiffness, so that muscles can present a substantial stiffness to disturbances even at low background forces. Despite the presence of feedback, the muscle may present a stiffness no greater than that provided by intrinsic properties when all motor units are recruited, but this level of stiffness is independent of operating point on the force-length plane. The length-tension curves illustrated in Fig. 3.1 represent those of fixed populations of motor units in the absence of feedback, and those of the muscle when feedback was present (two cases with different slopes are shown). For any given operating point in the force-length plane, the stiffness with feedback

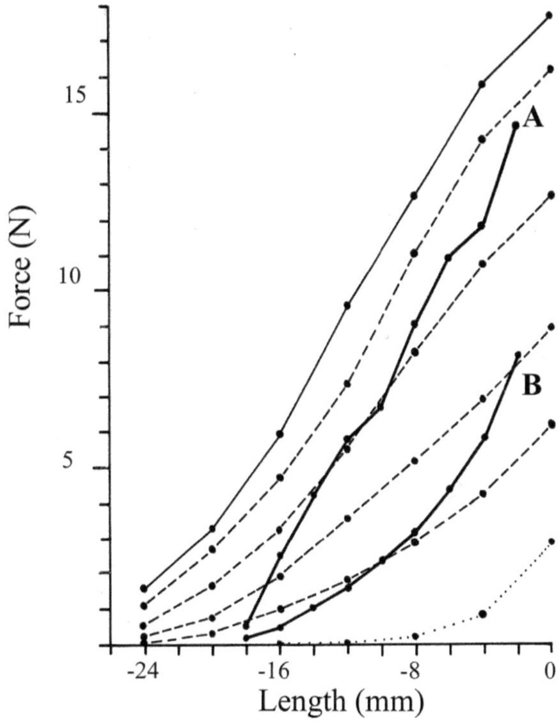


**Fig. 3.1** *Transformation of steady-state mechanics of the feline soleus muscle by autogenic feedback.* The two heavy solid lines (A, B) represent force-length relationships obtained at different times in one animal in the unanesthetized decerebrate state. The muscle was held at each length for 30 s. The intrinsic force-length relationships of fixed populations of motor units were obtained by stimulating progressively larger groups of ventral root filaments at 8 pps after transecting the ventral roots. This rate was selected as the firing rate observed during steady force production in this muscle (Grillner and Udo 1971). Intermediate groups are shown as *dashed lines* and 100% of the population is shown as the *light solid line*. Each data point was obtained by stimulating the muscle at a given length, and length was then changed during the inactive period between stimuli. The *dotted line* denotes the relationship for the inactive muscle. Zero length denotes the maximum physiological extension determined while the muscle was still connected to the calcaneus. Note that all force-length relationships have positive slopes across the physiological range of the muscle lengths. Autogenic feedback increases the stiffness of the muscle at each operating point in the force-length plane by the recruitment of additional motor units. The relative contributions of intrinsic stiffness and autogenic feedback can be estimated as the difference in slopes at the points of intersection of *dashed* and *heavy solid lines*. The stiffness of the regulated muscle can change under different states of the spinal cord, as illustrated by the different slopes of curves A and B. In different preparations (Nichols 1974) or with different levels of stimulation of Dieter's nucleus (Feldman and Orlovsky 1972), the threshold of activation of the muscle can change. Threshold and slope can therefore be controlled independently of background force, so that the muscle can present substantial stiffness even if background force is low. Data were adapted from a thesis (Nichols 1974)

exceeds the intrinsic stiffness. At longer lengths and higher levels of recruitment, however, the intrinsic properties make a larger contribution to the total stiffness. The force-length relationships in the presence of feedback can also shift along the length axis (Feldman and Orlovsky 1972), representing different thresholds of activation due to different command signals to the motor neuron pool, with no substantial changes in slope. The slope can change, however, under different behavioral conditions and in response to descending control signals (Nichols and Steeves 1986), as also shown in Fig. 3.1. The length-tension characteristic with lower slope was observed during the same experiment due to spontaneous changes in the state of the preparation. While intrinsic muscular stiffness depends on the level of activation and consequently force, the stiffness and threshold of the muscle with local feedback can vary independently of background force. This principle is a basis for the Equilibrium Point Hypothesis of motor control (Feldman and Levin 1995). As motor units are recruited, the rate of firing of recruited units also increases (Monster and Chan 1977). The intrinsic properties of muscle are therefore somewhat more complicated than portrayed in Fig. 3.1, but the transformation brought about by feedback is similar to what is shown.

When the length of an active muscle is changed, the initial response is determined by the intrinsic properties. This initial response coincides with short-range stiffness (Joyce et al. 1969), followed by a complex interaction between the intrinsic properties and autogenic feedback. The importance of short-range stiffness is that this mechanical response occurs before any changes in motor unit recruitment due to feedback and is therefore responsible for the initial response of the body to disturbing forces. Short-range stiffness is subject, therefore, to the scaling of stiffness with background force. During quiet standing, background force is low, so the response to postural perturbations is dominated by contributions of feedback. At high forces, the intrinsic properties including short-range stiffness contribute substantially to the mechanical response (Nichols and Houk 1976).

Short-range stiffness and the remainder of the intrinsic response are dependent upon movement history, as illustrated in Fig. 3.2, and this dependence has a profound influence on the contributions of feedback to the mechanical properties of muscle (Huyghues-Despointes et al. 2003b). When a contracting muscle is stretched following a period of isometric contraction, the stiffness decreases following the short range, a phenomenon known as yielding (Joyce et al. 1969; Nichols and Houk 1976). In the presence of autogenic feedback, the yield is compensated by additional recruitment of motor units (Fig. 3.2, upper left panel), resulting in a response that more closely resembles a linear viscoelastic response (Nichols and Houk 1976). If stretch is preceded by release, the yield is reduced in relation to the magnitude of the shortening. If the release is large enough, no yield is observed. The contribution of feedback is complementary, preserving the viscoelastic character of the intrinsic response. In this case, the contribution of feedback is nil, and there is no real distinction between the short and longer-range stiffnesses. It appears that this remarkable control strategy arises from the history dependence of muscle spindle receptors (Haftel et al. 2004; Nichols and Cope 2004).

In the above examples, the "spring constant" of the muscle is determined by feedback, with intrinsic properties contributing variable amounts depending upon conditions, including level of recruitment, length and previous mechanical history. As stated above, the available evidence suggests that the critical autogenic feedback

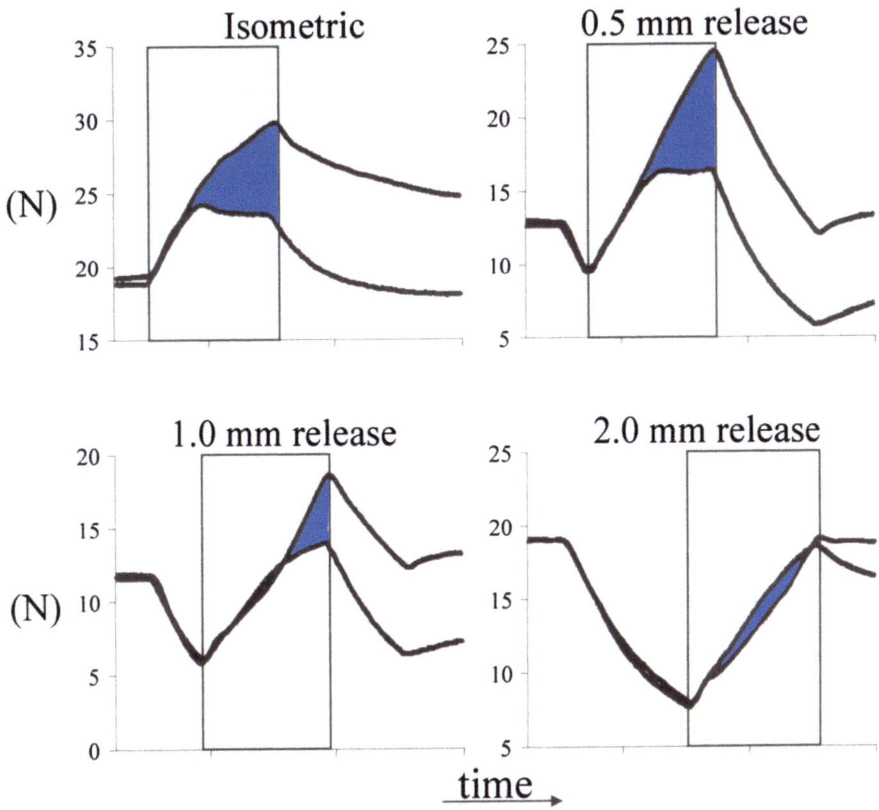


**Fig. 3.2** *Transformation of the transient mechanical properties of muscle by autogenic feedback.* The intrinsic response of the deafferented right feline soleus muscle in a decerebrate cat is compared to the response of the muscle with intact autogenic feedback during stretch. Intrinsic responses were obtained by reinnervation of the right muscle approximately 1 year prior to the experiment. Motor units were reinnervated during this time, but autogenic feedback was blocked by a process of synaptic stripping (Bullinger et al. 2011). The data in each panel represents the responses of the two muscles during activation by crossed-extension reflexes. For the responses illustrated in each successive panel, muscle stretch (2 mm) was preceded by muscle shortening of the given amplitude. The differences between intrinsic responses and those with intact feedback are indicated by the shaded areas, allowing an estimate of the relative contributions of intrinsic properties and feedback. With no prior stretch, autogenic feedback increased muscular stiffness and compensated for the yield in the intrinsic response. As the prior release was increased, the yield diminished and the contribution of feedback progressively diminished, conserving the resultant response of the muscle. The feedback compensated for amplitude and history dependent nonlinearities of the muscle. Adapted from Huyghues-Despointes et al. (2003b)

is from the muscle spindle receptor. During different behaviors, other sources of feedback contribute to the mechanical response. During locomotion for example, inertial loads become greater, and greater stiffness is desirable. As stated above, excitatory force feedback is enabled and projects primarily autogenically (Ross 2006). This feedback serves to increase muscular stiffness, but based on force rather than length. This arrangement is advantageous for muscles in which tendon compliance is such that the length changes of muscle fascicles do not correspond to the length changes of the muscle-tendon unit. Indeed, during level walking, the fascicles of the medial gastrocnemius muscle actually shorten during weight acceptance while the muscle tendon unit is lengthening (Hoffer et al. 1989). Since forces in the tendon are increasing, excitatory force feedback still reinforces muscular contraction even in the face of decreasing length feedback. There is also evidence that in muscles not exhibiting excitatory force feedback during locomotion, autogenic inhibition (Granit 1950) is expressed (Ross 2006), presumably reducing stiffness slightly. It is not clear what functional advantage this would have, but may simply be a byproduct of mobilization of the force feedback system.

These data indicate that autogenic length feedback transforms the properties of the muscle to present spring-like characteristics (with nonlinear damping) with variable length threshold, while autogenic force feedback may be responsible for determining the magnitude of muscular stiffness in a task dependent manner. Some of the complex mechanical properties of muscle, such as short-range stiffness and thixotropy, are incorporated into the response, while others, such as the dependence of stiffness on force, are compensated. In order to represent these interactions, suitably sophisticated models of muscle are needed. The available computational models of muscle systems utilize various modifications of “Hill-type” muscle models that in some cases capture important steady-state and transient muscle properties (Lin and Crago 2002a, b). However, these models by and large do not represent the history-dependent properties of muscles and spindle receptors described above. More mechanistically based models would help to overcome this difficulty. Moreover, Hill-type models are commonly used to represent the dependencies of force on length and velocity, but these models, with some exceptions (for example Lin & Crago), are often based on unphysiological conditions of activation. Hill-type models also do not usually represent short-range stiffness (but see Lin & Crago). Hill-type models are a useful starting point for ongoing locomotion, but for more complex behaviors involving changes in speed and direction or sudden initiation of movement, more realistic, mechanistic models will be necessary.

## 3.4 Organization of Heterogenic Feedback

The functions of proprioceptive feedback can best be appreciated by considering the attachments and moment arms of individual muscles. The interpretation of autogenic and intermuscular (heterogenic) pathways is critically dependent upon an understanding of the anatomical context. In this section we will discuss proprioceptive pathways in this anatomical context. Insight into the anatomical context can be obtained from two types of studies that were initiated in order to understand the actions of muscles in three dimensions.[2] For the distal musculature, the torques

[2] In the analysis of proprioceptive feedback for muscles crossing the hip and the knee (Eccles and Lundberg 1958a), actions out of the sagittal plane were acknowledged, but functions of specific

exerted by the major distal muscles during electrical stimulation (Lawrence and Nichols 1999a, b) and their moment arms (Young et al. 1993) have been documented. In another approach, the muscular actions during electrical simulation were measured as the reaction forces on the ground (Honeycutt and Nichols 2014). These latter measurements agree substantially with predictions from the Burkholder computational model (Burkholder and Nichols 2004). A table providing a qualitative description of the actions of muscles crossing the hip and knee based on several sources is available (Nichols et al. 1999a).

It should be noted that connective tissues, including tendon and fascia, are important components of the peripheral motor apparatus and can also influence limb mechanics. Tendons provide mechanical buffering during locomotion (Griffiths 1991; Roberts and Konow 2013), and in some cases result in shortening of the muscle fibers while the muscle-tendon unit is lengthening (Hoffer et al. 1989; Prilutsky et al. 1996; Maas et al. 2009; Konow et al. 2012). Fascia provides an additional route for force transmission (Maas et al. 2005; Stahl 2010). Connective tissue, including tendon and fascia, together with muscle and bone forms a complex mechanical network that has been described as a tensegrity structure (Silva et al. 2010). Muscle fibers work within this mechanical network and produce movement by altering the patterns of force within it.

Individual muscles or portions of muscles (English and Weeks 1987; Carrasco et al. 1999) influence joint stiffness by virtue of their attachments and routes of force transmission between these attachments. Individual muscles can cross one or more joints and one or more axes of rotation. Muscles contribute to the stiffness of joints by virtue of their resultant stiffness (the sum of intrinsic stiffness and contributions from proprioceptive pathways) and the moment arm of the muscle at the spanned joint. For example, the FHL muscle exerts substantial plantarflexion torque and therefore stiffness at the ankle joint, while its strong synergist FDL (Eccles et al. 1957a; Bonasera and Nichols 1994) exerts relatively little due to the differences in moment arms at the ankle (Lawrence et al. 1993; Lawrence and Nichols 1999a). The greater moment arm for ankle plantarflexion by FHL can explain the very different activation patterns of these two muscles (O'Donovan et al. 1982), where FHL presumably contributes to ankle stiffness during stance. This example also illustrates the fact that the presence of strong connections through Ia afferents does not underlie patterns of muscular activity, since both muscles are strong Ia synergists with very different patterns of activity.

Multiarticular muscles provide mechanical coupling across joints. The gastrocnemius muscles (MG and LG) have greater moment arms for plantarflexion at the ankle and therefore contribute preferentially to ankle joint stiffness, but also impart a flexor moment to the knee. During locomotion, the expression of excitatory force feedback combined with length feedback greatly enhances this mechanical coupling in addition to increasing the stiffness of the ankle. The distal attachment of MG also couples plantarflexion with abduction, promoting postural stability and

muscles were still expressed in terms of flexion and extension. Later work underscored the necessity of incorporating muscular actions in three dimensions (Macpherson 1988b).

contributing to the initiation of turning (Lawrence et al. 1993; Nichols et al. 1993; Lawrence and Nichols 1999a, b). Several proximal muscles produce combinations of multiarticular and biaxial torques. PSOAS, ADF and GRAC provide combinations of hip extension or flexion and adduction, while BFA and BFP provide abduction in addition to knee flexion. The tensor fascia lata muscle, which is not included in our model as yet, provides flexion and abduction of the hip. Examples of relatively uniaxial actions include the vastus muscles, providing knee extension due to the patellar mechanism.

In order to represent the relationships between muscles and to understand the functions of heterogenic feedback, we utilized a model constructed in Neuromechanic (Bunderson et al. 2012; Bunderson and Bingham 2015) to generate a "similarity matrix" similar to that originally constructed by Dr. Thomas Burkholder (Nichols et al. 2002). The similarity matrix provides a quantification of the mechanical similarity between any two muscles. In our model, the mechanics of the feline hindlimb are described by seven degrees of freedom (DOF), including three at the hip, two at the knee and two at the ankle. 31 muscles are represented in the model and the mechanical action of each muscle is represented by a seven element moment arm vector corresponding to these seven DOF. The $a_{ij}$ element of the similarity matrix A (Fig. 3.3) is calculated as the angle between the moment arm vectors for muscle $i$ and muscle $j$, shifted and scaled to vary between −1 (for perfect mechanical antagonists) and 1 (for perfect mechanical agonists)

$$a_{ij} = 1 - \frac{1}{\pi}\cos^{-1}\left(\frac{\vec{v}_i \cdot \vec{v}_j}{\left\|\vec{v}_i\right\|\left\|\vec{v}_j\right\|}\right).$$

The rationale for the order in which muscles are listed on the rows and columns is given by the dendrogram at the right of Fig. 3.3. The dendrogram is the result of a cluster analysis performed in Matlab (Mathworks, Natick MA). The 31 similarity comparisons between a single muscle and all other muscles were used as the observation set and the Euclidean distance between observation sets for each muscle was used to form the hierarchical cluster tree shown in the dendrogram in Fig. 3.3. These clusters denote the mechanically synergistic and antagonistic groups. Although the vector directions are biased toward the dominant actions of the muscles, it should be kept in mind that most muscles have complex actions, and in some cases a given pair may have both synergistic and antagonistic relationships (for example, ST and BFP).

The similarity matrix shows the synergistic groupings near the diagonal and the antagonistic groupings mainly distant from the diagonal. Note that the triceps surae muscles (LG, MG, SOL) group with PLAN and FHL. Even though FHL and FDL insert into the same tendons to the toes, they fall into two different mechanical synergies since their actions at the ankle differ substantially. The group of biarticular and biaxial muscles crossing the hip and ankle form the largest mechanically synergistic group. Note that ST and BFP are represented as synergistic, although they do have opposing actions in the frontal plane (Nichols et al. 1999a). Within this group,

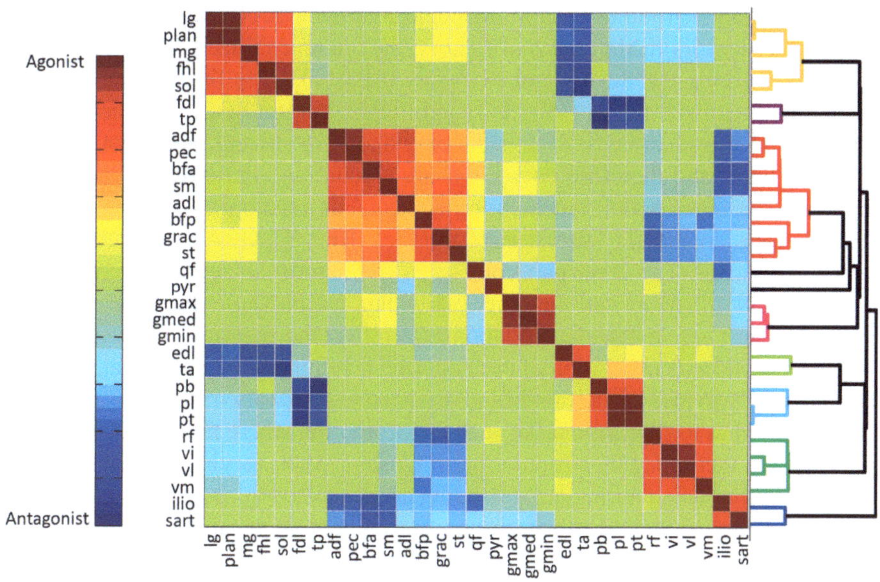


**Fig. 3.3** *The similarity between the moment arms of any two muscle pairs of the feline hindlimb model at a quiet standing posture is shown here in a symmetric matrix*. The colors depict a continuum ranging from completely agonistic muscles (*dark red*) to completely antagonistic muscles (*dark blue*) with a midpoint where muscle pairs have no shared joint actions (*light green*). The order of the muscles in the rows and columns was chosen based on a cluster analysis (dendrogram at *right*) that groups muscles according to their normalized moment arm vector. The different colors of the terminal branches of the dendrogram show ten clusters identified in the analysis. Four clusters can be identified with primarily ankle actions including a plantar flexor cluster (LG, PLAN, MG, FHL, SOL), ankle adductor cluster (FDL, TP), ankle abductor cluster (PB, PL, PT), and ankle extensor cluster (EDL, TA). The quadriceps (RF, VI, VL, VM) and gluteal (GMAX, GMED, GMIN) muscles form two additional clusters. Two muscles, PYR muscle and QF, had sufficiently different moment arm vectors to each warrant their own individual "cluster". ILIO and SART formed a hip flexor group and the remaining muscles (ADF, PEC, BFA, SM, ADL, BFP, GRAC, ST) formed the largest group as hip extensors

one can discern at least two subgroups, namely, the "hamstrings muscles" (BFP, ST, GRAC) with strong actions at the knee and another subgroup acting predominantly at the hip (ADF, PEC, BFA, SM, ADL) (cf. Rossignol 1996). The five remaining synergistic groups are the gluteus muscles (Gmax, Gmed, Gmin), the pretibial flexors (TA, EDL), the peroneus muscles (PB, PL, PT), the quadriceps (RF, VI, VL, VM) and the group consisting of ILIO and SART. Note that TA and EDL are synergistic with PL as both produce dorsiflexion, but not with PB since PB is neutral with respect to the flexion/extension direction (Lawrence et al. 1993; Lawrence and Nichols 1999a). Therefore the synergistic groups of pretibial flexors and peroneus muscles have some overlap. There is a region of weak synergism between the biarticular ankle extensors (MG, LG, PLAN) and the hamstrings muscles (BFP, GRAC, ST). Our model does not yet distinguish the two divisions of SART (Eccles and Lundberg 1958b; Hoffer et al. 1987).

The antagonist groupings fall in two regions of the matrix. One grouping relates the ankle extensors to the pretibial flexors, the ankle stabilizers (TP and FDL to PB), and the long toe flexors and extensors (FHL, FDL) to the pretibial flexors. The gastrocnemius muscles (MG, LG) and PLAN, but not SOL, are mechanically dissimilar to the quadriceps. SOL is the only one of the triceps surae that does not cross the knee, and therefore does not share a common joint with the quadriceps muscles. The other region representing antagonistic relationships can be divided into two parts. In one, the "hamstrings" muscles (BFP, ST, GRAC) are antagonistic to the quadriceps. The other part consists of mechanical antagonism between hip flexors (ILIO & SART) and the other biarticular and biaxial muscles crossing the hip and knee. These mechanical relationships reflect the anatomy considering both articulation and axis of rotation of each muscle. It is important to note that the similarity matrix provides a snapshot of the mechanical similarity of muscles at a given posture and that for dynamic tasks such as locomotion where the posture can vary substantially there may be a reordering of clusters. Such varying relationships for different limb spacing during quiet standing are suggested by changes in the organization of directionally tuned force responses to perturbations of the support surface (Macpherson 1994; Honeycutt and Nichols 2010).

### *3.4.1 Heterogenic Length Feedback*

The distribution of length feedback from group Ia spindle afferents (Fig. 3.4a) generally corresponds to the synergistic and antagonistic mechanical groupings, although the neural connections are more restricted than the mechanical connections (Fig. 3.3). The illustrated connections represent a compendium of results from the papers of Eccles et al. (1957a, b; Eccles and Lundberg 1958b) and more recent papers (Nichols and Koffler-Smulevitz 1991; Bonasera and Nichols 1994, 1996; Nichols 1999; Wilmink and Nichols 2003; Ross and Nichols 2009). As is well known, the triceps surae muscles form a synergistic group (Eccles et al. 1957a; Nichols 1999). Interestingly, MG provides substantially greater Ia input to LG than the reverse, which has consequences for directional tuning of LG during postural responses. MG torque at the ankle has a strong abduction component (Lawrence and Nichols 1999a). Therefore, the directional tuning of LG corresponds approximately to the directional tuning of MG (Honeycutt and Nichols 2014). Although a powerful ankle plantarflexor, PLAN has only weak Ia connections to LG and SOL perhaps related to its additional actions at the toes. FHL is somewhat paradoxical in that it has no known Ia connections to the triceps surae or PLAN (Eccles et al. 1957a), despite the fact that it shares mechanical actions with these muscles; FDL and FHL both flex the toes and are strong Ia synergists, but do not share action at the ankle. Although the activation patterns of these two muscles are very different during locomotion, the proprioceptive link suggests that they would be coactivated in response to dorsiflexion of the toes at the onset of stance.

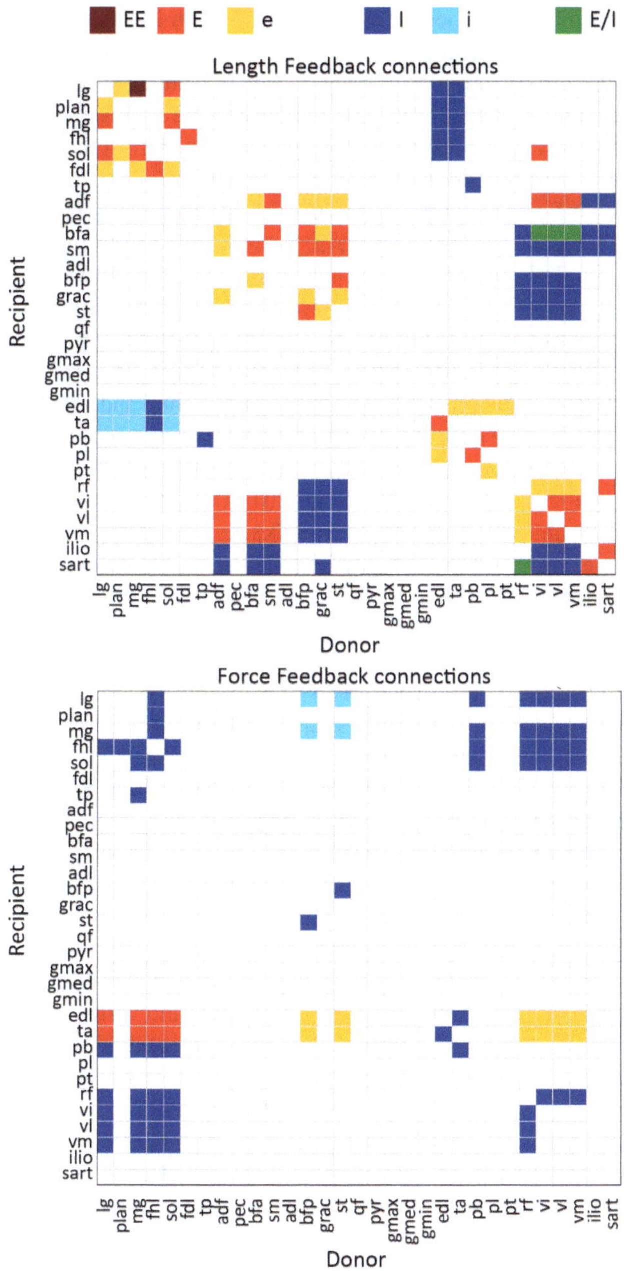


**Fig. 3.4** *The short latency reflex interactions between muscles of the feline hindlimb described in these sections are summarized in two matrices dividing the interactions between length and force dependencies*. Very strong, strong, and weak excitatory interactions are depicted in *dark red, red*, and *yellow* respectively. Strong and weak inhibitory interactions are depicted in *dark blue* and cyan respectively. The interactions marked in *green* (from RF to SART and from the vastus muscles to BF) represent data from Eccles et al. (1957b) where different motoneurons in the same motor pool received either inhibitory or excitatory inputs from the stimulated nerves. Note that some muscles received both length and inhibitory force feedback from other muscles. The dominant effect is then force dependent

Other mechanical synergies are also associated with Ia linkages, such as the pretibial flexors, the peroneus muscles, and the group consisting of ILIO and SART. The quadriceps group forms a Ia synergy, but the linkages between RF and the vastus muscles are somewhat weaker than between the vastus muscles. The weaker synergies between vastus muscles and RF presumably reflect the differences in activation patterns and functions of these two groups (Engberg and Lundberg 1969) rather than any differences in the direction of endpoint forces (Honeycutt and Nichols 2014). A large Ia synergy (Fig. 3.4a) corresponds to the mechanical synergy linking the complex muscles causing hip extension, knee flexion, adduction and abduction at the hip (Fig. 3.3). One can also discern subgroupings corresponding to the hamstrings muscles (BFP, GRAC, ST) and to muscles that primarily extend the hip (ADF, PEC, BFA, SM). Members of these two subgroups correspondingly show two distinct patterns of endpoint forces in both horizontal and sagittal planes (Honeycutt and Nichols 2014). Identification of the hamstrings as a distinct subgroup is further justified by the connection of these muscles through the crural fascia to the calcaneus (van Ingen Schenau 1994; Stahl 2010). The predominantly hip extensor subgroup also forms a Ia synergy. The Ia linkages among the members of the gluteus group have not been studied to our knowledge, and therefore these linkages are not represented.

Although most Ia linkages for hind limb muscles represent close mechanical synergies (Fig. 3.3), there is one prominent example of a Ia linkage that crosses joints, namely, a unidirectional link from VI to SOL (Eccles et al. 1957a; Wilmink and Nichols 2003). VI and SOL are both single joint extensors, crossing the knee and ankle, respectively, and show no mechanical synergy (Fig. 3.3). The vastus muscles also share Ia feedback with ADF and receive excitation from SM and BFA, all hip extensors (Eccles and Lundberg 1958b). Apparently due to technical limitations, it was not specified whether all three vasti project to ADF, or only VI ("crureus").

Muscles are also linked by reciprocal Ia inhibition (Fig. 3.4a), although these linkages do not represent all the examples of mechanical dissimilarity (Fig. 3.3). These linkages include the triceps surae muscles and FHL with the TA & EDL, hamstrings with the quadriceps muscles, and BFA & SM with ILIO & SART. Linkages are also apparent for ADF with ILIO and SART, and vastus muscles with ILIO and SART. These linkages all correspond to mechanical dissimilarity, and involve joints common to both muscle groups. The linkage between RF and SART is represented as mixed. RF is linked by excitation to lateral SART, which extends the knee, and by inhibition to medial SART, that flexes the knee. Note that inhibition between the vastus muscles and ILIO represents an example of cross-joint inhibition. The quadriceps muscles are mechanically dissimilar to MG, LG and PLAN for action at the knee (Fig. 3.3), but these groups are not linked by reciprocal inhibition. Further, adductors and abductors of the hip are not linked by reciprocal inhibition (Eccles and Lundberg 1958b).

At the ankle, the reciprocal inhibition between triceps surae muscles and TA & EDL is stronger in the direction TA & EDL to triceps surae, but is balanced between TP and PB (Bonasera and Nichols 1994), two important ankle stabilizing muscles that are coactivated during stance. This reciprocal inhibition has the effect of stiffening the ankle in the frontal plane during stance. Although FDL and FHL are strong Ia synergists, only FHL shares reciprocal inhibition with TA & EDL, presumably because FHL, but not FDL, has a strong action at the ankle (Lawrence and Nichols 1999a).

### 3.4.2 Implications for the Myotatic Unit Hypothesis

The distribution of excitatory and inhibitory Ia feedback generally reflects mechanical synergism and antagonism, respectively, but the connections in many cases are inconsistent with the myotatic unit hypothesis (cf. Eccles and Lundberg 1958b). The presence of monosynaptic linkages or inhibition across joints and the projection of excitation and inhibition to more than one muscle group are examples of patterns not consistent with the hypothesis. Even in the distal hind limb, one can find examples of muscles that form strong Ia linkages with a synergist (e.g. FHL & FDL) but do not share inhibition with direct antagonists (e.g. FDL).

The myotatic unit has also been considered a basis for synergistic activation of muscles (Caicoya et al. 1999). However, there are numerous examples of muscular activation patterns that do not correspond to patterns of length feedback. For example, the ankle stabilizers are linked by reciprocal inhibition, but yet are coactivated during stance. The strong reciprocal inhibition during cocontraction provides increased stiffness in the frontal plane (Nichols and Koffler-Smulevitz 1991; Bonasera and Nichols 1996). Furthermore, cutaneous pathways (e.g. sural nerve) can simultaneously activate MG and inhibit LG and SOL, even though these muscles are closely linked by Ia afferents (LaBella et al. 1989; LaBella and McCrea 1990; Nichols et al. 1993). Finally, FDL and FHL are recruited according to very different patterns during locomotion (O'Donovan et al. 1982) and yet are strong Ia synergists. We conclude from these observations that length feedback helps to regulate limb mechanics rather than providing a basis for muscular activation patterns, and that this regulatory system operates at the level of the limb rather than at the level of single joints.

### 3.4.3 Heterogenic Force Feedback

These results have been drawn from the studies of Eccles et al. (1957b) and more recent studies (Bonasera and Nichols 1994, 1996; Wilmink and Nichols 2003). Most heterogenic pathways from Golgi tendon organs are inhibitory during conditions of steady force production and during locomotion (Ross and Nichols 2009), but some excitatory pathways exist also (Fig. 3.4b). Heterogenic inhibition between major muscle groups is found principally between the quadriceps muscles, the triceps surae muscles, and FHL. FHL is a particularly powerful source of inhibitory feedback (Bonasera and Nichols 1994). Eccles et al. (1957b) reported that FDL was a powerful source of inhibition to other muscles, but more recent results indicate that this inhibition emerges from FHL while FDL provides relatively little force feedback (Bonasera and Nichols 1994). Since the studies of Eccles et al. depended upon electrical stimulation of muscle nerves, it may have been difficult to distinguish contributions of these two synergists. There is also some inhibition from the hamstrings muscles to MG and LG. Therefore, force dependent inhibition appears to link the major antigravity muscles across the joints of the limb. Inhibition across axes of rotation is found between PB and the triceps surae. Force related excitation was found in a single direction from the triceps surae and FHL to TA & EDL, which is complementary to the relatively weak reciprocal inhibition in this direction. Observation of these excitatory connections may have been responsible for the suggestion of the "inverse myotatic reflex mechanism" (Laporte and Lloyd 1952). According to Eccles et al. (1957b), however, TA and EDL receive excitation from a number of extensor muscles across the limb, so this system is not localized in the manner of length feedback. Therefore, the concept of the "inverse myotatic reflex mechanism" proposed by Laporte and Lloyd is not generalizable to the whole limb and does not parallel the organization of length feedback.

Heterogenic inhibition was also found within muscle groups. The weak Ia excitation between the vastus muscles and RF is complemented by mutual force dependent inhibition. Within the triceps surae group, inhibition is found projecting from MG and to some extent from LG to SOL, but not generally in the reverse direction (Nichols 1999). Finally, force-related inhibition was found between EDL and TA and between ST and BFP. These pathways are force-dependent, so the inhibitory linkages across joints and axes of rotation are likely to become more important during movements than during quiet standing.

### *3.4.4 Magnitude and Directionality*

Recent studies, consistent with an earlier report (see Table 3.1 from Eccles et al. (1957b)), indicate that the magnitude of heterogenic inhibition varies considerably across preparations (Bonasera and Nichols 1994; Lyle, Niazi, Tuthill and Nichols, unpublished), in different tasks (Nichols et al. 2014) and following spinal cord injury (Niazi et al. 2012), while the strength of length feedback remains relatively constant over tasks. In addition, inhibitory force feedback appears to have a directional bias. During locomotion on a level treadmill, heterogenic inhibition follows a proximal to distal gradient (Ross and Nichols 2009) in which inhibition is stronger from the quadriceps to the triceps surae muscles to FHL. In various non-locomoting preparations, this gradient, its reverse or balanced inhibition may be observed (Bonasera and Nichols 1994; Lyle and Nichols 2014). Following spinal cord injury, the greater strength of force feedback is from FHL to the triceps surae muscles or the quadriceps. Returning to the original hypothesis of Houk (1972a) in which length and force feedback are integrated for the regulation of muscular stiffness, and its extrapolation to the regulation of limb stiffness, the different gradients of force feedback that have been observed suggest the stiffness of the limb measured at the endpoint as well as the distribution of stiffnesses across the joints are regulated by the central nervous system. Our recent results suggest further that force feedback constitutes an important mechanism for modulating limb stiffness.

## 3.5 Summary and Discussion

In this chapter we have attempted to assemble information about the organization of proprioceptive circuits for the cat spinal cord where connections to specific muscles could be identified. The main sources of this information are the extensive studies of Eccles et al. on group I pathways using electrically evoked afferent responses recorded intracellularly from motoneurons in anesthetized animals, and the more recent mechanographic studies using unanesthetized decerebrate animals. These two series of studies, where the same muscle combinations were used, are largely compatible. We propose that an important advantage of the mechanographic method—which involves evoking afferent responses with muscle stretch (i.e. physiological input) and recording the net population effect as force responses (i.e. physiological output)—is the closer approximation of reality affording more salient functional inferences. However, it remains difficult to interpret the function of individual pathways in many cases, and even more difficult to perceive how these pathways function together as an integrated whole. Current work utilizing robotic perturbations to calculate endpoint stiffness ellipsoids is anticipated to provide insight concerning the integrated role of proprioceptive pathways at the whole limb level. We believe that a more complete picture of the integrated pathways will come through mathematical modeling of the neuromusculoskeletal system, where the contributions of various components can be systematically tested. In a neuromechanical modeling environment, de novo limb controllers can be constructed based on theories derived from experimental observations and the features of that controller as well as the performance of the limb under that control can be compared with experimentally observed features and behavior. For example, by extending the insights of James Houk (that the integration of length and force feedback act as regulators of muscle stiffness) to the entire limb we can construct an integrated feedback system that regulates whole limb stiffness and compare the performance and features of that feedback system with the experimental observations.

### *3.5.1 Summary of Intrinsic Properties of the Musculoskeletal System*

Understanding the actions of proprioceptive pathways requires knowledge of the intrinsic properties of the musculoskeletal system. Muscles have complex mechanical properties, such as nonlinear viscosity, and stiffness that depends on amplitude of perturbation, background force, and movement history. Properties such as intrinsic stiffness are advantageous for posture and movement, while others appear to be compensated (yielding and the dependence of stiffness on background force level and length). Muscles transmit forces through an elastic network consisting largely of tendons and fascia that provide filtering and distribution of forces across different joints. Energy storage and dissipation occurs through the elasticity of connective tissue elements and viscous muscle. The mechanical actions of each muscle depend on these routes of force transmission and on the location of their attachments through fascia or tendon. Finally, each limb segment has inertial properties that have a major impact on motor coordination.

### *3.5.2 General Principles of Organization and Action of Individual Pathways: Length Feedback*

Pathways arising from muscle spindle receptors are stimulated by length changes in muscles, and primarily by muscle stretch that occurs during tasks involving eccentric muscular work such as downslope walking. Autogenic length feedback decouples force and stiffness so that muscles can respond vigorously to length change when starting with minimal levels of muscular recruitment, as occurs in response to postural disturbances during quiet standing, and also compensate for certain other nonlinear properties of muscle. The excitatory length feedback shared by close synergists would seem to have similar actions to autogenic length feedback, and function to increase the stiffness of shared joints or mechanically coupled joints spanned by the muscles.

There are also several examples of short latency excitation between muscles crossing different joints. In the case of VI to SOL, there are not parallel pathways from VL and VM to SOL, even though all these muscles are technically single joint. This might suggest that this short-latency connection could be related to motor unit type. However, ADF and other hip extensor muscles linked to the quadriceps by short-latency excitation are heterogeneous in fiber-type composition, and it has been argued elsewhere that proprioceptive linkages are related to articulation rather than motor unit type (Wilmink and Nichols 2003). It is therefore not clear what principle determines the VI to SOL connection. However, it is interesting that VL and VM are connected to fascia and therefore have more possibility of myofascial force transmission than VI. This neuronal pathway from VI to SOL therefore might constitute a neural pathway that parallels the mechanical coupling of VM and VL to the distal limb.

Inhibitory effects of length feedback, usually referred to as reciprocal inhibition, appear to have the effect of increasing joint stiffness. The reciprocal inhibition shared by PB and TP, two ankle stabilizers in the frontal plane, clearly enhances stiffness in the frontal plane since these muscles are coactivated during stance. For multi-joint muscles, the strict reciprocal relationships are not so clear. In the distal limb, TA & EDL receive inhibition from several muscles crossing various combinations of joints (MG, LG, PLAN, SOL, FHL), even though the different pairings share some antagonistic action. Although reciprocal inhibition generally links muscles having antagonistic actions, this is not always the case as illustrated by inhibitory pathways from the quadriceps to BFA and other hip extensors, in addition to the expected inhibitory linkages of quadriceps with knee flexors. These heterogenic connections of length feedback illustrate that the idea of the myotatic unit is rarely realized in the feline hind limb.

By virtue of the actions of length feedback to mediate vigorous muscle reflex responses at low background forces, the associated pathways are important for balance and stability during tasks such as quiet standing. Indeed, the directional tuning of initial muscular activations in response to perturbations of the support surface observed in intact animals (Macpherson 1988a, b) can be explained by short-latency length feedback and limb anatomy, both experimentally and computationally (Honeycutt et al. 2009, 2012; Bunderson et al. 2010; Honeycutt and Nichols 2010). That is, the principal direction of the tuning curve for each muscle opposes the direction of force produced by stimulation of that muscle (Honeycutt and Nichols 2014). The synergy structure during postural perturbations (Torres-Oviedo et al. 2006) is therefore explained in part by the anatomical organization of the limb. The magnitude of muscular responses is, however, likely to be determined also by integrated feedback from other limbs (Ting et al. 1998; Zehr et al. 2009; Hayes et al. 2012; Stevenson et al. 2013), as well as by the relative strengths of length feedback between muscles and length feedback across joints.

### *3.5.3 General Principles of Organization and Action of Individual Pathways: Force Feedback*

Force feedback is expected to influence muscular action in response to perturbations and also in response to active contractile force due to central commands. Force feedback can be quite powerful at low background forces (FHL to triceps surae and quadriceps), but is generally force and task dependent, unlike length feedback. Autogenic excitation from Golgi tendon organs (MG and LG) is observed only during locomotion, and given that it is observed mainly in the biarticular members of the triceps surae, presumably functions to increase mechanical coupling between the knee and ankle by increasing the stiffness of these muscles. Autogenic inhibition in muscles not exhibiting excitation is also amplified during locomotion (Ross 2006). Although heterogenic inhibition is observed under static conditions as well as locomotion, its magnitude is subject to the specific task (Tuthill and Nichols 2009; Nichols et al. 2014).

The distribution of force feedback also differs considerably from the distribution of length feedback. The major anti-gravity groups are linked across joints and axes of rotation by inhibitory force feedback (quadriceps, triceps surae and FHL; triceps surae, PB and FHL) and force dependent inhibition links muscles with different articulation or axes of rotation within the major groups (RF and vastus muscles; MG and SOL; TA and EDL; TA and PB; ST and BFP). Within the major groups, one finds mixtures of length dependent excitation and force dependent inhibition, where the balance between the two sources of feedback depends on background tension (e.g. MG contributes excitation to SOL at lower forces that is gradually replaced by inhibition at higher background forces) (Nichols 1999). TA & EDL also receive heterogenic excitation from many muscles in the limb (e.g. triceps, Q, PB & ST) (Eccles et al. 1957b). The functions of the excitatory pathways onto TA & EDL are unknown, but it is interesting that reciprocal inhibition from the triceps surae to TA & EDL is considerably weaker than the reverse direction, and it is in this weaker direction that the force dependent excitation is found. The force dependence of these pathways suggests that these muscle groups are more tightly linked under dynamic conditions. Work is in progress to incorporate these pathways into the feline hind limb model to better understand their integrative action. It should be born in mind that confining the analysis to individual limbs provides an incomplete picture of the neuromechanical system. These proprioceptive pathways also influence muscular activation in other limbs (Ting et al. 1998; Zehr et al. 2009; Hayes et al. 2012; Stevenson et al. 2013). Interlimb communication will be important to include in future versions of the model.

### *3.5.4 Integration of Intrinsic Mechanical Properties, Length and Force Feedback During Functional Tasks*

The relative contributions of mechanical and proprioceptive pathways to the regulation of limb dynamics are challenging to determine under most experimental conditions. It is fruitful nonetheless to seek this critical knowledge if we are to understand the causes of movement impairment in the presence of musculoskeletal and neurological disorders. We propose, based on the principles discussed previously, some hypotheses concerning the regulation of limb dynamics for the contrasting locomotor tasks of locomotion and landing from a fall. Locomotion occurs over a range of speeds over which inertial influences and overall loading varies, whereas landing is characterized by exaggeration of segment inertial disparity and high loading. Sense can be made of the actions of many of the individual proprioceptive pathways previously discussed during these tasks, but there are some seemingly conflicting actions as well.

There are several problems associated with managing the mechanics of multi-segmented limbs during locomotion, besides provision of the appropriate sequencing of muscles. Movements of the individual joints should be coordinated so as to maintain muscles within ranges of motion that optimize force and torque production. Indeed, disruption of proportional coordination between the ankle and knee is observed following the loss of proprioceptive feedback from the triceps surae muscles (Abelew et al. 2000; Maas et al. 2007). In addition, the disparity in inertia between the proximal and distal limb segments, and intersegmental dynamics need to be compensated when inertial effects become important, such as during trotting and running. For example, during interaction with the ground, the lighter distal segments potentially would absorb more of the perturbation than the heavier proximal segments. During swing, the inertia of the distal segments can lead to unwanted relative motions of the distal and proximal segments. The limb stiffness must also be appropriate to the task for shock absorption and efficient use of energy (Ferris and Farley 1997; Farley et al. 1998; Ferris et al. 1998; Daley et al. 2006).

The roles of some pathways in the regulation of stiffness, in the regulation of gravitational forces, and in inertial compensation are apparent. During the weight acceptance phase of stance (E2), TP and PB are coactive, stiffening the ankle in the transverse plane, and the reciprocal inhibition between these two muscles reinforces this stabilization. The knee and the ankle are mechanically coupled by the biarticular MG and LG, reinforced by excitatory force feedback. This mechanical coupling helps to reduce the effects of inertial disparity between the limb segments. The moment arms for MG and LG are greater at the ankle than at the knee (Lan and Crago 1992), so these muscles contribute to weight support, propulsion and ankle stiffness. Ankle stiffness also receives a contribution from the substantial reciprocal inhibition from TA & EDL. Unlike the balanced reciprocal inhibition between TP and PB, there is relatively little reciprocal inhibition from the triceps surae onto TA & EDL. The stronger reciprocal inhibition from the pretibial flexors onto triceps surae seems to be correlated with the direction of gravity: it is focused on the muscles that are stretched under the weight of the body (Nichols and Koffler-Smulevitz 1991). Stiffness at the knee is presumably similarly regulated by autogenic feedback and reciprocal inhibition between the quadriceps muscles and knee flexors (BFP, GRAC, ST). Since the knee flexors and RF cross the hip as well, mechanical coupling between knee and hip is provided.

Superimposed on this system of length feedback, force feedback provides an additional source of mechanical coupling, stiffness regulation and inertial compensation under dynamic conditions, since force feedback is itself force dependent. As discussed earlier, autogenic force feedback can be excitatory during locomotion. Force dependent excitation in MG and LG during locomotion (Ross and Nichols 2009) contributes to ankle stiffness and coupling between ankle and knee. In contrast, heterogenic force feedback appears to be predominantly inhibitory. The integration of this more global feedback with length feedback helps to determine limb stiffness and the distribution of stiffness across the component joints.

Heterogenic force feedback between major muscle groups can contribute to inertial compensation and the distribution of limb stiffness. The distal limb is the site of direct interaction with the environment, so one might expect that the impedance of the distal joints of the limb might be less than that of the proximal joints in order to provide a moderately compliant interface. Since the distal limb segments have smaller mass than the more proximal segments, impedance due to inertia is therefore non-uniform in the appropriate direction based on limb mechanics. As discussed previously, the strong Ib inhibition from what we believe is FHL (i.e., reported as FDL) onto the triceps surae and quadriceps muscles in the anesthetized state (Eccles et al. 1957b) would tend to compensate for this non-uniformity in impedance when the anti-gravity muscles, including FHL, are activated during stance. The regulatory mechanism(s) mediating the balance between limb inertia and the gradient of force feedback is unknown, but is anticipated to have an important influence on task-specific limb mechanics. In the unanesthetized, non-locomoting decerebrate animal, the directionality of inhibitory force feedback was distal to proximal in some preparations, the opposite in others and more balanced in the remainder (Bonasera and Nichols 1994), so it may be that the effects of inertial non-uniformity on impedance at different joints are reinforced for some tasks and compensated for others by variations in the strength of Ib inhibition.

Interestingly, strong distal to proximal inhibition has been consistently observed after acute and chronic spinal hemisection (Niazi et al. 2012). This observation, along with the findings of Eccles et al. (1957b), suggests that the distal to proximal gradient of inhibition is the default state of the spinal cord that is then modulated according to the specific task, such as walking (Ross and Nichols 2009). During stepping on a level treadmill in the unanesthetized decerebrate animal, inhibitory feedback was found to be stronger from proximal to distal muscles (Ross and Nichols 2009). During relatively slow walking, inertial effects should be relatively small so that the proximal to distal gradient of force feedback would ensure that the distal limb segment would be appropriately compliant. It is predicted then that during rapid locomotion, or landing from jump, inhibitory feedback would be reweighted so as to influence proximal muscles to compensate for the effects of non-uniform inertia. As noted above, proximal to distal inhibitory feedback can be upregulated during downslope walking to reduce the impedance of distal joints even more. This hypothesis that *force feedback is engaged for inertial compensation and limb stiffness* remains to be validated using computational modeling and experiment.

A potential role for the excitatory force feedback observed between muscles of propulsion and the pretibial flexors (Eccles et al. 1957b; Nichols 1989) is suggested by a consideration of limb mechanics during locomotion. During E3 and into the initial stages of swing, especially for more rapid locomotion, there is coactivation of RF, BFP, ST and TA (Rossignol 1996). It is possible that the activity of TA is reinforced or even mediated during that time by the force-related excitation from the hamstrings muscles and RF. The hamstrings muscles may aid in propulsion through the crural fascia during E3 (Stahl 2010), but then continue with knee flexion as swing progresses. Their activity would also reinforce the action of TA to flex the ankle during the subsequent swing phase (Rossignol 1996). This role of force feedback may be viewed as an "assistance" rather than resistance and would promote the reversal in the direction of limb movement between stance and swing.

As discussed above, intrinsic mechanical properties of the limb and proprioceptive pathways have an important role in regulating joint dynamics in a task dependent manner for locomotion. Landing from a jump, which involves rapid deceleration of total body center of mass through eccentric muscular control of the limb segments in a distal to proximal sequence, is an additional motor task that highlights the complex motor control synergy between intrinsic musculotendon properties and proprioceptive circuits. The act of landing from a height has been studied in animals (Prochazka et al. 1977; McKinley et al. 1983; Abraham and Loeb 1985; Konow et al. 2012) and in humans (Ferris and Farley 1997; McDonagh and Duncan 2002; Santello 2005; Galindo et al. 2009; Lyle et al. 2013). Landing involves higher loads and angular velocities of the limb segments (particularly the distal segments with lower inertia) that necessitate rapid control of joints as compared to locomotion. The rapidity by which the peak forces are experienced (typically within 50–60 ms) indicates that intrinsic musculotendon properties, feedforward muscle activation and sensory feedback mediated by spinal pathways is primarily responsible for deceleration of body center of mass.

The assumed global goal for the nervous system during landing is to smoothly decelerate the center of mass by regulating whole limb stiffness to act as a brake. Animals and humans possess a remarkable ability to modulate limb stiffness across a wide spectrum of loading and surface conditions (Ferris and Farley 1997; Ferris et al. 1998; Dickinson et al. 2000; Daley et al. 2006; Perreault et al. 2008; Krutky et al. 2013). While the details of how this is accomplished remains an open question, it is clear that intrinsic musculotendon properties coupled with feedforward activation of muscles prior to foot contact functions to resist the initial impact due to delays in sensory feedback (Prochazka et al. 1977; McKinley et al. 1983; Abraham and Loeb 1985; Konow et al. 2012). In addition to the complex mechanical properties of muscle, evidence indicates that tendons can act as a mechanical buffer (Griffiths 1991; Prilutsky et al. 1996; Roberts and Konow 2013) by delaying and reducing the velocity of active lengthening of muscle fascicles in response to the rapid joint flexion during landing (Konow et al. 2012). This intrinsic property has been proposed to protect muscles from damage (Griffiths 1991; Konow et al. 2012; Roberts and Konow 2013) and perhaps preserves a favorable force-velocity relation for muscle action (Griffiths 1991; Prilutsky et al. 1996). The rapid flexion of the distal segment (e.g. ankle) additionally imposes mechanical coupling of the ankle and knee joints due to stretching the gastrocnemius for example (Zajac 1993; Prilutsky and Zatsiorsky 1994).

Rapid flexion and high loads characteristic of landing indicate proprioceptive length and force feedback contribute to regulation of whole limb stiffness with short delays occupied by short-range stiffness. As previously discussed, length feedback from muscle spindles contribute to limb stiffness primarily at the joint level but also reinforce segment coupling in cases such as autogenic feedback to gastrocnemius and likely rectus femoris (Prilutsky and Zatsiorsky 1994). Interestingly, a burst of activation from FHL and FDL are both observed after foot contact (Abraham and Loeb 1985) likely attributed to rapid toe dorsiflexion and the strong Ia connections between them. The synergy observed by these muscles during landing is in contrast to dissimilar activations of these muscles during locomotion (see above).

We propose that inhibitory force feedback would best match the task goal of attenuating impact forces by being distributed such that inhibition is strongest from distal to proximal muscles. The clear advantage of this organization is an explicit compensation for the effects of non-uniform inertia. Functionally, this would serve the purpose of increasing the compliance of the knee and hip and thus facilitate energy absorption by the more massive proximal knee and hip extensor muscles. In support of the distal to proximal concept, Prochazka et al. (1977), examining landing in cats, proposed that a reduction in lateral gastrocnemius activation for a short period immediately after impact was due to force dependent inhibition from toe flexors. Alternatively, landing and other rapid deceleration maneuvers could achieve a distal to proximal inhibitory force feedback gradient simply from the mechanics of the task, which is characterized by a distal to proximal sequencing of joint flexion and peak net joint moments. That is, landing involves rapid flexion of the toe and ankle such that toe and ankle flexor muscles presumably could begin inhibiting the proximal muscles at the knee prior to or coincident with rising force in the quadriceps muscles for example.

It argued that functions of proprioceptive feedback may be understood by considering their role in regulating the magnitude and distribution of limb stiffness through the interaction of length and force feedback. Evidence also suggests that the balance of length and force feedback can be altered in a task-specific manner by modulation of force feedback. Force feedback becomes more important under dynamic conditions and can compensate or reinforce the effects of non-uniform inertia of the limb segments. These hypotheses are qualitative and deduced by a review of the experimental evidence, but remain to be directly validated by further experiment and simulation. Work is currently in progress utilizing Neuromechanic to validate these hypotheses. It should be noted that the forgoing analysis assumes that the actions of these pathways are focused on the limb of origin of the proprioceptive signals. It may well be that projections of these pathways to other limbs must be considered to provide a more complete understanding of their *integrative actions*.

**Acknowledgements** This work was supported by NIH grants NS20855 and HD 32571 to TRN and 1F32NS080393-01A1 to MAL.

## References

Abelew TA, Miller MD, Cope TC, Nichols TR (2000) Local loss of proprioception results in disruption of interjoint coordination during locomotion in the cat. J Neurophysiol 84:2709–2714

Abraham LD, Loeb GE (1985) The distal hindlimb musculature of the cat. Patterns of normal use. Exp Brain Res 58:583–593

Bonasera SJ, Nichols TR (1994) Mechanical actions of heterogenic reflexes linking long toe flexors with ankle and knee extensors of the cat hindlimb. J Neurophysiol 71:1096–1110

Bonasera SJ, Nichols TR (1996) Mechanical actions of heterogenic reflexes among ankle stabilizers and their interactions with plantarflexors of the cat hindlimb. J Neurophysiol 75:2050–2070

Boskov D, Heckman CJ (1996) Motor unit recruitment patterns during reflex compensation of muscle yield investigated by computer simulations. Biol Cybern 75:211–217

Brink EE, Suzuki I, Timerick SJ, Wilson VJ (1985) Tonic neck reflex of the decerebrate cat: a role for propriospinal neurons. J Neurophysiol 54:978–987

Bullinger KL, Nardelli P, Pinter MJ, Alvarez FJ, Cope TC (2011) Permanent central synaptic disconnection of proprioceptors after nerve injury and regeneration. II. Loss of functional connectivity with motoneurons. J Neurophysiol 106:2471–2485

Bunderson N, Bingham J (2015) Better science through predictive modeling: numerical tools for understanding neuromechanical interactions. In: Prilutsky BI, Edwards DH Jr (eds) Neuromechanical modeling of posture and locomotion. Springer, New York (in press)

Bunderson NE, Burkholder TJ, Ting LH (2008) Reduction of neuromuscular redundancy for postural force generation using an intrinsic stability criterion. J Biomech 41:1537–1544

Bunderson NE, McKay JL, Ting LH, Burkholder TJ (2010) Directional constraint of endpoint force emerges from hindlimb anatomy. J Exp Biol 213:2131–2141

Bunderson NE, Bingham JT, Sohn MH, Ting LH, Burkholder TJ (2012) Neuromechanic: a computational platform for simulation and analysis of the neural control of movement. Int J Numer Methods Biomed Eng 28:1015–1027
Burkholder TJ, Nichols TR (2000) The mechanical action of proprioceptive length feedback in a model of cat hindlimb. Motor Control 4:201–220
Burkholder TJ, Nichols TR (2004) Three-dimensional model of the feline hindlimb. J Morphol 261:118–129
Caicoya AG, Illert M, Janike R (1999) Monosynaptic Ia pathways at the cat shoulder. J Physiol 518(Pt 3):825–841
Campbell KS, Moss RL (2000) A thixotropic effect in contracting rabbit psoas muscle: prior movement reduces the initial tension response to stretch. J Physiol 525(Pt 2):531–548
Carrasco DI, English AW (1999) Mechanical actions of compartments of the cat hamstring muscle, biceps femoris. Prog Brain Res 123:397–403
Carrasco DI, Lawrence J 3rd, English AW (1999) Neuromuscular compartments of cat lateral gastrocnemius produce different torques about the ankle joint. Motor Control 3:436–446
Chang YH, Auyang AG, Scholz JP, Nichols TR (2009) Whole limb kinematics are preferentially conserved over individual joint kinematics after peripheral nerve injury. J Exp Biol 212:3511–3521
Cope TC, Bonasera SJ, Nichols TR (1994) Reinnervated muscles fail to produce stretch reflexes. J Neurophysiol 71:817–820
Cui L, Perreault EJ, Maas H, Sandercock TG (2008) Modeling short-range stiffness of feline lower hindlimb muscles. J Biomech 41:1945–1952
Daley MA, Usherwood JR, Felix G, Biewener AA (2006) Running over rough terrain: guinea fowl maintain dynamic stability despite a large unexpected change in substrate height. J Exp Biol 209:171–187
Daley MA, Voloshina A, Biewener AA (2009) The role of intrinsic muscle mechanics in the neuromuscular control of stable running in the guinea fowl. J Physiol 587:2693–2707
Dickinson MH, Farley CT, Full RJ, Koehl MA, Kram R, Lehman S (2000) How animals move: an integrative view. Science 288:100–106
Donelan JM, Pearson KG (2004) Contribution of force feedback to ankle extensor activity in decerebrate walking cats. J Neurophysiol 92:2093–2104
Eccles JC, Eccles RM, Lundberg A (1957a) The convergence of monosynaptic excitatory afferents on to many different species of alpha motoneurones. J Physiol 137:22–50
Eccles JC, Eccles RM, Lundberg A (1957b) Synaptic actions on motoneurones caused by impulses in Golgi tendon organ afferents. J Physiol 138:227–252
Eccles RM, Lundberg A (1958a) Integrative pattern of Ia synaptic actions on motoneurones of hip and knee muscles. J Physiol 144:271–298
Eccles RM, Lundberg A (1958b) Integrative pattern of Ia synaptic actions on motoneurones of hip and knee muscles. J Physiol 144:271–298
Engberg I, Lundberg A (1969) An electromyographic analysis of muscular activity in the hindlimb of the cat during unrestrained locomotion. Acta Physiol Scand 75:614–630
English AW, Weeks OI (1987) An anatomical and functional analysis of cat biceps femoris and semitendinosus muscles. J Morphol 191:161–175
Farley CT, Houdijk HH, Van Strien C, Louie M (1998) Mechanism of leg stiffness adjustment for hopping on surfaces of different stiffnesses. J Appl Physiol 85:1044–1055
Feldman AG, Levin MF (1995) The origin and use of positional frames of reference in motor control. Behav Brain Sci 18:723–806
Feldman AG, Orlovsky GN (1972) The influence of different descending systems on the tonic stretch reflex in the cat. Exp Neurol 37:481–494
Ferris DP, Farley CT (1997) Interaction of leg stiffness and surfaces stiffness during human hopping. J Appl Physiol (1985) 82:15–22; discussion 13–14
Ferris DP, Louie M, Farley CT (1998) Running in the real world: adjusting leg stiffness for different surfaces. Proc Biol Sci 265:989–994
Galindo A, Barthelemy J, Ishikawa M, Chavet P, Martin V, Avela J et al (2009) Neuromuscular control in landing from supra-maximal dropping height. J Appl Physiol (1985) 106:539–547

Gordon J, Ghilardi MF, Ghez C (1995) Impairments of reaching movements in patients without proprioception. I. Spatial errors. J Neurophysiol 73:347–360
Gottschall JS, Nichols TR (2007) Head pitch affects muscle activity in the decerebrate cat hindlimb during walking. Exp Brain Res 182:131–135
Gottschall JS, Nichols TR (2011) Neuromuscular strategies for the transitions between level and hill surfaces during walking. Philos Trans R Soc Lond B Biol Sci 366:1565–1579
Granit R (1950) Reflex self-regulation of muscle contraction and autogenetic inhibition. J Neurophysiol 13:351–372
Griffiths RI (1991) Shortening of muscle fibres during stretch of the active cat medial gastrocnemius muscle: the role of tendon compliance. J Physiol 436:219–236
Grillner S, Udo M (1971) Motor unit activity and stiffness of the contracting muscle fibres in the tonic stretch reflex. Acta Physiol Scand 81:422–424
Haftel VK, Bichler EK, Nichols TR, Pinter MJ, Cope TC (2004) Movement reduces the dynamic response of muscle spindle afferents and motoneuron synaptic potentials in rat. J Neurophysiol 91:2164–2171
Hayes HB, Chang YH, Hochman S (2012) Stance-phase force on the opposite limb dictates swing-phase afferent presynaptic inhibition during locomotion. J Neurophysiol 107:3168–3180
Hoffer JA, Loeb GE, Sugano N, Marks WB, O'Donovan MJ, Pratt CA (1987) Cat hindlimb motoneurons during locomotion. III. Functional segregation in sartorius. J Neurophysiol 57:554–562
Hoffer JA, Caputi AA, Pose IE, Griffiths RI (1989) Roles of muscle activity and load on the relationship between muscle spindle length and whole muscle length in the freely walking cat. Prog Brain Res 80:75–85; discussion 57–60
Honeycutt CF, Nichols TR (2010) The decerebrate cat generates the essential features of the force constraint strategy. J Neurophysiol 103:3266–3273
Honeycutt CF, Nichols TR (2014) The mechanical actions of muscles predict the direction of muscle activation during postural perturbations in the cat hindlimb. J Neurophysiol 111(5):900–907
Honeycutt CF, Gottschall JS, Nichols TR (2009) Electromyographic responses from the hindlimb muscles of the decerebrate cat to horizontal support surface perturbations. J Neurophysiol 101:2751–2761
Honeycutt CF, Nardelli P, Cope TC, Nichols TR (2012) Muscle spindle responses to horizontal support surface perturbation in the anesthetized cat: insights into the role of autogenic feedback in whole body postural control. J Neurophysiol 108:1253–1261
Houk J (1972a) The phylogeny of muscular control configurations. In: Drischel P, Dettmar P (eds) Biocybernetics, vol 4. Fisher, Jena, pp 125–44
Houk JC (1972b) On the significance of various command signals during voluntary control. Brain Res 40:49–58
Houk JC (1979) Regulation of stiffness by skeletomotor reflexes. Annu Rev Physiol 41:99–114
Houk JC, Rymer WZ, Crago PE (1981) Dependence of dynamic response of spindle receptors on muscle length and velocity. J Neurophysiol 46:143–166
Houk JC, Fagg AH, Barto AG (2002) Fractional power damping model of joint motion. In: Latash ML (ed) Structure-function relations in voluntary movements, vol 2. Human Kinetics, Champaign
Huyghues-Despointes CM, Cope TC, Nichols TR (2003a) Intrinsic properties and reflex compensation in reinnervated triceps surae muscles of the cat: effect of activation level. J Neurophysiol 90:1537–1546
Huyghues-Despointes CM, Cope TC, Nichols TR (2003b) Intrinsic properties and reflex compensation in reinnervated triceps surae muscles of the cat: effect of movement history. J Neurophysiol 90:1547–1555
Hyde LA, Burkholder TJ, Nichols TR (1999) Reflex action of the hamstrings muscles at the feline ankle mediated by the crural fascia. Soc Neurosci Abstr 25:1151
Jankowska E, Edgley SA (2010) Functional subdivision of feline spinal interneurons in reflex pathways from group Ib and II muscle afferents; an update. Eur J Neurosci 32:881–893
Joyce GC, Rack PM, Westbury DR (1969) The mechanical properties of cat soleus muscle during controlled lengthening and shortening movements. J Physiol 204:461–474

Kirsch RF, Boskov D, Rymer WZ (1994) Muscle stiffness during transient and continuous movements of cat muscle: perturbation characteristics and physiological relevance. IEEE Trans Biomed Eng 41:758–770

Konow N, Azizi E, Roberts TJ (2012) Muscle power attenuation by tendon during energy dissipation. Proc Biol Sci 279:1108–1113

Krouchev N, Kalaska JF, Drew T (2006) Sequential activation of muscle synergies during locomotion in the intact cat as revealed by cluster analysis and direct decomposition. J Neurophysiol 96:1991–2010

Krutky MA, Trumbower RD, Perreault EJ (2013) Influence of environmental stability on the regulation of end-point impedance during the maintenance of arm posture. J Neurophysiol 109:1045–1054

LaBella LA, McCrea DA (1990) Evidence for restricted central convergence of cutaneous afferents on an excitatory reflex pathway to medial gastrocnemius motoneurons. J Neurophysiol 64:403–412

LaBella LA, Kehler JP, McCrea DA (1989) A differential synaptic input to the motor nuclei of triceps surae from the caudal and lateral cutaneous sural nerves. J Neurophysiol 61:291–301

Lafreniere-Roula M, McCrea DA (2005) Deletions of rhythmic motoneuron activity during fictive locomotion and scratch provide clues to the organization of the mammalian central pattern generator. J Neurophysiol 94:1120–1132

Lan N, Crago PE (1992) A noninvasive technique for in vivo measurement of joint torques of biarticular muscles. J Biomech 25:1075–1079

Laporte Y, Lloyd DP (1952) Nature and significance of the reflex connections established by large afferent fibers of muscular origin. Am J Physiol 169:609–621

Latash ML, Zatsiorsky VM (1993) Joint stiffness: myth or reality? Human Mov Sci 12:653–692

Lawrence JH 3rd, Nichols TR (1999a) A three-dimensional biomechanical analysis of the cat ankle joint complex: I. Active and passive postural muscles. J Appl Biomech 15:95–105

Lawrence JH 3rd, Nichols TR (1999b) A three-dimensional biomechanical analysis of the cat ankle joint complex: II. Effects of ankle joint orientation on evoked isometric joint torque. J Appl Biomech 15:106–119

Lawrence JH 3rd, Nichols TR, English AW (1993) Cat hindlimb muscles exert substantial torques outside the sagittal plane. J Neurophysiol 69:282–285

Liddell EGT, Sherrington C (1924) Reflexes in response to stretch (Myotatic Reflexes). Proce Roy Soc Lond B, Contain Pap Biol Character 96:212–242

Lin CC, Crago PE (2002a) Neural and mechanical contributions to the stretch reflex: a model synthesis. Ann Biomed Eng 30:54–67

Lin CC, Crago PE (2002b) Structural model of the muscle spindle. Ann Biomed Eng 30:68–83

Lloyd DP (1946) Integrative pattern of excitation and inhibition in two-neuron reflex arcs. J Neurophysiol 9:439–444

Lyle MA, Valero-Cuevas FJ, Gregor RJ, Powers CM (2013) Control of dynamic foot-ground interactions in male and female soccer athletes: females exhibit reduced dexterity and higher limb stiffness during landing. J Biomech 47(2):512–517

Lyle MA, Nichols TR (2014) Variable gradient of intermuscular inhibition as preliminary evidence for spinal mediated modulation of task dependent limb behavior. In: Society for Neuroscience Meeting. Program No. 827.19. Neuroscience Meeting Planner. Washington, DC

Maas H, Meijer HJ, Huijing PA (2005) Intermuscular interaction between synergists in rat originates from both intermuscular and extramuscular myofascial force transmission. Cells Tissues Organs 181:38–50

Maas H, Prilutsky BI, Nichols TR, Gregor RJ (2007) The effects of self-reinnervation of cat medial and lateral gastrocnemius muscles on hindlimb kinematics in slope walking. Exp Brain Res 181:377–393

Maas H, Gregor RJ, Hodson-Tole EF, Farrell BJ, Prilutsky BI (2009) Distinct muscle fascicle length changes in feline medial gastrocnemius and soleus muscles during slope walking. J Appl Physiol (1985) 106:1169–1180

Macpherson JM (1988a) Strategies that simplify the control of quadrupedal stance. I. Forces at the ground. J Neurophysiol 60:204–217
Macpherson JM (1988b) Strategies that simplify the control of quadrupedal stance. II. Electromyographic activity. J Neurophysiol 60:218–231
Macpherson JM (1994) Changes in a postural strategy with inter-paw distance. J Neurophysiol 71:931–940
Malamud JG, Godt RE, Nichols TR (1996) Relationship between short-range stiffness and yielding in type-identified, chemically skinned muscle fibers from the cat triceps surae muscles. J Neurophysiol 76:2280–2289
Marchand AR, Manzoni D, Pompeiano O, Stampacchia G (1987) Effects of stimulation of vestibular and neck receptors on Deiters neurons projecting to the lumbosacral cord. Pflugers Arch 409:13–23
Markin SN, Lemay MA, Prilutsky BI, Rybak IA (2012) Motoneuronal and muscle synergies involved in cat hindlimb control during fictive and real locomotion: a comparison study. J Neurophysiol 107:2057–2071
McDonagh MJ, Duncan A (2002) Interaction of pre-programmed control and natural stretch reflexes in human landing movements. J Physiol 544:985–994
McKinley PA, Smith JL, Gregor RJ (1983) Responses of elbow extensors to landing forces during jump downs in cats. Exp Brain Res 49:218–228
Merton PA (1953) Speculations on the servo-control of movement. In: Wolstenhome GEW (ed) The spinal cord. Churchill, London, pp 247–255
Monster AW, Chan H (1977) Isometric force production by motor units of extensor digitorum communis muscle in man. J Neurophysiol 40:1432–1443
Munson JB, Sypert GW, Zengel JE, Lofton SA, Fleshman JW (1982) Monosynaptic projections of individual spindle group II afferents to type-identified medial gastrocnemius motoneurons in the cat. J Neurophysiol 48:1164–1174
Niazi IF, Nichols TR, Rising A, Little L, Mondello S, Howland D (2012) Reorganization of heterogenic inhibitory force feedback between hindlimb extensors after both acute and chronic time points post spinal cord injury. In: Society for Neuroscience Meeting. Program No. 885.16. Neuroscience Meeting Planner. New Orlean, LA
Nichols TR (1974) Soleus muscle stiffness and its reflex control[Thesis]. Type, Harvard University.
Nichols TR (1989) The organization of heterogenic reflexes among muscles crossing the ankle joint in the decerebrate cat. J Physiol 410:463–477
Nichols TR (1994) A biomechanical perspective on spinal mechanisms of coordinated muscular action: an architecture principle. Acta Anat (Basel) 151:1–13
Nichols TR (1999) Receptor mechanisms underlying heterogenic reflexes among the triceps surae muscles of the cat. J Neurophysiol 81:467–478
Nichols TR, Cope TC (2004) Cross-bridge mechanisms underlying the history-dependent properties of muscle spindles and stretch reflexes. Can J Physiol Pharmacol 82:569–576
Nichols TR, Houk JC (1976) Improvement in linearity and regulation of stiffness that results from actions of stretch reflex. J Neurophysiol 39:119–142
Nichols TR, Koffler-Smulevitz D (1991) Mechanical analysis of heterogenic inhibition between soleus muscle and the pretibial flexors in the cat. J Neurophysiol 66:1139–1155
Nichols TR, Ross KT (2009) The implications of force feedback for the lambda model. Adv Exp Med Biol 629:663–679
Nichols TR, Steeves JD (1986) Resetting of resultant stiffness in ankle flexor and extensor muscles in the decerebrate cat. Exp Brain Res 62:401–410
Nichols TR, Lawrence JH 3rd, Bonasera SJ (1993) Control of torque direction by spinal pathways at the cat ankle joint. Exp Brain Res 97:366–371
Nichols TR, Cope TC, Abelew TA (1999a) Rapid spinal mechanisms of motor coordination. Exerc Sport Sci Rev 27:255–284
Nichols TR, Lin DC, Huyghues-Despointes CM (1999b) The role of musculoskeletal mechanics in motor coordination. Prog Brain Res 123:369–378

Nichols TR, Wilmink RJH, Burkholder TJ (2002) The multidimensional and temporal regulation of limb mechanics by spinal circuits. In: Latash ML (ed) Progress in motor control, vol 2. Human Kinetics, Champaign, pp 179–94
Nichols TR, Gottschall JS, Tuthill C (2014) The regulation of limb stiffness in the context of locomotor task. In: Levin MF (ed) Progress in motor control IX. Springer, New York, pp 41–54
Nishikawa K, Biewener AA, Aerts P, Ahn AN, Chiel HJ, Daley MA et al (2007) Neuromechanics: an integrative approach for understanding motor control. Integr Comp Biol 47:16–54
O'Donovan MJ, Pinter MJ, Dum RP, Burke RE (1982) Actions of FDL and FHL muscles in intact cats: functional dissociation between anatomical synergists. J Neurophysiol 47:1126–1143
Pearson KG (1995) Proprioceptive regulation of locomotion. Curr Opin Neurobiol 5:786–791
Perreault EJ, Kirsch RF, Crago PE (2004) Multijoint dynamics and postural stability of the human arm. Exp Brain Res 157:507–517
Perreault EJ, Chen K, Trumbower RD, Lewis G (2008) Interactions with compliant loads alter stretch reflex gains but not intermuscular coordination. J Neurophysiol 99:2101–2113
Prilutsky BI, Zatsiorsky VM (1994) Tendon action of two-joint muscles: transfer of mechanical energy between joints during jumping, landing, and running. J Biomech 27:25–34
Prilutsky BI, Herzog W, Leonard TR, Allinger TL (1996) Role of the muscle belly and tendon of soleus, gastrocnemius, and plantaris in mechanical energy absorption and generation during cat locomotion. J Biomech 29:417–434
Prochazka A, Schofield P, Westerman RA, Ziccone SP (1977) Reflexes in cat ankle muscles after landing from falls. J Physiol 272:705–719
Proske U, Morgan DL (1999) Do cross-bridges contribute to the tension during stretch of passive muscle? J Muscle Res Cell Motil 20:433–442
Rack PM, Westbury DR (1969) The effects of length and stimulus rate on tension in the isometric cat soleus muscle. J Physiol 204:443–460
Rack PM, Westbury DR (1974) The short range stiffness of active mammalian muscle and its effect on mechanical properties. J Physiol 240:331–350
Roberts TJ, Konow N (2013) How tendons buffer energy dissipation by muscle. Exerc Sport Sci Rev 41:186–193
Ross KT (2006) Quantitative analysis of feedback during locomotion[Thesis]. Type, Georgia Institute of Technology, Atlanta
Ross KT, Nichols TR (2009) Heterogenic feedback between hindlimb extensors in the spontaneously locomoting premammillary cat. J Neurophysiol 101:184–197
Rossignol S (1996) Neural control of stereotypic limb movements. In: Rowell LB, Shepherd JT (eds) Handbook of physiology, section 12: Exercise: Regulation and integration of multiple systems, vol 12. Oxford, New York
Rymer WZ, Hasan Z (1980) Absence of force-feedback regulation in soleus muscle of the decerebrate cat. Brain Res 184:203–209
Sainburg RL, Poizner H, Ghez C (1993) Loss of proprioception produces deficits in interjoint coordination. J Neurophysiol 70:2136–2147
Sainburg RL, Ghilardi MF, Poizner H, Ghez C (1995) Control of limb dynamics in normal subjects and patients without proprioception. J Neurophysiol 73:820–835
Santello M (2005) Review of motor control mechanisms underlying impact absorption from falls. Gait Posture 21:85–94
Silva PL, Fonseca ST, Turvey MT (2010) Is tensegrity the functional architecture of the equilibrium point hypothesis? Motor Control 14:e35–e40
Smith JL, Carlson-Kuhta P (1995) Unexpected motor patterns for hindlimb muscles during slope walking in the cat. J Neurophysiol 74:2211–2215
Smith JL, Chung SH, Zernicke RF (1993) Gait-related motor patterns and hindlimb kinetics for the cat trot and gallop. Exp Brain Res 94:308–322
Sponberg S, Spence AJ, Mullens CH, Full RJ (2011) A single muscle's multifunctional control potential of body dynamics for postural control and running. Philos Trans R Soc Lond B Biol Sci 366:1592–1605
Stahl VA (2010) Contributions of fascia to force transmission in the cat hindlimb [Thesis]. Type, Georgia Institute of Technology, Atlanta

Stauffer EK, Watt DG, Taylor A, Reinking RM, Stuart DG (1976) Analysis of muscle receptor connections by spike-triggered averaging. 2. Spindle group II afferents. J Neurophysiol 39:1393–1402

Stein R (1982) What muscle variable (s) does the nervous system control in limb movements. Behav Brain Sci 5:535–577

Stevenson AJ, Geertsen SS, Andersen JB, Sinkjaer T, Nielsen JB, Mrachacz-Kersting N (2013) Interlimb communication to the knee flexors during walking in humans. J Physiol 591:4921–4935

Stuart DG, Mosher CG, Gerlach RL, Reinking RM (1970) Selective activation of Ia afferents by transient muscle stretch. Exp Brain Res 10:477–487

Sypert GW, Fleshman JW, Munson JB (1980) Comparison of monosynaptic actions of medial gastrocnemius group Ia and group II muscle spindle afferents on triceps surae motoneurons. J Neurophysiol 44:726–738

Ting LH, Raasch CC, Brown DA, Kautz SA, Zajac FE (1998) Sensorimotor state of the contralateral leg affects ipsilateral muscle coordination of pedaling. J Neurophysiol 80:1341–1351

Torres-Oviedo G, Macpherson JM, Ting LH (2006) Muscle synergy organization is robust across a variety of postural perturbations. J Neurophysiol 96:1530–1546

Tuthill C, Nichols TR (2009) The asymmetric organization of force feedback in the decerebrate cat during varied motor tasks. Soc Neurosci. In: Society for Neuroscience Meeting. program No.658.10. Neuroscience Meeting Planner. Chicago, IL

van Ingen Schenau GJ (1994) Proposed action of bi-articular muscles and the design of hindlimbs of bi- and quadrupeds. Human Mov Sci 13:665–681

Wilmink RJ, Nichols TR (2003) Distribution of heterogenic reflexes among the quadriceps and triceps surae muscles of the cat hind limb. J Neurophysiol 90:2310–2324

Young RP, Scott SH, Loeb GE (1993) The distal hindlimb musculature of the cat: multiaxis moment arms at the ankle joint. Exp Brain Res 96:141–151

Zajac FE (1993) Muscle coordination of movement: a perspective. J Biomech 26(Suppl 1):109–124

Zehr EP, Hundza SR, Vasudevan EV (2009) The quadrupedal nature of human bipedal locomotion. Exerc Sport Sci Rev 37:102–108

# Chapter 4
# Model-Based Approaches to Understanding Musculoskeletal Filtering of Neural Signals

**Thomas J. Burkholder**

**Abstract** The musculoskeletal system lies between the nervous system and its effects on the external world. Strong constraints imposed by joints, ligaments, and muscle attachments shape both the effect the nervous system can have on the external world and the proprioceptive information conveyed to the nervous system. The integrative nature of the neuro-musculo-skeletal system makes it difficult to examine the effects of this structure experimentally, and a well-developed mathematical framework allows analysis of otherwise inseparable aspects of the chain of control. Structure in the limbs seems to separate aspects of limb performance, such as weight support and propulsion, to different degrees of freedom, which may simplify neural control processes. The bi-directional nature of the musculoskeletal filter amplifies signals that the nervous system can most strongly influence and suppresses feedback of task parameters that are poorly controllable.

**Keywords** Musculoskeletal model · Motor control · Posture · Rigid body model · Muscle model

## 4.1 Introduction

This chapter is based on the idea that musculoskeletal structure and dynamics serve to separate and focus different aspects of the flow of control and information between the central nervous system and the physical world. The underlying question is: Can we separate motor control into separate domains of mechanics and neural drive? Mathematical modeling will be used to illustrate system properties closely associated with structural features, like muscle attachments, joint articulations, and posture. It is also important to recognize the limitations of mathematical approximations of biological systems, and consideration will be given to potential misdirection.

T. J. Burkholder (✉)
School of Applied Physiology and Interdisciplinary Bioengineering Program,
Georgia Institute of Technology, 555 14th St NW, Atlanta, GA 30332-0356, USA
e-mail: thomas.burkholder@ap.gatech.edu

 
B. I. Prilutsky, D. H. Edwards (eds.), *Neuromechanical Modeling of Posture and Locomotion,* Springer Series in Computational Neuroscience,
DOI 10.1007/978-1-4939-3267-2_4

The nervous, muscular, and skeletal systems all work together to produce movement, but one tends to attribute control of the movement to the nervous system—and particularly to the brain, active production of power to the muscles, and passive impedance to skeletal system. We attribute intent to often nebulous higher neural centers and view the musculoskeletal system as a responsive structure interposed between that intent and its realization in the external world. This is sometimes described as the nervous system planning a behavior, while the musculoskeletal system represents a problem that must be solved in order to bring the intent into actuality (Latash and Gottlieb 1991; Gomi and Kawato 1997; Gribble et al. 1998). It does seem that structures within the nervous system, even within the spinal cord, are capable of producing free-running cyclical behaviors, and these central pattern generators (CPGs) can be activated with very little encouragement. The physical consequence of CPG activity depends on muscular performance and the physics of interaction with the world, and success of the motion is communicated back to the CPG through myriad sensory systems, particularly muscle-resident spindles and Golgi tendon organs, that may set the cadence of the CPG (Lam and Pearson 2001; Rybak et al. 2006). In this context, one might consider the CPG as a source of movement intent, processed through the musculoskeletal filter to drive real-world motion. One might also consider the resulting motion as a source of information subjected to filtering through musculoskeletal dynamics before driving the CPG or other neural intent. Both movement and postural control depend on the musculoskeletal and environmental response to neural activity and on the sensory perception of that response. For signals passing in either direction, musculoskeletal dynamics sit between source and sink and change the nature of the signal.

The musculoskeletal system can also be considered to separate wanted from unwanted signals. Several studies have suggested that the motor system is capable of selectively controlling task relevant performance, while leaving an uncontrolled manifold of task-irrelevant parameters (Scholz and Schoner 1999). That is, the combined neuro-musculo-skeletal system is capable of damping perturbations that affect the intended task, while leaving other perturbations undamped. Within the same system, the nervous system is capable of adjusting the relative stability of different task parameters to meet changing instruction or demand (Hsu et al. 2007; Auyang and Chang 2013). Tuning of this filter depends at least partly on neural modulation in two ways: selection of a motor plan to weight the contribution of intrinsic muscle mechanics, and selection of sensory feedback gains to amplify the muscular contribution.

## 4.2 Models to Decipher Neuro-Musculo-Skeletal Performance

Because of its central position between a neural program and its environmental impact, it is difficult to isolate the behavior of the musculoskeletal system experimentally. Attempts at this generally include imposing a recorded or idealized length trajectory on an isolated muscle while imposing electrically stimulated contractions, such as the "work loop" experiments (Josephson 1985) or reconstruction of locomotion (Lutz and Rome 1996; Sandercock and Heckman 1997). These methods suffer from non-physiological nature of electrical stimulation and the sensitivity of muscle (and muscle models) to small variations in velocity trajectory that may produce indistinguishable length trajectories. It can be difficult to distinguish a neural "excitation" signal from muscular "activation." It can be difficult to distinguish descending neural drive from superimposed reflex action. Mathematical modeling is a convenient method either to estimate unmeasureable parameters from experimental data or to screen hypothetical control rules for functional plausibility.

This chapter is primarily interested in high parameter count models that intend to represent the detailed behavior of neuromechanical systems. This is specifically to distinguish from simplified models better suited to examining or illustrating general physical principles (Full and Koditschek 1999). Simplified models explicitly limit the level of detail and the extent to which they represent the physical reality, and this helps to limit inferences to those appropriate to the model.

Models with many parameters are also simplified representations of reality, but the simplifications and their implications are less explicit. Musculoskeletal models are generally rigid body representations with mathematically rigid constraints between segments and dynamically independent muscle actuators (Delp et al. 1990, 2007). This contrasts with the physical reality, in which soft tissues—especially muscle and adipose—deform and move relative to the bones, and even the bones flex. In reality, articular surfaces deform and constraining ligaments stretch, so the motions of gentle flexion and forceful flexion will differ. Intermuscular connective tissue results in force production spreading across muscles (Huijing et al. 1998; Maas and Huijing 2009).

Mechanical models are very coarse and not individually representative, even if one accepts the reduction to mathematical precision. This has been demonstrated by comparison of subject-specific models with scaled generic models (Scheys et al. 2008; Correa et al. 2011), but it is apparent from the experimental data used to define the generic models. The measured muscle attachment points have substantial variability (Brand et al. 1982; Burkholder and Nichols 2004). Experimentally measured moment arms often have variability between specimens exceeding 20% (Loren et al. 1996; Smutz et al. 1998), which is not much less than deviations reported between subject-specific and generic models (Correa et al. 2011). Even within purpose-bred animals of nearly identical body size, estimates of muscle parameters like optimal fiber length have variance of 10% or more (Burkholder et al. 1994; Eng et al. 2008). These parameters are relatively easy to measure and require little post-processing. More difficult parameters, like joint centers can have much greater variability or uncertainty (Hollister et al. 1993; Burkholder and Nichols 2004). This variability reflects a combination of true inter-individual differences, experimental error, and the reduction of a complex physical structure to a simple, mathematically rigid approximation. Although the model mathematics are rigid and precise, they are based on estimation and approximation which is often imprecise, and model results are best interpreted for their qualitative features.

Several studies have examined the sensitivity of model predictions to the mechanical parameters of the model. Specific parameter choices, or errors in their determination, can have substantial impact on predicted moment arms (Brand et al. 1982), joint loads (Southgate et al. 2012), muscle induced accelerations (Correa et al. 2011), and functional performance (Ackland et al. 2012). It should not be surprising that the effect of parameter variability or error is relatively small for individual muscle parameters, and relatively large for parameters that influence many muscles.

Model behavior is also sensitive to choices of neural parameters. Experimentally, these are most difficult to define, because it is generally not possible or practical to record from all of the muscles involved in a task. To overcome this limitation, several optimization methods have been developed to predict the activity of all muscles based on limited recorded data (Crowninshield and Brand 1981; McKay and Ting 2008), but these methods are difficult to validate and often contentious (see (Prilutsky 2000) and related commentaries). Model performance is clearly sensitive to these parameters, both the explicit execution of a task (McKay and Ting 2012) and the derivatives or stability (Bunderson et al. 2008).

Fortunately, the computational framework provides exactly the platform to separate properties that might be features of the biological system from features that might be limitations of the modeling paradigm. For models with relatively few parameters, it may be possible to evaluate the stability and sensitivity of model performance within a parameter neighborhood directly (Bingham and Ting 2013). For more complex models, it may be possible to distinguish between behaviors that are qualitatively resistant to the artificial constraints of the modeling structure and those that depend on an unrealistic precision by repeated simulation with perturbed parameters. This is slightly different from a sensitivity analysis, because a model may be very sensitive to a particular parameter, but maintain the same qualitative behavior. In neuromuscular control, there are two aspects to consider: the extent to which musculoskeletal parameter choices influence the inputs and outputs of the nervous system and the extent to which neural parameter choices influence the behavior of the mechanical system.

Mathematical models allow fantastic access to intervention within parameter space. This means the investigator is not limited to one specific parameter set, but can survey model performance using a wide range of plausible parameter sets to find which qualitative features are conserved. It also means that experimentally inaccessible parameters can be manipulated and the effects compared with known parameter variability. In general, parameter spaces are much larger than can be directly measured or systematically sampled, so optimization and Monte Carlo methods are often used (Anderson and Pandy 2001; Valero-Cuevas et al. 2003; Ackland et al. 2012). It is important to be aware of structural dependencies within the model and within the parameter space being sampled. For example, randomizing synergistic muscle activations independently will result in force production concentrated near the group mean and limiting actual kinetic variance. Likewise, if one makes independent, proportional variations in two parts of a whole, for example, muscle fiber length and tendon length, which make up the origin-insertion distance (Scovil and Ronsky 2006; Ackland et al. 2012), then system behavior will appear to be more sensitive to the larger part (*eg.* tendon length).

**Table 4.1** Muscles included in the model, abbreviations, and major actions

| Muscle | Abbreviation | Primary action |
|---|---|---|
| Adductor femoris | ADF | Hip adduction |
| Adductor longus | ADL | Hip adduction |
| Biceps femoris, anterior head | BFA | Hip extension |
| Biceps femoris, posterior head | BFP | Knee flexion |
| Extensor digitorum longus | EDL | Ankle flexion |
| Flexor digitorum longus | FDL | Ankle extension |
| Flexor hallicus longus | FHL | Ankle extension |
| Gluteus maximus | GMAX | Hip extension/ Abduction |
| Guteus medius | GMED | Hip extension/ Abduction |
| Gluteus minimus | GMIN | Hip extension/ Abduction |
| Gracilis | GRAC | Knee flexion |
| Lateral gastrocnemius | LG | Ankle extension |
| Medial gastrocnemius | MG | Ankle extension |
| Peroneus brevis | PB | Ankle eversion |
| Pectineus | PEC | Hip adductor |
| Peroneus longus | PL | Ankle eversion |
| Plantaris | PLAN | Ankle extension |
| Iliopsoas | PSOAS | Hip flexion |
| Peroneus tertius | PT | Ankle eversion |
| Pyriformis | PYR | Hip extension |
| Quadratus femoris | QF | Hip extension |
| Rectus femoris | RF | Knee extension |
| Sartorius | SART | Knee flexion |
| Semimembranossus | SM | Knee flexion |
| Soleus | SOL | Ankle extension |
| Semitendinosus | ST | Knee flexion |
| Tibialis anterior | TA | Ankle flexion |
| Tibialis posterior | TP | Ankle inversion |
| Vastus intermedius | VI | Knee extension |
| Vastus lateralis | VL | Knee extension |
| Vastus medialis | VM | Knee extension |

## 4.3 The Model System

Much of the following is based on our work with a model of the cat hindlimb (Burkholder and Nichols 2004; Bunderson et al. 2010). This model has seven purely rotational kinematic degrees of freedom and 31 muscles (Table 4.1), each acting at discrete attachment points. Muscle activity is controlled by a combination of assigned program and length feedback. Forward simulations are performed using Neuromechanic (Bunderson et al. 2012; Bunderson and Bingham 2015), but substantial analysis is based on linearizations of the model in specific configurations.

The combination of rigid segmental model, with pointwise muscle connections, and architecture-based muscle function estimates is common to many modeling platforms (Zajac 1989; Delp et al. 1990; Todorov 2007). We generally assume muscles are near their optimal length in the neutral posture (Burkholder and Lieber 2001), so quasi-static stiffness due to the length-tension relationship is small. Our applications of this model have emphasized postural tasks or perturbations of relatively low acceleration, in which we believe the mechanical response is dominated by musculoskeletal forces, rather than inertial and Coriolis forces.

It is important to recognize that the dimensional mismatches among endpoint, joint, and muscle spaces means that the transformations between these spaces are one directional. That is, the conversion is unique in one direction, but indeterminate in the other. The joint angles uniquely determine both the endpoint position and all muscle lengths. However, there is a multi-dimensional subspace of joint angles that satisfy any endpoint position. To determine the joint displacements resulting from an imposed endpoint displacement or force requires additional knowledge of the mechanical impedance of the system, the stiffness, viscosity, and inertia that resist displacement. Because these properties are strongly influenced by muscle activity, the inverse filter and the sensory perception of perturbations may be highly dependent on neural parameters and behavior. If the musculoskeletal system is considered a filter between the nervous system and the physical environment, then in one direction the filter depends only on geometry, and in the other it depends on both geometry and the neural state. If the neural state and muscular forces are highly dependent on configuration, then the forward and inverse filters may not be complimentary.

## 4.4 Conversion of Neural Signals to Mechanical Output

One of the challenges facing neuroscientists is to understand how the nervous system selects a specific muscle activation pattern from the multidimensional subspace producing identical endpoint forces. There are several assumptions built into phrasing the problem in this way: (1) an exact solution must be found (2) muscles act independently (3) endpoint force or motion fully describes the goal (4) task planning is primarily a feedforward process in muscle space. In this context, the arrangement of the musculoskeletal system defines the manifold of equivalent performance and may influence other constraints or considerations the nervous system uses to determine its preferred activation. If the musculoskeletal system acts as a filter, then the nervous system must know the properties of that filter in order to plan its behaviors accurately. Alternatively, the nervous system may plan its behaviors in a space based on proprioceptive signals representing a version of the external task space distorted by the impedance of the body.

The first sense in which the musculoskeletal system behaves as a filter for the nervous system is in the determination of mechanical output from neural input. At least some movements are planned entirely as feedforward programs, minimally altered by perturbation or distortion during performance (Burrows and Morris 2001).

These movements tend to be highly dynamic motions, and their brief duration may simply not allow time for delayed feedback signals to influence the ongoing motor program. Planning an open-loop movement requires that the system performance be predictable and consistent. If the performance of the musculoskeletal system is highly sensitive to small changes in configuration, then task planning is greatly complicated. On the other hand, if the musculoskeletal system imposes strong constraints on endpoint behavior, then task planning is simplified, because the biomechanics reject errors in the neural program.

The influence of the musculoskeletal filter on the neural output can be seen by exploring the mechanical parameter space. Because those parameters are easily measurable, the magnitude of their variability or uncertainty is known. Many of the neural parameters are not measured or measurable, and we can only speculate on the biological variability. The mechanical parameters define mathematically rigid constraints, such as joint axes, and potentially impose limits. Knowledge of the system sensitivity to mechanical parameter variation will give us some sense of whether the nervous system can strongly affect those same system performance measures.

The lowest-level musculoskeletal transformation to consider is the relationship between joint angle and endpoint position or posture. The Jacobian (**J**) of this transformation, the matrix of partial derivatives of position with respect to joint angle, also defines the relationship between endpoint forces and joint torques. If this transformation is uniquely invertible, then the system is kinematically constrained, and a great deal of system behavior will resolve to geometry. In general, musculoskeletal systems and models have more independent joints than endpoint coordinates and are not uniquely invertible. This is important because a number of analyses require inversion of the Jacobian ($\mathbf{J}^{-1}$). If $\mathbf{J}^{-1}$ is not unique, a pseudoinverse can be calculated, but the pseudoinverse is a strictly geometrical construction that discards any system physics.

The Jacobian is a transcendental function of the joint angles. Small changes in the definition of joints or in joint angles may have a large effect on the Jacobian components. This would substantially complicate control, because the mapping between joint angles, which are most directly controlled, and endpoint trajectory, the space in which motions appear to be planned, might depend strongly on the precise posture. The computational model can be easily manipulated to evaluate the effect of changing joint axes and posture. As can be seen in Fig. 4.1, the components of **J** lie close to anatomical planes with relatively little change in orientation due to small changes in posture. This reflects the condition that the joint axes are closely aligned with the anatomical axes in the horizontal plane, with small vertical components. The Jacobian seems to be insensitive to small changes in the axis definitions and orientations. The limb performance is more strongly influenced by the global limb orientation than by the individual joint positions or angles. The postural variation illustrated resulted in the limb axis (line from hip center to toe) bounded by a cone with 18° apex angle, and the knee extension components of **J** correlated more strongly with limb axis ($R^2=0.90$) than with knee angle ($R^2=6e-3$).

This demonstrates that over a range of postures near the stance configuration, the Jacobian components are segregated along the cardinal directions of the animal.

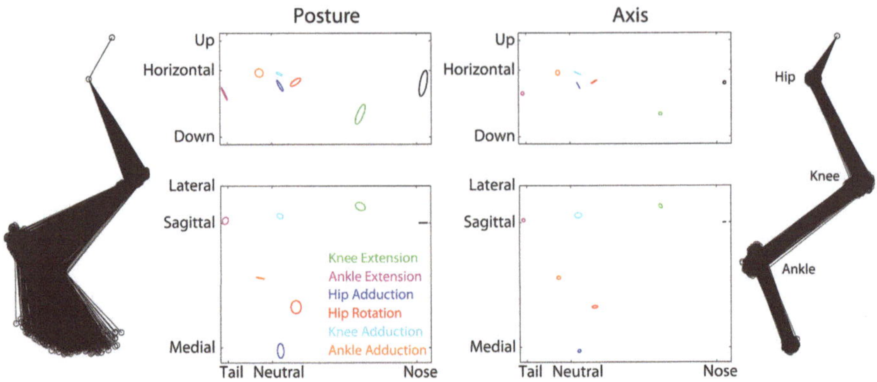


**Fig. 4.1** The Jacobian focuses endpoint motion onto specific joint motions. Ellipses represent one standard deviation of the respective joint components. Postural perturbations result from joint angle changes up to 12°. Axis perturbations result from displacements up to 5 mm and orientation changes up to 6°. Stick figures represent joint-center to joint-center illustrations of the models examined

The major flexion joints are associated with Jacobian components aligned near the cranial-caudal axis where their motion contributes strongly to propulsion. The non-sagittal axes have Jacobian components near the medial-lateral axis, and medial-lateral action is primarily restricted to the hip joint. The mechanics of the limb therefore focus motion into the sagittal plane and focus control of lateral positioning into the hip. Interestingly, no joints have substantial vertical components, which means that vertical motion of the endpoint can only be accomplished by relatively large displacements of multiple joints. This provides substantial structural rigidity near the stance posture and minimizes the muscular force required to reject vertical perturbations. It also isolates vertical support of the body from joint displacements required for propulsion. The largest horizontal components are found in the hip joint, indicating that reorientation of the limb is most easily achieved at the hip.

The structural concentration of horizontal displacement onto hip joint and distribution of vertical displacement across joints is consistent with a number of modeling and experimental observations. Both the feasible force set (McKay and Ting 2008) and the response to horizontal perturbation (Winter et al. 1998; Henry et al. 2001) rotate with the limb axis. The vertical components of restoring forces are much greater than the horizontal components, even for horizontal plane perturbations (Macpherson 1988; Henry et al. 2001). The notion that the limb may have strut-like properties underlies both the spring-mass and inverted pendulum models of locomotion (Chow and Jacobson 1972; Alexander 1992). Segregation of support and direction control is also consistent with the observation that turning relies strongly on changes in hip muscle activity and foot placement (Hase and Stein 1999). This mechanical filter may allow the nervous system to separate the control of direction onto the proximal musculature and control of speed onto the distal musculature.

Another low level characteristic is the transformation between joint angles (q) and muscle lengths (L). The muscle moment arms ($\mathbf{R} = \mathbf{dL/dq}$) define this transformation, and is generally not invertible. That is, a given set of joint angle changes

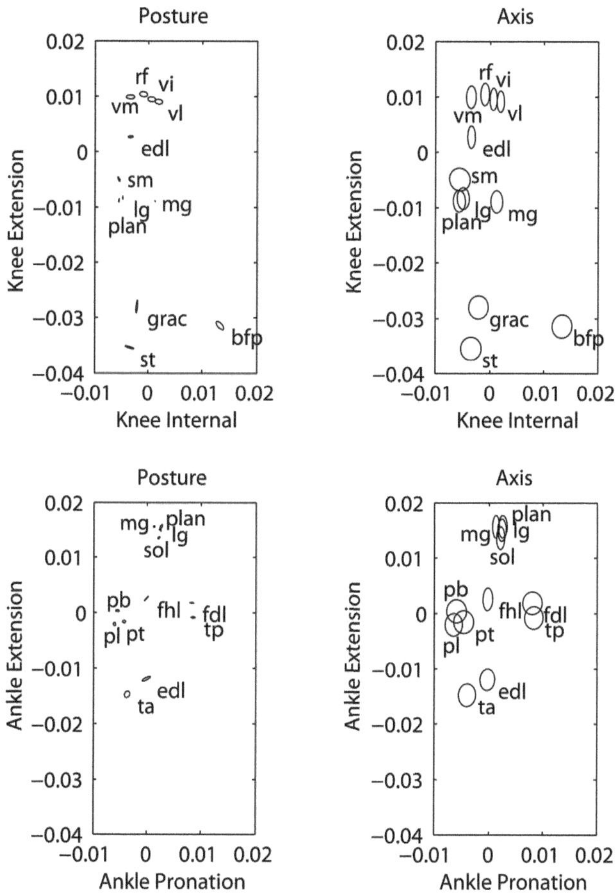


**Fig. 4.2** Knee and ankle moment arms (in meters) emphasize sagittal plane torques. Ellipses represent one standard deviation of moment arm distribution with the same structural perturbations as Fig. 4.1

will uniquely determine the resulting muscle length changes, but a set of muscle length changes do not necessarily define a compatible set of joint angle changes. Moment arms depend on the choice of attachment points or structures and on the choice of joint axes. Variations in attachments affect each muscle individually, where variations in the axes affect whole groups of muscles.

It is not surprising, then, that the gross performance of models with many synergist muscles is relatively insensitive to changes in the moment arms of individual muscles (Ackland et al. 2012; O'Neill et al. 2013). The main effect of changing individual moment arms appears to be in the particular pattern of force sharing among synergist muscles. In contrast to the Jacobian, which was more sensitive to changes in posture than changes in axis definition, the moment arms are highly dependent on axis definition, but relatively insensitive to posture (Fig. 4.2). The range of moment arms is greater in the flexion-extension axis than the non-sagittal axis, indicating a structural bias towards weight support and propulsion. At the ankle, muscles with large non-sagittal action have very small extension moment arms, suggesting compartmentalization of propulsion and stabilization, but no such separation exists at the knee. The small effects of posture on moment arms suggests that the nervous system can expect consistent joint-level function from the musculature, but also minimizes any intrinsic stability provided by the moment arms (Young et al. 1992).

The musculoskeletal system seems to separate sagittal and frontal plane movement, to separate control of limb orientation from length, and to be qualitatively insensitive to moderate perturbations in key parameters. This is imposed on the mathematical model by the geometrical constraints of joint axes and fixed muscle attachments, and on the biological system by the elastic constraints of ligament and cartilage. It supports the intuitive impression that limbs are largely planar structures, but the hip and shoulder are qualitatively different.

## 4.5 Conversion of Mechanical Signals to Neural Input

The transformation of forces acting on the body to changes in configuration and proprioceptive recognition depends on the mechanical impedance of the system. For slow motions, the impedance is strongly influenced by the activity and properties of the spanned muscles. Therefore, the properties of the musculoskeletal filter will be influenced by the nervous system. Estimating the response to endpoint perturbation is more complicated than estimating the geometrical properties discussed above, because dimensional redundancy generally means the relationship between endpoint position or velocity and joint angles or muscle lengths is not unique.

The modeling approach offers great potential for insight in the context of neural parameters. These are difficult or impossible to characterize fully *in vivo*, because of technical limitations on the number of muscles or neurons one can sample, because the model parameters (e.g. activation) differ from the experimental measures (e.g., EMG), or other discrepancies. A model platform offers a number of opportunities for exploring the effect of neural parameters on the performance and properties of the system, from optimization (Crowninshield and Brand 1981; Lockhart and Ting 2007; Bunderson et al. 2008) to manifold mapping (Buchanan et al. 1993; McKay et al. 2007; Bunderson et al. 2010). Essentially, in the face of limited experimental data, models provide an opportunity to invent and test hypotheses of neural organization, and evaluate their consistency with both neural and performance data.

The impedance of the hindlimb model can be characterized by examining the apparent endpoint stiffness, as estimated from a linear approximation. The linear approximation begins with a quasi-static estimate of the joint stiffness ($\mathbf{K_J}$):

$$\mathbf{K_J} = \frac{\partial \vec{G}}{\partial \vec{\theta}} + \left( \frac{\partial \mathbf{R}}{\partial \vec{\theta}} \vec{F}_M + \mathbf{R} \frac{\partial \vec{F}_M}{\partial \vec{\theta}} \right) + \frac{\partial \mathbf{J}^T}{\partial \vec{\theta}} \vec{F}_{END}$$

where $\mathbf{G}$ are the torques due to gravity, $\vec{F}_{END}$, the balancing endpoint force. In practice, the joint stiffness is dominated by $\mathbf{R}\, \partial \vec{F}_M / \partial \vec{\theta}$, and highly dependent on the

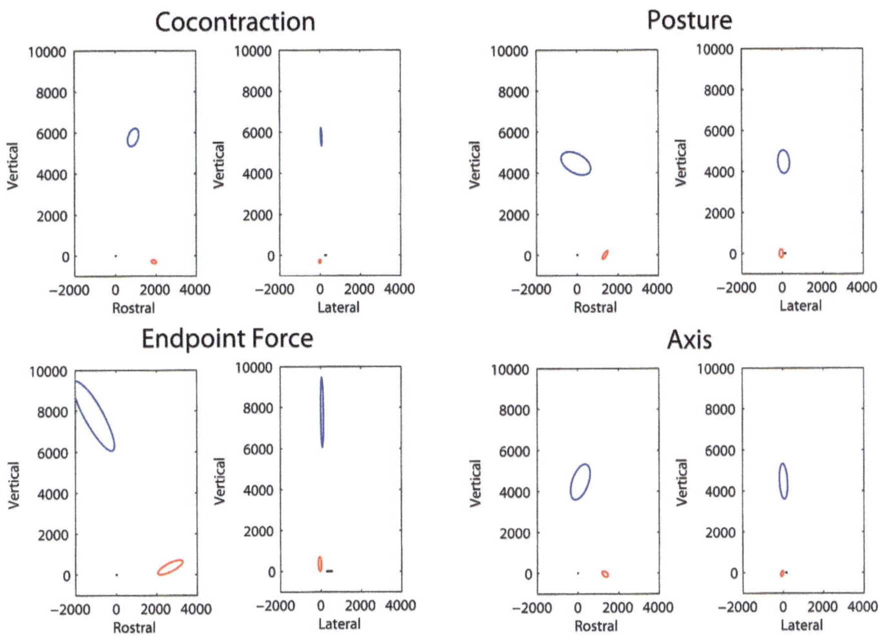


**Fig. 4.3** Effect of parameter variance on endpoint stiffness (in N/m). Ellipses represent one standard deviation of the largest (*blue*), intermediate (*red*), and smallest (*black*) semi-principal axes of stiffness. For cocontraction, muscle activation pattern was allowed to vary within the null space capable of generating a stance-like endpoint force, rather than the minimum squared-activation pattern. For endpoint, force variations represent the limb in static equilibrium against random endpoint forces, rather than the endpoint force produced by a cat during quiet standing. Posture and Axis represent, as above, models with mechanical parameters altered within the range of experimental variability

muscle properties and activation pattern. Note that in dynamic systems, it would be appropriate to replace $\mathbf{K}_\mathrm{J}$ with a generalized joint impedance incorporating viscous and inertial terms. This term depends on all the properties of the limb: Jacobian, muscle moment arms, muscle forces, and muscle stiffnesses, and allows calculation of the endpoint stiffness ($\partial \vec{F}_{END} / \partial x$) as $(\mathbf{J}\ \mathbf{K}_\mathrm{J}^{-1}\ \mathbf{J}^\mathrm{T})^{-1}$ without resorting to pseudo-inversion. This stiffness estimate is symmetric and results in a stiffness ellipse that can be represented by its three semi-principal axes. Figure 4.3 shows the sensitivity of the stiffness structure to perturbations in model parameters.

Qualitatively, endpoint stiffness is aligned close to the anatomical axes, with greatest stiffness near vertical, and this result in essentially independent of parameter perturbations. This structure largely results from the Jacobian (or square of the Jacobian, $\mathbf{J}\ \mathbf{J}^\mathrm{T}$), which admits only small vertical endpoint motions for large joint motions. That is, to vertically displace the endpoint, several joints must go through large displacements, with correspondingly large changes in muscle length. The size and orientation of the stiffness ellipsoid varied most when the endpoint force was manipulated. These manipulations changed both the magnitude and direction of endpoint force. Where the reference model produced an 8 N, primarily vertical endpoint force, the random forces ranged from 0.02 N to 43 N and had substantial

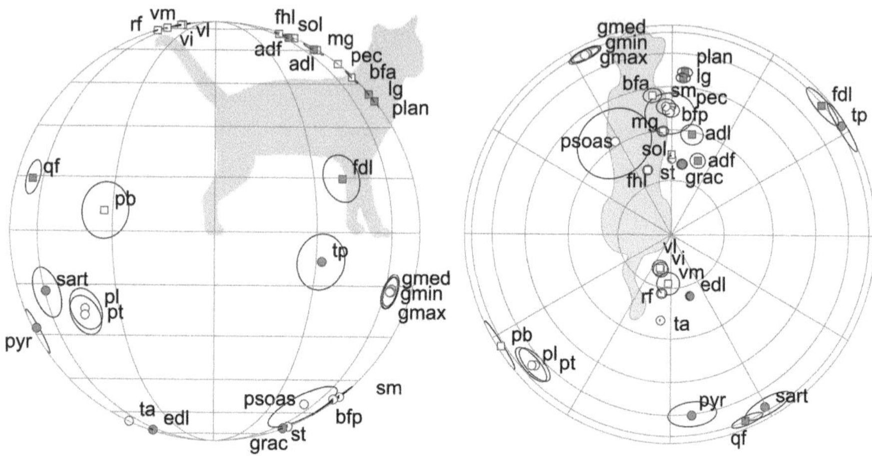


**Fig. 4.4** Pulling directions for *right* hindlimb muscles in the stance configuration

caudal-directed and upward-directed components, rather than downward vertical components. These dramatic changes in endpoint force and muscle activation still retained the fundamentally vertical endpoint stiffness. It is worth noting that none of these parameter perturbations was specifically designed to manipulate endpoint stiffness, and it may be possible to find muscle activation patterns that do alter the shape of the intrinsic stiffness ellipsoid.

The biases in endpoint stiffness influence the transduction of force perturbations into the nervous system. That is, because of the small medial-lateral stiffness, forces applied in this direction will result in large displacements concentrated at the hip. Conversely, vertical force perturbations will produce much smaller displacements and offer a smaller proprioceptive stimulus. The strong vertical mechanical stiffness compliments a muted proprioceptive response, while the muted medial-lateral stiffness compliments an exaggerated proprioceptive response. This is a consequence of the same physical structure serving both a forward, nerve-to-endpoint and a backward, endpoint-to-nerve filter. The constraints that limit the nervous system output will necessarily isolate its inputs in the same way.

We have argued that musculoskeletal mechanics result in each muscle being associated with a characteristic endpoint direction in which displacement of the endpoint causes the greatest muscle stretch, the "pulling direction" (Bunderson et al. 2010). These directions are highly dependent on the joint stiffness ($\mathbf{K_J}$), and the direction of endpoint displacement giving the greatest muscle length change is

$$\vec{L}_{MTL}^{\max} = \mathbf{R}\mathbf{K}_{\mathbf{J}}^{-1}\mathbf{J}^{T}(\mathbf{J}\mathbf{K}_{\mathbf{J}}^{-1}\mathbf{J}^{T})^{-1}$$

Sampling the activation patterns that produce a stance-like force reveals that the muscle pulling directions are generally insensitive to the activation pattern and lie in a few clusters in space (Fig. 4.4). The majority of pulling directions lie near a

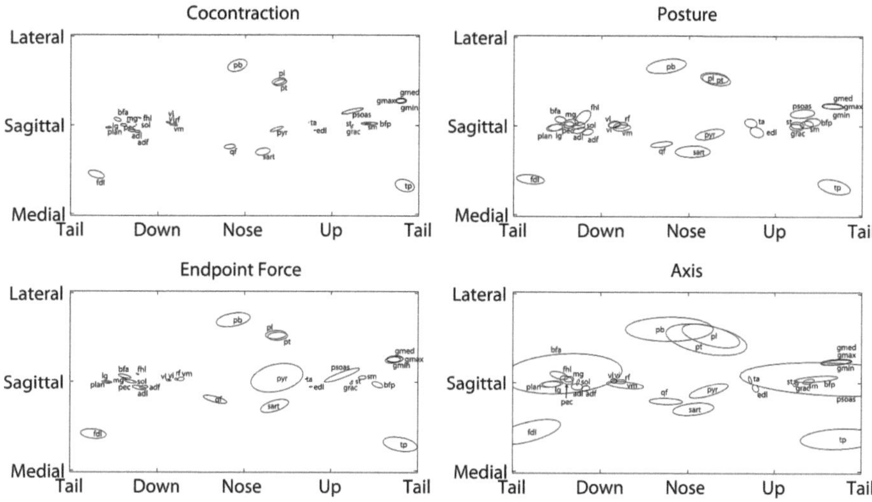


**Fig. 4.5** Ellipses represent one standard deviation of muscle pulling directions drawn from repeated simulations. In *Cocontraction*, muscle activation was allowed to increase above the minimum energy solution, but endpoint force vector and limb mechanics were held constant. In *Endpoint Force*, the limb generated arbitrary endpoint forces. In *Posture*, muscle activation was minimized while joint angles were perturbed by up to 12°. In Axis, joint axes were perturbed by 5 mm and 6°

plane rotated 15° from the parasagittal plane. Those muscles with pulling directions away from this plane (e.g. FDL, PB) also lie near a plane not quite perpendicular to the major plane. This result is not surprising, given the similar nearly orthogonal structure in the Jacobian (Fig. 4.1). Although many of these clusters represent classical, joint-level synergists, some of the clusters span multiple joints. Notably, the cluster of muscles with cranial-dorsal direction includes ankle extensors like soleus, but also hip muscles like adductor longus and biarticular muscles like medial gastrocnemius. This suggests that one function of the musculoskeletal filter may be to focus feedback from muscle proprioceptors into behaviorally relevant signals, independent of variations or noise in muscle activation patterns.

This has two consequences for proprioception. First, if the information represented by each of the muscle proprioceptors is functionally equivalent, then the muscle clusters become redundant estimates of that information, and their aggregate signal will have substantially reduced error. Second, the nervous system might create arbitrary directional sensors from appropriately weighted combinations of muscles. It is worth noting that muscle length change shows cosine tuning relative to its pulling direction as a consequence of the biomechanics, and cosine-tuned directional responses can be found in many neurons and behaviors (Bosco and Poppele 1993; Nozaki et al. 2005). That is, the nervous system itself need not perform any transcendental transformations to produce cosine tuning of gross mechanical performance because the system mechanics produce that tuning directly.

Pulling directions are sensitive to model parameters. In Fig. 4.5, the pulling direction sphere has been mapped to a 2-D projection, with the sagittal plane forming the equator. Pulling directions are sensitive to changes in limb posture and joint axes, but also to the muscle activation pattern, indicating that the nervous system is capable of tuning the mechanical properties of the limb. Regardless of the perturbation, the general alignment of pulling directions near the sagittal plane is preserved, with the same five muscles lying outside the plane. Individual muscles are differentially affected, depending on the mode of perturbation.

Perturbations of muscle activation within the null space of the stance-like postural force had very little influence on muscle pulling direction. This suggests that, within a specific task, where the joint torques are constrained, variations in the neural strategy for achieving that task have little influence on the distribution of mechanical impedance of the limb. That is, the limb as a whole may become stiffer, but this happens without disrupting the relative stiffness among the joints, such that the kinematics of limb deformation are minimally altered. The muscular component of impedance is the sum of muscle force-weighted derivative of moment arm and the moment arm-weighted derivative of muscle force with respect to joint angle (i.e., $F\, dR/dq + R\, dF/dq$). In these simulations, the static stiffness of individual muscles is quite low, and it was seen in Fig. 4.2 that the variation in moment arm with joint angle is also small, so this likely represents a low estimate of the contribution of neural activation. The nervous system can affect the system impedance, and much greater variability in pulling direction is seen if the static endpoint force or posture is varied.

Muscle pulling directions were highly sensitive to changes in joint axis definition. This reflects the quadratic reliance of pulling direction on muscle moment arms: once to define the muscle length changes imposed by joint displacement, and once to define the torque produced by each muscle. The muscles which were most sensitive to joint displacement also had the smallest moment arms, so perturbations of the joint location are relatively exaggerated. For some muscles, particularly at the hip, small perturbations in joint definition essentially eliminated the directional consistency, which raises the question of whether the notions of pulling direction or induced acceleration reflect the biological system or the reduction to mathematical rigidity (Zajac and Gordon 1989; Bunderson et al. 2010).

Regardless of the perturbation, the qualitative emphasis on the sagittal plane persists, but the precise focusing of individual muscle directions is lost. This suggests that the filtering properties of the musculoskeletal system are insensitive to the neural strategy for performing a given task, and more sensitive to the mechanical conditions. It has already been suggested that properties or behaviors that are highly sensitive to specific choice of mechanical parameters may represent artifacts of the mathematical approximation, rather than features of the biological system. The narrow focusing of pulling directions, with high sensitivity to mechanical parameters suggests that the task-level function of these muscles may be less precise than the nominal pulling directions imply. However, the limited influence of changes in neural activity, within the task null space, to alter pulling directions implies that they are more mechanical than neural phenomena.

## 4.6 Dynamic Models

Many activities, such as walking and running involve periodic motion and substantial muscle-dependent energy storage in non-contractile connective tissue (Morgan et al. 1978; Alexander 1991), and the musculoskeletal system can strongly influence the damping or amplification of such actions (Gasser and Hill 1924; Josephson 1985; Lin and Rymer 2000). The importance of intrinsic properties and geometry to dynamic behaviors has been well demonstrated by the success of passive dynamic walkers (McGeer 1990a, b) and their elastic extensions (Collins et al. 2005; Geyer et al. 2006; Whittington et al. 2008). From the perspective of understanding the musculoskeletal filter, its frequency response is at least as important as its equilibrium response.

Dynamic behavior has proven to be challenging to high parameter count modeling approaches. Kinematic changes are large, so linearized models diverge from the non-linear formulation. Forward simulation is a stiff problem, and integration over time periods required to analyze repetitive motions are subject to mathematical instability. Muscle activation changes through the performance, and finding activity patterns across dozens of muscles that produce periodic behavior is extremely challenging.

## 4.7 Conclusion

Models of the musculoskeletal system have very rigid structure and this structure shapes both the outputs from and the inputs to the nervous system. This rigidity is closely tied to joint mechanics, particularly as characterized by the system Jacobian. The mathematically rigid kinematic constraints may not accurately represent biological kinematics, but a few observations are consistent, even with substantial variance in parameters.

The limb separates vertical and horizontal actions. Individual segments are primarily vertical, so joint rotation moves the endpoints primarily within the horizontal plane. Conversely, this structure allows much of weight support forces to be transmitted through bone-bone contact. In the postural setting, this means that the limb is structured to reject vertical perturbations. Changes in vertical force induce displacements distributed across joints with opposing horizontal plane actions, focusing the perturbation onto muscles aligned with the limb axis (Fig. 4.4). This exaggerates the length change imposed on these muscles, providing a strong proprioceptive signal. In contrast, horizontal plane forces, particularly medial-lateral forces, are focused at the hip. This structure may allow spinal and lower centers of the nervous system to effectively manage vertical posture, while simplifying higher control of horizontal plane disturbances by limiting effective responses to a single joint.

The mechanical properties of the limb are relatively independent of the neural strategy for accomplishing a particular task, but do change in different task settings. Although many different patterns of muscle co-contraction can produce identical endpoint forces, the muscle redundancy results in little change to endpoint stiffness (Fig. 4.3). Task performance and response to perturbations of muscle strength or external force should therefore be consistent and independent of the exact muscle activation pattern used to accomplish the task. The nervous system thus receives consistent proprioceptive input, independent of the descending control signal. Conversely, altering the mechanical task does have a large effect on endpoint stiffness and consequently on the proprioceptive response to perturbation.

The musculoskeletal system filters neural control signals, focusing its effort into sagittal plane motion and force. It filters mechanical signals from the external world, exaggerating those which can be directly and easily countered while suppressing those that require more complex responses.

**Acknowledgements** This work was supported by NIH grants HD032571 and HD46922. The NIH had no role in the design, performance or interpretation of the study.

## References


Ackland DC, Lin YC, Pandy MG (2012) Sensitivity of model predictions of muscle function to changes in moment arms and muscle-tendon properties: a Monte-Carlo analysis. J Biomech 45:1463–1471

Alexander RM (1991) Energy-saving mechanisms in walking and running. J Exp Biol 160:55–69

Alexander RM (1992) A model of bipedal locomotion on compliant legs. Philos Trans R Soc Lond B Biol Sci 338:189–198

Anderson FC, Pandy MG (2001) Dynamic optimization of human walking. J Biomech Eng 123:381–390

Auyang AG, Chang YH (2013) Effects of a foot placement constraint on use of motor equivalence during human hopping. PLoS One 8:e69429

Bingham JT, Ting LH (2013) Stability radius as a method for comparing the dynamics of neuromechanical systems. IEEE Trans Neural Syst Rehabil Eng 21:840–848

Bosco G, Poppele RE (1993) Broad directional tuning in spinal projections to the cerebellum. J Neurophysiol 70:863–866

Brand RA, Crowninshield RD, Wittstock CE, Pedersen DR, Clark CR, van Krieken FM (1982) A model of lower extremity muscular anatomy. J Biomech Eng 104:304–310

Buchanan TS, Moniz MJ, Dewald JP, Zev Rymer W (1993) Estimation of muscle forces about the wrist joint during isometric tasks using an EMG coefficient method. J Biomech 26:547–560

Bunderson N, Bingham J (2015) Better science through predictive modeling: numerical tools for understanding neuromechanical interactions. In: Prilutsky BI, Edwards DH, Jr (eds) Neuromechanical modeling of posture and locomotion. Springer, New York (in press)

Bunderson NE, Burkholder TJ, Ting LH (2008) Reduction of neuromuscular redundancy for postural force generation using an intrinsic stability criterion. J Biomech 41:1537–1544

Bunderson NE, McKay JL, Ting LH, Burkholder TJ (2010) Directional constraint of endpoint force emerges from hindlimb anatomy. J Exp Biol 213:2131–2141

Bunderson NE, Bingham JT, Hongchul Sohn M, Ting LH, Burkholder TJ (2012) Neuromechanic: a computational platform for simulation and analysis of the neural control of movement. Int J Numer Meth Biomed Eng 28:1015–1027

Burkholder TJ, Lieber RL (2001) Sarcomere length operating range of vertebrate muscles during movement. J Exp Biol 204:1529–1536

Burkholder TJ, Nichols TR (2004) Three-dimensional model of the feline hindlimb. J Morphol 261:118–129

Burkholder TJ, Fingado B, Baron S, Lieber RL (1994) Relationship between muscle fiber types and sizes and muscle architectural properties in the mouse hindlimb. J Morphol 221:177–190
Burrows M, Morris G (2001) The kinematics and neural control of high-speed kicking movements in the locust. J Exp Biol 204:3471–3481
Chow CK, Jacobson DH (1972) Further studies of human locomotion: postural stability and control. Math Biosci 15:93–108
Collins S, Ruina A, Tedrake R, Wisse M (2005) Efficient bipedal robots based on passive-dynamic walkers. Science 307:1082–1085
Correa TA, Baker R, Graham HK, Pandy MG (2011) Accuracy of generic musculoskeletal models in predicting the functional roles of muscles in human gait. J Biomech 44:2096–2105
Crowninshield RD, Brand RA (1981) A physiologically based criterion of muscle force prediction in locomotion. J Biomech 14:793–801
Delp SL, Loan JP, Hoy MG, Zajac FE, Topp EL, Rosen JM (1990) An interactive graphics-based model of the lower extremity to study orthopaedic surgical procedures. IEEE Trans Biomed Eng 37:757–767
Delp SL, Anderson FC, Arnold AS, Loan P, Habib A, John CT et al (2007) OpenSim: open-source software to create and analyze dynamic simulations of movement. IEEE Trans Biomed Eng 54:1940–1950
Eng CM, Smallwood LH, Rainiero MP, Lahey M, Ward SR, Lieber RL (2008) Scaling of muscle architecture and fiber types in the rat hindlimb. J Exp Biol 211:2336–2345
Full RJ, Koditschek DE (1999) Templates and anchors: neuromechanical hypotheses of legged locomotion on land. J Exp Biol 202:3325–3332
Gasser HS, Hill AV (1924) The dynamics of muscular contraction. Proc R Soc Lond B Biol Sci 96:398–437
Geyer H, Seyfarth A, Blickhan R (2006) Compliant leg behaviour explains basic dynamics of walking and running. Proc Biol Sci 273:2861–2867
Gomi H, Kawato M (1997) Human arm stiffness and equilibrium-point trajectory during multi-joint movement. Biol Cybern 76:163–171
Gribble PL, Ostry DJ, Sanguineti V, Laboissiere R (1998) Are complex control signals required for human arm movement? J Neurophysiol 79:1409–1424
Hase K, Stein RB (1999) Turning strategies during human walking. J Neurophysiol 81:2914–2922
Henry SM, Fung J, Horak FB (2001) Effect of stance width on multidirectional postural responses. J Neurophysiol 85:559–570
Hollister AM, Jatana S, Singh AK, Sullivan WW, Lupichuk AG (1993) The axes of rotation of the knee. Clin Orthop Relat Res 290:259–268
Hsu WL, Scholz JP, Schoner G, Jeka JJ, Kiemel T (2007) Control and estimation of posture during quiet stance depends on multijoint coordination. J Neurophysiol 97:3024–3035.
Huijing PA, Baan GC, Rebel GT (1998) Non-myotendinous force transmission in rat extensor digitorum longus muscle. J Exp Biol 201:683–691.
Josephson RK (1985) Mechanical power output from striated muscle during cyclic contraction. J Exp Biol 114:493–512
Lam T, Pearson KG (2001) Proprioceptive modulation of hip flexor activity during the swing phase of locomotion in decerebrate cats. J Neurophysiol 86:1321–1332
Latash ML, Gottlieb GL (1991) Reconstruction of shifting elbow joint compliant characteristics during fast and slow movements. Neuroscience 43:697–712
Lin DC, Rymer WZ (2000) Damping actions of the neuromuscular system with inertial loads: soleus muscle of the decerebrate cat. J Neurophysiol 83:652–658.
Lockhart DB, Ting LH (2007) Optimal sensorimotor transformations for balance. Nat Neurosci 10:1329–1336
Loren GJ, Shoemaker SD, Burkholder TJ, Jacobson MD, Friden J, Lieber RL (1996) Human wrist motors: biomechanical design and application to tendon transfers. J Biomech 29:331–342
Lutz GJ, Rome LC (1996) Muscle function during jumping in frogs. II. Mechanical properties of muscle: implications for system design. Am J Physiol 271:C571–C578

Maas H, Huijing PA (2009) Synergistic and antagonistic interactions in the rat forelimb: acute effects of coactivation. J Appl Physiol (1985) 107:1453–1462
Macpherson JM (1988) Strategies that simplify the control of quadrupedal stance. I. Forces at the ground. J Neurophysiol 60:204–217
McGeer T (1990a) Passive bipedal running. Proc R Soc Lond B Biol Sci 240:107–134
McGeer T (1990b) Passive dynamic walking. Int J Robot Res 9:62–82
McKay JL, Ting LH (2008) Functional muscle synergies constrain force production during postural tasks. J Biomech 41:299–306
McKay JL, Ting LH (2012) Optimization of muscle activity for task-level goals predicts complex changes in limb forces across biomechanical contexts. PLoS Comput Biol 8:e1002465
McKay JL, Burkholder TJ, Ting LH (2007) Biomechanical capabilities influence postural control strategies in the cat hindlimb. J Biomech 40:2254–2260
Morgan DL, Proske U, Warren D (1978) Measurements of muscle stiffness and the mechanism of elastic storage of energy in hopping kangaroos. J Physiol 282:253–261
Nozaki D, Nakazawa K, Akai M (2005) Muscle activity determined by cosine tuning with a nontrivial preferred direction during isometric force exertion by lower limb. J Neurophysiol 93:2614–2624
O'Neill MC, Lee LF, Larson SG, Demes B, Stern JT, Jr., Umberger BR (2013) A three-dimensional musculoskeletal model of the chimpanzee (Pan troglodytes) pelvis and hind limb. J Exp Biol 216:3709–3723
Prilutsky BI (2000) Coordination of two- and one-joint muscles: functional consequences and implications for motor control. Motor Control 4:1–44
Rybak IA, Stecina K, Shevtsova NA, McCrea DA (2006) Modelling spinal circuitry involved in locomotor pattern generation: insights from the effects of afferent stimulation. J Physiol 577:641–658
Sandercock TG, Heckman CJ (1997) Force from cat soleus muscle during imposed locomotor-like movements: experimental data versus Hill-type model predictions. J Neurophysiol 77:1538–1552
Scheys L, Spaepen A, Suetens P, Jonkers I (2008) Calculated moment-arm and muscle-tendon lengths during gait differ substantially using MR based versus rescaled generic lower-limb musculoskeletal models. Gait Posture 28:640–648
Scholz JP, Schoner G (1999) The uncontrolled manifold concept: identifying control variables for a functional task. Exp Brain Res 126:289–306
Scovil CY, Ronsky JL (2006) Sensitivity of a hill-based muscle model to perturbations in model parameters. J Biomech 39:2055–2063
Smutz WP, Kongsayreepong A, Hughes RE, Niebur G, Cooney WP, An KN (1998) Mechanical advantage of the thumb muscles. J Biomech 31:565–570
Southgate DF, Cleather DJ, Weinert-Aplin RA, Bull AM (2012) The sensitivity of a lower limb model to axial rotation offsets and muscle bounds at the knee. Proc Inst Mech Eng H 226:660–669
Todorov E (2007) Probabilistic inference of multijoint movements, skeletal parameters and marker attachments from diverse motion capture data. IEEE Trans Biomed Eng 54:1927–1939
Valero-Cuevas FJ, Johanson ME, Towles JD (2003) Towards a realistic biomechanical model of the thumb: the choice of kinematic description may be more critical than the solution method or the variability/uncertainty of musculoskeletal parameters. J Biomech 36:1019–1030
Whittington B, Silder A, Heiderscheit B, Thelen DG (2008) The contribution of passive-elastic mechanisms to lower extremity joint kinetics during human walking. Gait Posture 27:628–634
Winter DA, Patla AE, Prince F, Ishac M, Gielo-Perczak K (1998) Stiffness control of balance in quiet standing. J Neurophysiol 80:1211–1221
Young RP, Scott SH, Loeb GE (1992) An intrinsic mechanism to stabilize posture–joint-angle-dependent moment arms of the feline ankle muscles. Neurosci Lett 145:137–140
Zajac FE (1989) Muscle and tendon: properties, models, scaling, and application to biomechanics and motor control. Crit Rev Biomed Eng 17:359–411
Zajac FE, Gordon ME (1989) Determining muscle's force and action in multi-articular movement. Exerc Sport Sci Rev 17:187–230

# Chapter 5
# Modeling the Organization of Spinal Cord Neural Circuits Controlling Two-Joint Muscles

**Natalia A. Shevtsova, Khaldoun Hamade, Samit Chakrabarty, Sergey N. Markin, Boris I. Prilutsky and Ilya A. Rybak**

**Abstract** The activity of most motoneurons controlling one-joint muscles during locomotion are locked to either extensor or flexor phase of locomotion. In contrast, bifunctional motoneurons, controlling two-joint muscles such as posterior biceps femoris and semitendinosus (PBSt) or rectus femoris (RF), express a variety of activity patterns including firing bursts during both locomotor phases, which may depend on locomotor conditions. Although afferent feedback and supraspinal inputs significantly contribute to shaping the activity of PBSt and RF motoneurons during real locomotion, these motoneurons show complex firing patterns and variable behaviors under the conditions of fictive locomotion in the immobilized decerebrate cat, i.e., with a lack of patterned supraspinal and afferent inputs. This suggests that firing patterns of PBSt and RF motoneurons are defined by neural interactions inherent to the locomotor central pattern generator (CPG) within the spinal cord. In this study, we use computational modeling to suggest the architecture of spinal circuits representing the locomotor CPG and the connectivity pattern of spinal interneurons defining the behavior of bifunctional PBSt and RF motoneurons. The proposed model reproduces the complex firing patterns of these motoneurons during

N. A. Shevtsova (✉) · K. Hamade · S. N. Markin · I. A. Rybak
Department of Neurobiology and Anatomy, Drexel University College of Medicine, 2900 W. Queen Lane, Philadelphia, PA 19129, USA
e-mail: Natalia.Shevtsova@drexelmed.edu

K. Hamade
e-mail: Khaldoun.Hamade@drexelmed.edu

S. N. Markin
e-mail: smarkin@drexelmed.edu

S. Chakrabarty
School of Biomedical Sciences, Faculty of Biology, University of Leeds, Garstang Building, Leeds, West Yorkshire LS2 9JT, UK
e-mail: s.chakrabarty@leeds.ac.uk

B. I. Prilutsky
School of Applied Physiology, Center for Human Movement Studies, Georgia Institute of Technology, 555 14th Street NW, Atlanta, GA 30332, USA



B. I. Prilutsky, D. H. Edwards (eds.), *Neuromechanical Modeling of Posture and Locomotion,* Springer Series in Computational Neuroscience,
DOI 10.1007/978-1-4939-3267-2_5

fictive locomotion under different conditions including spontaneous deletions of flexor and extensor activities and provides insights into the organization of spinal circuits controlling locomotion in mammals.

**Keywords** Spinal cord · Two-joint muscles · Fictive locomotion · Central pattern generator · Computational modeling

**Abbreviations**

| | |
|---|---|
| CPG | Central pattern generator |
| EMG | Electromyoagam |
| ENG | Electroneurogram |
| GS | Gastrocnemius combined with soleus |
| LGS | Lateral gastrocnemius combined with soleus |
| MG | Medial gastrocnemius |
| MLR | Mesencephalic locomotor region |
| PB | Posterior biceps femoris |
| PBSt | PB combined with semitendinosus |
| PF | Pattern formation |
| Plant | Plantaris |
| RF | Rectus femoris |
| RG | Rhythm generator |
| Sart | Sartorius |
| SmAB | Semimembranosus combined with anterior biceps femoris |
| St | Semitendinosus |
| TA | Tibialis anterior |
| UBG | Unit burst generator |

## 5.1 Introduction

The bipartite half-center organization of the spinal locomotor central pattern generator (CPG) was originally proposed by T. Graham Brown (1914) and then expanded by Lundberg, Jankowska and their colleagues (e.g., Jankowska et al. 1967a, b; Lundberg 1981). According to this concept, the locomotor rhythmic activity is generated by the alternating activity of two populations of excitatory interneurons (the "half-centers") mutually inhibiting each other via inhibitory interneurons. This alternating half-center activity directly controls alternating activation of extensor and flexor motoneurons whose activity is locked to one of the two locomotor phases. However, the activity of motoneuron pools controlling the muscles spanning more than one joint (e.g., the two-joint muscles) is more complicated and often depends on locomotor conditions. In particular, motoneuron pools controlling such muscles as posterior biceps femoris and semitendinosus (PB and St or PBSt, hip extensor and knee flexor) and rectus femoris (RF, hip flexor and knee extensor) express a variety of activity patterns depending on gait, speed, and slope of locomotion, and/or other locomotor conditions (Halbertsma 1983; Pratt et al. 1996; Carlson-Kuhta et al. 1998; Smith et al. 1998a). The activity of PBSt during real locomotion can be characterized as flexor- or extensor-related, depending on the primary muscle group with which it is co-active, or as biphasic (Grillner 1981; Halbertsma 1983; Smith et al. 1998b). RF also shows similar variability in activity patterns during locomotion that also depends on gait and locomotor conditions (Pratt et al. 1996; Smith et al. 1998b).

The complex activity patterns generated by motoneurons controlling two-joint muscles have been considered as a strong argument against a half-center (bipartite) organization of the CPG (discussed by Grillner 1981; Stein and Smith 1997). In contrast, Perret and colleagues (Perret 1983; Orsal et al. 1986; Perret et al. 1988) suggested that the PBSt and RF motoneuron pools may receive excitatory and inhibitory inputs from both flexor and extensor CPG half-centers and proposed several schematics of spinal circuits that could potentially generate the complex PBSt and RF patterns within a framework of the bipartite locomotor CPG organization. The principal point of the proposed architecture was the existence of additional interneurons interposed between the rhythm generator and PBSt and RF motoneurons that could provide shaping of their firing patterns under control of supraspinal and/or afferent signals (Perret 1983). In this architecture (see Fig. 5.1), intermediate neuron populations are controlled by excitatory inputs from the CPG and afferent pathways and shape PBSt and RF activities through excitatory and inhibitory influences.

However, although afferent feedback and supraspinal inputs may significantly contribute to shaping activity of PBSt and RF motoneurons during real locomotion, these motoneurons express a wide repertoire of firing patterns during fictive locomotion in decerebrate immobilized cats in the absence of sensory feedback and patterned supraspinal inputs (Grillner and Zangger 1979; Perret and Cabelguen 1980; Perret 1983; Orsal et al. 1986; Guertin et al. 1995; Markin et al. 2012). Moreover, with afferent stimulation, the spontaneous activity of PBSt and RF motoneurons was reportedly altered from being flexor-related to extensor-related (Perret 1983) or biphasic (McCrea and Chakrabarty 2007; Shevtsova et al. 2007; Hamade et al. 2008). Such a wide repertoire of PBSt and RF motoneuron firing behaviors during fictive locomotion have never been explained or reproduced with computational models.

Rybak et al. (Rybak et al. 2006a, b; McCrea and Rybak 2007, 2008) have recently proposed a two-level organization of the locomotor CPG in which a bipartite rhythm generator (RG) controls a specially organized pattern formation (PF) network that in turn projects to motoneurons and controls their behavior (Fig. 5.2). The two-level CPG architecture permits separate control of the locomotor rhythm and motoneuron activity. The first level, RG, defines the locomotor rhythm and the durations of the flexor and extensor phases, and the second level, the PF network, transforms the RG activity to activity patterns of different motoneuron pools. The two-level CPG model was able to reproduce multiple effects of afferent stimulation during fictive locomotion as well as the specific behavior of flexor and extensor motoneurons during spontaneous deletions (missing bursts) of motoneuron activity (Rybak et al. 2006a, b; McCrea and Rybak 2007, 2008; Shevtsova 2015). However, this model included only two antagonist motoneuron pools (flexor and extensor, see Fig. 5.2) and the generation of more complex activity patterns, such as patterns of PBSt and RF, was not considered.

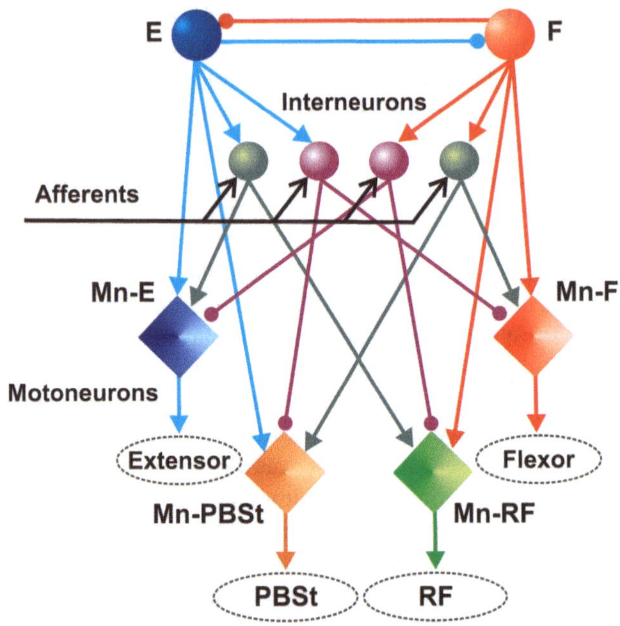


**Fig. 5.1** Hypothetical network capable of generating the PBSt and RF activity patterns. The central rhythm generator (*F*—*E*) provides the alternating flexor-extensor activity. The flexor half-center, *F*, directly excites the flexor and RF motoneuron pools while the extensor half-center, *E*, directly excites the extensor and PBSt motoneurons. Additional interneuron populations of alternative reflex pathways are influenced by afferent inputs and provide complimentary excitatory/inhibitory signals to motoneuron populations. In this and following figures, populations of interneurons are represented by *spheres*; excitatory and inhibitory synaptic connections are indicated by lines ended with arrows and small circles, respectively; populations of motoneurons are represented by diamonds. Modified from (Perret 1983)

Here we propose a possible organization of spinal circuits in the frameworks of a two-level CPG organization allowing the model to generate different patterns of motoneuron activity including the complex patterns of the PBSt and RF observed during fictive locomotion, as well as to reproduce the variety of changes of these patterns during deletions.

## 5.2 Patterns of PBSt and RF Activity During fictive Locomotion

### *5.2.1 Data*

Experimental data used in this study represent series of recordings of activity of different spinal nerves from decerebrate immobilized cat preparations during fictive locomotion induced by electrical stimulation of the brainstem mesencephalic

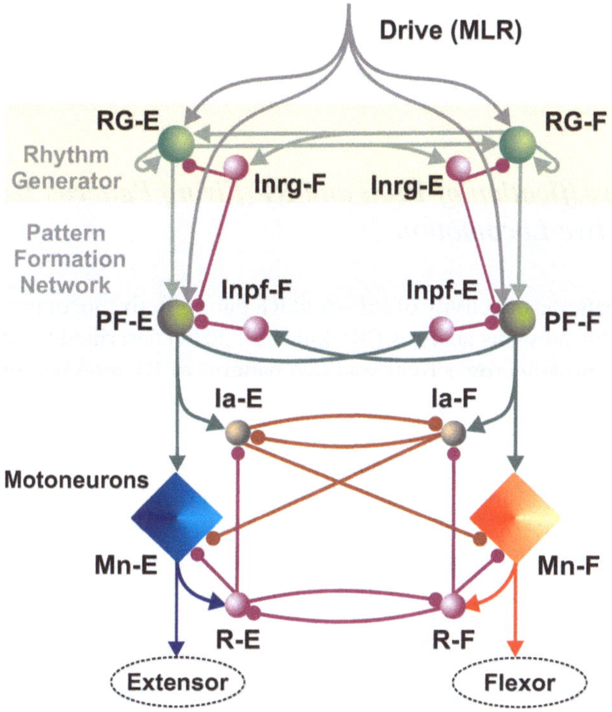


**Fig. 5.2** Schematic of the two-level locomotor CPG model by (Rybak et al. 2006a). Note that for consistency with the current model description we slightly changed the names of the inhibitory interneuron populations Inrg-E, Inrg-F, Inpf-E and Inpf-F from those in the basic model given in (Rybak et al. 2006a, b). The extension E or F in a population name corresponds now to the phase in which this population is active. Modified from (Rybak et al. 2006a)

locomotor region (MLR). These data were collected for many years in the laboratories of Drs. McCrea and Jordan at the Spinal Cord Research Centre, University of Manitoba, Winnipeg, Canada. All experiments were performed in compliance with the guidelines set out by the Canadian Council on Animal Care and the University of Manitoba. No new animal experiments were performed for the present study.

Motoneuron activity during fictive locomotion was recorded extracellularly as electroneurograms (ENGs) from multiple hindlimb nerves. Nerve recordings were rectified and low-pass filtered before digitization (for details see Guertin et al. 1995; Lafreniere-Roula and McCrea 2005). Data were analyzed from experiments which included records from RF nerve as well as PB or St nerves or the combined PB and St (PBSt) nerves. As described above, the locomotor activities of two-joint muscles can vary widely depending upon behavioral or experimental conditions. However, the particular pattern of motoneuron activity of PBSt or RF occurring in a given fictive locomotion preparation generally remained unchanged during the course of an experiment despite changes in locomotor period or variations in MLR stimulus intensity employed during the experiment. When recorded separately during fictive locomotion, the ENGs from PB and St (both muscles are hip extensors and knee flexors) exhibited similar firing pattern regardless of whether their activity was flexor-phase or extensor-phase related. This similarity justified the common practice of using recordings from the combined PBSt nerves.

### 5.2.2 Classification of PBSt and RF Firing Patterns During Fictive Locomotion

The typical patterns of activity of PBSt and RF during fictive locomotion are shown in Fig. 5.3. Our previous analysis (Markin et al. 2012) described three distinct patterns of PBSt motoneuron activity and two patterns of RF motoneuron activity occurring during fictive locomotion (Fig. 5.4, panels b and c, respectively). In Fig. 5.4, the typical patterns of flexor (Sart) and extensor (SmAB) activity are also shown for reference. This figure shows that during fictive locomotion, PBSt usually exhibits one of the following three ENG activity patterns: (1) a single burst at the beginning of the flexor phase (*type 1*, observed in 73 % of PBSt records, Fig. 5.4b, upper trace; see examples in Fig. 5.3a1, a2, d1, d2) that is usually short in duration (less than 30 % of the flexor phase, *type 1a,* Fig. 5.3a1, d1, d2) but sometimes longer (up to 70 % of the flexor phase, *type 1b*, Fig. 5.3a2); (2) a typical extensor activity, i.e., firing throughout the entire extensor phase (*type 2*, 9 % of cases, Fig. 5.4b, middle trace; see example in Fig. 5.3b); and (3) a biphasic pattern, consisting of a short first burst at the beginning of the flexor phase and a longer second burst throughout the extensor phase (*type 3*, 18 % of cases, Fig. 5.4b, bottom trace, examples are in Fig. 5.3c1, c2).

RF usually exhibited one of the following two ENG patterns: (1) a single burst at the end of the flexor phase (*type 1*, observed in 53 % of RF records, Fig. 5.4c, upper trace, example is shown in Fig. 5.3b, d1) or (2) a biphasic pattern consisting of a burst at the end of the flexor phase and an additional burst at the end of the extensor phase (*type 2*, 47 % of cases, Fig. 5.4c, bottom trace, see example in Fig. 5.3d2). In all experiments with simultaneous recordings, PBSt and RF never were active simultaneously (see for example, see Fig. 5.3b, d1, d2). Possible combinations of PBSt and RF ENG activities recorded simultaneously in fictive locomotion experiments are shown in Fig. 5.4d1–d4.

### 5.2.3 Activity of PBSt and RF Motoneurons During Spontaneous Deletions

Deletions represent brief periods of inactivity (missing one or several consecutive bursts) occurring spontaneously during generation of locomotor activity simultaneously in multiple synergist (e, g., flexor or extensor) motoneuron pools. During deletions, the maintained activity of antagonist motoneuron pools usually becomes tonic or continue to be rhythmic (Lafreniere-Roula and McCrea 2005; Rybak et al. 2006a;

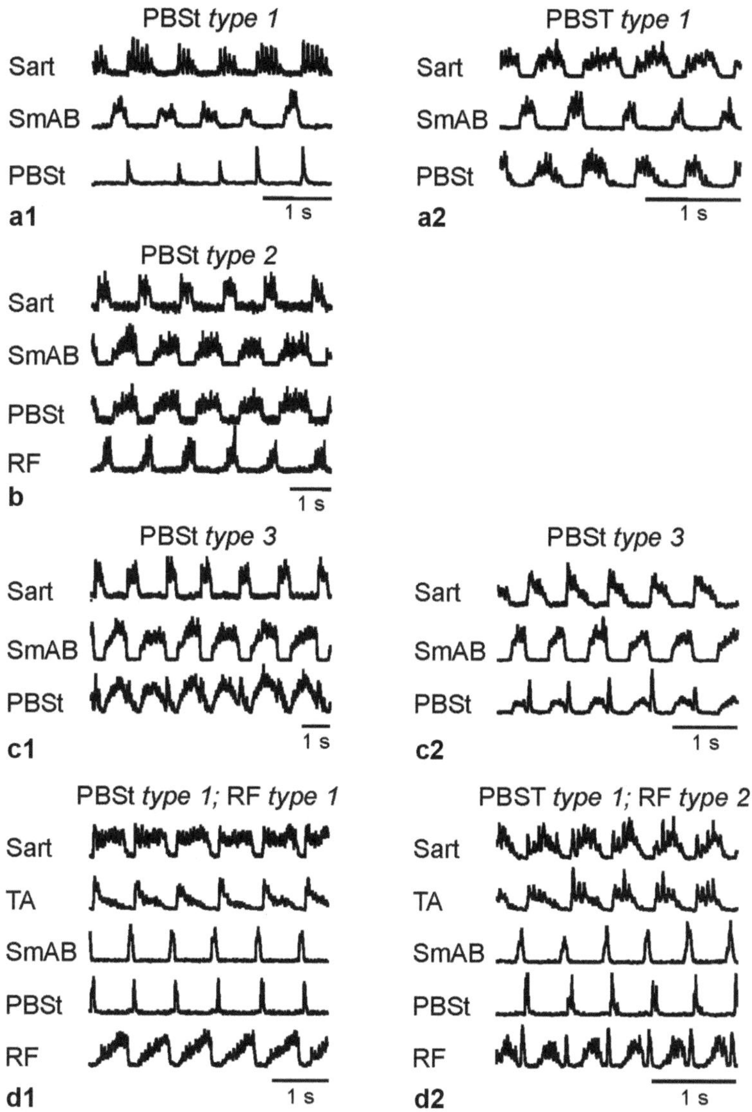


**Fig. 5.3** Activities of the PBSt and RF motoneuron pools in selected fictive locomotion experiments. **a1, a2** The PBSt motoneuron pool (type 1) is active either at the onset of the flexor phase or during most of the flexor phase. **b** The PBSt motoneuron pool (type 2) is active in the extensor phase. **c1, c2** The biphasic PBSt ENG (type 3). **d1** The motoneuron pools of PBSt (type 1) and RF (type 1) are active in the flexor phase. **d2** The PBSt motoneuron pool (type 1) is active in the flexor phase, and the RF ENG (type 2) is biphasic

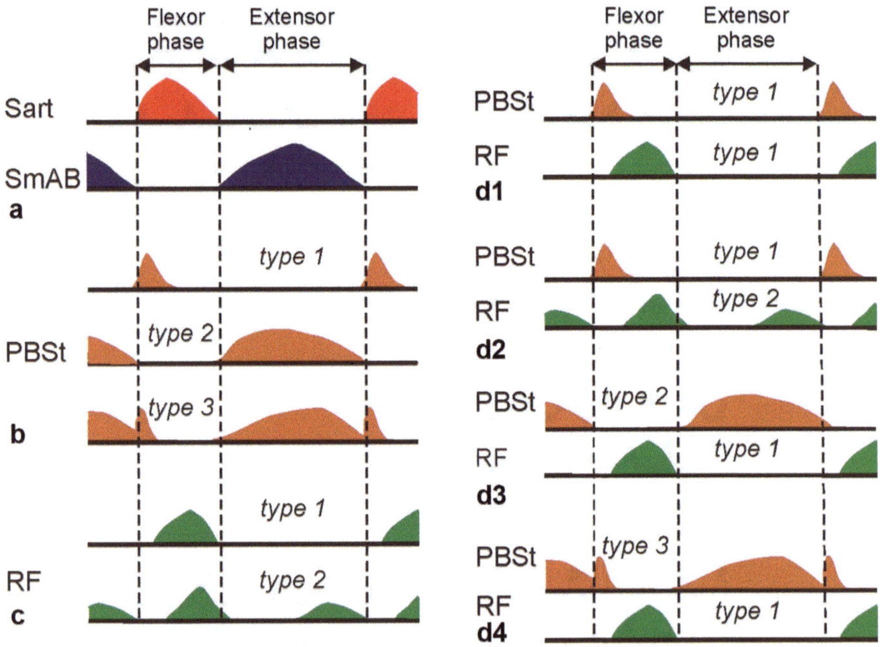


**Fig. 5.4** Averaged and normalized patterns of PBSt and RF ENG activities with respect to flexor and extensor activity profiles during fictive locomotion. **a** Typical flexor (Sart) and extensor (SmAB) patterns. **b** Three patterns of PBSt motoneuron activities: type 1, the PBSt motoneuron pool is active at the beginning of the flexor phase; type 2, the PBSt motoneuron pool is active during the extensor phase; and type 3, the PBSt motoneuron pool is active in both phases. **c** Two typical RF ENG patterns: type 1, the RF motoneuron pool is active in the flexor phase, and type 2, the RF motoneuron pool is active in both phases. **d1–d4** Possible combinations of PBSt and RF ENG activities recorded simultaneously. Modified from (Markin et al. 2012)

Shevtsova 2015). Deletions have been previously classified into two types: *resetting* deletions, characterized by a shift in the phase of post-deletion rhythmic activity relative to the pre-deletion rhythm, and *non-resetting* deletions, after which the post-deletion rhythmic activity re-appears with no phase shift relative to a pre-deletion rhythm (Lafreniere-Roula and McCrea 2005; Rybak et al. 2006a; Shevtsova 2015).

One would expect that if before deletions PBSt or RF was active in one phase only (i.e., showed activity like a flexor or extensor), its behavior during the deletion would resemble the activity of the corresponding motoneurons (flexor or extensor). However, experimental data indicate that PBSt and RF behaviors during deletions have no direct correspondence with their pre- or post-deletion activity. For example, in three experiments shown in Fig. 5.5a–c, each episode contains a resetting extensor deletion (silence of SmAB, MG, Plant, LGS) accompanied by tonic activity of flexor motoneuron pools (Sart, TA). In all three episodes before and after deletion, PBSt demonstrates flexor-related activity (*type 1*). However, the behavior of PBSt during deletions in these episodes is completely different. In Fig. 5.5a, the PBSt motoneuron pool becomes tonically active during the deletion, i.e., behaves as the

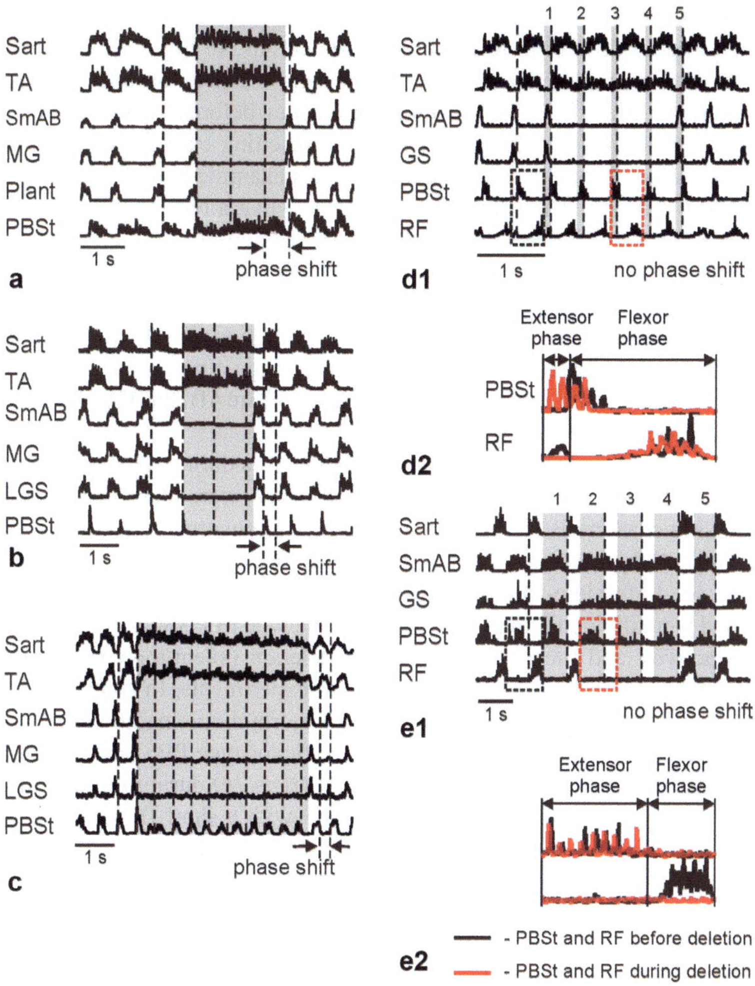


**Fig. 5.5** **a, b, c** Examples of variable behavior of PBSt ENGs during resetting deletions of extensor activity (SmAB, MG, Plant, and LGS) accompanied by sustained activity in flexors (Sart and TA). In this and the following figures, the *vertical dashed lines* plotted at the intervals of an average locomotor period indicate the beginning of the flexor phases before and after deletions and the beginning of the expected flexor phases during deletions. In all three deletions shown in **a–c**, there is an obvious phase shift of the post-deletion rhythm with respect to the pre-deletion rhythmic activity (see arrows at the bottom) indicating that these deletions are resetting. In all three recordings in **a–c** PBSt ENG is of the flexor type before and after deletions. During deletions, the PBSt motoneuron pool demonstrates tonic activity like flexors in **a**, is silent like extensors in **b**, and expresses rhythmic activity in **c**. Note that frequency of PBSt ENG oscillations in **c** differs from the locomotor frequency before and after the deletion. **d1, e1** Examples of PBSt (**d1, e1**) and RF (**e1**) ENG activity during extensor (**d1**) and flexor (**e1**) deletions. The lack of phase shift of the post-deletion rhythmic activity with respect to the pre-deletion rhythm indicates that the deletions shown in panels **d1** and **e1** are non-resetting. **d2, e2** Enlarged and overlapped traces of PBSt and RF ENG activities during one locomotor period before (*black*) and during (*red*) the deletions outlined in **d1** and **e1** by *dashed* rectangles. Shaded rectangles in **d1** and **e1** highlight the extensor phases before and after the deletions and the expected extensor phases during the deletions

flexor motoneuron pools. In Fig. 5.5b, the PBSt motoneuron pool is silent during the deletion episode similar to extensor motoneurons. In Fig. 5.5c, the PBSt motoneuron pool expresses rhythmic activity during the deletion and the frequency of this activity differs from the locomotor frequency before and after the deletion.

Figure 5.5d1 shows an episode with a non-resetting extensor deletion (SmAB and GS are silent for some time interval). In this example before and after the deletion, PBSt is active in the flexor phase (*type 1*) like a typical flexor. During the deletion, however, PBSt remains rhythmic, and its activity becomes biphasic (see an expanded insert d2 showing two overlapped locomotor periods, one taken before and the other taken during the deletion).

Figure 5.5e1 shows an episode of fictive locomotion, in which PBSt normally expresses extensor-related activity (*type* 2), but during a non-resetting flexor deletion (Sart is silent) with tonically active extensors (SmAB and GS) PBSt remains rhythmically active.

To analyze a full specter of PBSt and RF behaviors during deletions, the experimental recordings containing PBSt and/or RF ENGs were classified according to the following characteristics: (a) type of agonist motoneuron pool (flexor or extensor) whose activity was missing during deletion; (b) deletion type (resetting or non-resetting, see Lafreniere-Roula and McCrea 2005; Rybak et al. 2006a; Shevtsova 2015), and were divided into groups based on the type of PBSt or RF pre- and post-deletion activity and their behavior during deletion (silent/tonic/rhythmic). This analysis (see Table 5.1) included data from 36 experiments. In some experiments, there were several deletion episodes separated by rhythmic locomotor activity. We noticed obvious differences in behavior of the flexor-type PBSt ENG with a short flexor burst (*type 1a*) vs. a longer flexor burst (*type 1b*) during most of deletions. Similarly, the behaviors of biphasic PBSt ENG with a short flexor burst and that with a longer flexor burst were also different during resetting extensor deletions. Though no data were found demonstrating biphasic PBSt ENG with a longer flexor burst for other types of deletions, we separated biphasic PBSt ENG patterns into two subgroups, *type 3a* and *3b* depending on the length of the flexor burst similar to the flexor-type PBSt. The results of our classification are summarized in Table 5.1. Numbers in parentheses in Table 5.1 indicate the numbers of episodes where the indicated behavior was observed during particular deletion types for each type of PBSt or RF ENG patterns. The re-appearing post-deletion activity patterns were the same as the pre-deletion patterns in the absolute majority of experiments.

## 5.3 Constructing the Extended CPG Model

### *5.3.1 Shaping Profiles of PBSt and RF Activity*

Analysis of intracellular recordings from PBSt and RF motoneurons shows that they often exhibit depolarization in both locomotor phases, even if this depolarization is sufficient for firing during only one phase, flexor or extensor. This observa-

**Table 5.1** Behavior of PBSt and RF ENGs during spontaneous deletions

| | Deletion type | | | | |
|---|---|---|---|---|---|
| | Extensor deletions | | | Flexor deletions | |
| | Tonic flexors | | Rhythmic flexors | Tonic extensors | |
| | Resetting | Non-resetting | | Resetting | Non-resetting |
| *PBSt* | | | | | |
| Flexor, short burst (*type 1a*) | Silent (7) | Rhythmic[a] (1) | Silent or rhythmic[a] (1) | Silent (4) | |
| Flexor, long burst (*type 1b*) | Tonic (4) or rhythmic[b] (1) | Rhythmic[c] (4) | Rhythmic (3) | Silent (4) | |
| Extensor (*type 2*) | Silent (9) | | Silent (4) | Tonic (8) | Silent (1) or rhythmic[a] (1) |
| Biphasic, short flexor burst (*type 3a*) | Silent (7) | | Silent (1) | Tonic (5) | Not found |
| Biphasic, long flexor burst (*type 3b*) | Tonic (2) | Not found | | | |
| *RF* | | | | | |
| Flexor (*type 1*) | Tonic (5) | Rhythmic (1) | Rhythmic (3) | Silent (7) | |
| Biphasic (*type 2*) | Tonic (5) | Rhythmic[d] (7) | Rhythmic[d] (9) | Not found | |

[a] Amplitude of rhythmic activity is markedly reduced and some bursts are missing
[b] Frequency of PBSt ENG oscillations differs from the frequency before/after deletion
[c] Flexor-type PBSt ENG pattern becomes biphasic during deletions
[d] RF ENG pattern loses its extensor component

tion led to the suggestion that PBSt and RFs motoneurons receive excitatory inputs from both flexor and extensor half-centers of the CPG (Perret 1983; Orsal et al. 1986; Perret et al. 1988). Another suggestion of Perret et al. was the existence of additional excitatory and inhibitory interneuron populations interposed between the CPG and PBSt and RF motoneurons that shape their firing patterns. These suggestions were explicitly used in the construction of our model.

Figure 5.6 shows schematically the possible neural connectivity allowing shaping the activity profiles of PBSt and RF motoneurons. As suggested above, they receive excitatory inputs from both half-centers and additional inhibitory inputs from hypothetical interneuron populations shaping their activity. Panel A shows typical profiles of flexor (Sart) and extensor (SmAB) activities that may represent the profiles of excitatory input from flexor (red) and extensor (blue) half-centers, respectively. Figure 5.6b1–b5 schematically shows the net excitation that the PBSt and RF would receive from flexor and extensor half-centers (filled gray areas) superimposed with the typical PBSt (Fig. 5.6b1–b3, upper panels) or RF (Fig. 5.6b4, b5, upper panels) profiles. For each particular PBSt or RF pattern, we can see "extra excitation" that should be eliminated by some inhibitory inputs to the corresponding populations (lower panels in Fig. 5.6b1–b5, flexor- and extensor related components of the inhibitory signals are shown in red or blue, respectively). Figure 5.6c1–c5 schematically shows how each of PBSt and RF patterns can be sculptured by additional

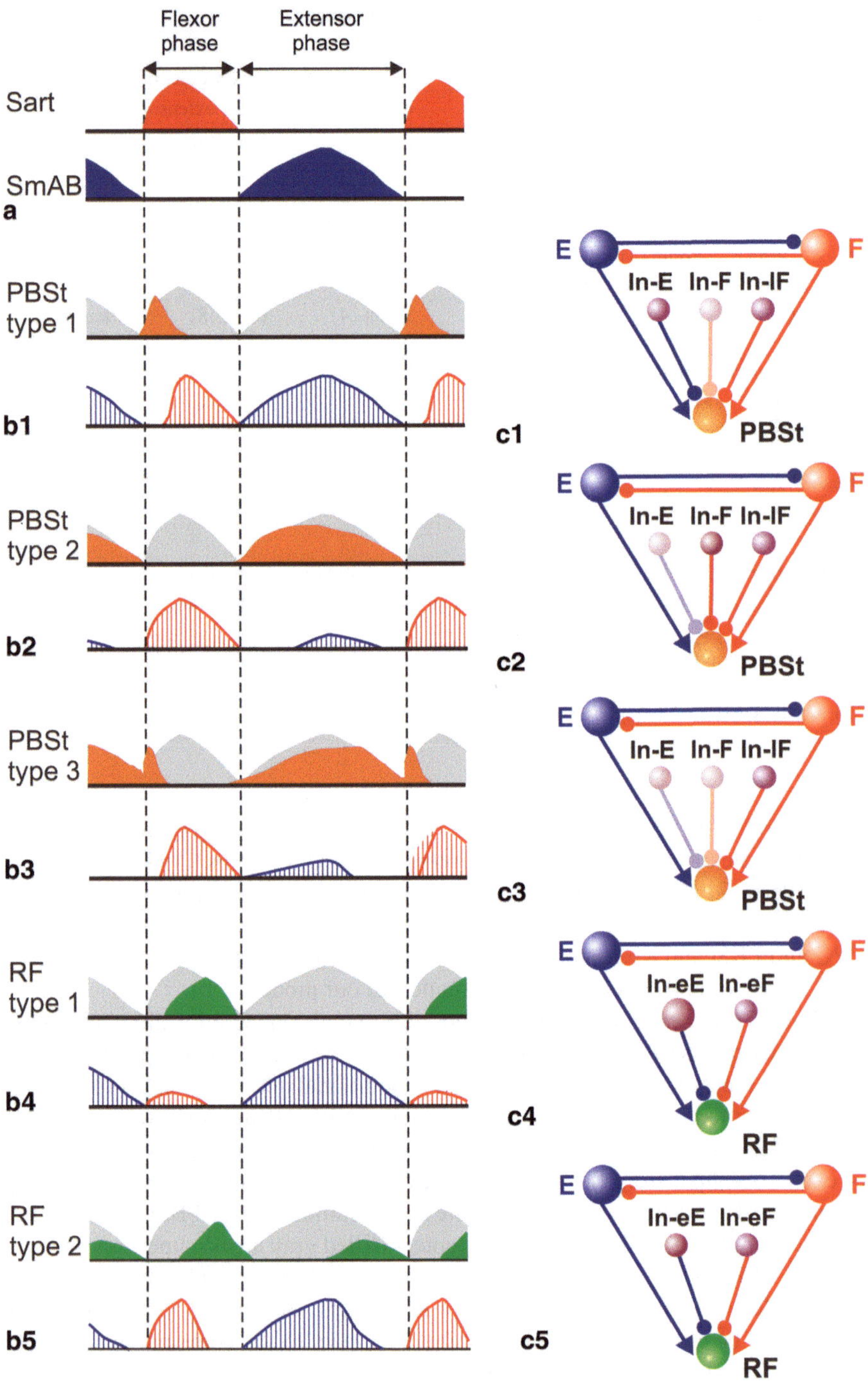


**Fig. 5.6** Shaping PBSt and RF motoneuron activities of different types in fictive locomotion. **a** Averaged and normalized activities of typical flexor (Sart) and extensor (SmAB) motoneuron pools. **b1–b5** *Upper* traces, typical PBSt (**b1–b3**, *brown filled areas*) or RF (**b4** and **b5**, *green filled*

inhibitory interneuron populations. Specifically, PBSt activity in the extensor phase can be shaped by an extension-related inhibitory population, In-E, which is active throughout the extensor phase. If activation of this population is strong, the extensor component of PBSt activity will be fully inhibited (Fig. 5.6b1, c1). If activation of this population is moderate or weak, PBSt will maintain an extensor component and exhibit an extensor (Fig. 5.6b2, c2) or biphasic pattern (Fig. 5.6b3, c3). PBSt activity in the flexor phase can be controlled by an inhibitory population firing throughout the flexor phase, In-F. Strong activation of this population will result in complete suppression of the flexor component of PBSt activity (Fig. 5.6b2, c2); if the In-F population is silent or weakly activated, the PBSt motoneuron pool will demonstrate activity in the flexor phase (Fig. 5.6b1, c1, b3, c3). The length of the PBSt flexor burst can be regulated by an additional inhibitory population, active in a later part of the flexor phase, In-lF (Fig. 5.6b1, c1, b3, c3).

Similarly, the flexor component of RF profile can be shaped by two inhibitory populations: In-eF that shapes the RF activity during the flexor phase, and In-eE that shapes the extensor component of RF activity (Fig. 5.6b4, c4, b5, c5). Note that if the In-eE population is strongly activated, it becomes active during the whole extensor phase (see Fig. 5.6b4, c4). In this case the extensor component of RF activity will be fully suppressed and RF will exhibit a flexor-type activity (Fig. 5.6b4, c4); otherwise RF activity will be biphasic (Fig. 5.6b5, c5).

### *5.3.2 Basic Concept and PF Network Architecture*

The ultimate goal of our modeling effort was to extend our basic two-level model of the spinal locomotor CPG (Rybak et al. 2006a) so that it could reproduce complex patterns of bifunctional motoneurons such as PBSt and RF. Similar to the preceding model, the extended model includes a bipartite RG and PF network. Also, similar to that model the locomotor activity is initiated by external tonic "MLR" drive to the excitatory neural populations of the CPG. This way to initiate the rhythm allowed us to perform direct simulation of fictive locomotion conditions.

While extending the basic two-level CPG model, we have suggested that the model reproduction of complex patterns of PBSt and RG motoneurons can be achieved by a special construction of pattern formation (PF) network without changing the

---

*areas*) activity patterns overlapped with summarized flexor and extensor activities (*gray filled areas*). Bottom traces, hypothetical inhibitory inputs shaping each type of PBSt and RF activity patterns and representing the difference between summarized flexor and extensor activities and the corresponding PBSt or RF profile. *Red* and *blue patterned filled areas* correspond to inhibitory influence on the PBSt and RF motoneurons in the flexor or extensor phase, respectively. **c1–c5** Hypothetical neural circuitries participating in shaping particular PBSt and RF patterns. Spheres marked by *E* (*blue*) and *F* (*red*) schematically represent the flexor and extensor parts of the CPG. *PBSt* (*brown*) and *RF* (*green*) populations are shown as larger spheres. Smaller spheres represent hypothetical inhibitory interneuron populations sculpting PBSt and RF activity profiles. Strength of neuron activation in populations is symbolically shown by level of brightness of the corresponding sphere. Larger size of the In-eE population in **c4** indicates a strong activation of this population during the whole extensor phase to shape the flexor-type RF. See text for details bipartite organization of the RG. So, similar to the preceding model (Rybak et al. 2006a, see Fig. 5.2) the locomotor rhythm generation in the RG is based on a combination of intrinsic (persistent sodium current dependent) properties of the excitatory RG neurons, mutual excitation within each half-center (RG-F and RG-E populations) and reciprocal inhibition between the half-centers via the inhibitory interneuron populations (Inrg-F and Inrg-E, see Fig. 5.2). The alternating bursting activities of the RG-F and RG-E populations (half-centers) define the extensor and flexor phases of the locomotor cycle, respectively. The PF network contains the two principal PF populations (PF-F and PF-E) that also receive the MLR drive. They also receive excitatory inputs from the homonymous RG populations and inhibitory inputs from the interneuron populations (Inrg-F and Inpf-F, or Inrg-E and Inpf-E, correspondingly, see Fig. 5.2). The PF-F and PF-E populations transmit rhythmic activities to the flexor (Mn-F) and extensor (Mn-E) motoneuron populations, respectively, and to the homonymous inhibitory Ia populations, providing reciprocal inhibition between the flexor and extensor motoneuron populations. There are also populations of Renshaw cells (R-F and R-E) that receive collateral excitatory input from the corresponding motoneuron populations (Mn-F and Mn-E) and provide feedback inhibition to the homonymous motoneuron population and the population of Ia inhibitory neurons (Ia-F or Ia-E), rhythmically inhibiting motoneuron populations during the inactive phase of the locomotor cycle (Fig. 5.2). To control the populations of PBSt and RF motoneurons (Mn-PBSt and Mn-RF, respectively) we needed to incorporate in the PF network the principle PF-PBSt and PF-RF populations that would project to the Mn-PBSt and Mn-RF motoneuron populations and to organize interactions between the PF-PBSt and PF-RF and other CPG populations (Fig. 5.7).

According to this idea (and similar to the PF-F and PF-E populations), the PF-PBSt and PF-RF populations should receive the "MLR" drive. They should also receive excitatory inputs from both extensor and flexor parts of the CPG that could come from the RG and/or PF levels (i.e., from the RG-E or PF-E, and from the RG-F or PF-F populations). Additional inhibitory interneuron populations, that are necessary to sculpt the PBSt and RF activity (In-E, In-F, In-lF, In-eF, and In-eE, see Fig. 5.6), should be also included in the PF network and shape the activity of the PF-PBSt and PF-RF (and hence Mn-PBSt and Mn-RF) populations (purple small spheres and connections in Fig. 5.7a1, b1). To be active in a particular phase of the locomotor cycle (flexor or extensor), these additional populations should receive rhythmic excitatory inputs from either flexor or extensor side of the CPG at different levels (RG or PF). We suggested that to be active in a certain locomotor phase, these populations receive tonic excitatory drive and are inhibited in the opposite phase by inhibition from the corresponding Inrg or Inpf populations. To determine possible organization of inputs from RG and PF populations to the PF-PBSt, PF-RF and additional interneuron populations, we analyzed the activity of PBSt and RF during two non-resetting deletions shown in Fig. 5.5d1, d2. During this analysis, we took into account a previous suggestion (Rybak et al. 2006a) that non-resetting deletions result from perturbations (additional drives or increase of excitability) affecting only one side (flexor or extensor) of the PF network, while keeping rhythmic activity of the RG populations (see Table 5.2). The results of such logical approach are illustrated in Fig. 5.9.

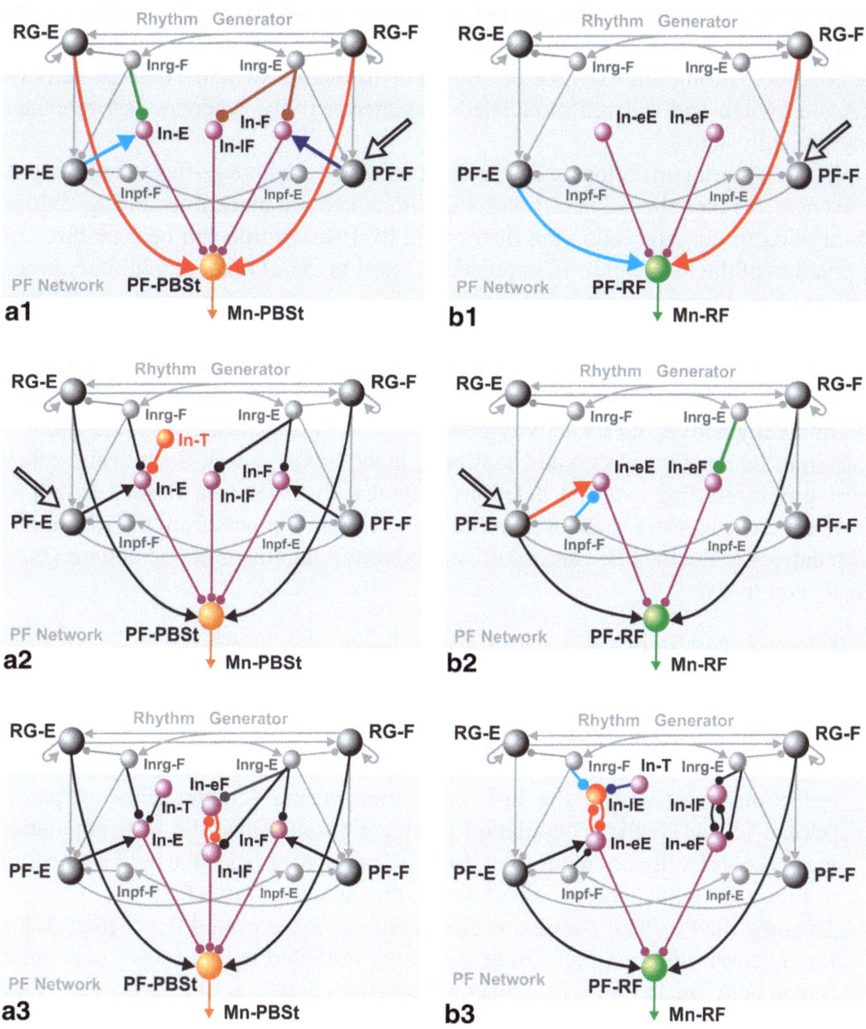


**Fig. 5.7** Incorporating circuits controlling PBSt (**a1, a2**) and RF (**b1, b2**) motoneuron activities in the PF network on the base of analysis of fictive locomotion episodes with particular deletions shown in Fig. 5.5d1, e1. The *left column* (panels **a1** and **a2**) illustrates sequential building of PBSt circuitry. The *right column* (panels **b1** and **b2**) shows sequential construction of RF circuitry. At each step of circuitry construction newly introduced network elements are highlighted by particular color. **a3, b3** Extension of PF circuitries controlling the PF-PBSt (**a3**) and PF-RF (**b3**) populations behavior in the extended model to provide complimentary patterns of PBSt and RF activity. See details and explanations in the text

Let us first consider fictive locomotion recordings shown in Fig. 5.5d1. These recordings show a fictive locomotion episode with a non-resetting deletion of extensor activity. Since this deletion is non-resetting it may result from an increased excitation of the PF-F population that fully inhibits the PF-E population and hence produces a non-resetting deletion of extensor activity, while the RG-E and RG-F populations maintain rhythmic activity (Rybak et al. 2006a). In Fig. 5.7a1, b1 this increase of excitation is schematically shown by large unfilled arrows to the PF-F population. During this extensor deletion, not all flexors switched to tonic activity: TA did switch and showed modulated sustained activity, whereas Sart remained rhythmically active.

In this experiment, before and after deletion, PBSt exhibited a flexor activity profile (*type 1b*) (see shaded bars 1 and 5 in Fig. 5.5d1 and black trace in Fig. 5.5d2). According to our suggestion the flexor-type PF-PBSt profile can be sculptured by activation of the In-E and In-lF populations (see Fig. 5.6c1). These inhibitory populations would inhibit the PF-PBSt population during the extensor phase and later in the flexor phase, respectively. In contrast, the In-F population that inhibits the PF-PBSt population in the flexor phase should be silent or weakly activated. As seen in Fig. 5.5d1, d2, during the deletion, the PBSt motoneuron pool remained rhythmically active, but its activity profile turned into biphasic; it started firing in the expected extensor phases and continued in the flexor phases. Such firing pattern during non-resetting extensor deletions is typical for PBSt (see Table 5.1) and allows several suggestions about possible organization of inputs from the RG and PF populations to the PF-PBSt and additional inhibitory interneuron populations (In-E, In-lF, and In-F):

1. Biphasic activity of PBSt during this deletion may indicate that the PF-PBSt population receives excitatory inputs from the RG-F and RG-E populations (see red bold connections in Fig. 5.7a1) that continue to be rhythmically active during this deletion (see Table 5.2).
2. PBSt activity that appears in the expected extensor phases may result from a reduction in activity of the In-E population during deletion allowing partial release of the PF-PBSt population from inhibition. Thus, the In-E population may receive excitatory input from the PF-E population (see blue bold connection in Fig. 5.7a1) that becomes silent during this deletion (Table 5.2).
3. To allow the PF-PBSt population to be active in the expected flexor phases during deletion, the In-E population should be inhibited by the Inrg-F population (green bold line in Fig. 5.7a1) that remains rhythmically active during non-resetting extensor deletions (Table 5.2).
4. Similarly, to allow PBSt activity during the expected extensor phases during this deletion, the In-F and In-lF populations should both be inhibited by the Inrg-E population (brown bold lines in Fig. 5.7a1) that remains rhythmically active during non-resetting extensor deletions (see Table 5.2).
5. During this deletion, PBSt may remain active at the beginning of the flexor phase but with reduced amplitude and duration (see the traces of PBSt activity before and during deletion in Fig. 5.5d2). This indicates that in addition to excitatory tonic input drive, the In-F population may receive a weak excitatory input from the PF-F population (dark blue bold line in Fig. 5.7a1) that increases its activity during non-resetting extensor deletion.

In experiment shown in Fig. 5.5d1, RF expresses biphasic activity (*type 2*) before and after deletion. According to our suggestion (see Fig. 5.6c5), the activity of

**Table 5.2** Activity of CPG populations in the basic model during spontaneous deletions

| Population name | Deletion type | | | | |
|---|---|---|---|---|---|
| | Extensor deletions | | | Flexor deletions | |
| | Tonic flexors | | Rhythmic flexors | Tonic extensors | |
| | Resetting | Non-resetting | Non-resetting | Resetting | Non-resetting |
| RG-E/Inrg-E | Silent | Rhythmic | Rhythmic | Tonic | Rhythmic |
| RG-F/ Inrg-F | Tonic | Rhythmic | Rhythmic | Silent | Rhythmic |
| PF-E/ Inpf-F | Silent | Silent | Silent | Tonic | Tonic |
| PF-F/ Inpf-F | Tonic | Tonic | Rhythmic | Silent | Silent |

biphasic RF can be sculptured by activation of the In-eE and In-eF populations, inhibiting the PF-RF population during the beginning of the extensor and flexor phases, respectively. During this deletion (and other similar deletions, see Table 5.1), RF remains rhythmically active in the expected flexor phases, but loses its extensor component. Such firing pattern of biphasic RF during non-resetting extensor deletions allows the following suggestions about organization of inputs from the RG and PF populations to the PF-RF population and interneuron populations incorporated to shape RF activity (In-eE and In-eF):

1. The PF-RF population may receive an excitatory input from the RG-F population (red bold line in Fig. 5.7b1) that continues to be rhythmically active during this deletion (see Table 5.2).
2. The PF-RF population may receive excitatory input from the PF-E population (blue bold line in Fig. 5.7b1) that becomes silent during non-resetting extensor deletions (Table 5.2).

Let us now consider a fictive locomotion episode with a non-resetting deletion of flexor activity (missing activity of Sart) shown in Fig. 5.5e1. During such deletions, the RG-E and RG-F populations may maintain rhythmic activity, while the PF-E population would show sustained tonic activity and the PF-F population would become silent (Rybak et al. 2006a). Such deletion in the model may result from increasing excitation of the PF-E population (Rybak et al. 2006a) shown by the large unfilled arrows to the PF-E population in Fig. 5.9a2, b2.

In episode shown in Fig. 5.5e1, before and after deletion PBSt was active during the extensor phase and silent during the flexor phase (*type 2*). To provide shaping of such PBSt pattern, the In-F population that inhibits the PF-PBSt population during the flexor phase should be strongly activated, while the In-E population inhibiting the PF-PBSt population in the extensor phase should be weakly active or silent. According to our suggestion above, the In-E population receives excitatory input from the PF-E population (see Fig. 5.7a1, blue bold connection). Thus, to suppress activity of the In-E population and allow PBSt to be active in the extensor phase, an In-T population is included in the model that receives tonic excitatory input and inhibits the In-E population (shown by red in Fig. 5.7a2). Hence, the activity of the In-T population controls the In-E (and hence PF-PBSt) population activity during the extensor phase. Specifically, if the In-T is silent, the In-E population is strongly activated and inhibits the PF-PBSt population in the extensor phase. In contrast, if the In-T population is strongly activated, it inhibits the In-E population, thus allowing the PF-PBSt population activity in the extensor phase. Finally, if the In-T population is moderately activated, the In-E population expresses a moderate activity in the extensor phase and the PF-PBSt population would exhibit a reduced extensor activity.

During the deletion shown in Fig. 5.5e1, PBSt rhythmic activity persists in the expected extensor phases (see shaded bars 2–3 in Fig. 5.5e1 and overlapped traces in Fig. 5.5e2), although reduced relative to its activity before and after deletion. This supports our suggestion that the In-E population receives an excitatory synaptic input from the PF-E population (Fig. 5.7a1, blue bold connection) whose activity increases during this and similar deletions. Evidently, during the deletion shown in Fig. 5.5e1, the In-F population remains rhythmically active and inhibits the PF-PBSt population activity during the expected flexor phase, supporting our previous suggestion that in addition to the excitatory tonic drive this population may receive an inhibitory synaptic input from the Inrg-E population (Fig. 5.7a1, dark blue bold connection).

In the experimental recordings shown in Fig. 5.7e1, RF expressed flexor activity (*type 1*) before and after a non-resetting extensor deletion. Such RF pattern may be sculptured by activation of the In-eF populations at the beginning of the flexor phase and by the In-eE population during the extensor phase (Fig. 5.6c4). During the deletion, RF became silent, which allows the following suggestions:

1. The In-eE population inhibiting the PF-RF population during the extensor phase may be excited by the PF-E population (see red bold connection in Fig. 5.7b2) that becomes tonic and increases its activity during such deletions (see Table 5.2).
2. The In-eE population is inhibited by the Inpf-F population (blue bold connection in Fig. 5.7b2) that becomes silent during such deletions (see Table 5.2).

To complete organization of connections to populations shaping RF activity we suggested that the In-eF population is inhibited by the Inrg-E population (green bold connection in Fig. 5.7b2).

Inhibitory interneuron populations (In-lF, In-eF, and In-eE) incorporated in the PF network to shape PF-PBSt and PF-RF activity (Figs. 5.6 and 5.7) should be active either in the beginning ("early" type, e.g., In-eF or In-eE) or in the later part of the corresponding phase ("late" type, In-lF). To shape such activity patterns, we suggested that neurons with an "early" pattern of activity have intrinsic adaptive properties and that neurons with the "late" pattern of activity in the In-lF population are inhibited by the neurons of the "early" type in the In-eF populations. In addition, we suggested that the In-lF and In-eF populations mutually inhibit each other, thus providing complimentary activity patterns of the In-eF and In-lF populations and hence, complimentary shaping of the of PBSt and RF activity profiles during the flexor phase (Fig. 5.7a3, red connections). To provide complimentary shaping PF-PBSt and PF-RF activities in extension, we incorporated in the PF network an In-lE population (shown in red in Fig. 5.7b3) that is active in late extension, receives an excitatory tonic drive, and is inhibited by the Inrg-F populations (blue connection in Fig. 5.7b3). We also suggested that similarly to the In-eF and In-lF populations, the In-eE and In-lE populations mutually inhibit each other (red connections in Fig. 5.7b3). Finally, we suggested that in addition to tonic drive, activity of the In-lE population is regulated by inhibition from the In-T population (dark blue connection in Fig. 5.7b3) which controls activity of the PF-PBSt population in the extensor phase (see Fig. 5.7a2). If this inhibition is strong, it "prohibits" activation of the PF-RF population (by disinhibition of the In-eE population which inhibits the PF-RF activity) and "allows" the PF-PBSt activity (by inhibition of the In-E population and hence disinhibition of the PF-PBSt population) during the extensor phase. Thus, the In-T population provides complimentary shaping of the profiles of PF-PBSt and PF-RF activities during the extensor phase.

## *5.3.3 Extended Model*

### 5.3.3.1 Model Architecture

Figure 5.8 shows the schematic of the extended model that includes all populations and circuits that were incorporated in the PF network of the CPG (see Figs. 5.6 and 5.7) to provide the full repertoire of PBSt and RF ENG activity profiles observed during fictive locomotion and their variable behavior during spontaneous deletions. In this model, activation of several inhibitory interneuron populations (In-E, In-T, In-F, In-eF, In-lF, In-eE, and In-lE) by tonic drives controls shaping of the PBSt and RF activity patterns and their behavior during deletions. The extended model also incorporates the motoneuron populations, Mn-PBSt and Mn-RF, and the populations of Renshaw cells, R-PBSt and R-RF, which receive collateral excitatory inputs from, and provide inhibitory feedback to, the homonymous motoneuron populations (Mn-PBSt and Mn-RF). Little data are available to conclude which motoneuron pools can be considered the direct antagonists of PBSt or RF motoneuron pools and if reciprocal interactions exist between the PBSt or RF motoneurons and some other motoneuron populations. Therefore, Ia-mediated reciprocal inhibition to and from PBSt and RF motoneurons was not included in the current version of the extended model.

### 5.3.3.2 Modeling Single Neurons and Neural Populations

All neurons were modelled in the Hodgkin-Huxley style. Interneurons were simulated as single-compartment models. Motoneurons had two compartments: soma and dendrite. The single neuron models were previously described in detail (Rybak et al. 2006a). A brief description of these neuron models, ionic currents included in each neuron type, and all model parameters are provided in the Appendix.

Each interneuron population included 20 neurons. Each motoneuron population had 40 neurons. Connections between populations were established such that, if

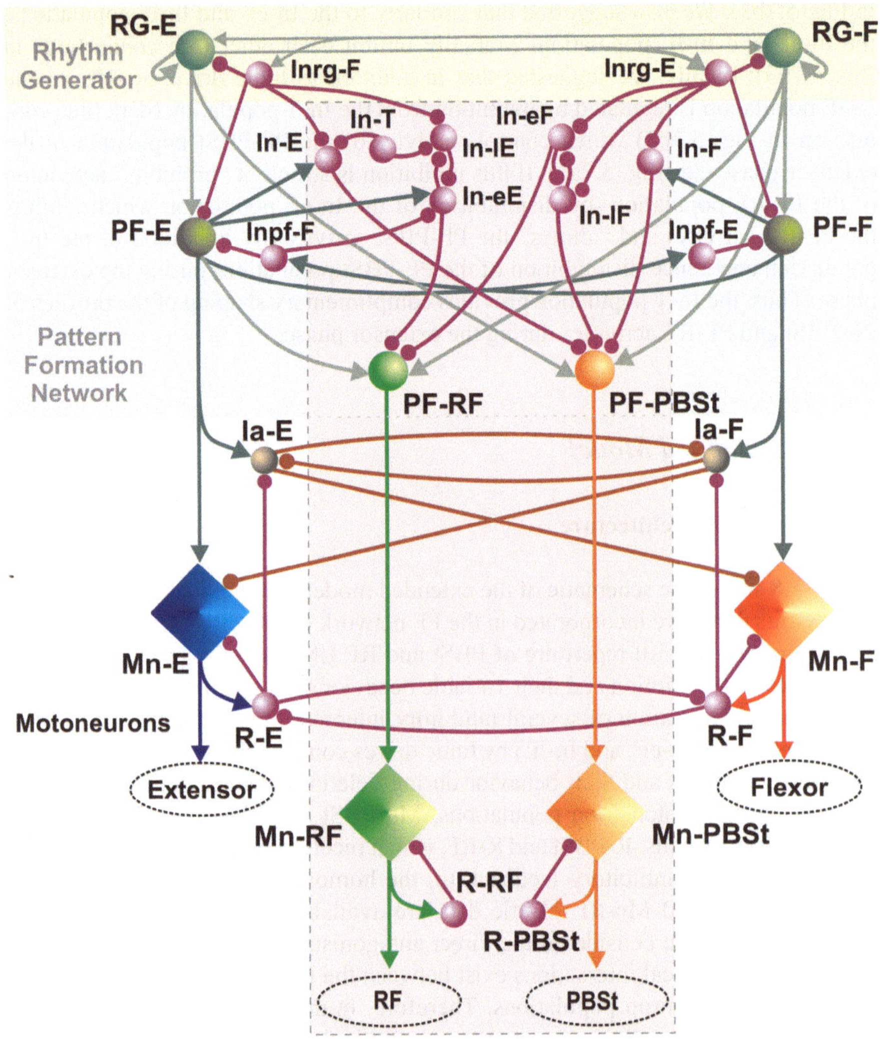


**Fig. 5.8** Schematic of the extended model of the locomotor CPG. The model includes all populations of the basic CPG model (see Fig. 5.2) and an extension to simulate generation of PBSt and RF motoneuron activities (outlined by *pink rectangle*) that includes hypothetical circuitry incorporated into the PF network to control PBSt and RF behavior (see Figs. 5.6 and 5.7), two bifunctional motoneuron populations, Mn-PBSt and Mn-RF, and two populations of Renshaw cells, R-PBSt and R-RF, which receive collateral excitatory input from and provide feedback inhibition to the Mn-PBSt and Mn-RF populations, respectively. Note that all populations in the Rhythm Generator and Pattern Formation Network receive excitatory tonic drives (not shown)

a population *A* was assigned to receive an excitatory or inhibitory input from a population *B* or external drive *d*, then each neuron of population *A* received the corresponding excitatory or inhibitory synaptic input from each neuron of population *B* or from drive *d*, respectively. All connections were randomly distributed to provide heterogeneity of individual neuron behavior in populations. The mean weights of synaptic connections between the neural populations in our model are shown in the Appendix (Table 5.A2). The standard deviation (SD) of synaptic weights varied from 10–15 % of the mean value.

The heterogeneity of neurons within each population was also provided by random distribution of the reversal potential of leak channel, $E_L$ (see mean values ± SD in Appendix), initial conditions for values of the membrane potential, calcium concentrations and some channel conductances. In each simulation, a settling period of 20 s was allowed before data were collected. Each simulation was repeated 20–30 times, and demonstrated qualitatively similar behavior for particular mean values and standard deviations of distributed parameters.

All simulations were performed on a Dual Core Opteron(tm), 2.61 GHz/2.0 GB (DELL) with a Windows XP operating system using a special simulation package NSM 2.1 RC2, developed at Drexel University by I. A. Rybak, S. N. Markin, and N. A. Shevtsova using Microsoft Visual C++. Differential equations were solved using the exponential Euler integration method (MacGregor 1987) with a step of 0.1 ms (for details see Rybak et al. 2006a).

## 5.4 Model Performance

### *5.4.1 Sculpting PBSt and RF Activity Patterns*

Similar to the original model by (Rybak et al. 2006a), the locomotor rhythm in the extended model is generated by the bipartite rhythm generator (RG). The alternating rhythmic bursts in flexor and extensor RG half-centers (RG-F and RG-E traces in Fig. 5.9a) define the durations of the extensor and flexor phases and hence the locomotor cycle period. At the PF level, the PF-F and PF-E populations follow the activity of the corresponding RG populations (see corresponding traces in Fig. 5.9). As shown in Fig. 5.8, the PF-PBSt and PF-RF populations project to the corresponding motoneuron populations (Mn-PBSt or Mn-RF, respectively) and receive excitatory synaptic inputs from both the RG-F and RG-E (for PBSt) or from the PF-E and RG-F (for RF) populations. The total synaptic drive to PF-PBST and PF-RF is shaped by a number of inhibitory interneuron populations (In-E, In-T, In-F, In-eF, In-lF, In-eE, and In-lE). For each type of PBSt and RF patterns, the activity of inhibitory interneuron populations (five upper traces in Fig. 5.9b1–b5) depends on tonic drive to each of these populations ($d_E$, $d_T$, $d_F$, $d_{eF}$, $d_{lF}$, $d_{eE}$, and $d_{lE}$, respectively, see Fig. 5.9c1–c5). The resulting activities of the PF-PBSt or PF-RF populations (and, respectively, Mn-PBSt or Mn-RF populations shown at the bottom in Fig. 5.9b1–b5) for each particular activity pattern depend on relative strength of the corresponding input drives ($d_E$, $d_T$, $d_F$, $d_{eF}$, $d_{lF}$, $d_{eE}$, $d_{lE}$).

*Flexor-type Profile of PBSt Activity* Figure 5.9b1, c1 illustrates sculpting the flexor-type PBSt activity pattern (type 1) in the extended model. To generate this

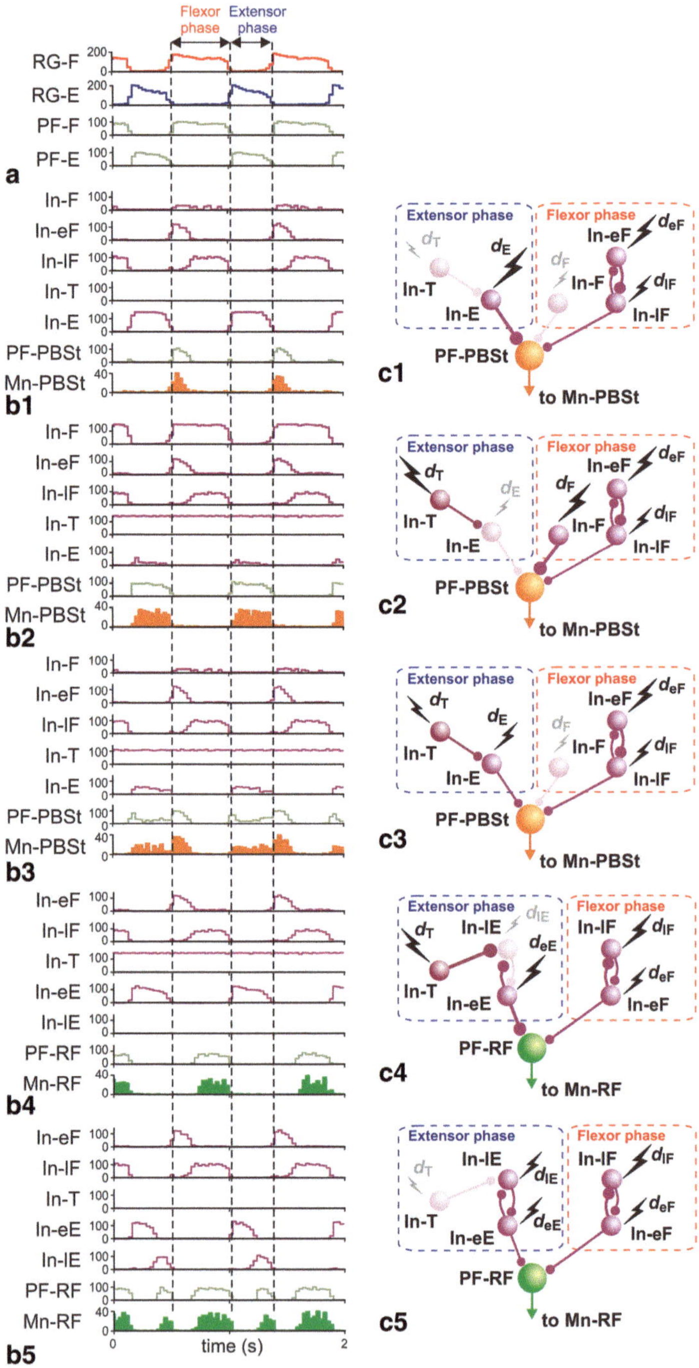


**Fig. 5.9** Sculpting PBSt and RF motoneuron activity patterns in the extended model. The left column shows results of computer simulation for different PBSt and RF ENG activity types. Each trace represents a histogram of average neuron activity in a particular population (spikes

pattern, the PF-PBSt population should be inhibited during the extensor phase by the strongly activated In-E population and during later part of the flexor phase by the In-lF population. Activity of the In-E population depends on the excitatory tonic drive, $d_E$, and inhibition from the In-T population controlled by the tonic drive, $d_T$. Thus, if $d_T$ is weak and $d_E$ is strong enough (see Fig. 5.9c1), the In-E population will be strongly activated and fully suppress the activity of PF-PBSt (and hence Mn-PBSt) population during the extensor phase. At the same time, the In-F population, controlled by the tonic drive, $d_F$, should be silent or weakly activated to allow activity in PBSt motoneurons in the flexor phase, i.e., $d_F$ should be weak. Duration of PBSt flexor bursts is defined by the activity of the In-eF and In-lF populations, inhibiting each other, and finally by tonic drives to these populations, $d_{eF}$ and $d_{lF}$, respectively (see Fig. 5.9c1): the stronger $d_{lF}$ and/or the weaker $d_{eF}$ is, the shorter is the duration of PBSt flexor burst. Figure 5.9b1 shows activities of the In-E, In-T, In-F, In-eF, and In-lF populations shaping the flexor-type of PF-PBSt and hence Mn-PBSt profiles of activity.

*Extensor-type Profile of PBSt Activity* Figure 5.9b2, c2 shows shaping of the extensor type PBSt activity pattern (type 2). In case of the extensor type of PBSt profile, activity of PF-PBSt during the extensor phase should not be inhibited, i.e., the In-E population should be weakly activated or silent, which may occur when $d_E$ is weak and/or $d_T$ is strong enough (see Fig. 5.9c2). At the same time, the PF-PBSt activity during the flexor phase should be completely inhibited by the In-F population, i.e., $d_F$ should be strong (Fig. 5.9c2). Figure 5.9b2 shows activities of the In-E, In-T, In-F, In-eF, and In-lF populations shaping the extensor-type of PF-PBSt and hence Mn-PBSt profiles of activity.

*Biphasic Profile of PBSt Activity* Figure 5.9b3, c3 illustrates shaping the biphasic type PBSt pattern (type 3). The biphasic PBSt profile includes a persistent activity (usually reduced) during the extensor phase followed by a short burst at the begin-

---

per neuron per second, bin=30 ms). The right column shows neural circuits shaping different activity patterns. The rectangles outlined by *dashed lines* indicate populations participating in shaping activity of the PF-PBSt and PF-RF populations (*large green spheres*) during the flexor (*red*) or extensor (*blue*) phase. Small spheres represent inhibitory populations sculpting the PBSt and RF activity profiles. The bolts of different sizes schematically show strengths of excitatory input drives to inhibitory interneuron populations, the larger the bolt size, the stronger the input drive. Strength of neuron activation in a particular population is symbolically shown by thickness of output connection, the greater the level of excitation, the thicker the connection. Silent or weakly activated populations and their outputs are shown by lower brightness of the corresponding sphere or connection. **a** Alternating rhythmic bursts of the RG-F and RG-E populations (*two upper traces*) define locomotor cycle period and durations of the flexor and extensor phases (indicated by *vertical dashed* lines for one locomotor period). Two lower traces represent activities of the PF-F and PF-E populations, following activity of the corresponding RG populations. **b1–b5** Upper five traces represent activities of inhibitory interneuron populations shown in **c1–c5** and sculpting the flexor-type PBSt, type 1 (**b1, c1**); extensor-type PBSt, type 2 (**b2, c2**); biphasic PBSt, type 3 (**b3, c3**); flexor-type RF, type 1 (**b4, c4**); and biphasic RF, type 2 (**b5, c5**) activities. The last four traces in **b1–b5** show activity of the PF-PBSt or PF-RF populations and resulting activity of the corresponding motoneuron population (Mn-PBSt or Mn-RF, respectively). For values of input drives to the RG-F, RG-E, and hypothetical interneuron populations see Table 5.A3 in Appendix ning of the flexor phase. In case of this profile, the reduced activity during the extensor phase can be provided by moderate inhibition of the PF-PBSt population by the In-E population; the level of inhibition here can be defined by the relative strengths of the tonic drives $d_E$ and $d_T$ to the In-E and In-T populations, respectively (see Fig. 5.9c3). The In-F population in this case should be silent or weakly activated to allow PF-PBSt activity in the flexor phase, i.e., $d_F$ should be weak or absent (see Fig. 5.9c3). The duration of PBSt flexor bursts for the biphasic PBSt pattern is defined by relative activation of the In-eF and In-lF populations similar to the flexor-type PBSt pattern, and finally by the tonic drives $d_{eF}$ and $d_{lF}$ (see Fig. 5.9c3). Activities of the In-E, In-T, In-F, In-eF, and In-lF populations shaping the biphasic PBSt motoneuron activity pattern and the resulting PF-PBSt and Mn-PBSt activity profiles are shown in Fig. 5.9b3.

*Flexor-type Profile of RF Activity* Shaping the flexor-type RF profile of activity (type 1) is shown in Fig. 5.9b4, c4. In this case, the PF-RF population should be inhibited during the extensor phase and at the beginning of the flexor phase. In the model, whether or not the PF-RF population is active in the extensor phase is determined by activation of the In-eE population that depends on relative strengths of the tonic input drives $d_{eE}$ and $d_{lE}$ to the In-eE and In-lE populations, respectively (see Fig. 5.9c4), and on activation of the In-T population, i.e., on the tonic drive $d_T$. If $d_{eE}$ and/or $d_T$ are strong enough and $d_{lE}$ is weak or absent, the In-eE population is active throughout the whole extensor phase and inhibits PF-RF activity during the extensor phase. Simultaneously, the PF-RF population is inhibited by the In-eF population at the beginning of the flexor phase, which depends on tonic drive $d_{eF}$ and interaction between In-eF and In-lF populations, i.e., on the interplay between the tonic drives $d_{eF}$ and $d_{lF}$. Figure 5.9b4 shows activities of the In-T, In-eE, In-lE, In-eF, and In-lF populations shaping the flexor-type PF-RF and Mn-RF activity profiles.

*Biphasic Profile of RF Activity* Figure 5.9b5, c5 illustrates shaping of the biphasic profile of RF activity (type 2). In this case, the PF-RF population should be inhibited at the beginning of the extensor phase and at the beginning of the flexor phase. Shaping RF activity later in the flexor phase for the biphasic RF activity is similar to that for the flexor-type RF activity (see above). To allow RF to generate an additional burst in the extensor phase, the tonic drive $d_T$ to the In-T population should be weak (if present), and at the same time, the tonic drive $d_{lE}$ to the In-lE population should be strong enough to overcome inhibition of this population by the In-eE population. In other words, if $d_{lE}$ is strong enough, at some moment during the extensor phase the In-lE population escapes inhibition from the In-eE population and suppresses its activity by the end of the extensor phase. This allows the emergence of a short extensor burst in the PF-RF (and hence in Mn-RF) population during the later part of the extensor phase. Activities of the In-T, In-eE, In-lE, In-eF, and In-lF populations shaping the biphasic RF activity pattern and the resulting PF-RF and Mn-RF activities are shown in Fig. 5.9b5.

Figure 5.10 shows the results of our simulations for different types of PBSt and RF activities recorded during fictive locomotion experiments. Experimental recordings are shown for one flexor (Sart), one extensor (SmAB), PBSt, and RF (panels a1, b1, c1, and d1). The simulated traces (in panels a2, b2, c2, and d2, respectively) show averaged frequency of spiking including (from top to bottom): the flexor (Mn-F), extensor (Mn-E), PBSt (Mn-PBSt), and RF (Mn-RF) motoneuron populations. In Fig. 5.10a1, PBSt has a flexor-type profile and RF shows a biphasic profile. The corresponding simulation is shown in Fig. 5.10a2. In Fig. 5.10b1, PBSt is also active in the flexor phase, but demonstrates a longer flexor burst, which is simulated in Fig. 5.10b2. In this simulation, the Mn-RF population also shows flexor-type activity profile, complimentary to the activity profile of the Mn-PBSt population. Figure 5.10c1 shows the experimental records, in which the PBSt demonstrates extensor-type of activity profile and the RF motoneuron pool is active in the flexor phase. This is simulated in Fig. 5.10c2. In Fig. 5.10d1, PBSt is biphasic and the RF ENG is of the flexor type. This is simulated in Fig. 5.10d2.

## *5.4.2 Modeling PBSt and RF Behavior During Deletions*

### 5.4.2.1 Resetting Deletions

While developing the extended model, the challenge was not only to reproduce the full repertoire of PBSt and RF activity patterns observed in experimental studies but also explain their variable behavior during deletions (see Table 2.1). This specifically concerns the PBSt with a typical flexor-type activity profile that nevertheless during resetting extensor deletions can demonstrate tonic, silent, or rhythmic activity (Fig. 5.5a–c). According to the two-level concept of the CPG organization, resetting extensor deletions can be produced by either spontaneous increase of excitation of the flexor half-center (RG-F) or decrease in excitation of the extensor half-center (RG-E) (Rybak et al. 2006a). In the model, such deletion can be produced by a temporal change in the MLR drive to the corresponding RG population. During resetting extensor deletion, all major extensor-related populations become silent while flexor-related populations demonstrate sustained activity (see Table 5.2). Figure 5.11 demonstrates our simulations of variable behavior of the flexor-type PBSt during resetting extensor deletions shown in Fig. 5.5a–c. In order to have a flexor-type activity profile before and after deletions, the PF-PBSt population should be strongly inhibited during extension by the In-E population (see Fig. 5.9b1, c1). Besides, PF-PBSt activity during the flexor phase is controlled by inhibition from the In-lF population. The length of the PBSt flexor bursts before and after deletions depends on relative activation of the In-eF and In-lF populations, inhibiting each other, i.e. on the interplay between the excitatory tonic drives to these populations ($d_{eF}$ and $d_{lF}$, respectively). During resetting extensor deletions, the PF-PBSt population receives a sustained excitation from the RG-F population shown by large unfilled arrows in Fig. 5.11a1–a3. The activity of In-E population is weak because it loses an excitatory input from the PF-E population and becomes tonically inhibited by the Inrg-F population (see Table 5.2). Both In-eF and In-lF populations lose inhibition from the Inrg-E population that becomes silent during resetting extensor deletions (see Table 5.2) and their behavior depends on the tonic drives to these populations ($d_{eF}$ and $d_{lF}$, respectively) and mutual inhibition between

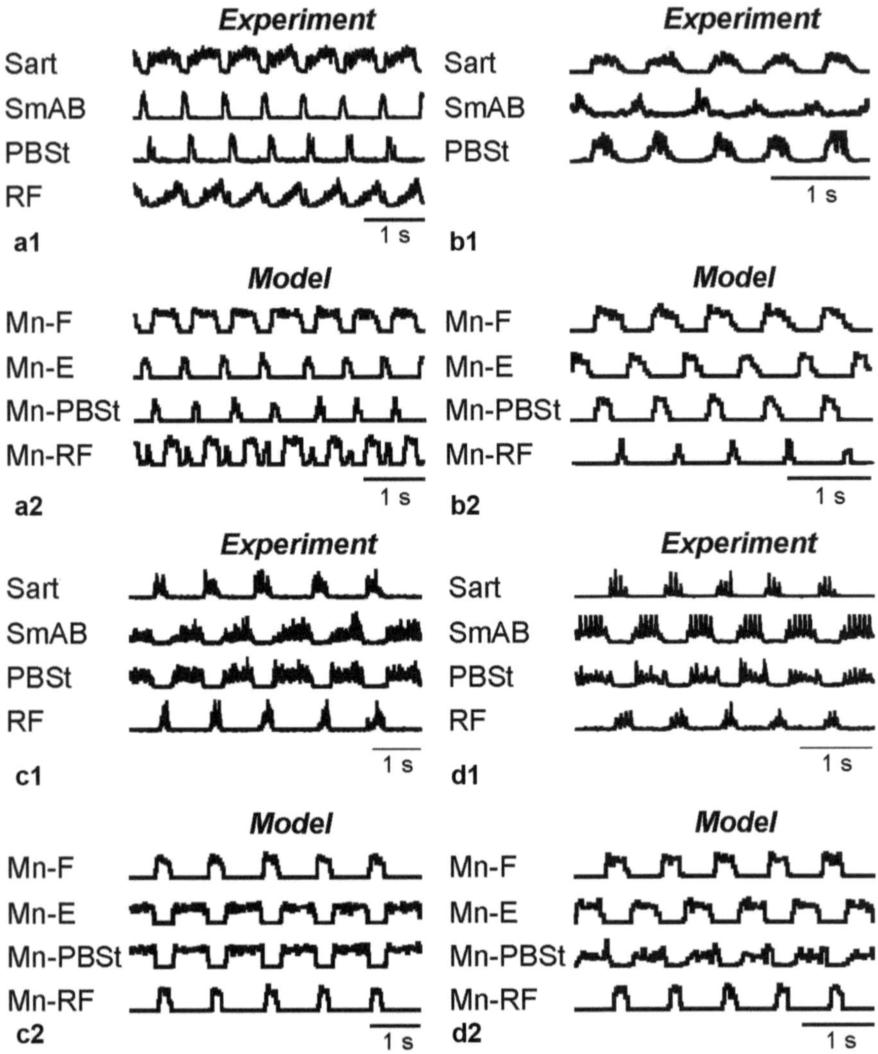


**Fig. 5.10** Simulation of different types of the PBSt and RF ENG activities recorded during fictive locomotion experiments in the extended CPG model. Panels **a1**, **b1**, **c1**, and **d1** show recordings of the flexor (Sart), extensor (SmAB), PBSt, and RF ENGs. Panels **a2**, **b2**, **c2**, and **d2** show simulated activities of the Mn-F, Mn-E, Mn-PBSt, and Mn-RF populations as histograms of average neuron activity in a particular population (spikes per neuron per second, bin = 30 ms). **a1, a2** Flexor-type PBSt (type 1) and biphasic RF (type 2) motoneuron activity patterns. **b1, b2** Flexor-type PBSt and RF motoneuron activity patterns (both type 1, RF is shown in **b2** only). **c1, c2** Extensor-type PBSt (type 2) and flexor-type RF (type 1) motoneuron activity patterns. **d1, d2** Biphasic PBSt (type 3) and flexor-type RF (type 1) motoneuron activity patterns. For values of input drives to the RG-E, RG-E, and hypothetical interneuron populations see Table 5.A3 in Appendix

them. Mini-circuitries in Fig. 5.11a1–a3 schematically show how the interplay between $d_{eF}$ and $d_{lF}$ affects the PF-PFSt and Mn-PBSt behavior during resetting extensor deletions. Figure 5.11b1–b3 shows the corresponding simulations.

In Fig. 5.11b1, the flexor-type PBSt motoneuron population is active during most of the flexor phase (type 1b) before and after the deletion and demonstrates sustained activity during the deletion, similar to the experimental recordings shown in Fig. 5.5a. To demonstrate a prolonged flexor burst, the PF-PBSt population should be inhibited by the In-lF population at the very end of the flexor phase. This can be achieved if the In-lF population receives a weaker excitatory input drive $d_{lF}$ compared to the drive $d_{eF}$ to the In-eF population (see Fig. 5.11a1) that in this case is active during most of the flexor phase. During the deletion the In-eF population becomes tonically active and inhibits the In-lF population allowing tonic activity of the PF-PBSt population (and hence Mn-PBSt) as can be seen in Fig. 5.11b1. Alternatively, if $d_{lF}$ is strong compared to $d_{eF}$ (see Fig. 5.11a2), which is expected for the short flexor-type PBSt activity profile (type 1a), the In-lF population inhibits the In-eF population during most of the flexor phase before and after the deletion. During the deletion, this population loses inhibitory input from the Inrg-E population, becomes tonically active, and inhibits the PF-PBSt population, resulting in the silent Mn-PBSt population (Fig. 5.11b2). This behavior is similar to the experimental recordings shown in Fig. 5.5b. Finally, in some situations the mutual inhibition between the In-eF and In-lF populations can produce rhythmic activity with a frequency independent of the locomotor frequency before and after deletion (see the corresponding traces in Fig. 5.11a3). In this case, the resulting activity of the PF-PBSt and Mn-PBSt populations becomes rhythmic as seen in Fig. 5.11b3. This can explain rhythmic PBSt activity in experimental records shown in Fig. 5.5c.

Figure 5.12 shows experimental recordings (panels a1, b1, c1, and d1) and our simulations (panels a2, b2, c2, and d2) of PBSt with typical extensor-type (type 2) and biphasic (type 3) activity profiles during resetting flexor and extensor deletions. In simulations, the traces represent histograms of average neuron activity in the motoneuron populations (Mn-F, Mn-E, Mn-PBSt, and Mn-RF). Shaded rectangles highlight the flexor, extensor, PBSt, and RF motoneuron population behavior during deletions.

In Fig. 5.12a1, before and after the flexor deletion, PBSt exhibits the extensor type activity profile (type 2) and RF shows the flexor type of activity (type 1). During the deletion, PBSt demonstrates sustained tonic activity similar to the extensor SmAB while RF becomes silent as the flexors (Sart and TA). Such PBSt and RF behavior was reproduced in our simulation shown in Fig. 5.13a2. In Fig. 5.12b1 before and after deletion, PBSt demonstrates an extensor-type activity profile (type 2). During the resetting extensor deletion, it becomes silent similar to extensors, which is reproduced in our simulation shown in Fig. 5.12b2. In Fig. 5.12c1, d2, before and after deletions, PBSt has a biphasic activity profile with a short flexor burst (type 3a) and becomes tonically active during the flexor deletion (Fig. 5.12c1) or silent during the extensor deletion (Fig. 5.12d1) (see also Table 5.1). Such behavior of the biphasic PBSt was reproduced in our simulations during flexor (Fig. 5.12c2) and extensor (Fig. 5.12d2) deletions.

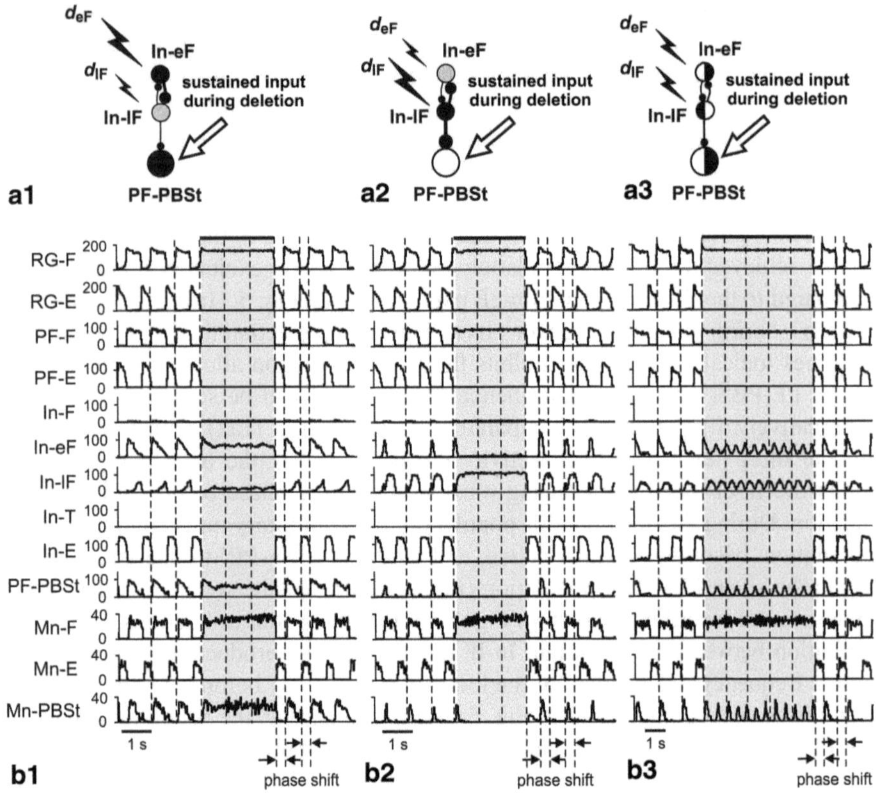


**Fig. 5.11** Modeling of flexor-type PBSt behavior during resetting extensor deletions. **a1–a3** Circuits shaping the PF-PBSt population activity during deletions for: PBSt with a long flexor burst before and after the deletion (**a1**), a short flexor burst before and after the deletion (**a2**), and a flexor burst of about half of the flexor phase before and after the deletion (**a3**). Bolts of different sizes schematically show excitatory input drives of different strengths received by the In-eF and In-lF inhibitory populations, the larger the bolt size the stronger the input drive. Inhibition provided by a population is schematically represented by thickness of the corresponding output connection, the greater the strength of inhibition, the thicker the connection. Unfilled arrows illustrate sustained activation of the PF-PBSt populations during the deletions by the RG-F population (see Table 5.2). *Black circles* represent populations highly activated during deletions; *gray circles* illustrate a low level of activation in populations; unfilled circle indicates the inactive population; *half-filled circles* indicate phasic activity of populations. **b1–b3** Results of simulation of PBSt behavior during deletions shown in Fig. 5.5a–c. Simulated deletions were produced by a temporal decrease (by 50 %) of the MLR tonic excitatory drive to the RG-E population indicated by horizontal black bars at the top of traces. Each trace represents a histogram of average neuron activity in a particular population (spikes per neuron per second, bin = 30 ms). Two upper traces show activities of the RG-F and RG-E populations followed by activities of the PF-F and PF-E populations (third and fourth traces). Next five traces show activities of the hypothetical interneuron populations shaping the PF-PBSt population activity (see Figs. 5.6–5.9). The fourth trace from the bottom shows activity of the PF-PBSt population. Three lower traces show activities of the Mn-F, Mn-E, and Mn-PBSt populations, respectively. *Shaded rectangles* highlight behaviors of neuron populations during the deletions. An obvious phase shift of the post-deletion rhythm with respect to the pre-deletion rhythm (see arrows at the bottom of traces) indicates that the deletions are resetting

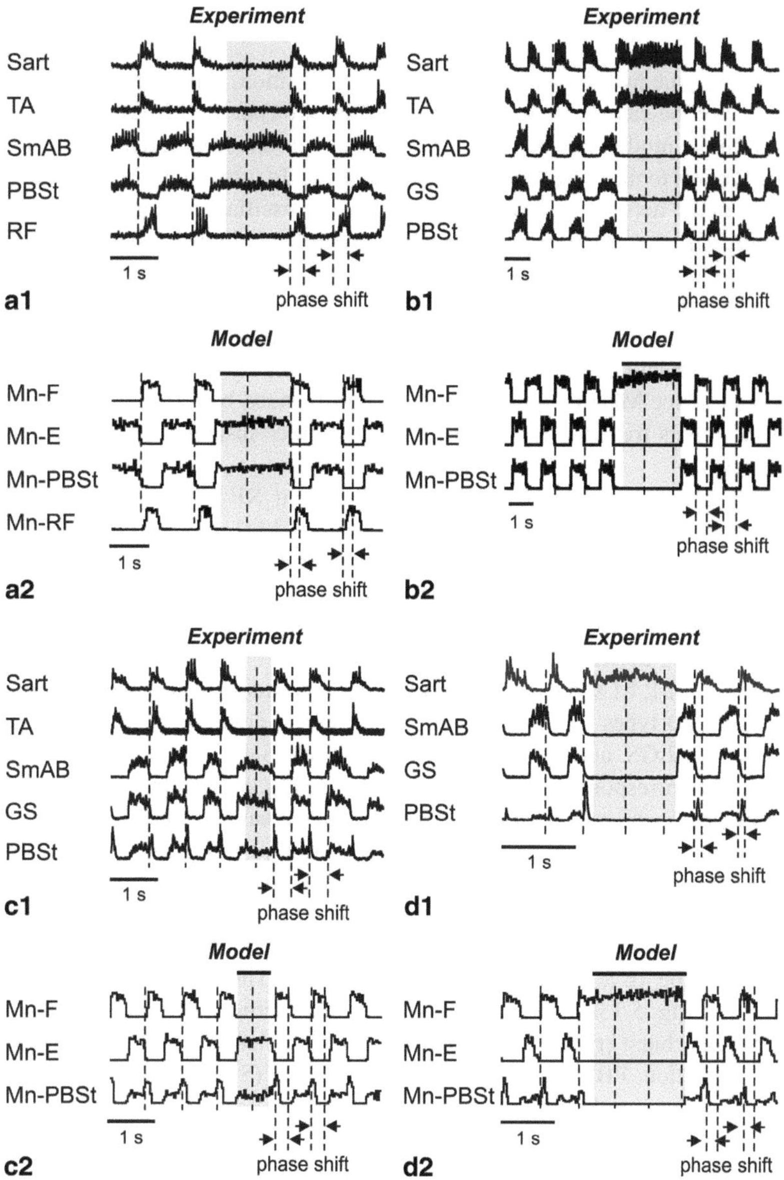


**Fig. 5.12** Simulation of the extensor-type (type 2) and biphasic (type 3) PBSt and flexor-type (type 1) RF behavior during resetting deletions (see an obvious phase shift of the post-deletion rhythm with respect to the pre-deletion rhythm). Shaded rectangles highlight flexor, extensor, PBSt and RF behaviors during deletions. **a1, b1, c1, d1** Experimental recordings demonstrating the PBSt and RF (in **a1**) behavior during the extensor (**a1, c1**) and flexor (**b1, d1**) deletions. **a2, b2, c2, d2** Simulation of experimental results shown in **a1, b1, c1, d1.** The traces represent histograms of average neuronal activity in the Mn-F, Mn-E, Mn-PBSt, and Mn-RF (in **a2**) populations. Horizontal bars above the traces indicate temporal increase of the MLR drive to the RG- or RG-F populations (by 20 % in **a2**, **b2**, **c2** and by 10 % in **d2**). For values of input drives to the RG-E, RG-E, and hypothetical interneuron populations see Table 5.A3 in Appendix.

#### 5.4.2.2 Non-Resetting Deletions

According to the two-level locomotor CPG organization, non-resetting extensor deletions could be caused by spontaneous perturbations at the PF level (Rybak et al. 2006a). In the model, this deletion type can be reproduced by a temporal change in the MLR drive to the corresponding PF population. In turn, the RG populations continue rhythmic activity and maintain the phase of oscillations during non-resetting deletions. The perturbation producing deletion affects the principal populations at the PF level, so that either PF-F or PF-E becomes silent, while the opposite principal population (PF-E or PF-F) switches to sustained activity or remains rhythmic (PF-F) (see Table 5.2). Thus, during non-resetting deletions, behaviors of the interneuron populations shaping activity profiles of the PF-PBSt and PF-RF populations (and hence Mn-PBSt and Mn-RF) are influenced by two factors: continuing rhythmic inputs from the RG populations and either sustained or lost inputs from the PF level.

Figure 5.13 shows our modelling of PBSt and RF typical behaviors during non-resetting extensor deletions with tonic (Fig. 5.13a1, a2, b1, b2) and rhythmic (Fig. 5.13c1, c2, d1, d2) activity in flexors. Upper panels (a1, b1, c1, and d1) show the experimental recordings and the lower panels (a2, b2, c2, and d2) demonstrate our simulation results. Shaded rectangles highlight flexor, extensor, PBSt and RF behavior during deletions.

Figure 5.13a1, b1 shows the behavior of flexor- and biphasic (with a short flexor burst, type 3a) types of PBSt during non-resetting deletions of extensor activity (SmAB, MG, LGS, and LG) when flexors (TA and Sart) switched to sustained activity. In the corresponding simulations shown in Fig. 5.13a2, b2, the non-resetting extensor deletions were produced by temporal increase of the MLR drive to the PF-F population (Rybak et al. 2006a). In Fig. 5.13a1, PBSt is active during most of the flexor phase (type 1b) before and after the deletion and demonstrates activity in both flexor and extensor expected phases during deletion. This resembles PBSt behavior during the non-resetting deletion shown in Fig. 5.5d1 and is typical for PBSt having the type 1b activity pattern (see Table 5.1). The corresponding simulation is shown in Fig. 5.13a2. In Fig. 5.13b1, the PBSt activity pattern is biphasic with a short flexor burst (type 3a) before and after the deletion. During a non-resetting extensor deletion, PBSt becomes silent as extensors (SmAB and GS) which is reproduced in our simulation shown in Fig. 5.13b1.

Figure 5.13c1, c2, d1, d2 shows two examples (panels c1 and d1) of non-resetting deletions of extensor activity (SmAB, MG, LGS, and LG) while flexors (TA and Sart) remain rhythmically active and the corresponding simulations (panels c2 and d2). In simulations, the non-resetting deletions of extensor activity were produced by temporal suppression of the PF-E population activity by application of an additional inhibitory drive to this population, while the PF-F population continued receiving the rhythmic input from the RG-F population and transferring it to the Mn-F population (Rybak et al. 2006a). In Fig. 5.13c1, PBSt is active in the beginning of the flexor phase (type 1) before and after the deletion and continues rhythmic activity during extensor deletion together with the flexors (Sart and TA). This is

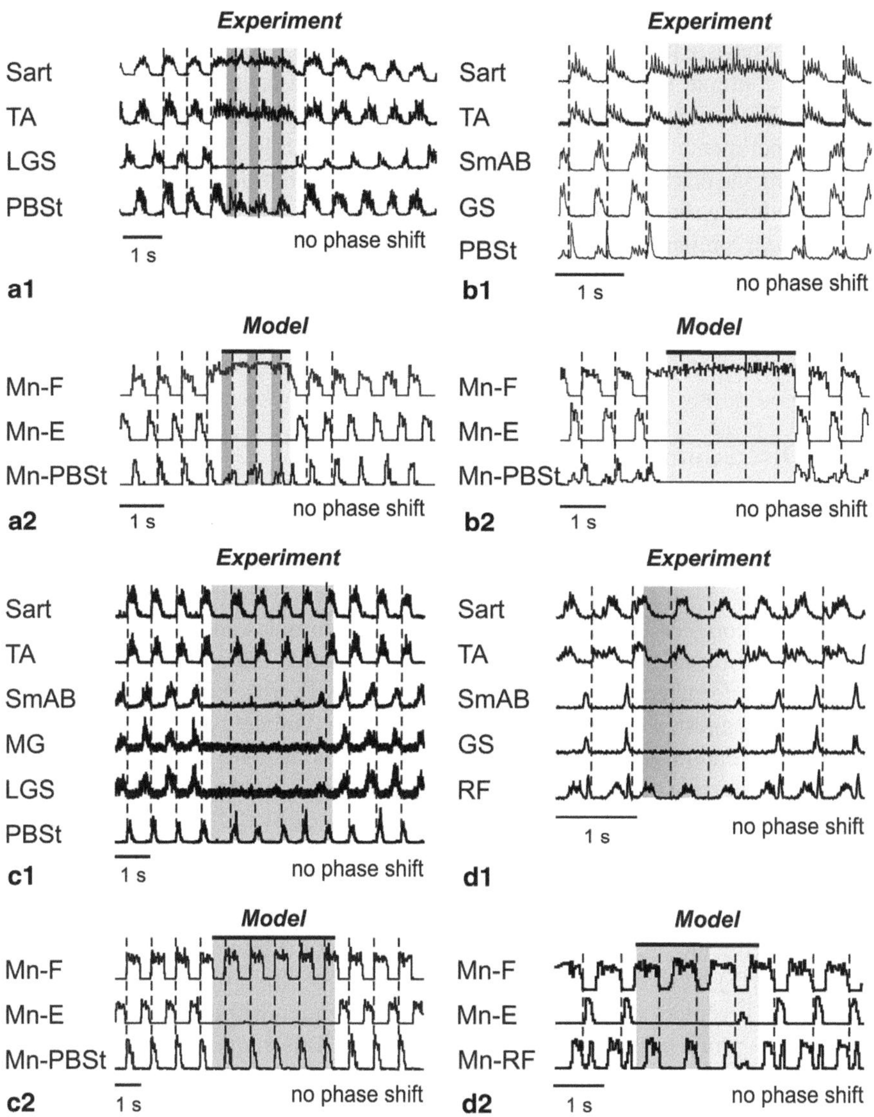


**Fig. 5.13** Modeling of PBSt with flexor (type 1) and biphasic (type 2) behavior during non-resetting extensor deletions with tonically active or rhythmic flexors. Shaded rectangles highlight flexor, extensor, PBSt and RF activity during deletions. **a1** and **b1** Experimental recordings demonstrating flexor-type PBSt activity during non-resetting deletions with tonically active flexors (Sart and TA). **a2, b2** Simulation of experimental results shown in **a1** and **b1**. *Horizontal bars* in **a2** and **b2** above the traces indicate temporal increase of the MLR drive to the PF-F population by 30 % (both in **a2** and **b2**). **c1, d1:** Experimental recordings demonstrating biphasic PBSt (**c1**, modified from Lafreniere-Roula and McCrea 2005) or RF (**d1**) behavior during non-resetting extensor deletions with rhythmically active flexors (Sart and TA). **a2, b2** Simulation of experimental results shown in **c1** and **d1**. Horizontal bars above the traces indicate temporal suppression of the PF-E population activity by an inhibitory drive ($d_{inh}$=0.7 in both **a2** and **b2**). In **b2**, changing level of shading schematically shows reduction of the additional inhibitory drive to the PF-E population by 70 %. See text for details. The traces in **a2, b2, c2,** and **d2** represent histograms of average neuronal activity in the Mn-F, Mn-E and Mn-PBSt or Mn-RF populations. For values of input drives to the RG-E, RG-E, and hypothetical interneuron populations see Table 5.A3 in Appendix

reproduced in our simulation shown in Fig. 5.13c2. Figure 5.13d1 shows the behavior of the biphasic RF during the non-resetting extensor deletion while flexors (Sart and TA) continue rhythmic activity. Interestingly, similar to the behavior of the biphasic RF during non-resetting deletions with tonic flexor activity (Fig. 5.5d1), in the experimental recording shown in Fig. 5.13d1, RF loses its extensor component during the deletion. The corresponding simulation reproducing the biphasic RF behavior during the extensor deletion with rhythmically active flexors is shown in Fig. 5.13d2. Similar to the experimental records, the Mn-RF population during the deletion maintains rhythmic activity but loses the extensor component of its activity profile. Our simulation shows a striking similarity to the experimental recordings including the low-amplitude extensor and RF bursts by the end of the deletion.

## 5.5 Discussion

The bipartite half-center organization of the locomotor CPG originally proposed by T. Graham Brown (1914) and expanded by Lundberg, Jankowska and their colleagues (Jankowska 1967a, b; Lundberg 1981; Stuart and Hultborn 2008) could not explain the complex activity patterns of two-joint muscles and the corresponding motoneurons recorded during real or fictive locomotion. This motivated researchers to search for alternative concepts for CPG organization. For example, Grillner (1981) has suggested a Unit Burst Generators (UBGs) model consisting of multiple coupled oscillators each of which operates at each joint. The UBG architecture is much more flexible and functionally richer than the original half-center CPG architecture. However, no UBG-based CPG computational model has been developed so far that could reproduce the complex patterns of bifunctional motoneurons. In contrast, Perret and colleagues (Perret and Cabelguen 1980; Perret 1983; Orsal et al. 1986; Perret et al. 1988) have proposed a solution for this problem in the framework of the bipartite locomotor CPG. They suggested that the bifunctional motoneurons (PBSt and RF) receive excitatory and inhibitory inputs from both flexor and extensor CPG half-centers and that special circuit organization within the CPG can shape the activity profiles of bifunctional motoneurons. This suggestion was used as a basis for our model development.

Our approach was based on a recently proposed concept of a two-level organization of the spinal locomotor CPG (Rybak et al. 2006a, b; McCrea and Rybak 2007, 2008). According to this concept, a bipartite half-center RG controls operation of PF networks that in turn coordinate rhythmic excitation and inhibition of multiple synergist motoneuron pools during locomotion. The proposed two-level CPG architecture has a number of advantages including the possibility for independent control of locomotor rhythm and phase durations at the RG level and control of patterns of motoneuron activations at the PF level. The two-level CPG organization proposed explanations for many experimental findings observed during fictive locomotion, including the specific alteration of activity of flexor and extensor motoneurons during afferent stimulations (Rybak et al. 2006b; McCrea and Rybak 2007; Shevtsova

2015) and their behavior during spontaneous deletions of motoneuron activity occurring with and without resetting of the locomotor cycle (Rybak et al. 2006a; McCrea and Rybak 2008; Shevtsova 2015). This two-level organization had been implemented in the reduced computational model of the CPG (Fig. 5.3) that could generate a realistic locomotor rhythm with alternating activity profiles of flexor and extensor motoneuron pools and reproduce various changes of these profiles under different conditions, such as afferent stimulations and deletions (Rybak et al. 2006a, b; Shevtsova 2015). However, this reduced model considered only two antagonist motoneuron populations (flexors and extensors) and did not include bifunctional motoneuron pools which are known to express more complicated firing patterns during both fictive and real locomotion. The major objective of this study was to extend the basic two-level CPG model by suggesting a special organization of the PF circuits, while keeping the bipartite organization of the RG. To this end, the proposed extended model reproduces the full repertoire of PBSt and RF motoneuron activity patterns recorded during fictive locomotion as well as all changes of these patterns observed during various deletions.

### *5.5.1 The Extended CPG Model Circuitry*

The idea that excitatory and inhibitory inputs from both extensor and flexor parts of the bipartite locomotor rhythm generator provide shaping of PBSt and RF motoneuron activity patterns suggested by Perret and colleagues (Perret and Cabelguen 1980; Perret 1983; Orsal et al. 1986; Perret et al. 1988) has not been implemented previously in a computational model. Moreover, the Perret schematic, through incorporating some additional neuronal networks shaping PBSt and RF motoneuron activities, can provide explanation for only the biphasic PBSt and RF patterns and cannot explain the variety of PBSt and RF activity profiles observed during fictive locomotion and the specific behaviors of PBSt and RF motoneuron populations during spontaneous deletions. Perret and colleagues (Perret et al. 1988) believed that the variability of PBSt and RF activity patterns is a result of interactions between the central locomotor drive and afferent influences. However, the analysis of fictive locomotion experiments in decerebrate cats (Markin et al. 2012) has shown that the full repertoire of non-trivial activity profiles of PBSt and RF can be generated in the absence of sensory feedback and hence may represent an inherent property of the locomotor CPG. By analyzing the PBSt and RF activity patterns we proposed a hypothetical network of interneuron populations incorporated in the pattern formation (PF) level of the two-level spinal CPG that can provide for the full repertoire of the PBSt and RF motoneuron activity profiles and explain their behaviors during deletions.

The key elements of the proposed network are (1) the principal neuron populations (PF-PBSt and PF-RF, see Fig. 5.8) that directly control activation of the PBSt and RF motoneuron pools, and (2) additional interneuron populations (In-E, In-T, In-F, In-lF, In-eF, In-eE, and, In-lE, see Fig. 5.8) that shape the activity profiles of the PF-PBSt and PF-RF populations during the flexor and extensor phases.

In the model, the activation of these interneuron populations (or the drives they receive from MLR) explicitly defines the activity profiles expressed during particular fictive locomotion experiment (from the repertoire of possible profiles). We suggest that these are the populations that during real locomotion receive excitatory input from sensory afferents and/or descending signals which, therefore, can select and shape the appropriate activity profiles of PBSt and RF motoneuron pools to adjust their activity to gait and speed of locomotion, and particular locomotor conditions.

It appears that activity of PB and RF (and other two-joint muscles) is tightly regulated by descending and afferent signals, so that the phasing and magnitude of their activity corresponds precisely to the mechanical demands. During skilled locomotor behaviors (walking, running, cycling (Wells and Evans 1987; Prilutsky et al. 1998a; Prilutsky and Gregor 2000) and other skilled non-rhythmic behaviors (load lifting or isometric exertion of external forces by the leg (Wells and Evans 1987; Prilutsky et al. 1998b), two-joint muscles typically exhibit their highest activity when they can simultaneously contribute to the extensor or flexor actions of all synergists at both joints the two-joint muscle crosses; the two-joint muscles are typically inhibited, when their anatomical antagonists at both joints produce the actions opposite to the ones of the two-joint muscle (Prilutsky 2000). During overground normal level and slope walking different bursts of PB and RF can be associated with actions of the corresponding flexor and extensor muscles at the knee and hip joints. For example, electromyogram (EMG) bursts of the PB muscle (knee flexor and hip extensor) are closely associated with the knee flexor-hip extensor combination of the resultant muscle moments in early swing and late swing-early stance phases of level and slope walking (Carlson-Kuhta et al. 1998; Smith et al. 1998a; Gregor et al. 2006). Similarly, the stance related EMG burst of the RF muscle (knee extensor-hip flexor) can be associated with the knee extensor-hip flexor combination of the resultant muscle moments (Carlson-Kuhta et al. 1998; Smith et al. 1998a; Gregor et al. 2006). Such organization of neural control of two-joint muscles appears to minimize total muscle stress, effort, fatigue and other related physiological variables as demonstrated by similarity between recorded EMG patterns of two-joint muscles and the muscle activity or muscle force patterns computed by minimizing the above cost functions in a musculoskeletal model during the same motor task (Anderson and Pandy 2001; Prilutsky and Zatsiorsky 2002).

## 5.6 Conclusion

Construction of a CPG model that would be able to realistically reproduce the behavior of the PBSt and RF during fictive locomotion required a comprehensive analysis of firing profiles of these motoneuron pools during fictive locomotion (Markin et al. 2012) and their behaviors during spontaneous deletions. This analysis allowed us to construct an extended model of the locomotor CPG by incorporating additional neuronal network to shape PBSt and RF activity profiles. To our knowledge this is the first computational model reproducing behavior of PBSt and RF during locomotion.

We have demonstrated that a two-level locomotor CPG with the bipartite half-center rhythm generator and special organization of neural circuits at the pattern formation level can generate complex activity patterns of bifunctional motoneurons, such as PBSt and RF. The proposed CPG model reproduces the full repertoire of PBSt and RF activity patterns recorded during fictive locomotion. The model also suggests explanations for the observed variability of PBSt and RF activity and may predict their behavior under different conditions. We hypothesize that sensory inputs to selected interneuron populations within the pattern formation level of the CPG provide a mechanism for the proprioceptive control of the activity of bifunctional motoneurons during real locomotion.

**Acknowledgements** This work was supported by the NIH grants R01 NS048844, R01 NS081713, R01 NS090919, R01 NS095366 and R01 IB012855. The authors would like to thank Dr. David McCrea for useful comments and discussions. We would also like to thank Drs. Angel, Gosgnach, Guertin, Jordan, Lafreniere-Roula, McCrea, Perreault, Stecina, and Quevedo for collection of experimental data used in this modeling study.

## Appendix

All neurons were modelled in the Hodgkin-Huxley style. Motoneurons had two compartments: soma and dendrite and were described based on the previous models (Booth et al. 1997; Rybak et al. 2006a). The membrane potentials of motoneuron soma ($V_{(S)}$) and dendrite ($V_{(D)}$) obey the following differential equations:

$$C \times \frac{dV_{(S)}}{dt} = -I_{Na(S)} - I_{K(S)} - I_{A(S)} - I_{CaN(S)} - I_{K,Ca(S)} - I_{L(S)} - I_{C(S)}$$

$$C \times \frac{dV_{(D)}}{dt} = -I_{NaP(D)} - I_{CaN(D)} - I_{CaL(D)} - I_{K,Ca(D)} - I_{L(D)} - I_{C(D)} - I_{SynE} - I_{SynI} \tag{5.A1}$$

where $C$ is the membrane capacitance and $t$ is time ($C=1\ \mu\text{F/cm}^2$), subscripts $S$ and $D$ indicate the soma or dendrite compartments, respectively.

The dendrite-soma coupling currents (with conductance $g_C$) for soma $I_{C(S)}$ and dendrite $I_{C(D)}$ are described as follows:

$$I_{C(S)} = \frac{g_C}{p} \times (V_{(S)} - V_{(D)});$$

$$I_{C(D)} = \frac{g_C}{1-p} \times (V_{(D)} - V_{(S)}), \tag{5.A2}$$

where $p$ is the parameter defining the ratio of somatic surface area to total surface area ($p=0.1$); $g_C=0.1$ mS/cm$^2$.

The following currents (with the corresponding maximal channel conductances) are included into motoneuron soma compartments (Booth et al. 1997; Rybak et al. 2006a): fast sodium, $I_{Na}$ (maximal conductance $\bar{g}_{Na}=120$ mS/cm$^2$); persistent sodium, $I_K$ ($=\bar{g}_{NaP}$ 100 mS/cm$^2$); calcium-N, $I_{CaN}$ ($\bar{g}_{CaN}=14$ mS/cm$^2$); calcium-dependent potassium, $I_{K,\,Ca}$ ($\bar{g}_{CaL}=2$ mS/cm$^2$), and leakage, $I_L$ ($g_L=0.51$ mS/cm$^2$) currents. In addition, based on evidence of the presence of the transient (rapidly inactivating) potassium current in the spinal cord interneurons and motoneurons (Safronov and Vogel 1995) this current has been also included in our motoneuron models ($I_A$ with maximal conductance $\bar{g}_A=200\pm40$ mS/cm$^2$). The following currents (with the corresponding maximal channel conductances) are included into motoneuron dendritic compartment: persistent sodium, $I_{NaP}$ ($\bar{g}_{NaP}=0.1$ mS/cm$^2$); calcium-N, $I_{CaN}$ ($\bar{g}_{CaN}=0.3$ mS/cm$^2$); calcium-L ($I_{CaL}$, $\bar{g}_{CaL}=0.33$ mS/cm$^2$), calcium-dependent potassium, $I_{K,\,Ca}$ ($\bar{g}_{K,Ca}=0.8$ mS/cm$^2$), and leakage, $I_L$ ($g_L=0.51$ mS/cm$^2$) currents.

Interneurons are simulated as single-compartment models. The neurons within the RG-F, RG-E, PF-F, PF-E, In-eF, and In-eE populations contain fast sodium, $I_{Na}$; persistent sodium, $I_{NaP}$; delayed-rectifier potassium, $I_K$; and leakage, $I_L$ currents:

$$C\times\frac{dV}{dt}=-I_{Na}-I_K-I_L-I_{SynE}-I_{SynI} \tag{5.A3}$$

The maximal channel conductances for neurons in theses populations are as follows: $g_L=0.51$ mS/cm$^2$; $\bar{g}_{Na}=150$ mS/cm$^2$ in RG-F and RG-E neurons and 120 mS/cm$^2$ in the PF-F, PF-E, In-eF, and In-eE populations; $\bar{g}_{NaP}=1.25$ mS/cm$^2$ in RG-F, RG-E, In-eF, and In-eF neurons and 0.1 mS/cm$^2$ in the PF-F and PF-E populations; $\bar{g}_K=5$ mS/cm$^2$ in the RG-F and RG-E populations and 10 mS/cm$^2$ in the PF-F, PF-E, In-eE, and In-eF populations.

For simplicity, all other interneurons contain only minimal set of ionic currents:

$$C\times\frac{dV}{dt}=-I_{Na}-I_K-I_L-I_{SynE}-I_{SynI} \tag{5.A4}$$

with the following maximal conductances: $\bar{g}_{Na}=120$ mS/cm$^2$; $\bar{g}_K=10$ mS/cm$^2$; $g_L=0.51$ mS/cm$^2$.

The ionic currents included into the modelled neurons are described as follows:

$$\begin{aligned}
I_{Na} &= \bar{g}_{Na}\times m_{Na}^3\times h_{Na}\times(V-E_{Na});\\
I_{NaP} &= \bar{g}_{NaP}\times m_{NaP}\times h_{NaP}\times(V-E_{Na});\\
I_K &= \bar{g}_K\times m_K^4\times(V-E_K);\\
I_A &= \bar{g}_A\times(0.6\times m_{A1}^4\times h_{A1}+0.4\times m_{A2}^4\times h_{A2})\times(V-E_K);\\
I_{CaN} &= \bar{g}_{CaN}\times m_{CaN}^2\times h_{CaN}\times(V-E_{Ca});\\
I_{CaL} &= \bar{g}_{CaL}\times m_{CaL}\times(V-E_{Ca});\\
I_{K,Ca} &= \bar{g}_{K,Ca}\times m_{K,Ca}\times(V-E_K);\\
I_L &= g_L\times(V-E_L),
\end{aligned} \tag{5.A5}$$

where $V$ is the membrane potential of the corresponding neuron compartment (soma, $V_{(S)}$, or dendrite, $V_{(D)}$) in two-compartment models, or the neuron membrane potential $V$ in one-compartment models; $E_{Na}$, $E_K$, $E_{Ca}$, and $E_L$ are the reversal potentials for sodium, potassium, calcium, and leakage current respectively; variables $m$ and $h$ with indexes indicating ionic currents represent, respectively, the activation and inactivation variables of the corresponding ionic channels.

The reversal potential values in the model are as follows: $E_{Na}=55$ mV; $E_K=-80$ mV; $E_{Ca}=80$ mV; $E_L=-64\pm0.64$ mV in RG-F and RG-E neurons, $E_L=-65\pm0.325$ mV in Inrg-F and Inrg-E interneurons and motoneurons, and $E_L=-68\pm0.34$ mV in all other neurons.

Activation $m$ and inactivation $h$ of voltage-dependent ionic channels (e.g., Na, NaP, K, A, CaN, CaL) are described by the following differential equations:

$$\tau_{mi}(V)\times\frac{d}{dt}m_i = m_{\infty i}(V)-m_i;$$
$$\tau_{hi}(V)\times\frac{d}{dt}h_i = h_{\infty i}(V)-h, \quad (5.A6)$$

where $i$ identifies the name of the channel, $m_{\infty i}(V)$ and $h_{\infty i}(V)$ represent the voltage-dependent steady-state activation and inactivation respectively, and $\tau_{mi}(V)$ and $\tau_{hi}(V)$ define the corresponding time constants (see their descriptions in Table 5.A1). Activation of the sodium channels is considered to be instantaneous ($\tau_{mNa}=\tau_{mNaP}=0$, see (Booth et al. 1997; Butera et al. 1999)).

Activation of the $Ca^{2+}$ -dependent potassium channels is also considered instantaneous and described as follows (Booth et al. 1997):

$$m_{K,Ca} = \frac{Ca}{Ca+K_d}, \quad (5.A7)$$

where $Ca$ is the $Ca^{2+}$ concentration within the corresponding compartment of motoneuron, and $K_d$ defines the half-saturation level of this conductance.

The kinetics of intracellular $Ca^{2+}$ concentration ($Ca$, described separately for each compartment) is modelled according to the following equation (Booth et al. 1997):

$$\frac{d}{dt}Ca = f\times(-\alpha\times I_{Ca}-k_{Ca}\times Ca), \quad (5.A8)$$

where $f$ defines the percent of free to total $Ca^{2+}$; $\alpha$ converts the total $Ca^{2+}$ current, $I_{Ca}$, to $Ca^{2+}$ concentration; $k_{Ca}$ represents the $Ca^{2+}$ removal rate.

The synaptic excitatory ($I_{synE}$ with conductance $g_{synE}$ and reversal potential $E_{SynE}=-10$ mV) and inhibitory ($I_{synI}$ with conductance $g_{synI}$ and reversal potential $E_{SynI}=-70$ mV) currents are described as follows:

$$I_{SynE} = g_{SynE}\times(V-E_{SynE});$$
$$I_{SynI} = g_{SynI}\times(V-E_{SynI}). \quad (5.A9)$$

**Table 5.A1** Steady state activation and inactivation variables and time constants for voltage-dependent ionic channels

| Ionic channels | $m_\infty(V)$, $V$ is in mV<br>$h_\infty(V)$, $V$ is in mV | $\tau_m(V)$, ms<br>$\tau_h(V)$, ms |
|---|---|---|
| Na | $m_{\infty Na} = (1+\exp(-(V+35)/7.8))^{-1}$ | $\tau_{mNa} = 0$ |
| | $h_{\infty Na} = (1+\exp((V+55)/7))^{-1}$ | $\tau_{hNa} = 30/(\exp((V+50)/15)+\exp(-(V+50)/16))$ |
| NaP | $m_{\infty NaP} = (1+\exp(-(V+47.1)/3.1))^{-1}$ | $\tau_{mNaP} = 0$ |
| | $h_{\infty NaP} = (1+\exp((V+59)/8))^{-1}$ | $\tau_{hNaP} = 800/\cosh((V+59)/16)$ |
| K | $m_{\infty K} = (1+\exp(-(V+28)/15))^{-1}$ | $\tau_{mK} = 7/(\exp((V+40)/40)+\exp(-(V+40)/50))$ |
| | $h_K = 1$ | N/A |
| A | $m_{\infty A1} = (1+\exp(-(V+60)/8.5))^{-1}$ | $\tau_{mA1} = 1/(\exp((V+35.82)/19.69)+\exp(-(V+79.69)/12.7)+0.37)$ |
| | $h_{\infty A1} = (1+\exp((V+78)/6))^{-1}$ | $\tau_{hA1} = 1/(1+\exp((V+46.05)/5)+\exp(-(V+238.4)/37.45))$<br>If $V<-63$, otherwise $\tau_{hA1}=19.0$ |
| | $m_{\infty A2} = (1+\exp(-(V+36)/20))^{-1}$ | $\tau_{mA2} = 1/(\exp((V+35.82)/19.69)+\exp(-(V+79.69)/12.7)+0.37)$ |
| | $h_{\infty A2} = (1+\exp((V+78)/6))^{-1}$ | $\tau_{hA2} = 1/(1+\exp((V+46.05)/5)+\exp(-(V+238.4)/37.45))$<br>If $V<-73$, otherwise $\tau_{hA2}=60.0$ |
| CaN | $m_{\infty CaN} = (1+\exp(-(V+30)/5))^{-1}$ | $\tau_{mCaN} = 4$ |
| | $h_{\infty CaN} = (1+\exp((V+45)/5))^{-1}$ | $\tau_{hCaN} = 40$ |
| CaL | $m_{\infty CaL} = (1+\exp(-(V+40)/7))^{-1}$ | $\tau_{mCaL} = 40$ |
| | $h_{CaL} = 1$ | N/A |

All expressions and parameters, except those for the NaP and potassium A channels, are taken from (Booth et al. 1997). The expressions for the NaP channel are from (Rybak et al. 2003). The expressions for potassium A channel are from (Huguenard and McCormick 1991; Huguenard and McCormick 1992).

The excitatory ($g_{SynE}$) and inhibitory synaptic ($g_{SynI}$) conductances are equal to zero at rest and may be activated (opened) by the excitatory or inhibitory inputs to neuron $i$ respectively:

$$g_{SynEi}(t) = \bar{g}_E \times \sum_j \sum_{t_{kj}<t} S\{w_{ji}\} \times \exp(-(t-t_{kj})/\tau_{SynE}) + \bar{g}_{Ed} \times \sum_m S\{w_{dmi}\} \times d_{mi};$$
$$g_{SynIi}(t) = \bar{g}_I \times \sum_j \sum_{t_{kj}<t} S\{-w_{ij}\} \times \exp(-(t-t_{kj})/\tau_{SynI}) + \bar{g}_{Id} \times \sum_m S\{-w_{dmi}\} \times d_{mi}, \quad (5.A10)$$

**Table 5.A2** Weights of synaptic connections between populations in the network

| Target population | Source population (weight of synaptic input to one neuron) |
|---|---|
| RG-E | RG-E (0.00125); RG-F (0.00125); Inrg-E (−0.01125) |
| RG-F | RG-E (0.00125); RG-F (0.00125); Inrg-F (−0.01125) |
| Inrg-E | RG-F (0.03) |
| Inrg-F | RG-E (0.03) |
| PF-E | RG-E (0.005); Inrg-E (−0.0035); Inpf-E (−0.04) |
| PF-F | RG-F (0.005); Inrg-F (−0.0035); Inpf-F (−0.04) |
| Inpf-E | PF-F (0.025) |
| Inpf-F | PF-E (0.025) |
| PF-PBSt | RG-E (0.005); RG-F (0.005); In-E (−0.02); In-eF (−0.02) |
| PF-RF | PF-E (0.015); RG-F (0.005); In-eE (−0.05); In-eF (−0.02) |
| In-E | PF-E (0.05); Inrg-F (−0.02): In-T (−0.0125) |
| In-F | PF-F (0.05); Inrg-E (−0.02): |
| In-eF | Inrg-E (−0.02); In-lF(−0.0125) |
| In-lF | PF-F (0.005); Inrg-E (−0.0125); In-eF (−0.0125) |
| In-eE | PF-E (0.005); Inpf-F (−0.02) |
| In-lE | InrgF (−0.0125); In-T (−0.0125) |
| Ia-E | PF-E (0.0275); Ia-E (−0.02); R-E (−0.02) |
| Ia-F | PF-F (0.0275); Ia-F (−0.02); R-F (−0.02) |
| R-E | Mn-E (0.015); R-F (−0.015) |
| R-F | Mn-F (0.0015); R-E (−0.015) |
| R-PBSt | Mn-PBSt (0.015) |
| R-RF | Mn-RF (0.015) |
| Mn-E | PF-E (0.05); Ia-F (−0.04); R-E (−0.0025) |
| Mn-F | PF-F (0.05); Ia-E (−0.04); R-F (−0.0025) |
| Mn-PBSt | PF-PBSt (0.05); R-PBSt (−0.0025) |
| Mn-RF | PF-RF (0.05); R-RF (−0.0255) |

Values in brackets represent relative weights of synaptic inputs from the corresponding source populations ($w_{ji}$ or -$w_{ji}$).

where the function $S\{x\}=x$, if $x\geq 0$, and 0 if $x<0$. According to equation (5.A10), the excitatory and inhibitory synaptic conductances have two terms: the first term describes the effects of inputs from other neurons in the network (excitatory and inhibitory respectively), and the second one describes effects of inputs from external drives $d_{mi}$ (see also Rybak et al. 1997). Each spike arriving to neuron $i$ from neuron $j$ at time $t_{kj}$ increases the excitatory synaptic conductance by $\bar{g}_E \times w_{ji}$ if the synaptic weight $w_{ji}>0$, or increases the inhibitory synaptic conductance by $-\bar{g}_I \times w_{ji}$ if the synaptic weight $w_{ji}<0$. $\bar{g}_E=0.05$ mS/cm$^2$ and $\bar{g}_I=0.05$ mS/cm$^2$ are the parameters defining an increase in the excitatory or inhibitory synaptic conductance, respectively, produced by one arriving spike at $|w_{ji}|=1$. $\tau_{SynE}=5$ ms and $\tau_{SynE}=15$ ms are the decay time constants for the excitatory and inhibitory conductances respectively. In the second terms of equations (5.A10), $\bar{g}_{Ed}=\bar{g}_{Id}=1$ mS/cm$^2$ is the parameter defining the increase in the excitatory synaptic conductance, produced by external input drive $d_{mi}=1$ with a synaptic weight of $|w_{dmi}|=1$. All synaptic weights used in the model can be found in Table 5.A2. The values of input drives used in particular simulation are shown in Table 5.A3.

**Table 5.A3** Values of input drives in simulations

| Figure No | $d_{RG\text{-}E}$ | $d_{RG\text{-}F}$ | $d_E$ | $d_T$ | $d_F$ | $d_{eF}$ | $d_{lF}$ | $d_{eE}$ | $d_{lE}$ |
|---|---|---|---|---|---|---|---|---|---|
| 9B1 | 0.17 | 0.18 | 0.25 | 0 | 0 | 0.18 | 0.17 | N/A | N/A |
| 9B2 | 0.17 | 0.18 | 0 | 0.3 | 0.3 | N/A | N/A | N/A | N/A |
| 9B3 | 0.17 | 0.18 | 0.2 | 0.1 | 0 | 0.18 | 0.22 | N/A | N/A |
| 9B4 | 0.17 | 0.18 | N/A | 0.3 | N/A | 0.2 | 0.2 | 0.22 | 0.25 |
| 9B5 | 0.17 | 0.18 | N/A | 0 | N/A | 0.2 | 0.2 | 0.22 | 0.25 |
| 10A2 | 0.175 | 0.17 | 0.25 | 0 | 0 | 0.17 | 0.22 | 0.2 | 0 |
| 10B2 | 0.18 | 0.19 | 0.25 | 0 | 0 | 0.2 | 0.28 | 0.2 | 0 |
| 10C2 | 0.185 | 0.175 | 0 | 0.25 | 0.25 | 0.17 | 0.22 | 0.2 | 0 |
| 10D2 | 0.175 | 0.17 | 0.1 | 0 | 0 | 0.17 | 0.22 | 0.2 | 0 |
| 11B1 | 0.175 | 0.195 | 0.25 | 0 | 0 | 0.18 | 0.185 | N/A | N/A |
| 11B2 | 0.18 | 0.19 | 0.25 | 0 | 0 | 0.17 | 0.23 | N/A | N/A |
| 11B3 | 0.16 | 0.18 | 0.25 | 0 | 0 | 0.18 | 0.16 | N/A | N/A |
| 12A2 | 0.165 | 0.15 | 0 | 0.25 | 0.25 | 0.16 | 0.23 | 0.2 | 0 |
| 12B2 | 0.15 | 0.15 | 0 | 0.25 | 0.25 | 0.18 | 0.22 | N/A | N/A |
| 12C2 | 0.18 | 0.19 | 0.1 | 0 | 0 | 0.18 | 0.2 | N/A | N/A |
| 12D2 | 0.18 | 0.205 | 0.1 | 0 | 0 | 0.18 | 0.17 | N/A | N/A |
| 13A2 | 0.17 | 0.21 | 0.12 | 0 | 0 | 0.16 | 0.21 | 0.25 | 0.1 |
| 13B2 | 0.18 | 0.205 | 0.09 | 0 | 0 | 0.18 | 0.17 | 0.22 | 0 |
| 13C2 | 0.16 | 0.175 | 0.25 | 0 | 0 | 0.2 | 0.17 | N/A | N/A |
| 13C2 | 0.16 | 0.175 | N/A | 0.1 | N/A | 0.2 | 0.2 | 0.22 | 0.25 |

Values of MLR drives to the PF-E and PF-F populations in all simulations were $d_{PF\text{-}E} = d_{PF\text{-}F} = 0.1$; to the PF-PBSt and RF populations, $d_{PF\text{-}E} = d_{PF\text{-}F} = 0$.

## References

Anderson FC, Pandy MG (2001) Static and dynamic optimization solutions for gait are practically equivalent. J Biomech 34:153–161

Booth V, Rinzel J, Kiehn O (1997) Compartmental model of vertebrate motoneurons for Ca2+ -dependent spiking and plateau potentials under pharmacological treatment. J Neurophysiol 78:3371–3385

Brown TG (1914) On the nature of the fundamental activity of the nervous centres; together with an analysis of the conditioning of rhythmic activity in progression, and a theory of the evolution of function in the nervous system. J Physiol 48:18–46

Butera RJ, Jr., Rinzel J, Smith JC (1999) Models of respiratory rhythm generation in the pre-Botzinger complex. I. Bursting pacemaker neurons. J Neurophysiol 82:382–397

Carlson-Kuhta P, Trank TV, Smith JL (1998) Forms of forward quadrupedal locomotion. II. A comparison of posture, hindlimb kinematics, and motor patterns for upslope and level walking. J Neurophysiol 79:1687–1701

Gregor RJ, Smith DW, Prilutsky BI (2006) Mechanics of slope walking in the cat: quantification of muscle load, length change, and ankle extensor EMG patterns. J Neurophysiol 95:1397–1409

Grillner S (1981) Control of locomotion in bipeds, tetrapods, and fish. In: Brooks V (ed) Handbook of physiology. section I. The nervous system, Vol II. American Physiological Society, Bethesda, pp 1179–236

Grillner S, Zangger P (1979) On the central generation of locomotion in the low spinal cat. Exp Brain Res 34:241–261

Guertin P, Angel MJ, Perreault MC, McCrea DA (1995) Ankle extensor group I afferents excite extensors throughout the hindlimb during fictive locomotion in the cat. J Physiol 487(Pt 1):197–209

Halbertsma JM (1983) The stride cycle of the cat: the modelling of locomotion by computerized analysis of automatic recordings. Acta Physiol Scand Suppl 521:1–75

Hamade K, Shevtsova NA, Markin SN, Chakrabarty S, McCrea DA, Rybak IA (2008) How a bipartite CPG can control the activity of bifunctional motoneurons: a modeling study with insights from deletions during fictive locomotion. In 2008 Neuroscience Meeting Planner. Society for Neuroscience. Abstract 925.4. Washington, DC.

Huguenard JR, McCormick DA (1991) Vclamp and Cclamp. A computational simulation of single thalamic relay and cortical pyramidal neurons. Neural simulation instruction manual. Stanford University, Stanford

Huguenard JR, McCormick DA (1992) Simulation of the currents involved in rhythmic oscillations in thalamic relay neurons. J Neurophysiol 68:1373–1383

Jankowska E, Jukes MG, Lund S, Lundberg A (1967a) The effect of DOPA on the spinal cord. 5. Reciprocal organization of pathways transmitting excitatory action to alpha motoneurones of flexors and extensors. Acta Physiol Scand 70:369–388

Jankowska E, Jukes MG, Lund S, Lundberg A (1967b) The effect of DOPA on the spinal cord. 6. Half-centre organization of interneurones transmitting effects from the flexor reflex afferents. Acta Physiol Scand 70:389–402

Lafreniere-Roula M, McCrea DA (2005) Deletions of rhythmic motoneuron activity during fictive locomotion and scratch provide clues to the organization of the mammalian central pattern generator. J Neurophysiol 94:1120–1132

Lundberg A (1981) Half-centres revisited. In: Szentagothai J, Palkovits M, Hamori J (eds) Regulatory functions of the CNS. Motion and organization principles. Pergamon Akadem Kiado, Budapest, pp 155–167

MacGregor RI (1987) Neural and brain modelling. Academic Press, New York

Markin SN, Lemay MA, Prilutsky BI, Rybak IA (2012) Motoneuronal and muscle synergies involved in cat hindlimb control during fictive and real locomotion: a comparison study. J Neurophysiol 107:2057–2071

McCrea DA, Chakrabarty S (2007) Activity patterns in bifunctional PBSt motoneuron pools during fictive locomotion in decerebrate cats: clues to CPG organization. In 2007 Neuroscience Meeting Planner. Society for Neuroscience. Abstract 925.3. Washington, DC.

McCrea DA, Rybak IA (2007) Modeling the mammalian locomotor CPG: insights from mistakes and perturbations. Prog Brain Res 165:235–253

McCrea DA, Rybak IA (2008) Organization of mammalian locomotor rhythm and pattern generation. Brain Res Rev 57:134–146

Orsal D, Perret C, Cabelguen JM (1986) Evidence of rhythmic inhibitory synaptic influences in hindlimb motoneurons during fictive locomotion in the thalamic cat. Exp Brain Res 64:217–224

Perret C (1983) Centrally generated pattern of motoneuron activity during locomotion in the cat. Symp Soc Exp Biol 37:405–422

Perret C, Cabelguen JM (1980) Main characteristics of the hindlimb locomotor cycle in the decorticate cat with special reference to bifunctional muscles. Brain Res 187:333–352

Perret C, Cabelguen JM, Orsal D (1988) Analysis of the pattern of activity in "knee flexor" motoneurons during locomotion in cat. In: Gurfinkel VS, Ioffe ME, Massim J (eds) Stance and motion: facts and concepts. Plenum Press, New York, pp. 133–141

Pratt CA, Buford JA, Smith JL (1996) Adaptive control for backward quadrupedal walking V. Mutable activation of bifunctional thigh muscles. J Neurophysiol 75:832–842

Prilutsky BI (2000) Coordination of two- and one-joint muscles: functional consequences and implications for motor control. Motor Control 4:1–44

Prilutsky BI, Gregor RJ (2000) Analysis of muscle coordination strategies in cycling. IEEE Trans Rehabil Eng 8:362–370

Prilutsky BI, Zatsiorsky VM (2002) Optimization-based models of muscle coordination. Exerc Sport Sci Rev 30:32–38

Prilutsky BI, Gregor RJ, Ryan MM (1998a) Coordination of two-joint rectus femoris and hamstrings during the swing phase of human walking and running. Exp Brain Res 120:479–486

Prilutsky BI, Isaka T, Albrecht AM, Gregor RJ (1998b) Is coordination of two-joint leg muscles during load lifting consistent with the strategy of minimum fatigue? J Biomech 31:1025–1034

Rybak IA, Paton JF, Schwaber JS (1997) Modeling neural mechanisms for genesis of respiratory rhythm and pattern. II. Network models of the central respiratory pattern generator. J Neurophysiol 77:2007–2026

Rybak IA, Ptak K, Shevtsova NA, McCrimmon DR (2003) Sodium currents in neurons from the rostroventrolateral medulla of the rat. J Neurophysiol 90:1635–1642

Rybak IA, Shevtsova NA, Lafreniere-Roula M, McCrea DA (2006a) Modelling spinal circuitry involved in locomotor pattern generation: insights from deletions during fictive locomotion. J Physiol 577:617–639

Rybak IA, Stecina K, Shevtsova NA, McCrea DA (2006b) Modelling spinal circuitry involved in locomotor pattern generation: insights from the effects of afferent stimulation. J Physiol 577:641–658

Safronov BV, Vogel W (1995) Single voltage-activated Na+ and K+ channels in the somata of rat motoneurones. J Physiol 487(Pt 1):91–106

Shevtsova NA (2015) Two-level model of mammalian locomotor CPG. In: Jaeger D, Jung R (eds) Encyclopedia of Computational Neuroscience. New York, NY, Springer, pp. 2999–3017

Shevtsova NA, Chakrabarty S, Hamade K, Markin SN, McCrea DA, Rybak IA (2007) Computational model of mammalian locomotor CPG reproducing firing patterns of flexor, extensor and bifunctional motoneurons during fictive locomotion. In 2014 Neuroscience Meeting Planner. Society for Neuroscience. Abstract 925.4. Washington, DC.

Smith JL, Carlson-Kuhta P, Trank TV (1998a) Forms of forward quadrupedal locomotion. III. A comparison of posture, hindlimb kinematics, and motor patterns for downslope and level walking. J Neurophysiol 79:1702–1716

Smith JL, Carlson-Kuhta P, Trank TV (1998b) Motor patterns for different forms of walking: cues for the locomotor central pattern generator. Ann NY Acad Sci 860:452–455

Stein PSG, Smith JL (1997) Neural and biomechanical control strategies for different forms of vertebrate hindlimb motor tasks. In: Stein P, Grillner S, Selverston AI, Stuart DG (eds) Neurons, networks, and motor behavior. MIT Press, Cambridge, pp 61–73

Stuart DG, Hultborn H (2008) Thomas Graham Brown (1882–1965), Anders Lundberg (1920-), and the neural control of stepping. Brain Res Rev 59:74–95

Wells R, Evans N (1987) Functions and recruitment patterns of one-joint and 2-joint muscles under isometric and walking conditions. Hum Mov Sci 6:349–372

# Chapter 6
# Muscles: Non-linear Transformers of Motor Neuron Activity

**Scott L. Hooper, Christoph Guschlbauer, Marcus Blümel, Arndt von Twickel, Kevin H. Hobbs, Jeffrey B. Thuma and Ansgar Büschges**

**Abstract** Predicting movement from neural activity requires quantitative understanding of muscle response to motor neuron input. Muscles are sufficiently complicated that fulfilling this goal requires computer simulation. We therefore first explain in considerable detail one approach to modeling muscle. We then provide multiple examples of how muscle intrinsic properties and muscle diversity make straightforward predictions of how muscles transform neural input into movement impossible, including the dependence of muscle velocity on sarcomere number, the inadequacy of mean data in muscle modeling, the effects of muscle low-pass filtering, spike-number vs. spike frequency coding for contraction amplitude, how the role of passive muscle force in movement generation varies as a function of limb size, how muscles produce forces greater than their 'maximum force', energy conserving mechanisms, muscles that brake rather than produce movement, and how muscles can generate restoring responses (preflexes) to perturbing input in the absence of sensory feedback.

S. L. Hooper (✉) · K. H. Hobbs · J. B. Thuma
Department of Biological Sciences, Irvine Hall, Athens, OH 45701, USA
e-mail: hooper@ohio.edu

K. H. Hobbs
e-mail: hobbsk@ohio.edu

J. B. Thuma
e-mail: thuma@ohio.edu

C. Guschlbauer · A. von Twickel · A. Büschges
Zoologisches Institut, Universität zu Köln, Zülpicher Straße 47B, 50674 Cologne, Germany
e-mail: c.guschlbauer@uni-koeln.de

A. von Twickel
e-mail: arndt.twickel@uni-koeln.de

A. Büschges
e-mail: ansgar.bueschges@uni-koeln.de

M. Blümel
Institut für Biologie und ihre Didaktik, Kunzenweg 21, 79117 Freiburg, Germany
e-mail: marcus.bluemel@ph-freiburg.de



B. I. Prilutsky, D. H. Edwards (eds.), *Neuromechanical Modeling of Posture and Locomotion,* Springer Series in Computational Neuroscience,
DOI 10.1007/978-1-4939-3267-2_6

**Keywords** Muscle · Muscle model · Muscle intrinsic properties · Movement generation · Sarcomere · Data averaging · Low-pass filtering · Passive muscle force · Limb scaling · Preflex

## 6.1 Introduction

A central goal of neuroscience is to explain behavior. In pursuit of this goal neuromuscular systems have been intensively studied. Equation-based descriptions of muscle were early developed (Hill 1938, 1950; Jewell and Wilkie 1958; Wilkie 1950, 1956). Unstimulated muscles act as non-linearly damped springs. Muscle active force depends on muscle length, contraction velocity, and the pattern of motor neuron activity. Motor neuron spikes induce graded post-synaptic potentials (many invertebrate, some vertebrate muscles) or muscle action potentials (most vertebrate muscles). Transforming these electrical responses to contraction requires some time, and muscle relaxation is also slow. Muscles thus low-pass filter motor neuron activity.

The equations that describe muscle activity are sufficiently complex to allow large diversity in muscle response to motor neuron activity. This raises the issue of whether real muscles express this possible diversity. How large a role does passive force play in generating motor activity? Does motor neuron spike frequency always determine muscle force amplitude? What are the functional consequences of muscle low pass filtering? Must muscles be modeled individual animal by individual animal? What portions of their force-length (FL) and force-velocity (FV) curves do muscles use? Are muscles always motors? Detailed understanding of muscle is also important for understanding motor networks. Neurons and muscles likely evolved from a common precursor (Arendt 2008), and since their separation have co-evolved. Assuming that the motor neuron activity driving the muscle changes in an appropriately compensatory manner, presumably any given motor act can be produced by a wide variety of quantitatively different muscles. Understanding this interplay between motor network and muscle properties requires detailed description of both neural network and muscle (Chiel and Beer 1997).

Muscles are hierarchical systems in which proteins are arranged into filaments, filaments into sarcomeres, sarcomeres into myofibrils, myofibrils into fibers, and fibers into fascicles, with motor neuron activity being transformed into muscle contraction by a similarly long sequence of events beginning with transmitter release and ending with actomyosin activation. We cannot yet model these complex, non-linear systems from the molecular level up in any muscle. Most models consequently work at the whole muscle level. Even at this level muscle models are complicated. We therefore first present in some detail one model of one muscle. This is not to imply that this model is 'correct'. However, providing this level of detail should allow readers to work profitably with all the muscle model literature. We then provide examples demonstrating the importance of detailed understanding of muscle properties for understanding how nervous systems create motor behavior.

## 6.2 Building a Muscle Model

Unstimulated muscles respond to imposed length changes with an initial increase in force that subsequently declines. This activity is similar to the activity of a spring and damper in parallel, with another spring in series to both elements (Fig. 6.1a). This arrangement is the 'mechanical' center of the model presented here. Also shown is a contractile element (CE), which transforms motor neuron input into force and movement. The boundary between the parallel elements, all of which must always have the same length, and the serial element, is marked 'D'. We reproduce this scheme in panels 'a' of Figs. 6.1–6.4 with a white oval in each figure indicating which aspect of the model is being examined.

The task is to determine the equations that describe all model components. We explain this process using data and a model of the extensor tibiae muscle of the stick insect, *Carausius morosus* (Blümel et al. 2012a, b, c; Guschlbauer et al. 2007). The b1 panels of Figs. 6.1, 6.2, 6.3 and 6.4 show raw data. Figures 6.1, 6.3, and 6.4 show the same raw data, with different characteristics being measured in each figure. We perform this replication to emphasize that many characteristics can be measured in single experiments, an important consideration in systems as complicated as muscles. The b2 panels of Figs. 6.1–6.4 show the output of models constructed from experimental data. To avoid clutter we do not show the data in these panels; interested readers should see Blümel et al. 2012c. Model output always well agreed with the data ($R^2$ values $\geq 0.93$).

### 6.2.1 *Modeling the Series Elastic (SE) Element*

Parallel element activity (force produced, change in length) depends wholly or in part on parallel element length. This length cannot be measured experimentally, but must instead be calculated by subtracting series elastic element length ($L_{SE}$) from muscle fiber length. In pennate muscle such as stick insect extensor, muscle fiber length does not equal muscle length. The data presented here are therefore actually muscle fiber lengths. Because this detail does not affect model generality, we refer to this length as muscle length ($L_M$). $L_{SE}$ also cannot be measured directly but must be calculated from muscle force, which always equals series elastic element force ($F_{SE}$). The key to determining the relationship between $F_{SE}$ and $L_{SE}$ is that the damper prevents parallel element length from changing instantaneously. The initial $L_M$ change associated with a step change in force (Fig. 6.1b1) is therefore solely due to changes in $L_{SE}$.

These experiments are performed by first stimulating the motor nerve under isometric conditions until muscle force stops changing ($F_1$ in Fig. 6.1b1). The control system is then switched to isotonic conditions (transition to $F_2$). The muscle length is inappropriate for this new force (in the example shown, $L_{SE}$ is too long), and $L_{SE}$ therefore immediately changes (in this example, shortens) to make muscle

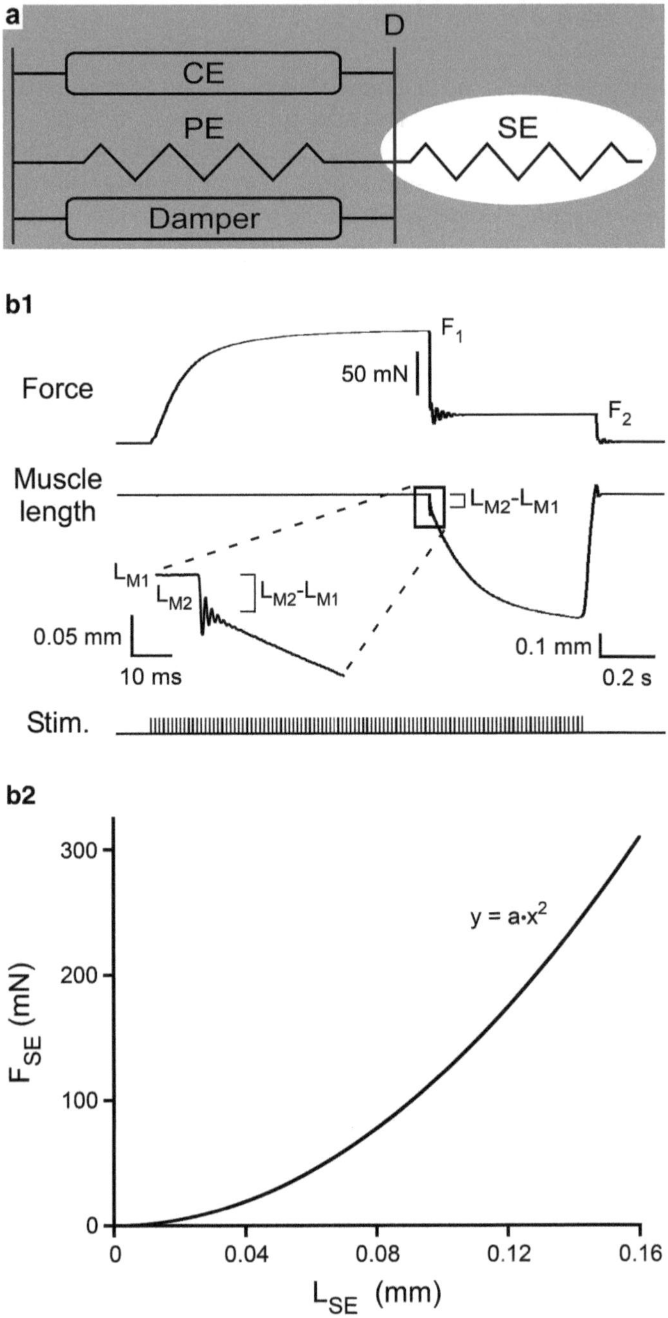


**Fig. 6.1** $F_{SE}$ dependence on $L_{SE}$ (*white oval* in panel a). **a** Schematic of muscle model used. **b1** Initial length change ($L_{M2} - L_{M1}$) after a step change in force is due solely to change in $L_{SE}$ because the damper prevents parallel element length from changing instantaneously. **b2** In stick insect extensor muscle $F_{SE}$ depends on $L_{SE}$ squared. (Modified from Blümel et al. 2012c)

and applied forces equal (inset; $\Delta L_{SE} = L_{M2} - L_{M1}$). Performing these steps from multiple F1 to multiple F2 gives a large number of $\Delta F$ vs. $\Delta L_M$ data points, from which the relationship between $F_{SE}$ and $L_{SE}$ can be determined (Blümel et al. 2012c; Guschlbauer et al. 2007). For the extensor muscle, this relationship is quadratic (Fig. 6.1b2).

In many models $L_{SE}$ can become negative, e.g., during imposed rapid shortenings before parallel element length has time to change, or during simulation oscillations. The biological relevance of negative $L_{SE}$ is uncertain. When a muscle's length is rapidly shortened to a length where it is at first slack, provided the new length is not too short, the slack will be slowly taken up as the parallel elements shorten, and a new, lower but non-zero, muscle force regained. It is unclear whether this taking up of muscle slack occurs because $F_{SE}$ becomes negative (the SE spring pushes on the parallel elements), or if $F_{SE}$ becomes zero (the parallel elements change length at the velocity appropriate for zero applied force). Regardless, modelers need to be aware of this difficulty, and make a decision of what $F_{SE}$ are given by negative $L_{SE}$.

### *6.2.2 Modeling the Parallel Elastic (PE) and Damper Elements*

PE characteristics are measured by ramp changes in muscle length of unstimulated muscles. Length increases cause a large force increase (due to $L_{SE}$ stretch) followed by a continuous decline (Fig. 6.2b1). These declines typically follow a power law (e.g., Thuma and Hooper 2010). As such, during the ramp relatively large relaxation would be expected to occur, one reason that force changes (Fig. 6.1b1) are instead used to determine SE characteristics.

Despite the power law relaxation, PE characteristics are typically analyzed as though the relaxation were exponential. In an exponential case, PE force ($F_{PE}$) equals a constant times PE length ($L_{PE}$) and damper force ($F_{damper}$) equals a constant times damper velocity ($dL_{damper}/dt$). Early in the relaxation $F_{damper}$ dominates $F_{PE} + F_{damper}$, but as relaxation proceeds $dL_{damper}/dt$, and thus $F_{damper}$, exponentially decline. After a few time constants the sum is essentially totally due to $F_{PE}$. $L_{PE}$ can be calculated from $L_M = L_{SE} + L_{PE}$, with $L_{SE}$ being calculated from the $F_{SE}$ equation derived above. $F_{PE}$'s dependence on $L_{PE}$ is therefore determined by performing multiple muscle stretches and measuring $F_{PE}$ when the change in muscle force becomes very small (measurement—asterisk—taken 40–100 s after stretch, note breaks in traces). $F_{PE}$ typically depends on $L_{PE}$ in a greater than linear fashion, in extensor muscle as an exponential (Fig. 6.2b2).

A concern with this approach is that relaxation is actually power law. Power laws that well match the beginning of an exponential match it much less well at later times. This raises the concern that, when $F_{PE}$ is measured, $F_{damper}$ may be large enough to prevent accurately determining $F_{PE}$'s dependence on $L_{PE}$. Fits to extensor muscle relaxations with $F = a + b \cdot (t + c)^d$, where $a$ is the force at which $F_{PE}$ equals $F_{SE}$ (the steady-state force at $t = \infty$), and $b$, $c$, and $d$ are parameters ($c$ allows the

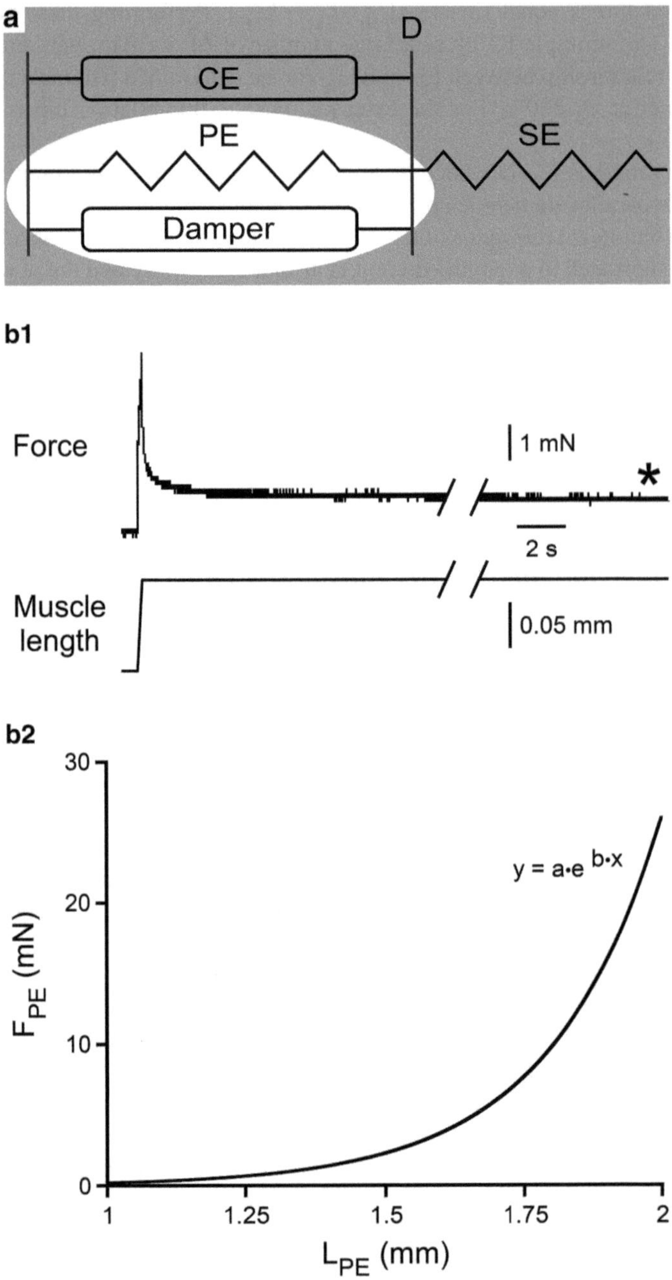


**Fig. 6.2** $F_{PE}$ dependence on $L_{PE}$ (*white oval* in panel a). **a** Schematic of muscle model used. **b1** A ramp change in muscle length induces a large force increase that slowly declines. $F_{PE}$ is measured (*asterisk*) after most of this decline has occurred. **b2** In extensor muscle, $F_{PE}$ is an exponential function of $L_{PE}$. (Modified from Blümel et al. 2012c)

power law to have finite values at $t=0$), show that for extensor muscle this concern is unlikely, as at measurement times the $b \cdot (t+c)^d$ term is small.

Another concern is modeling the power law decline. Because $F_{SE}$ and $F_{PE}$ are nonlinear functions of $L_{SE}$ and $L_{PE}$, determining what dependency of $F_{damper}$ on $dL_{damper}/dt$ (on the usual presumption that $F_{damper}$ depends on damper velocity) gives rise to a given $F_{PE}$ power law relaxation is difficult. A reasonable choice in muscles in which passive forces are small compared to active forces is to ignore passive force dynamics. However, in small limbs passive forces determine rest joint posture (see below), and dynamic passive forces can be considerable during rapid $L_M$ changes (Fig. 6.2b1). Including damper force could thus improve model performance. One work around is to use one or more linear dampers ($F_{damper} = -b \cdot dL_{CE}/dt$) and to fit the *b* parameter(s) to only physiologically relevant portions of the relaxation. Such approaches must use different damper parameters for lengthening vs. shortening to account for $F_{PE}$ being much larger during lengthening. A second parallel damper is sometimes necessary as a modeling construct to prevent excessively rapid D line movement (Blümel et al. 2012b).

### 6.2.3 *The Dependence of Contractile Element Force ($F_{CE}$) on CE Length ($L_{CE}$)*

In response to motor neuron activity, under isometric conditions, muscle force first increases and then reaches a steady state (asterisk, Fig. 6.3b1). The rise is modeled by a process (see below) that transforms motor neuron activity into a muscle activation parameter. We describe here how to model force once muscle activation has been calculated. Model functions and parameter values are determined from steady-state force data, and are often normalized to the maximum force ($F_{max}$) the muscle can produce under isometric conditions.

Force (Fig. 6.3b2) decreases with decreased motor neuron activity (decreased activation). It also varies with $L_{CE}$ because of the varying amounts of actin and myosin overlap that occur as $L_M$ changes, with $L_{CE}$ being calculated using the $F_{SE}$ expression above to obtain $L_{SE}$, and subtracting $L_{SE}$ from $L_M$. Curve peaks move to progressively longer $L_{CE}$ as activation decreases. The curves are well fit with sinusoidal functions, with logic functions used to make the model produce zero force at $L_{CE}$s outside the limits of a half-cycle (e.g., $L_{CE} < 1$ for the 0.15 act curve).

The general equation for these sinusoids is $F = amplitude \cdot \sin\left(\omega L_{CE} + \varphi\right)$, where *amplitude* is typically normalized to $F_{max}$, $\omega$ is cycle period, and $\varphi$ is waveform phase. The task is to express *amplitude*, $\omega$, and $\varphi$ in terms of activation (act) and $L_{CE}$. We do not reproduce this work in detail here (interested readers should see Blümel et al. 2012c). The general steps are to examine the dependence of *amplitude*, $\omega$, and $\varphi$ on activation, $L_{CE}$, and each other. Dependence of *amplitude*, $\omega$, or $\varphi$ on activation or $L_{CE}$ allows direct substitution of the derived equation in the general equation. Co-dependence of any two of *amplitude*, $\omega$, or $\varphi$ allows one of them to be replaced with the derived dependence (e.g., replacing $\omega$ with an equation in

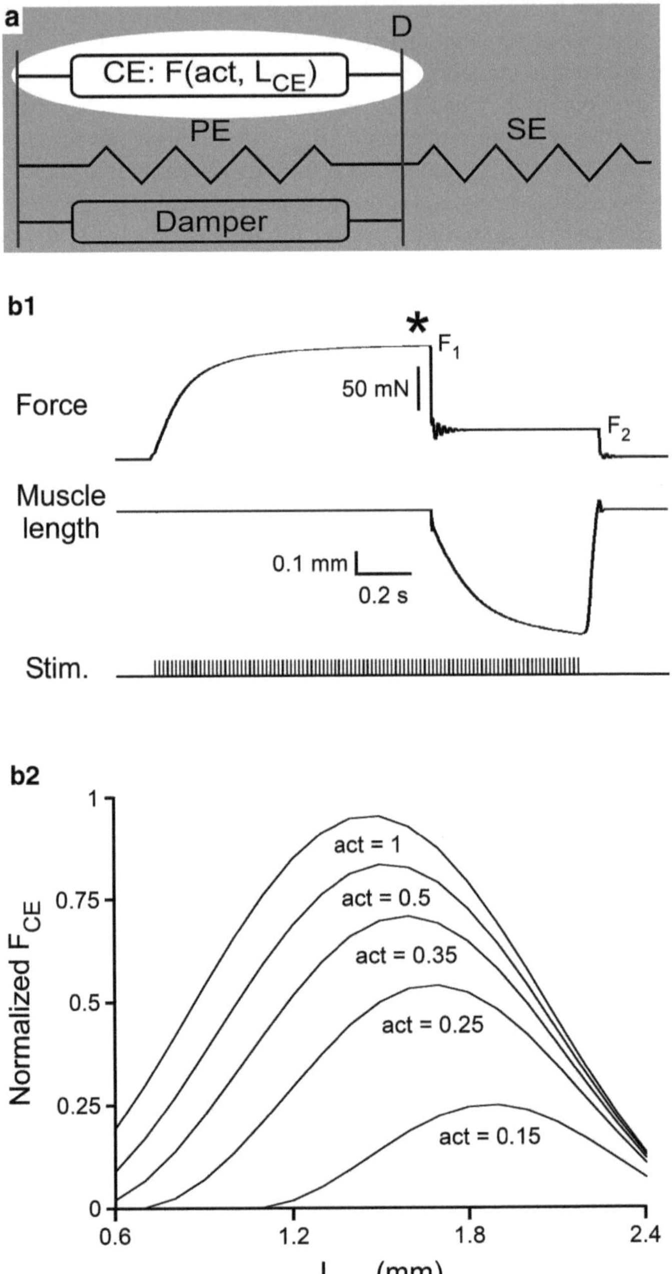


**Fig. 6.3** $F_{CE}$ dependence on $L_{CE}$ (*white oval* in panel a). **a** Schematic of muscle model used. **b1** Muscle activated under isometric conditions until force stabilizes (*asterisk*). $L_{CE}$ is calculated from muscle length and the known dependence on $F_{SE}$=measured force. **b2** $F_{CE}$ curves are sinusoidal functions of $L_{CE}$. $F_{CE}$ does not reach a maximum of 1 at act=1 because the fit found the best parameter set to reproduce all the curves. This best fit did not give a maximum of 1 for the act=1 curve. (Modified from Blümel et al. 2012c)

*amplitude*), after which if the remaining term (in this example, *amplitude*) can be expressed in terms of activation and $L_{CE}$, both terms can be replaced with equations in activation and $L_{CE}$.

For the data that gave the curves in Fig. 6.3b2, this process proceeded as follows. First, the descending arms of all the FL curves reached zero at approximately the same $L_{CE}$. In sinusoids related in this fashion, $\varphi = -(\pi/2 + g \cdot \omega)$, where $g$ is the $L_{CE}$ at which all the descending arms have zero force. Second, *amplitude* depended exponentially on $\omega$ ($amplitude = a \cdot e^{-b\omega}$), and $\omega$ depended on activation with an inverse square law ($\omega = c + (\alpha \cdot (\mathrm{act} + d))^{-2}$), where $a$, $b$, $c$, $d$, and $\alpha$ are parameters determined by fitting the data. Substituting into the general equation gave a single equation in $L_{CE}$ and activation for all the curves in Fig. 6.3b2:

$$F = a \cdot e^{b \cdot \left(c + (\alpha \cdot (act + d))^{-2}\right)} \cdot \frac{1}{2} \cdot \left(1 + sin\left[c + \left(\alpha \cdot (act + d)\right)^{-2} \cdot (L_{CE} - g) - \frac{\pi}{2}\right]\right).$$

The dependencies above (that all the descending arms reached zero at the same $L_{CE}$, that *amplitude* depended exponentially on $\omega$ and $\omega$ on the inverse square of activation) could be specific to extensor muscles. We therefore applied the same approach (Blümel et al. 2012c) to data on cat soleus muscle (Brown et al. 1999) and found the same dependencies (although, of course, with different values for $a$, $b$, $c$, $d$, $g$, and $\alpha$). Descending arm force reaching zero at a common $L_{CE}$ in both muscles is likely explained as this being the $L_{CE}$ at which actin and myosin can no longer interact. Being able to express $\varphi$ as a function of $\omega$ is thus likely common across muscles. The bases of the other dependencies are unclear. However, their presence in muscles as different as stick insect extensor and cat soleus suggests they may be generally present in muscle.

### 6.2.4 *The Dependence of $F_{CE}$ on CE Velocity ($V_{CE}$)*

$F_{CE}$'s dependence on $V_{CE}$ is characterized in the same quick release experiments used to determine SE properties, but by measuring the initial velocity of $L_M$ change after the SE length change has occurred (Fig. 6.4b1). Plotting $F_2$ vs. $V_{CE}$ (which equals $L_M$ velocity, $dL_M/dt$, because, when measured, $L_{SE}$ change has ceased) shows that force production decreases with CE shortening ($dL_{CE}/dt < 1$) and increases with CE lengthening ($dL_{CE}/dt > 1$), and that the shapes of these curves differ in shortening and lengthening contractions. These plots were originally expressed in terms of shortening velocity (Hill 1938; Jewell and Wilkie 1958), and thus positive $V_{CE}$ equals $-dL_{CE}/dt$ (note x axis label in Fig. 6.4b2). Although this convention is not always followed (Brown and Loeb 2000), we use it here.

Shortening and lengthening contractions are both well fit with hyperbolas of the forms

$$F = amplitude \cdot \left( \frac{a \cdot (1 + a)}{(-dL_{CE} / dt) / b + a} - a \right) (\text{shortening contractions}) \text{ and}$$

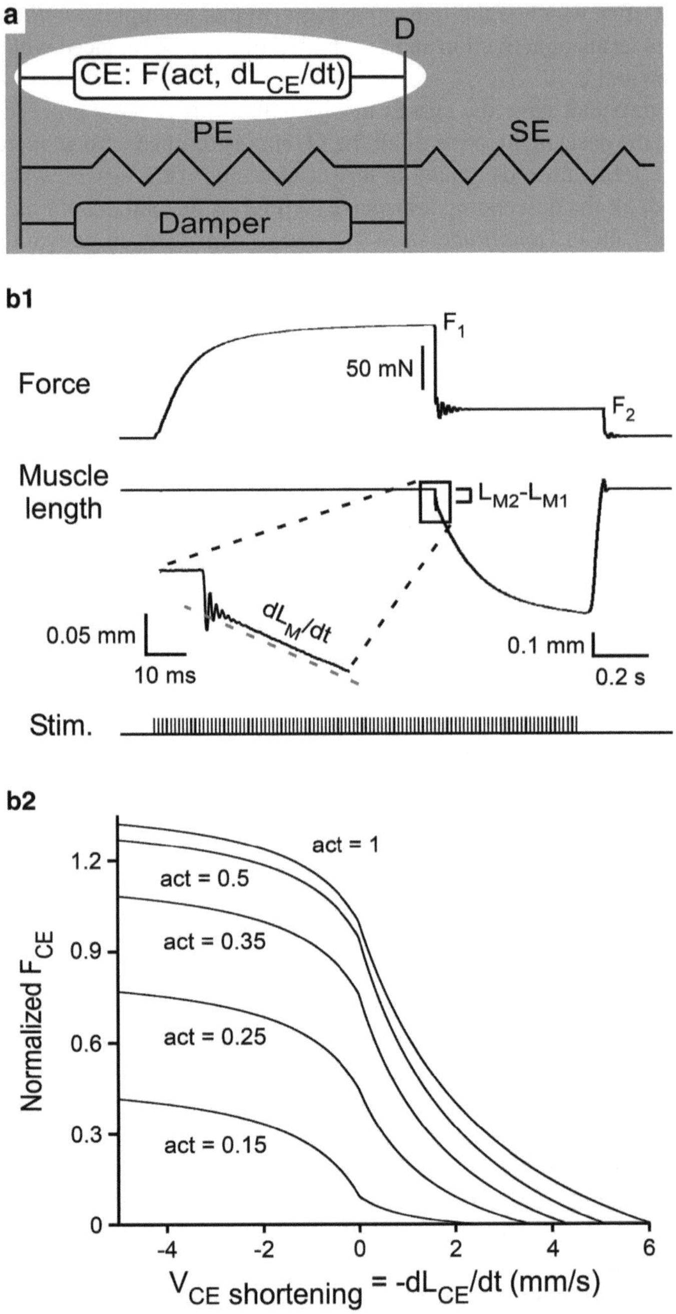


**Fig. 6.4** $F_{CE}$ dependence on $V_{CE}$ (*white oval* in panel a). **a** Schematic of muscle model used. **b1** The muscle is activated under isometric conditions until force stabilizes and then allowed to shorten (isotonic conditions) against a second force, $F_2$. The slope of the length change after the oscillation has ceased (*grey dashed line*) gives $V_{CE}$. **b2** $F_{CE}$ curves are hyperbolic functions of $V_{CE}$. (Modified from Blümel et al. 2012c)

$$F = \frac{f \cdot (1+f)}{(-dL_{CE}/dt)/g + f} - f - amplitude \text{ (lengthening contractions)}$$

where *a* and *f* are parameters, *amplitude* is typically normalized to $F_{max}$, and the differing treatment of *amplitude* in the equations is so the curves meet at $dL_{CE}/dt = 0$. *amplitude*, *b* and *g* are functions of activation (although *g* can sometimes be modeled as a parameter), each with a few parameters. We do not cover this process here; interested readers should see Blümel et al. 2012c. The above FV curves are independent of $L_{CE}$. This is not always the case (Edman 1979; Scott et al. 1996; Guschlbauer et al. 2007), a difficulty that can be overcome by including length dependence in the velocity equations (Brown et al. 1996). The importance of muscle FV properties has been investigated by comparing the effect of different FL and FV shapes on the performance of a periodic hopping model. Of the three shapes examined, only Hill-type hyperbolic FV curves gave rise to stable hopping (Haeufle et al. 2010).

### *6.2.5 Combining the Model Components*

One approach to combining the model components is to allow $V_{CE}$ to change instantaneously, in which case the activity of the model in the 'a' panels of Figs. 6.1–6.4 can be predicted from the functions in Figs. 6.1–6.4 (Brown et al. 1996). We ignore here the damper force that gives rise to $F_{PE}$ relaxation (Fig. 6.2). Total parallel module force is thus $F_{CE}+F_{PE}$ (where $F_{CE}$ and $F_{PE}$ are both either normalized to $F_{max}$ or actual force; actual $F_{CE} = F_{max} \cdot FL \cdot FV$ for normalized FL and FV curves). If included, parallel damper force would be added to $F_{CE}+F_{PE}$ to give total parallel module force. Since muscle passive forces are much greater with stretches than shortenings, this would primarily increase parallel module force at positive $dL_{CE}/dt$ (negative $V_{CE}$). In this approach the forces on the D line must always equal. $F_{SE}$ always equals the force the muscle is producing: in isotonic contractions, applied load; in isometric contractions, measured force. The relationship between $F_{SE}$ and $L_{SE}$ allows $L_{SE}$, and thus $L_{CE}$, to be calculated at all times.

First consider $F_{CE}+F_{PE}$ when $dL_{CE}/dt = 0$ (Fig. 6.5a). For many of these curves the same $F_{CE}+F_{PE}$ is present at three $L_{CE}$ (the three circles on the act = 1 curve mark these $L_{CE}$ for $F_{CE}+F_{PE} = 175$ mN). In the model the points on the descending (negative slope, dashed lines) arms of these curves are unstable. This point is demonstrated in Fig. 6.5c, which shows three isoforce lines as a function of V and $L_{CE}$ (for ease of presentation, these lines are with different parameters than those used in Fig. 6.5a and b1–b3). The black line is for a force $F_1$, the red a force $F_2$ less than $F_1$, and the blue a force $F_3$ greater than $F_1$. First consider the case where the system is originally at $F_1$ and the intermediate, unstable, stationary $L_{CE}$. If force instantaneously decreases to $F_2$, since $L_{CE}$ cannot instantaneously increase, parallel element V (upward red arrow) must decrease to match $F_2$. This is a shortening velocity, and thus so long as (no matter how briefly) force is at $F_2$, $L_{CE}$ decreases (leftward arrow on $F_2$ isoforce line). When force is instantaneously returned to $F_1$, again because

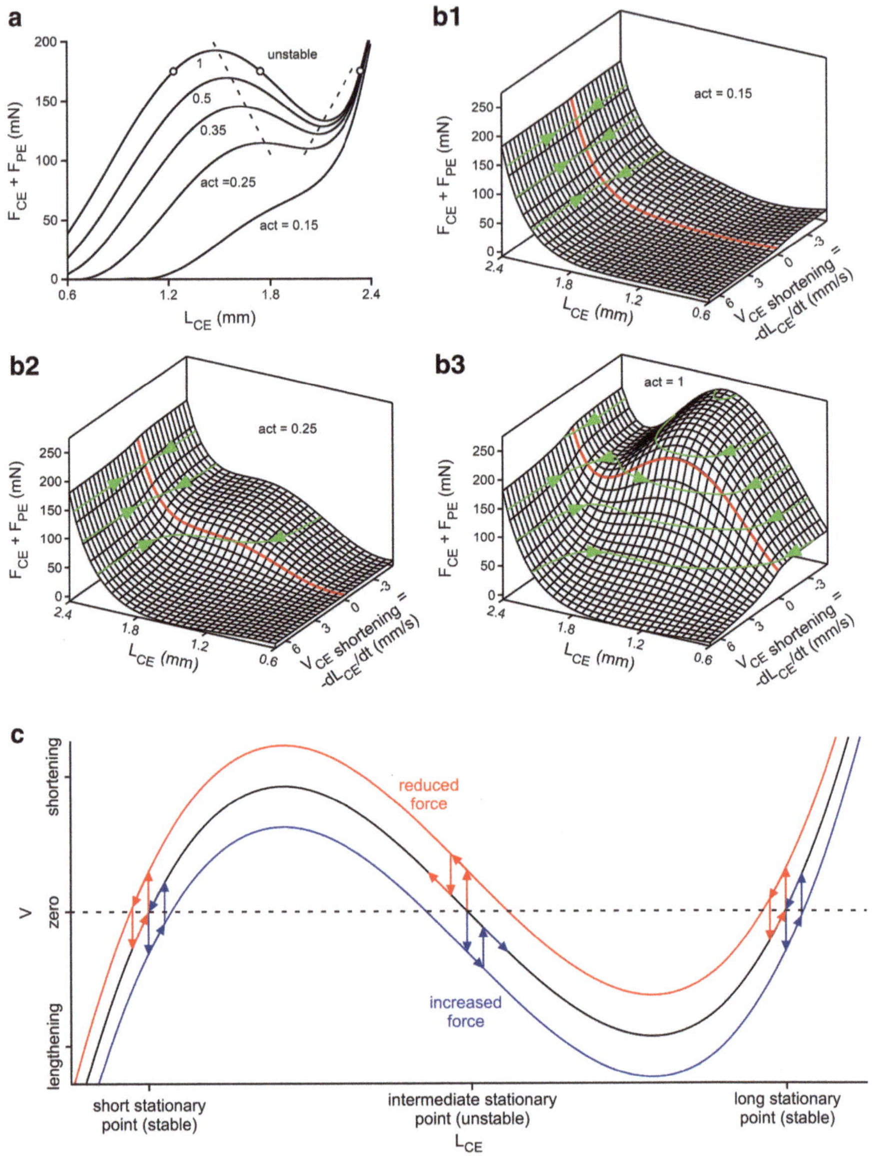


**Fig. 6.5** $L_{CE}$ activity in complete model. **a** Total parallel element force at zero $V_{CE}$ ignoring parallel damper force. Adding $F_{PE}$ to $F_{CE}$ increases force at long $L_{CE}$ and introduces, for most activation levels, a valley before the monotonic increase in $F_{CE}+F_{PE}$ at long $L_{CE}$. The negative slope regions of the $L_{CE}+L_{PE}$ curves are unstable to perturbation. **b1–b3** $F_{CE}+F_{PE}$ vs. $L_{CE}$ and $V_{CE}$ surfaces (note that $V_{CE}$ is positive for shortening contractions, and thus equals $-dL_{CE}/dt$). $L_{CE}$ axis is reversed relative to panel a. *Red lines* are zero $V_{CE}$, *green lines* isoforce contours. The intermediate intersection of the $V_{CE}=0$ and 150 mN isoforce lines is an unstable stationary point (note that *green arrows* point away from the intersection). These surfaces allow $V_{CE}$ prediction for any activation, force, and $L_{CE}$, and can be used, in combination with the $L_M$ and $F_{SE}$ equations, to model muscle activity (see text). **c** Briefly decreasing (*red arrows*) or increasing (*blue arrows*) applied force drives the system to the short or long (respectively) $L_{CE}$ stationary points when applied to the unstable intermediate stationary point, but returns to the point of origin when applied at the short or long stable stationary points

$L_{CE}$ cannot instantaneously increase, parallel element V must change to reach the $F_1$ isoforce line (downward red arrow). At this $L_{CE}$, V is still a shortening velocity. The system thus progresses along the $F_1$ isoforce line to the left stationary point. In an analogous fashion, brief force increases (blue arrows) move the system to the right stationary point.

Compare this response to the effect of the same perturbation at the right stationary point. Decreasing force again results in a shortening V (upward red arrow) and $L_{CE}$ decreases (red arrow on $F_2$ isoforce line). When $F_1$ is restored, however, V moves across the zero line to a lengthening V. $L_{CE}$ consequently increases and moves back along the $F_1$ isoforce line to the right stationary point. Analogous arguments show that this point is stable to force increases (blue arrows), and that the short $L_{CE}$ stationary point is also stable (red and blue arrows). These arguments do not prove that muscle could not work on the descending arms, as sensory feedback could counteract the muscle instability at these lengths. On what arm of the FL curves muscles operate has been surprisingly little investigated, but human soleus (Rubenson et al. 2012) and stick insect extensor (Guschlbauer et al. 2007) function on the ascending arm during locomotion.

Armed with this understanding, we can explain many aspects of model activity by considering $F_{CE}+F_{PE}$ vs. $L_{CE}$ and $dL_{CE}/dt$ plots (Fig. 6.5b1–b3 shows these surfaces at three activations). The green lines show model trajectories at various forces under isotonic conditions. Consider the 150 mN trajectory in Fig. 6.5b3. Since the extensor muscle works on the ascending arms of its FL curves, only the rightward portions of this surface ($L_{CE}$ <about 1.2) are physiological. However, we examine all portions of the 150 mN trajectory to show how the model works across the entire surface. If the initial condition is $L_{CE}=2.4$, the plot shows that at this $L_{CE}$, $-dL_{CE}/dt$ must be around 7 (rapid shortening). $L_{CE}$ therefore decreases (green arrow) until the intersection with the zero velocity line (red) at $L_{CE}$ about 2.3 (the long $L_{CE}$ stable stationary point) is reached. Initial conditions in the valley between $L_{CE}$'s of about 2.3 and 1.96 are in the unstable region at which $-dL_{CE}/dt$ is negative (lengthening). Initial conditions in this region thus also go to the long $L_{CE}$ intersection (opposite direction of relevant green arrow). For $L_{CE}$'s between 1.96 and 1.08, $-dL_{CE}/dt$ is again positive, and initial conditions in this region go to the short $L_{CE}$ stable stationary point. For shorter $L_{CE}$'s, $-dL_{CE}/dt$ is again negative, and from initial conditions in this region the system lengthens to also go to the short $L_{CE}$ intersection.

These surfaces also predict system activity under isometric conditions. Consider an increase in $L_M$. This stretches $L_{SE}$ and increases $F_{SE}$. $-dL_{CE}/dt$ must become negative ($L_{CE}$ lengthening) to equal the new $F_{SE}$. As $L_{CE}$ increases, $L_{SE}$ and $F_{SE}$ decrease. This process continues until $L_{CE}$ and $L_{SE}$ reach values at which $F_{CE}+F_{PE}$ equals $F_{SE}$, a point on the zero $dL_{CE}/dt$ line. Changes in activation shift the model between activation surfaces, which change smoothly as activation changes. Provided $V_{CE}$ is recalculated at each time step to keep D line net force zero, the $F_{CE}+F_{PE}$ vs. $L_{CE}$ and $dL_{CE}/dt$ surfaces thus provide all information needed to predict model activity for any combination of forces, lengths, and activations.

This approach, in which $V_{CE}$ changes in a non-history dependent fashion to always make $F_{CE}+F_{PE}=F_{SE}$, is not always used (Blümel et al. 2012b; Cheng et al. 2000). In these models $V_{CE}$ is updated by calculating $dV_{CE}/dt$ each time step ($\Delta t$)

and adding $\Delta t \cdot dV_{CE}/dt$ to the prior $V_{CE}$ to obtain a new $V_{CE}$. This $V_{CE}$ is then used to calculate new $L_{CE}$ and $L_{SE}$ and $F_{CE}$ and $F_{SE}$. The fact that in this approach $F_{SE}$ will not equal $F_{CE}+F_{PE}$, except at steady state, provides the $dV_{CE}/dt$ equation: one gives the D line mass (m) and calculates $dV_{CE}/dt$ from $F = m \cdot a = m \cdot -dV_{CE}/dt$ (the minus sign because $V_{CE}$ is defined as shortening velocity).

Which approach is better is unclear. In both one model element must change instantaneously ($V_{CE}$ in the first, acceleration in the second) and in reality neither can do so. D line mass complicates interpreting the FV data (Fig. 6.4), as the measured velocity would depend on both CE properties and D line acceleration over the measurement interval. Alternatively, including D line mass improves model stability, and such models can give very good results (Blümel et al. 2012b). Perhaps the most correct conclusion is that the model in the 'a' panels of Figs. 6.1–6.4 is extremely simplified. It lacks any details of the molecular basis of actomyosin force generation or of the giant muscle protein properties that give rise to $F_{PE}$ (presumably; $F_{PE}$ in unstimulated muscle, at least, is not due to 'background' actomyosin interaction (Thuma and Hooper 2010)). With such a simplified model, no approach may be able to reduce error below a certain minimum, in which case which approach is used may be primarily a matter of taste.

### 6.2.6 *Modeling Activation*

Individual motor neuron spikes induce relatively small force increases in non-spiking muscles and larger ones in spiking muscles. In both muscle types, if spike train frequency is high enough, the individual responses temporally summate, eventually reaching a steady-state (see Fig. 6.9). Force typically declines exponentially at spike train end. In both muscle types this response can be reasonably well modeled with a low pass filter. More accurate activation models can be obtained by modeling the responses to individual spikes and summing them. In muscles with history-dependence (e.g., facilitation), this history dependence must be measured and included in the activation process. A final complication is that, at least in some muscles, different filter parameter values must be used for isometric and isotonic conditions (Blümel et al. 2012b).

## 6.3 Consequences of Muscle Intrinsic Properties and Muscle Diversity

### 6.3.1 *Relating Whole Muscle and Individual Sarcomere FL and FV Curves*

The FL and FV curves in Figs. 6.3b2 and 6.4b2 arise from the summed activity of the muscle's sarcomeres. It is therefore important to understand the relationship between sarcomere intrinsic properties, the number of serially arranged sarcomeres per muscle fiber (here referred to as sarcomere number), and whole muscle properties (ignoring, for simplicity, pennation effects). The sliding filament theory of muscle contraction (Huxley and Niedergerke 1954) predicts that, all other factors being equal, muscle shortening (and, by the same logic, lengthening) velocity equals individual sarcomere shortening velocity times sarcomere number (Fig. 6.6a). Figure 6.6b shows data confirming this prediction (Thuma et al. 2007). It is important to stress this dependence of muscle velocity on sarcomere number because of the many articles (Atwood 1967, 1971, 1972, 1973, 1976; Costello and Govind 1983; Govind et al. 1975; Hoyle 1967, 1969; Jahromi and Atwood 1969a, b, 1971; Komuro 1981; Lang et al. 1977; Parnas and Atwood 1966; Royuela et al. 1998, 2000) stating that contraction velocity depends on sarcomere length. In muscles with equal sarcomere lengths, longer muscles, because of their greater sarcomere number, will therefore indeed have higher velocities, one functional consequence of which we examine below. However, sarcomere length, particularly in invertebrates, shows large variation. As such, in comparing muscles with different length sarcomeres, it is essential to recognize that the true dependence of contraction and relaxation velocity is on sarcomere number.

Muscle FL curves similarly depend on sarcomere FL characteristics. All other factors being equal, FL curve peaks will be at an $L_M$ equal to sarcomere number times sarcomere length at the peak force of the individual sarcomere FL curve, and FL curve width will equal sarcomere number times individual sarcomere curve width. In muscles with equal sarcomere lengths, longer muscles thus have broader FL curves. In muscles with equal sarcomere lengths (e.g., vertebrate muscle), the dependency on sarcomere number can be normalized away by dividing by muscle length, which results in each normalized unit length having an equal sarcomere number. In such muscles FV curves can be similarly normalized by expressing V as muscle length/s. Invertebrate muscles, alternatively, have highly variable sarcomere lengths. In these muscles one could attempt to normalize to sarcomere number, but muscles with different sarcomere lengths likely differ in other sarcomere intrinsic properties as well. Absent other justification, in invertebrate work it is therefore likely better to use real as opposed to normalized data.

### 6.3.2 *Mean Data are Unlikely Adequate for Accurate Muscle Modeling*

The equations obtained in Figs. 6.1–6.4 have many parameters. The values of these parameters are often determined by performing multiple experiments on the 'same' muscle (e.g., stick insect extensor) from different animals. One or a few parameters are measured in each experiment, the across-animal data are averaged, and the means are used to model an 'average' muscle. A difficulty with this approach is that means can be very unrepresentative of most individuals in the population. For instance, no member of the population given by $A^2+B^2=1$ has the population means of $A=B=0$ (Fig. 6.7a). An objection to this example is that A and B are not

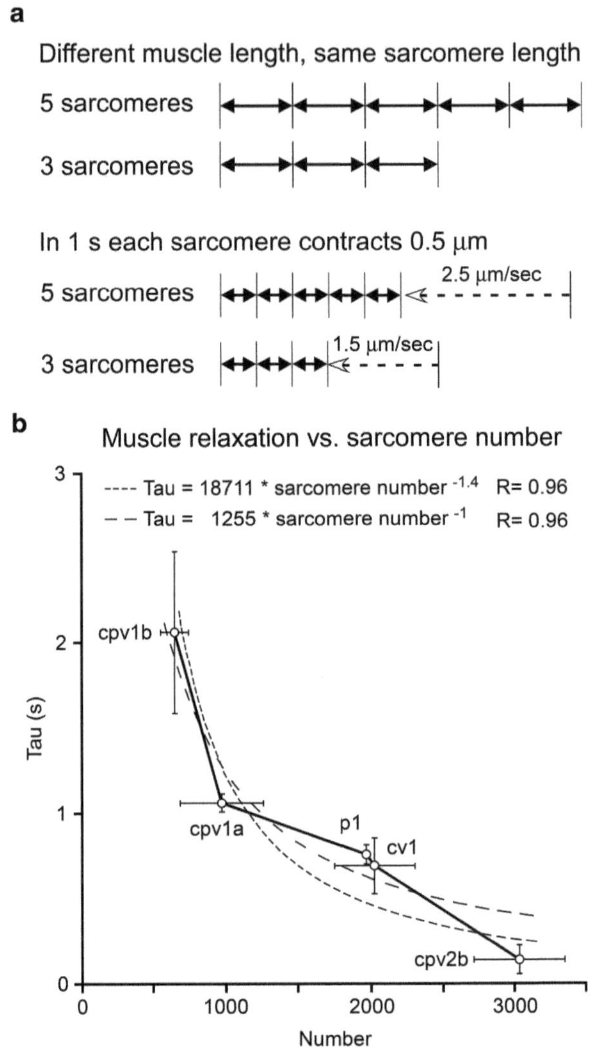


**Fig. 6.6** Muscle dynamics depends on sarcomere number. **a** Schematic showing that, all other factors being equal, increasing sarcomere number increases shortening velocity. **b** Data from lobster stomatogastric muscles, in which different muscles have both different sarcomere lengths and numbers, showing that relaxation dynamics (tau) depend on sarcomere number. Plotting tau vs. sarcomere length shows no dependence on sarcomere length (data not shown). The schematic shown in panel a predicts an inverse dependence of tau on sarcomere number (Thuma et al. 2007). Actual best fit has an exponent of −1.4, but a fit with the exponent set to −1 is essentially equally good. (Modified from Thuma et al. 2007)

independent. However, multi-dimensional systems with independent variables also have the property that individuals with mean values for each variable are extremely rare. The inset in Fig. 6.7b shows a variable that is Gaussian distributed with a mean of zero. In a system composed of eight such variables that freely distribute, only

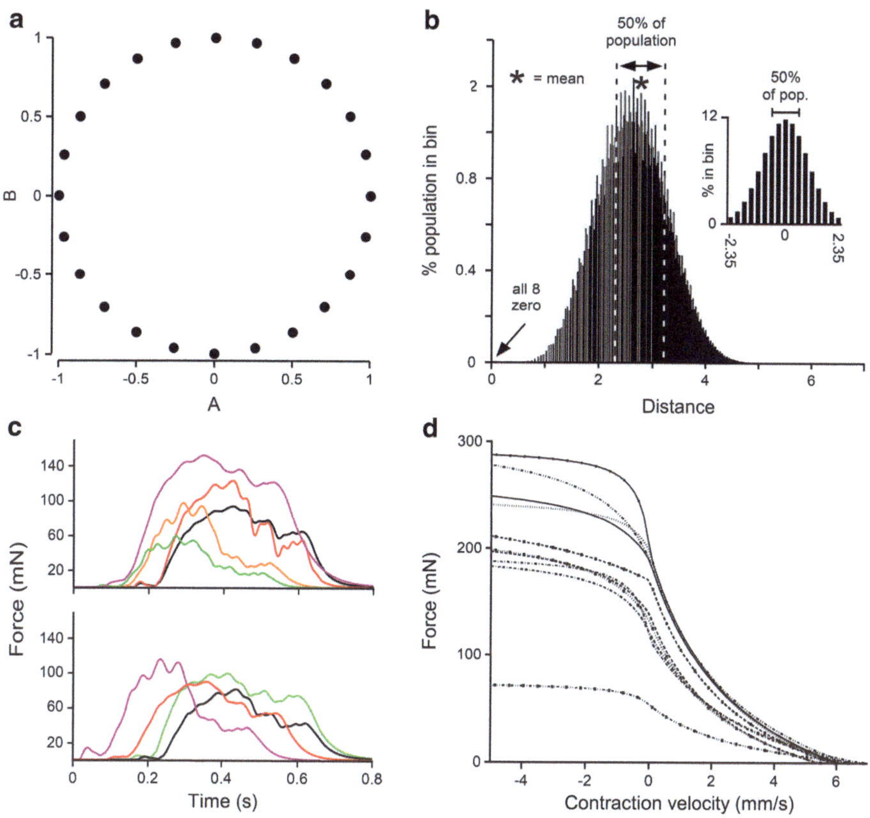


**Fig. 6.7** In populations characterized by multiple parameters, mean data can be unrepresentative of any member of the population and using population mean parameter values may be inadequate for accurate modeling. **a** A population whose members are defined by two parameters, A and B, that co-vary according to $A^2+B^2=1$. No member of the population has the population mean value of $A=B=0$. **b** A population defined by eight parameters, each of which varies from −2.35 to 2.35 in a Gaussian fashion (inset). If each member is defined by its Euclidean distance from the system mean (0,0,0,0,0,0,0,0), the mean distance of the population is about 2.75, and 50% of the population have distances between about 2.25 and 3.25. The percentage of the population with mean values for all eight parameters is vanishingly small (*arrow* pointing to a distance of zero). **c** Stick insect extensor muscles from nine animals respond differently to identical motor nerve stimulation (stimulation pattern, not shown, was motor neuron activity recorded previously during a step in another animal). Responses shown in two panels for clarity. Different colors in each panel activity of different muscles. **d** The nine muscles have different FV (and also $F_{SE}$, $F_{PE}$, and FL) curves. FV curves shown are for act=1. (Panels a, b, and c modified from Blümel et al. 2012c, panel d from Blümel et al. 2012a)

$3.5\times10^{-6}$% (arrow in Fig. 6.7b) of the population has the mean value of 0 in all eight variables (the amplitude of the center column of each variable's distribution, 11.7%, raised to the eighth power).

Each individual in this system can be characterized by the individual's Euclidean distance from the system mean, $\sqrt{\sum_i \left(x_i - \overline{x}_i\right)^2}$, where $x_i$ is variable value and $\overline{x}_i$ is variable mean (in this example, zero for each variable) (Fig. 6.7b). Almost all of the population's individuals are far from the system mean, with a mean distance of about 3, and individuals in the peak of the distribution can be very different, with the individuals (3,0,0,0,0,0,0,0), (0,0,0,0,0,0,0, −3), and (1.06, 1.06, −1.06, 1.06, −1.06, −1.06, 1.06, 1.06) all having distances of 3. In multi-dimensional systems, models using parameter means are thus very unlikely to represent any individual in the population from which the means were obtained, and individuals with the same value of an emergent property (in Fig. 6.7b, distance from system mean) can have very different values of the system's fundamental variables.

In neuromuscular systems the emergent property is to produce correct movement. This requirement can be met in many ways. One is that across-individual muscle parameter variation is small. In this case all muscles will respond similarly to identical motor neuron activity. Another is that muscles fundamentally differ but these differences are compensatory, analogous to how different individuals in the population in Fig. 6.7b can have the same distance if one variable decreases when another increases. In this case the fundamentally different muscles will again respond similarly to identical motor neuron activity. A third possibility is muscles that differ across individuals both in fundamental properties and on the whole muscle level. In this case muscles from different individuals would respond differently to identical motor neuron input, and different animals would therefore need to use different patterns of motor neuron activity to create identical movements.

In stick insect extensor the third possibility is the correct one: identical motor nerve stimulations induce different contractions in muscles from different animals (Fig. 6.7c) (Blümel et al. 2012a, b, c; Hooper et al. 2006, 2007a, b). These differences are not just in amplitude but are also qualitative, with contractions beginning at different times in the stimulation and having different shapes. These differences imply that in each animal different motor neuron firing must occur to generate any desired movement, and that each individual animal's locomotory neural networks must therefore be tuned to match the particular properties of that individual's muscles.

As expected, given these large differences in whole muscle response, stick insect muscle fundamental properties also differ (Fig. 6.7d). Modeling extensor muscles correctly therefore requires measuring, in each individual's muscle, the values of all individual-specific parameters in the model equations (Blümel et al. 2012c). Extensor across-animal parameter variation is large, ranging from 1.3 to 17-fold (Blümel et al. 2012a), and modeling individual muscles with their own parameter values, as opposed to mean parameter values, halves simulation error (Blümel et al. 2012b). A (non-exhaustive) review of the literature shows that large across-individual variation in muscle properties is widespread (spider: Siebert et al. 2010; cockroach: Ahn and Full 2002; locust: Wilson et al. 2010; rat: Bosboom et al. 2001; Gilliver et al. 2011; Grottel and Celichowski 1990; Hawkins and Bey 1997; cat: Brown and Loeb 2000; Herzog et al. 1992; human: Bottinelli et al. 1996; Gilliver et al. 2009;

Maughan et al. 1983; Rubenson et al. 2012; Wickiewicz et al. 1984). Individual by individual measurement of muscle properties and modeling may thus be necessary in most systems for accurate simulation.

### 6.3.3 *Muscles as Extreme Low Pass Filters*

Muscle low-pass filtering can be so large that muscles produce the activity of neural networks that do not innervate them. The pyloric, gastric mill, and cardiac sac neural networks are components of the crustacean stomatogastric nervous system (Selverston et al. 1976). The pyloric network cycles most rapidly, with a cycle period of about 1 s. Figure 6.8a (bottom trace) shows the rhythmic activity of one pyloric motor neuron, a Pyloric Dilator (PD). It would be natural to think (and was until the muscle work was performed) that this activity induced muscle contractions during the bursts and relaxations to rest length between them. The actual activity (top trace, Fig. 6.8a) of one PD-neuron innervated muscle is more complex, with a cycle period (time between equal amplitudes) of about 6 s in which PD neuron bursts first induce large responses that temporally summate to a peak followed by a return to near rest length in which PD neuron bursts induce very small contractions. This slow pattern matches gastric mill network cycle period (second trace, Fig. 6.8a). Detailed analysis reveals a very small gastric-timed modulation of PD neuron activity. The muscle slow filtering extracts this modulation and the muscle thus contracts in gastric mill time even though no gastric mill motor neuron innervates it.

Another PD neuron-innervated muscle has even slower dynamics than the muscle shown in Fig. 6.8a. This muscle's rhythm is even slower, with an approximate 60 s cycle period (top trace, Fig. 6.8b), and the contractions each PD neuron burst induces are very small compared to total amplitude variation (the tiny variations in the trace that look like noise are the contraction increase and decrease associated with each PD neuron burst). The long-period rhythm is in time with the cardiac sac rhythm (rectangles in second trace indicate cardiac sac bursts). PD neuron activity visibly changes during cardiac sac bursts, but the neuron always continues to cycle at approximately 1 Hz (expansion of PD neuron trace). The very slow muscle dynamics extract the cardiac-sac modulation of PD neuron activity and transform the modulation into the muscle's primary output. PD muscles are thus analogous to analog radios, in which the information is not in the carrier wave (the 1 s PD neuron bursting), but in the modulation of the carrier.

Without the muscle data, even something as fundamental as which neural network PD neuron-innervated muscles cycled with would thus be incorrect. Most vertebrate muscles have faster dynamics than the PD muscles. This type of gross misidentification is thus unlikely in these systems. However, slow muscles are much more common in invertebrates, and most of the best understood neural networks are invertebrate. These data thus emphasize the necessity of examining muscle properties in interpreting neural data.

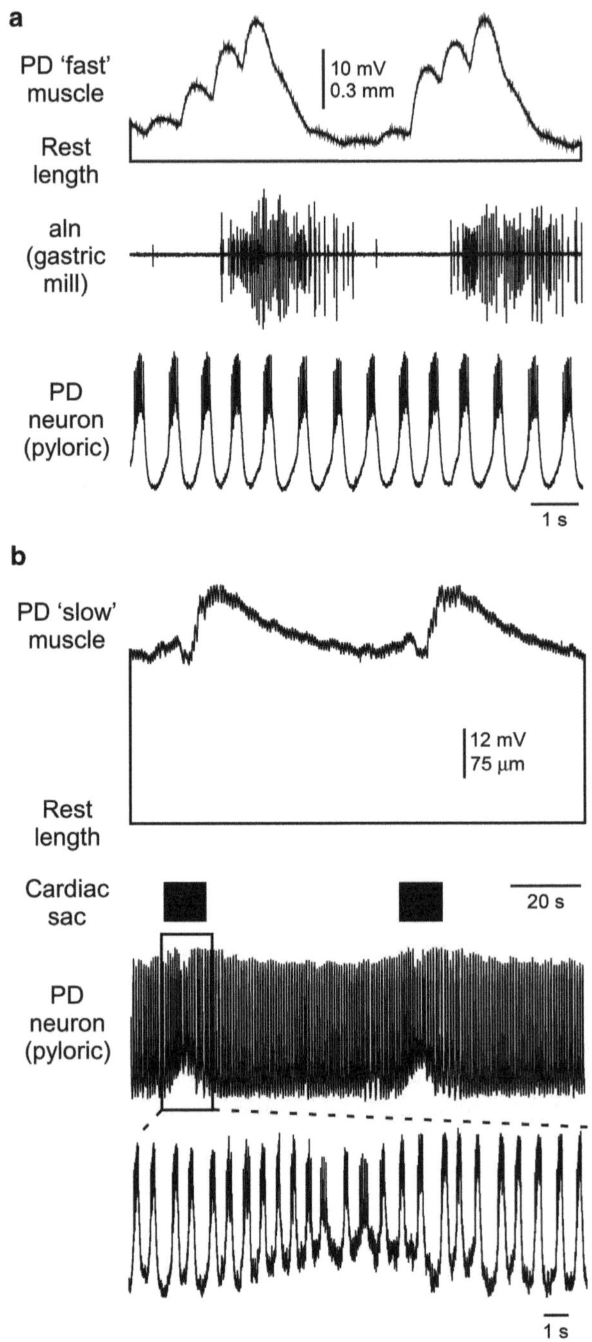


**Fig. 6.8** Slow temporal filtering can result in muscles primarily expressing motor patterns of neural networks none of whose motor neurons innervate the muscles. **a** PD muscle activity (*top trace*) does not fully relax between each PD neuron burst (*bottom trace*). The muscle instead extracts small (invisible at this time scale) variations in PD neuron activity induced by the gastric

### 6.3.4 Spike-Number vs. Spike-Frequency Contraction Amplitude Dependence

It is often stated that motor neuron spike frequency determines muscle contraction amplitude. However, this is not true in all muscles (Fig. 6.9a). Interspike relaxation slope is small at the beginning of spike trains but increases with contraction amplitude, eventually becoming large enough that contraction amplitude reaches a quasi-steady state (right arrow). Quasi-steady state contraction amplitude does depend on spike frequency. Alternatively, during the initial portion of the spike train, amplitude approximately equals spike number times the amplitude increase each spike induces (left arrow). That spike number determines contraction amplitude in real muscles was shown in stomatogastric pyloric muscles (Morris and Hooper 1997). Stimulating a pyloric muscle with 16 spikes at different spike frequencies induces contractions of very nearly equal amplitude (Fig. 6.9b1). Varying spike frequency and duration to create bursts with a variety of spike numbers shows that contraction amplitude depends linearly on spike number up to at least 32 spikes (far larger than physiological burst spike numbers) (Fig. 6.9b2). Spike number dependence also occurs in stick insect extensor muscle (Hooper et al. 2007a).

Controlling contraction amplitude is a primary task of motor neural networks. Spike number is determined by the interaction of burst duration and spike frequency, and thus places a different constraint on neural network output than changing spike frequency alone. Which control strategy is necessary depends in part on muscle response dynamics. Knowing which strategy a motor network uses is thus impossible without examining the muscles the network innervates.

### 6.3.5 The Role of Unstimulated (Passive) Muscle Force in Limb Posture and Movement Control Scales with Limb Size

Muscle passive force scales with muscle cross-section, and thus with animal dimension squared. Limb mass scales with muscle volume, and thus animal dimension cubed. These observations suggest that, in large limbs, muscle passive forces will be insignificant compared to the force of gravity and have small or negligible effects on the momentum of moving limbs. In small limbs, alternatively, passive forces will result in gravity-independent limb postures and, in the absence of motor neuron activity, rapidly return moving limbs to these postures.

mill network, and its activity thus has a cycle period equal to the gastric mill motor network's long cycle period (*middle trace*, approximately 6 s). **b** A slower pyloric muscle cycles almost completely in time with the very slow (cycle period approximately 60 s) stomatogastric system cardiac sac network (2nd trace). This network visibly modulates PD neuron activity (3rd trace), but never stops its approximately 1 s period bursting (time expansion in 4th trace). The tiny wiggles on the PD muscle trace that look like noise are the contractions each PD neuron burst induces. aln = anterior lateral nerve, which carries the output of a gastric mill network motor neuron. (Modified from Morris et al. 2000)

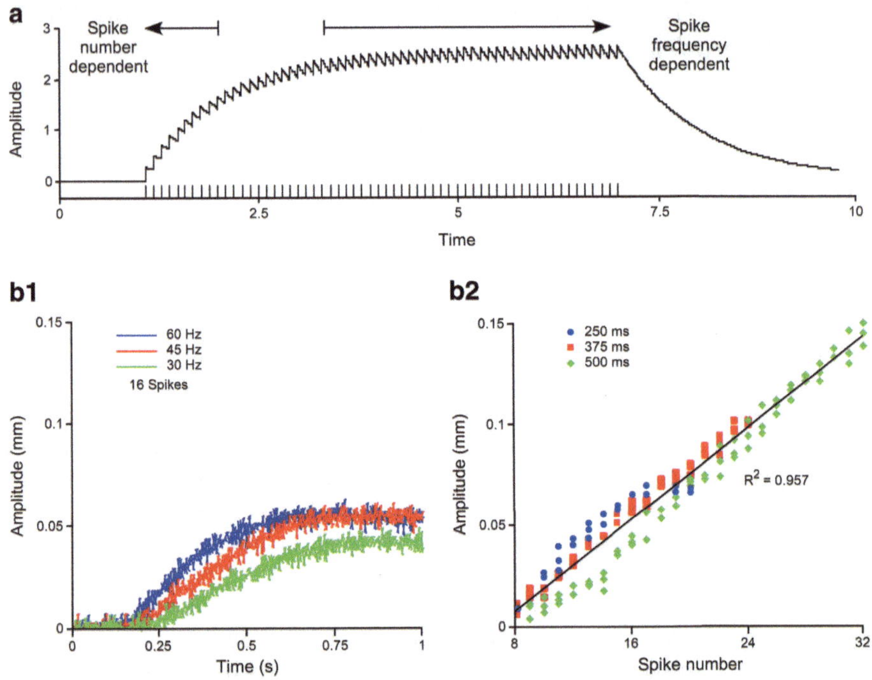


**Fig. 6.9** Contraction amplitude can depend on spike number, spike frequency, or a mixture of the two. **a** Schematic showing that early in a contraction there is little relaxation between motor neuron spikes, and thus achieved amplitude very nearly equals spike number times the amplitude increase induced by a single spike. Late in the contraction interspike relaxation amplitude increases sufficiently that amplitude reaches a quasi-steady state. Quasi-steady state amplitude depends on spike frequency. **b1** Lobster stomatogastric pyloric (PD) muscle contractions induced by 16 spike bursts at 30, 45, and 60 Hz (color code in inset). All three contractions are very similar. **b2** Plot of the amplitudes of contractions induced by motor neuron bursts of different durations and spike frequencies so as to achieve multiple spike numbers per burst. Spike number well predicts contraction amplitude (shown fit is to all data). Plots against spike frequency predict contraction amplitude much less well (data not shown). (Modified from Morris and Hooper 1997)

Data from small and large animals confirm these predictions (Hooper et al. 2009; Ache and Matheson 2012). In humans arm rest position is down regardless of whether the person is standing or held upside down. Gravity thus completely determines shoulder joint rest posture. In stick insects, alternatively, 'shoulder' angle is always approximately 90° relative to the body long axis (Fig. 6.10a) and this posture is maintained without any activity in leg muscle motor neurons. Work in stick insect 'knee' joints shows that these gravity-independent postures are those at which the muscles controlling the joint generate equal passive force. Human fingers at rest similarly assume a 'C'-shaped posture regardless of whether the palm is up or down relative to gravity. Neural networks controlling large limbs thus must always be 'aware of' and compensate for gravity when setting limb posture. Neural

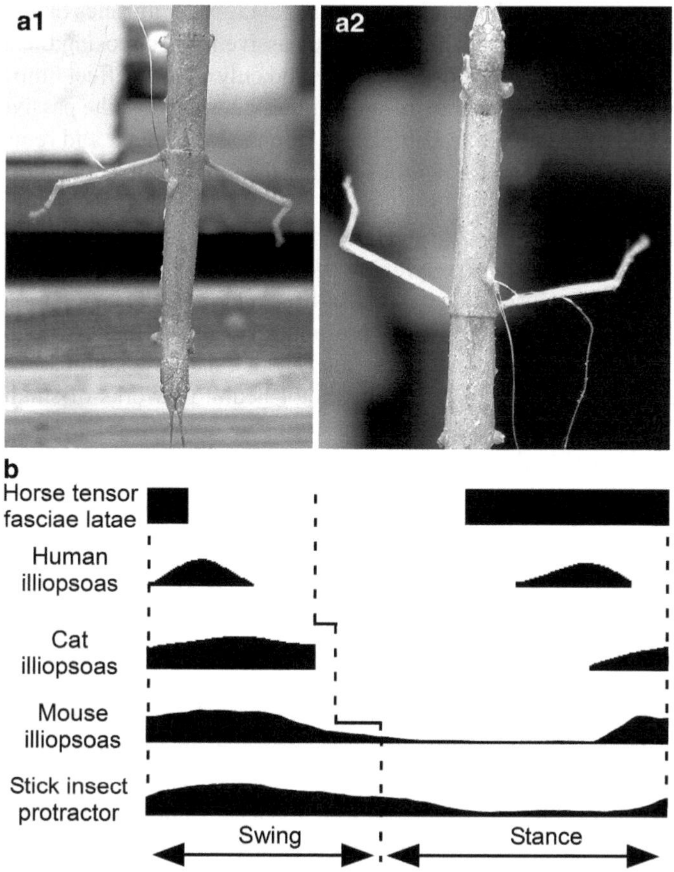


**Fig. 6.10** In small limbs passive (unstimulated) muscle force determines limb rest position in a gravity-independent manner, and likely explains differences in swing motor neuron activity as animal size varies. **a** Stick insect 'shoulder' angle remains constant when the animals are rotated to be head-down (**a1**) or head-up (**a2**). Simultaneous recordings from leg motor nerves show that this gravity-independent posture is maintained without motor nerve activity. This posture is thus due to agonist and antagonist passive muscle forces being at equal at this angle. These forces are large enough that changing the direction of gravity on the limb does not alter shoulder angle. **b** Large animals initiate swing with a brief burst of swing motor neuron activity, after which leg swing continues due to leg momentum alone. In small animals swing motor neurons fire throughout swing. This difference likely arises because, in small limbs, stopping swing muscle contraction would result in antagonist muscle passive force halting swing and moving the leg back to joint rest angle (in stick insect approximately 90° relative to the body's long axis, panel a). In large limbs, limb mass is so large that, even in the absence of continuing swing motor neuron activity, antagonist muscle passive force only slightly slows leg momentum over swing duration. (Modified from Hooper et al. 2009)

networks controlling small limbs, alternatively, can largely or completely ignore gravity for both rest posture and, since active forces are always larger than passive forces, active movement as well.

Limb movement stretches the movement's 'antagonist' muscles (e.g., flexing the elbow stretches the triceps), which generate passive force opposing the movement. In large limbs this passive opposing force would only slightly affect limb movement because limb momentum is large. In small limbs, alternatively, the passive opposing force could be large enough to completely stop the movement, and return the limb to its rest posture, if the motor neurons driving the movement stopped firing. Swing motor neuron firing during limb swing in animals of different sizes supports this hypothesis (Fig. 6.10b). In horse and human legs, swing motor neurons fire at swing beginning and then go silent, relying on the induced limb momentum to finish the movement. In cat, mouse, and stick insect legs, alternatively, swing motor neurons fire continuously throughout swing, presumably because if they stopped, stance muscle passive force would halt the swing (in stick insect, within 0.1 ms) and then return the limb to the rest posture. The motor neural networks of small and large limbs must thus differ not only with respect to generation of joint posture, but also to the generation of limb movement.

### 6.3.6 Generating Forces Greater than $F_{max}$

$F_{max}$ is the maximum force muscles can produce at zero velocity; i.e., the maximum force on the maximum activation curve in Fig. 6.3. Muscles produce forces greater than $F_{max}$ at negative shortening velocities (Fig. 6.4). Animals routinely use the negative velocity portions of the FV curves. For instance, when a human stands on tip toe, the heel raises because the gastrocnemius is activated to generate more force than body weight. If the heel is then lowered in a controlled fashion, the gastrocnemius lengthens while still activated, with the activation being such that $F_{gastrocnemius}$ is always less than or equal to body weight.

As such, in most movements, for a muscle to lengthen against a load it must first shorten against it. The $V_{pos}$ portions of the FV curves are always less, and the $V_{neg}$ portions of the curves always greater, than the force at $V=0$, and in general the same shortening and lengthening portions of the FL curve are traversed. It follows that in general muscle activation will be less during the shortening portion of a movement than the lengthening portion. This observation raises the question of whether animals ever use the portions of the FV curves generating forces greater than $F_{max}$. One example in which this would occur is leaping from heights, as these can be reached by climbing or occur naturally as terrain varies. In these cases, for a controlled landing, the gastrocnemius may indeed need to produce lengthening contractions with sufficient activation to generate forces greater than $F_{max}$.

Gastrocnemius forces as great as 175 % of $F_{max}$ also occur during startle-induced jumps in kangaroo rats (Biewener et al. 1988). Since jumps are caused by gastrocnemius shortening, these forces would seem impossible. The resolution to this difficulty is that shortening muscles can generate forces as great as twice $F_{max}$ if they are activated, stretched, and then allowed to shorten than if they are stretched while unactivated and then activated and allowed to shorten (Cavagna et al. 1965). Jumps with greater than $F_{max}$ forces always occur after prior jumps. During controlled landing from a prior jump the gastrocnemius would be activated to slow ankle flexion, and thus would be lengthening while activated. Once the landing was complete, if the rat immediately jumped again, the Cavagna et al. (1965) data show that the gastrocnemius would generate much more force than if it had passively lengthened during ankle flexion and then been activated. Thus, the gastrocnemius lengthening contractions that allow a controlled landing from a prior jump, if a subsequent jump happens quickly enough after, also allow the gastrocnemius to produce greater than $F_{max}$ forces during the shortening phase of the next jump.

### *6.3.7 Energy Storage in Neuromuscular Systems*

Minimizing energy consumption is an evolutionary pressure. Neuromuscular systems minimize energy consumption in two ways (Cavagna et al. 1964, 1977; Goslow et al. 1973, 1981; Heglund and Cavagna 1987). The first is pendular mechanics. A perfect pendulum swinging in vacuum dissipates no energy, with its total energy (if zero potential energy is defined as the potential energy at swing bottom) cycling smoothly from pure potential energy at the top of the swing to pure kinetic energy at swing bottom. A rigid limb supporting a mass can similarly work as an inverted pendulum. From heel strike to stance midpoint, mass height increases and mass momentum decreases. As the limb moves further back, mass height decreases and mass momentum increases. Swing then occurs, breaks the forward fall, and the cycle continues. The efficiency of this energy transfer is limited by the energy lost at heel strike, which can be minimized by keeping swing amplitude small. In walking this is the primary mechanism of energy conservation, accounting for as much as 70% of the energy changes occurring during a step cycle (Cavagna et al. 1977).

The second mechanism is analogous to the energy exchange of a perfect ball in vacuum, in which at each fall the ball elastically deforms and then, once downward momentum is zero, rebounds to power the next rise of the ball. In legs this elastic exchange occurs by energy storage by muscle and tendon stretch at stance beginning being re-used to power movement at stance end. If the muscles being stretched are activated, the increased force developed by stretching activated muscles (Cavagna et al. 1965) contributes to the energy exchange. This mechanism is used in hopping, trotting, and running (galloping uses both). Efficiencies (positive work done divided by the energetic equivalent of the oxygen consumed) are 35–50% in isolated muscles (provided they are activated before the stretch) and 40–50% during running in intact systems (in which pre-activation occurs naturally) (Cavagna et al. 1964; Heglund and Cavagna 1987). The importance of muscle activation before stretch can be demonstrated in isolated muscles, in which efficiency drops to 15–25% if the muscles are not pre-activated (Heglund and Cavagna 1987).

Some of this increased efficiency may be due to matching of muscle and load impedance. In mechanical systems impedance measures how much the system resists motion when subjected to rhythmic input. Impedance typically varies as a function of input frequency, with maximum energy transfer occurring at frequencies with the lowest impedance. In agonist:antagonist muscle systems, system impedance depends in part on the stiffness of the muscles. Muscle stiffness takes considerable time to decay after muscle contraction. Considering a simple agonist:antagonist pair of muscles (e.g., hip flexor and extensor), the agonist in one phase of the movement (hip flexor during flexion) works against residual stiffness in the antagonist (hip extensor) left over from the previous movement phase (hip extension). If this residual stiffness decreases system impedance at the movement's cycle frequency, this decrease would result in greater power being produced over the cycle period than if the two muscles had been rhythmically activated against antagonist muscles with zero stiffness. When tested with real muscles mechanically arranged to work as an agonist:antagonist pair, anti-phase rhythmic stimulation of the muscles resulted, at the best cycle frequency, in a 7-fold increase in power production relative to the sum of the power produced by each muscle when rhythmically stimulated individually (Farahat and Herr 2010).

### *6.3.8 Muscles that Act as Brakes*

Muscles are typically thought of as producing work—that is, acting as motors. Some muscles, however, absorb energy and thus act as brakes (Ahn and Full 2002; Full et al. 1998). During extension two cockroach hindleg extensor muscles are activated approximately equally and shorten at the same velocity. However, one muscle is approximately twice as long as the other. As explained above, this difference results in the muscles having different FV curves, with the long muscle developing more force than the short muscle at all shortening velocities (Fig. 6.11a). The short muscle consequently produces much more force during flexion (when it is passively lengthening) than it does during its 'active' shortening (actually driven by the long muscle) during extension, and thus produces negative work and power (Fig. 6.11b).

The energy-absorbing muscle is believed to help control and reverse flexion (Full et al 1998). The extensor muscles would thus have different functions, the longer powering extension and the shorter braking flexion. This raises the question of why a single muscle could not serve both functions, being activated in extension to power extension and in flexion to help end flexion. One answer may be the history-dependent nature of evolution, which can lead to solutions that while highly functional, are not what an engineer would have designed (e.g., the bones of the ear). Another is that, given the relatively slow dynamics of muscle and the high step frequencies of cockroach, it is impossible for a single muscle to fulfill both functions.

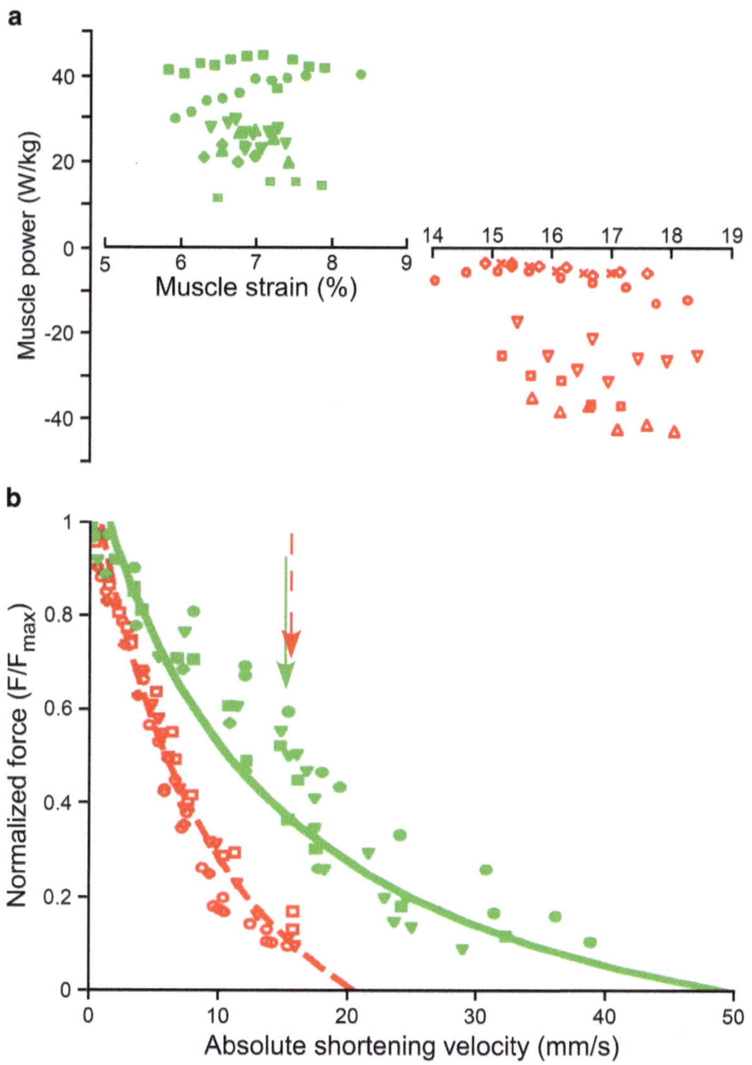


**Fig. 6.11** Muscles act as both motors (power producers) and brakes (power absorbers). **a** Power produced (positive values, *green* data) or absorbed (negative values, *red* data) by two cockroach leg extensor muscles (*green* data, muscle 177c; *red*, muscle 179) that span the same joint during movements with motor nerve stimulation mimicking *in vivo* activity. Different symbols are data from muscles from different animals. **b** The difference between the two muscles arises because the power-absorbing muscle is shorter than the power-producing muscle. During leg extension both muscles shorten with the same velocity (arrows), but because of their length difference the short muscle works near the end of its FV curve. The short muscle thus generates very little force, and the long muscle substantial force, when activated. The force the short muscle produces when it is stretched during flexion is greater than the force it produces during extension, and over a step cycle the muscle absorbs power. The long muscle, alternatively, produces much more force when it shortens than when it is being passively lengthened, and thus over a step cycle produces power. (Modified from Ahn and Full 2002)

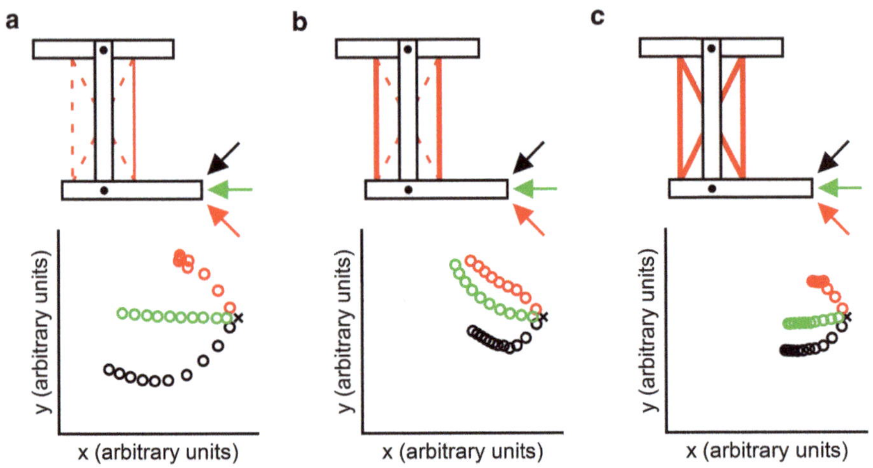


**Fig. 6.12** Muscle activation reduces perturbation-induced movement, and thus acts as a zero-delay restorative reflex (a preflex). In all three panels the top is a schematic of an arm (*rectangles*) with six muscles (*red lines*). *Dashed lines* are inactive muscles; increasing thickness of *full lines* shows increasing activation. **a** When only one muscle is activated, sufficient to keep the elbow at 90°, perturbing forces delivered to the hand at different angles (*colored arrows*, *top panel*) result in large changes in hand position (*bottom panel*, each circle is hand position at 10 ms intervals; circle color corresponds to perturbation direction in top panel). **b** Activating two limb muscles, to a greater but equal degree so that elbow angle is maintained at 90, decreases hand movement for the down perturbation. **c** Activating all six muscles, still such that elbow posture is maintained, decreases hand movement for all perturbations. In all *bottom panels* the 'x' indicates initial hand position. (Modified from Brown and Loeb 2000)

### 6.3.9 Preflexes

Preflexes are zero-delay, muscle intrinsic restorative responses to perturbation. The stable stationary points in Fig. 6.5 are an example. This concept was first developed from a model of human arm responses to perturbations applied to the hand (top panels in Fig. 6.12; arrows indicate perturbation direction) (Brown and Loeb 2000). The model has six muscles (red lines in Fig. 6.12; dashed lines, unstimulated; increasingly thick solid lines, increasing activation). When only one muscle is active, at only a level sufficient to keep the elbow at 90° (top panel Fig. 6.12a), the perturbations induce large hand movements (bottom panel Fig. 6.12a; each circle is hand position at 10 ms intervals after the perturbation). Activating both long muscles to a greater degree altered hand trajectories and, for the down perturbation, reduced hand movement amplitude (Fig. 6.12b). Activating all six muscles reduced hand movement amplitude for all perturbations (Fig. 6.12c).

Preflexes are much faster than neuron-based sensory-feedback loops, and so may play an important role in rapid restorative responses (Full and Koditschek 1999). In rapidly executed motor patterns, which preclude sensory-feedback based mechanisms because of their long intrinsic delays, preflexes may be the only restorative responses. Importantly, as in Fig. 6.12, in most motor systems many different muscle activation patterns can produce the same initial state (e.g., all three muscle contraction patterns in Fig. 6.12 induce a 90° elbow angle). System response to perturbation thus depends not only on the limb posture present at the time of perturbation, but also on the pattern of muscle activation. A pressure in motor neural network evolution may thus be not only to generate a desired movement or posture, but to do so with a muscle activation pattern that makes the system resistant to likely perturbations of the desired movement or posture. A seeming redundancy of neuromuscular systems, that they generally have more muscles than necessary to produce a limb's movement repertoire, may reflect this second function of muscles to promote movement and posture stability (Brown and Loeb 2000).

## 6.4 Summary

Muscles are complicated, non-linear, hierarchical systems that can produce widely different responses to identical driving input depending on variations in their fundamental properties. These properties vary not only between different muscle (biceps vs. triceps), but also between the 'same' muscle (biceps) across individuals. Accurate muscle modeling likely requires modeling muscles as individual entities, and motor networks are likely tuned to the specific properties of an individual's muscles. It is impossible to predict movement from motor neuron activity alone, and even such elementary characteristics as what motor pattern a muscle produces, or whether a muscle acts as a motor or a brake, cannot be known without detailed examination of the muscle in question. The importance of muscle passive properties in determining limb rest posture and how motor networks control movement scales with limb size. Muscles have evolved not only to generate movement, but to generate movement in energetically efficient manners and to create motor systems that have intrinsic restorative mechanisms. Understanding how nervous systems generate movement cannot be achieved without detailed understanding of the muscles the system innervates.

## References

Ache JM, Matheson R (2012) Passive joint forces are tuned to limb use in insects and drive movements without motor activity. Curr Biol 23:1418–1426
Ahn AN, Full RJ (2002) A motor and a brake: two leg extensor muscles acting at the same joint manage energy differently in a running insect. J Exp Biol 205:379–389
Arendt D (2008) The evolution of cell types in animals: emerging principles from molecular studies. Nat Rev Genet 9:868–882
Atwood HL (1967) Crustacean neuromuscular mechanisms. Am Zool 7:527–551
Atwood HL (1971) Z and T tubules in stomach muscles of the spiny lobster. J Cell Biol 50:264–268
Atwood HL (1972) Crustacean muscle. In: Bourne GH (ed) Structure and function of muscle. Academic Press, New York, pp 421–489

Atwood HL (1973) An attempt to account for the diversity of crustacean muscles. Am Zool 13:257–378

Atwood HL (1976) Organization and synaptic physiology of crustacean neuromuscular systems. Prog Neurobiol 7:291–391

Biewener AA, Blickhan R, Perry AK, Heglund NC, Taylor CR (1988) Muscle forces during locomotion in kangaroo rats: force platform and tendon buckle measurements compared. J Exp Biol 137:191–205

Blümel M, Guschlbauer C, Daun-Gruhn S, Hooper SL, Buschges A (2012a) Hill-type muscle model parameters determined from experiments on single muscles show large animal-to-animal variation. Biol Cybern 106:559–571

Blümel M, Guschlbauer C, Hooper SL, Buschges A (2012b) Using individual-muscle specific instead of across-muscle mean data halves muscle simulation error. Biol Cybern 106:573–585

Blümel M, Hooper SL, Guschlbauer C, White WE, Buschges A (2012c) Determining all parameters necessary to build Hill-type muscle models from experiments on single muscles. Biol Cybern 106:543–558

Bosboom EMH, Hesselink MKC, Oomens CWJ, Bouten CVC, Drost MR, Baaijens FPT (2001) Passive transverse mechanical properties of skeletal muscle under *in vivo* compression. J Biomech 34:1365–1368

Bottinelli R, Canepari M, Pellegrino MA, Reggiani C (1996) Force-velocity properties of human skeletal muscle fibres: myosin heavy chain isoform and temperature dependence. J Physiol 495:573–586

Brown IE, Loeb GE (2000) A reductionist approach to creating and using neuromusculoskeletal models. In: Winters JM, Crago PE (ed) Biomechanics and neural control of posture and movement. Springer, New York, pp 148–163

Brown IE, Scott SH, Loeb GE (1996) Mechanics of feline soleus: II. Design and validation of a mathematical model. J Muscle Res Cell Motil 17:221–233

Brown IE, Cheng EJ, Loeb GE (1999) Measured and modeled properties of mammalian skeletal muscle. II. The effects of stimulus frequency on force-length and force-velocity relationships. J Muscle Res Cell Motil 20:627–643

Cavagna GA, Saibene FP, Margaria R (1964) Mechanical work in running. J Appl Physiol 19:249–256

Cavagna GA, Saibene FP, Margaria R (1965) Effect of negative work on the amount of positive work performed by an isolated muscle. J Appl Physiol 20:157–158

Cavagna GA, Heglund N, Taylor R (1977) Mechanical work in terrestrial locomotion: two basic mechanisms for minimizing energy expenditure. Am J Physiol 233:243–261

Cheng EJ, Brown IE, Loeb GE (2000) Virtual muscle: a computational approach to understanding the effects of muscle properties on motor control. J Neurosci Methods 101:117–130

Chiel H, Beer R (1997) The brain has a body: adaptive behavior emerges from interactions of nervous system, body and environment. Trends Neurosci 20:553–557

Costello WJ, Govind CK (1983) Contractile responses of single fibers in lobster claw closer muscles: correlation with structure, histochemistry, and innervation. J Exp Zool 227:381–393

Edman KAP (1979) The velocity of unloaded shortening and its relation to sarcomere length and isometric force in vertebrate muscle fibers. J Physiol 291:143–159

Farahat WA, Herr HM (2010) Optimal workloop energetics of muscle-actuated systems: an impedance matching view. PLoS Comput Biol 6:e1000795. doi:10.1371/journal.pcbi.1000795

Full RJ, Koditschek DE (1999) Templates and anchors: neuromechanical hypotheses of legged locomotion on land. J Exp Biol 202:3325–3332

Full RJ, Stokes D, Ahn A, Josephson R (1998) Energy absorption during running by leg muscles in a cockroach. J Exp Biol 201:997–1012

Gilliver SF, Degens H, Rittweger J, Sargeant AJ, Jones DA (2009) Variation in the determinants of power of chemically skinned human muscle fibres. Exp Physiol 94:1070–1078

Gilliver SF, Jones DA, Rittweger J, Degens H (2011) Variation in the determinants of power of chemically skinned type I rat soleus muscle fibres. J Comp Physiol [A] 197:311–319

Goslow GE Jr, Reinking RM, Stuart DG (1973) The cat step cycle: hind limb joint angles and muscle lengths during unrestrained locomotion. J Morphol 141:1–42
Goslow GE Jr, Seeherman HJ, Taylor CR, McCuthin MN, Heglund NC (1981) Electrical activity and relative length changes of dog limb muscles as a function of speed and gait. J Exp Biol 94:15–42
Govind CK, Atwood HL, Maynard DM (1975) Innervation and neuromuscular physiology of intrinsic foregut muscles in the blue crab and spiny lobster. J Comp Physiol [A] 96:185–204
Grottel K, Celichowski J (1990) Division of motor units in medial gastrocnemius muscle of the rat in the light of variability of their principal properties. Acta Neurobiol Exp (Warsz) 50:571–588
Guschlbauer C, Scharstein H, Büschges A (2007) The extensor tibiae muscle of the stick insect: biomechanical properties of an insect walking leg muscle. J Exp Biol 210:1092–1108
Haeufle DFB, Grimmer S, Seyfarth A (2010) The role of intrinsic muscle properties for stable hopping–stability is achieved by the force-velocity relation. Bioinspir Biomim 5:016004 doi:10.1088/1748-3182/5/1/016004
Hawkins D, Bey M (1997) Muscle and tendon force-length properties and their interactions *in vivo*. J Biomech 30:63–70
Heglund NC, Cavagna GA (1987) Mechanical work, oxygen consumption, and efficiency in isolated frog and rat muscle. Am J Physiol 253:C22–C29
Herzog W, Leonard TR, Renaud JM, Wallace J, Chaki G, Bornemisza S (1992) Force-length properties and functional demands of cat gastrocnemius, soleus and plantaris muscles. J Biomech 25:1329–1335
Hill AV (1938) The heat of shortening and the dynamic constants of muscle. Proc R Soc Lond B Biol Sci 126:136–195
Hill AV (1950) The series elastic component of muscle. Proc R Soc Lond B Biol Sci 141:104–117
Hooper SL, Guschlbauer C, von Uckermann G, Büschges A (2006) Natural neural output that produces highly variable locomotory movements. J Neurophysiol 96:2072–2088
Hooper SL, Guschlbauer C, von Uckermann G, Büschges A (2007a) Different motor neuron spike patterns produce contractions with very similar rises in graded slow muscles. J Neurophysiol 97:1428–1444
Hooper SL, Guschlbauer C, von Uckermann G, Büschges A (2007b) Slow temporal filtering may largely explain the transformation of stick insect (*Carausius morosus*) extensor motor neuron activity into muscle movement. J Neurophysiol 98:1718–1732
Hooper SL, Guschlbauer C, Blümel M, Rosenbaum P, Gruhn M, Akay T, Büschges A (2009) Neural control of unloaded leg posture and of leg swing in stick insect, cockroach, and mouse differs from that in larger animals. J Neurosci 29:4109–4119
Hoyle G (1967) Specificity of muscle. In: Wiersma CAG (ed) Invertebrate muscle systems. University of Chicago Press, Chicago, pp 151–167
Hoyle G (1969) Comparative aspects of muscle. Annu Rev Physiol 31:43–84
Huxley AF, Niedergerke R (1954) Structural changes in muscle during contraction. Nature 173:971–973
Jahromi SS, Atwood HL (1969a) Correlation of structure, speed of contraction, and total tension in the fast and slow abdominal muscle fibers of the lobster (*Homarus americanus*). J Exp Zool 171:25–38
Jahromi SS, Atwood HL (1969b) Structural features of muscle fibres in the cockroach leg. J Insect Physiol 15:2255–2262
Jahromi SS, Atwood HL (1971) Structural and contractile properties of lobster leg-muscle fibers. J Exp Zool 176:475–486
Jewell BR, Wilkie DR (1958) An analysis of the mechanical components in frog's striated muscle. J Physiol 143:515–540
Komuro T (1981) Fine structural study of the abdominal muscle receptor organs of the crayfish (*Procambuarus clarkii*). Fast and slow receptor muscles. Tissue Cell 13:79–92
Lang F, Costello WJ, Govind CK (1977) Development of the dimorphic claw closer muscles of the lobster *Homarus americanus*. I Regional distribution of muscle fiber types in adults. Biol Bull 152:75–83

Maughan RJ, Watson JS, Weir J (1983) Strength and cross-sectional area of human skeletal muscle. J Physiol 338:37–49
Morris LG, Hooper SL (1997) Muscle response to changing neuronal input in the lobster (*Panulirus interruptus*) stomatogastric system: spike number- versus spike frequency-dependent domains. J Neurosci 17:5956–5971
Morris LG, Thuma JB, Hooper SL (2000) Muscles express motor patterns of non-innervating neural networks by filtering broad-band input. Nat Neurosci 3:245–250
Parnas I, Atwood HL (1966) Phasic and tonic neuromuscular systems in the abdominal extensor muscles of the crayfish and rock lobster. Comp Biochem Physiol 18:701–723
Royuela M, Meyer-Rochow VB, Fraile B, Paniagua R (1998) Ultrastructure of muscle cells in *Acetabulostoma* (Crustacea, Ostracoda)—mussel shrimp from the Ross Sea (Antarctica). Polar Biol 20:77–84
Royuela M, Fraile B, Arenas MI, Paniagua R (2000) Characterization of several invertebrate muscle cell types: a comparison with vertebrate muscles. Microsc Res Tech 48:107–115
Rubenson J, Pires NJ, Loi HO, Pinniger GJ, Shannon DG (2012) On the ascent: the soleus operating length is conserved to the ascending limb of the force-length curve across gait mechanics in humans. J Exp Biol 215:3539–3551
Scott SH, Brown IE, Loeb GE (1996) Mechanics of feline soleus: I. Effect of fascicle length and velocity on force output. J Muscle Res Cell Motil 17:207–219
Selverston AI, Russell DF, Miller JP, King DG (1976) The stomatogastric nervous system: structure and function of a small neural network. Prog Neurobiol 7:215–290
Siebert T, Weihmann T, Rode C, Blickhan R (2010) *Cupiennius salei*: biomechanical properties of the tibia-metatarsus joint and its flexing muscles. J Comp Physiol 180:199–209
Thuma JB, Hooper SL (2010) Direct evidence that stomatogastric (*Panulirus interruptus*) muscle passive responses are not due to background actomyosin cross-bridges. J Comp Physiol [A] 196:649–657
Thuma JB, Harness PI, Koehnle TJ, Morris LG, Hooper SL (2007) Muscle anatomy is a primary determinant of muscle relaxation dynamics in the lobster (*Panulirus interruptus*) stomatogastric system. J Comp Physiol [A] 193:1101–1113
Wickiewicz TL, Roy RR, Powell PL, Perrine JJ, Edgerton VR (1984) Muscle architecture and force-velocity relationships in humans. J Appl Physiol Respirat Environ Exercise Physiol 57:435–443
Wilkie DR (1950) The relationship between force and velocity in human muscle. J Physiol 110:249–280
Wilkie DR (1956) The mechanical properties of muscle. Br Med Bull 12:177–182
Wilson E, Rustighi E, Mace BR, Newland PL (2010) Isometric force generated by locust skeletal muscle: responses to single stimuli. Biol Cybern 102:503–511

# Part III
# Neuromechanical Modeling of Posture and Postural Control

# Chapter 7
# Why Is Neuromechanical Modeling of Balance and Locomotion So Hard?

**Jessica L. Allen and Lena H. Ting**

**Abstract** A goal and challenge in neuromechanical modeling is to develop validated simulations to predict the effects of neuromotor deficits and therapies on movements. This has been particularly challenging in balance and locomotion because they are inherently unstable, making it difficult to explore model parameters in a way that still coordinates the body in a functional way. Integrating realistic and validated musculoskeletal models with neural control mechanisms is critical to our ability to predict how human robustly move in the environment. Here we briefly review both human locomotion models, which generally focus on modeling the physical dynamics of movement with simplified models of neural control, as well as balance models, which model sensorimotor dynamics and processing with simplified biomechanical models. Combining complex neural and musculoskeletal models increases the redundancy in a model and allows us to study how motor variability and robustness are exploited to produce movements in both healthy and impaired individuals. To advance, the integration of neuromechanical modeling and experimental approaches will be critical in testing specific hypotheses concerning how and why neuromechanical flexibility is both exploited and constrained under various movement contexts. We give a few examples of how the close interplay between models and experiments can reveal neuromechanical principles of movement.

**Keywords** Balance control · Biomechanics · Musculoskeletal modeling · Neuromechanics · Locomotion · Postural control · Muscle synergies · Sensorimotor control · Sensorimotor integration · Sensorimotor feedback

L. H. Ting (✉) · J. L. Allen
Coulter Department of Biomedical Engineering at Georgia Institute of Technology and Emory University, 1760 Haygood Dr, Ste W 200, Atlanta, GA 30322-4350, USA
e-mail: lting@emory.edu

J. L. Allen
e-mail: jessica.allen@emory.edu



B. I. Prilutsky, D. H. Edwards (eds.), *Neuromechanical Modeling of Posture and Locomotion,* Springer Series in Computational Neuroscience,
DOI 10.1007/978-1-4939-3267-2_7

## 7.1 Introduction

Our ability to use models to simulate and predict motor behaviors in humans based on principles of neural control and biomechanics remains elusive. This is particularly true for whole-body behaviors that are inherently unstable and involve significant balance control, including locomotor behaviors such as walking and running. Balance and locomotor functions are impaired by a wide range of neurological and musculoskeletal disorders, negatively affecting quality of life (CDC 2008). Validated and predictive neuromechanical models have the potential to help diagnose, accurately characterize, and optimize treatments for a broad range of movement deficits. These models could allow researchers and clinicians to play "*what if*" by changing the neural control and biomechanical parameters of the model and observing the motor behaviors that emerge. Thus, a grand challenge in neuromechanical modeling is to develop predictive models that can be used to understand effects of neuromuscular deficits, predict outcomes of rehabilitation interventions, and develop individualized interventions and therapies optimized to the abilities of each participant.

There are a number of barriers to the development of validated neuromechanical models in balance and locomotion that we discuss in this chapter. In the first part of this chapter we review the current state of the art in human locomotion and balance models. Although walking and balance are integrated motor behaviors, there is a disciplinary divide in the investigation of balance control and locomotion, both experimentally and computationally. Neuromechanical models of human locomotion and balance thus reflect this divide and are not well integrated. Whereas the goal in locomotion is to move the body from one place to another resulting in large, measurable external movements amenable to biomechanical analysis, in standing balance the goal is to maintain a stable posture resulting in very small motions that belie the complex underlying sensorimotor processing. To date, much of the focus of locomotion modeling has been on reproducing the mechanics of the movement whereas balance modeling has typically focused on understanding the underlying neural control mechanisms, with very minimal overlap between the two modeling endeavors.

To advance, it will be necessary to develop integrated experiments and neuromechanical models to test fundamental sensorimotor principles for integration of balance and locomotion. A primary challenge is in validating both the neural and biomechanical elements of a neuromechanical model. Neuromechanical models may be a necessary tool to decompose and interpret the multifaceted ways in which balance and locomotor function can be achieved in different individuals and impairments. Experimental observations are difficult to interpret from a mechanistic level because internal signals such as muscle force and neural activity cannot be directly measured. Moreover, there is a great deal of redundancy in biomechanical and neural contributions to locomotion and balance, which may be a major contributor to the high level of variability observed in muscle and neural activity compared to measures of motor function. Using neuromechanical models tailored to a particular hypothesis and experimental paradigm will thus be critical for drawing conclusions about how balance and locomotion are jointly achieved. In the second part of the chapter we will give some examples from our work of how experimental and computational approaches can be combined to reveal fundamental principles of human balance and locomotion that could form a foundation for the development of predictive neuromechanical models.

## 7.2 Current Neuromechanical Models of Locomotion and Balance

### *7.2.1 Locomotion Models are Mechanics-Focused*

Neuromechanical models of locomotion, from the very simple to the very complex, typically focus on reproducing the mechanics of gait without emphasis on underlying neural control mechanisms. The simplest models demonstrate that locomotor-like motion can be generated almost entirely by passive. More complex models typically focus on identifying the necessary joint torques or muscle forces that replicate experimentally observed locomotor patterns, but without regard to how they are generated from a neural or sensorimotor perspective. Such models have generated a large body of literature regarding the mechanics of walking and how muscles can be coordinated to achieve the necessary biomechanical subtasks of locomotion. Both simple and complex models are essential for understanding the physical constraints and principles governing locomotion, defining the biomechanical tasks and functions critical for a successful neural controller to achieve. In some cases they can identify aspects of the movement that are governed by the properties of the musculoskeletal system as well as performance criteria governing kinematic and muscular patterns. However, to date, human locomotion models generally do not address specific neural control mechanisms that contribute to the generation of locomotor motor patterns and movement strategies.

### *7.2.2 Simple Locomotion Models Describe Body Mechanics*

Point mass models of locomotion are important because they may describe the targets of neural control (e.g., neural control variables) and demonstrate the reconfigurability and versatility of the body when subject to neural control. The simplest locomotion models are typically energetics based, motivated by experimental evidence of the different mechanisms for exchange of kinetic and potential energy during walking and running (Cavagna et al. 1964, 1977; Dickinson et al. 2000). These energetic exchanges resemble the patterns that occur in common mechanical elements subject to gravity, such as pendulums and springs. Simple locomotion models based on such elements require minimal control and instead rely on initial conditions and gravity to produce motion. They are thought to reflect energetically efficient modes of locomotion that result from the optimal tuning and interactions between complex neural and musculoskeletal systems (Srinivasan and Ruina 2006) and the overall movement goals of neuromechanical control (Full and Koditschek 1999).

Multisegmental models can reveal features of locomotion that are dominated by the mechanical properties of the body versus those that require more active neural regulation. Such models have facilitated the development of more energy efficient robots and revealed mechanisms of human gait. For example, walking motions can be achieved with minimal actuation by harnessing the passive dynamics of pendulum-like limbs. Machines built on such mechanisms can walk down a gentle slope without any active control of the joints (McGeer 1990a, b) and require only minimal actuation when walking on level ground (e.g., Collins et al. 2001, 2005; Collins and Ruina 2005). These passive walking models reveal that the swing phase of walking can be driven entirely by passive mechanics of the swinging limb resulting from gravity, which has been corroborated through comparison to experimental data (Mochon and McMahon 1980; Hoy and Zernicke 1986). Although muscle activity is often present during swing (Sutherland 1984; Perry 1992) the fact that passive dynamics can generate realistic swing kinematics suggests that neural control may be less critical within this region of the gait cycle. In contrast, appropriately timed hip or ankle torque in late stance can propel the center-of-mass (CoM) forward and is sufficient to generate steady state walking on level ground (Collins et al. 2005; Collins and Ruina 2005; Wisse et al. 2007). The need for actuation for propulsion provides evidence that neural mechanisms are necessary to produce the biomechanical function of forward propulsion. Running dynamics can be reproduced by passive spring-mass systems (Blickhan 1989; McMahon and Cheng 1990) and the addition of springs to pendulum-based models further improves the realism of gait features such as the characteristic M-shaped curve of the vertical ground reaction force in walking (Geyer et al. 2006). This suggests that the characteristics of the mechanical system are equally important to the neural controller in generating realistic movements. Different gaits can be achieved by modulating the leg stiffness or compliance, which could be achieved biologically through the active regulation of neural commands to muscles. However, neural control mechanisms cannot be explicitly examined in such models and require the addition of actively controlled muscle models with appropriate mechanical properties.

### *7.2.3 Complex Locomotion Models Describe Musculoskeletal Mechanics*

Neuromechanical models that incorporate muscles are necessary in order to examine the detailed neural excitation signals to multiple muscles necessary to drive locomotion. A primary focus of musculoskeletal modeling studies of locomotion has been to find the appropriate muscle excitation patterns to reproduce a particular gait that was measured experimentally. These studies have provided important information about the biomechanical functions of muscles during walking and how muscles can be coordinated to produce the necessary biomechanical subtasks of walking. However, because of the redundancy in the musculoskeletal system in production of actions, there are many feasible solutions to a particular motor task (Collins 1995; Martelli et al. 2013; Sohn et al. 2013; Simpson et al. 2015) and so optimization techniques are used to identify a single pattern of muscle activation that are sufficient to replicate experimentally observed kinematics and/or kinetics (e.g., Neptune et al. 2001; McLean et al. 2003; Liu et al. 2008). The criteria for optimization are typically based on minimizing muscle stress, which has been assumed to be a goal of the neural control system (Crowninshield and Brand 1981; Collins 1995; Erdemir et al. 2007). The choice of the cost function can have large effects on the selected optimal solution of muscle activations and thus realistic optimality criteria are essential for predictive, optimal control models that can be used to examine the effects of surgical interventions, rehabilitation, etc (Ackermann and van den Bogert 2010). However, even if realistic optimality criteria are chosen, the patterns of muscle activation are found without specific consideration of various neural mechanisms that may impose other constraints or features (Ting et al. 2012). Therefore, while the physics of the motion may be accurate in such models, the neural origin on muscle activity (e.g. feedforward motor pattern vs. feedback response to error) cannot be identified.

Another common focus of studies using detailed musculoskeletal models is to understand how altered muscle coordination in clinical populations (e.g., in cerebral palsy, stroke, amputation, etc.) affects walking performance (Higginson et al. 2006; Hall et al. 2011; Peterson et al. 2011; Silverman and Neptune 2012; Steele et al. 2012, 2013). Towards this goal, a variety of studies have examined how the contributions of specific muscles to important biomechanical subtasks of walking are altered. However, such models cannot provide information about *why* muscle coordination is changed because muscle coordination is usually determined only by reproducing experimentally observed data. Therefore, dissociation between the effect of an initial impairment versus compensation for that impairment is difficult and yet gait deficiencies resulting from impairment versus compensation may benefit from different rehabilitation strategies. It is also difficult to make predictions using the results of models that were found by tracking experimental measures and do not contain explicit representations of neural mechanisms that can be altered. Even if it were possible, the number of parameters in the model make it difficult to “emerge” new strategies based on various performance criteria and constraints compared to simpler models (e.g., Srinivasan and Ruina 2006).

### *7.2.4 Locomotion Models Designed to Incorporate Neural Control Are Unvalidated*

There are several examples of locomotion models that can produce stable human-like locomotion using biologically-inspired control mechanisms yet none have been rigorously tested as a predictive model of real human walking. Most neurally-driven models are based upon the hypothesis that spinal central pattern generators (CPGs) produce feedforward muscle excitation patterns that are modulated by sensory feedback. But, how the CPG is modeled, which types of sensory feedback are utilized and how sensory information is integrated differs widely across models. For example, there are CPG models based on the half-center hypothesis (Brown 1914) in which alternating flexor and extensor activity produces the basic muscle activity underlying gait (Taga 1995a, b; Ogihara and Yamazaki 2001; Paul et al. 2005), while others are based on a more complex two-layer CPG network composed of a rhythm generator that provides the basic locomotor rhythm and a pattern formation network that distributes this activity to the appropriate set of muscles (Jo and Massaquoi 2007; Aoi et al. 2010; Aoi 2015). Similarly, these models also use different methods of modifying the CPG and its outputs based on sensory feedback. Despite these differences, however, all of these models reproduce locomotion that is qualitatively similar to experimentally observed gait patterns. Although some validation of each control scheme has been performed, none of the models have been rigorously tested against experimental data, particularly data designed specifically to test the validity of specific components of the model. This is particularly important as these models typically incorporate a large number of parameters in both neural and biomechanical components, increasing the sources of redundancy. Thus, while current neuromechanical models of locomotion do provide some insight into the role of different neural mechanisms in locomotion, to date they represent more of a proof of concept such that their utility for addressing questions about the neural mechanisms contributing to locomotion and to predict the effect of lesions is currently limited.

### 7.2.5 Balance Models Are Focused on Understanding Neural Control

An obvious difference in standing balance compared to locomotion is that there is much less movement involved while standing still, and even when responding to perturbations that occur while standing. This may make balance tasks seem boring or trivial from a mechanics perspective and yet there are no feedforward models of balance that can stand independently without eventually going unstable. Thus, from a neural perspective balance is an interesting problem of neural sensorimotor control. The focus of a majority of studies on standing balance control have therefore been on understanding this underlying sensorimotor control system rather than the more detailed mechanics of the musculoskeletal system. A commonly assumed goal of standing balance control is to keep the CoM within the base of support. However, this can be achieved by a number of joint and muscle level strategies. As it is unclear how the sensorimotor system controls the specific patterns associated with different postural strategies, most studies of standing balance control have focused on the control of relatively simple mechanical variables such as body orientation or CoM dynamics without regard for the control of multi-muscle patterns that generate different biomechanical strategies.

### 7.2.6 Simple Models of Balance Reveal Principles of Sensorimotor Control

Torque-actuated, inverted pendulum models are commonly used to investigate the sensorimotor control of standing balance. Typically, torque at the base of the pendulum is representative of muscle actions at the ankle. Models usually include a physiologically-relevant time delay, are inherently unstable, and therefore require some form of neural control scheme to remain upright. There has been considerable debate about the role of feedforward versus feedback control in standing balance control, as well as which variables or error signals are necessary to reliably return the body to equilibrium given a perturbation. Further, which sensory signals are used to generate a feedback signal is unclear, as well as how this information is transformed into motor actions in response to perturbations.

Simple inverted pendulum postural models have been used on both sides of the debate of whether movement during quiet standing is generated from feedforward or feedback neural mechanisms. Even in the absence of external perturbations the body is in constant motion. This motion, typically referred to as postural or body sway, is a low-frequency movement that can easily be quantified using measures of center of pressure (CoP) displacement. It has been suggested the intrinsic stiffness of active muscles about the ankle is sufficient to stabilize the body during quiet stance (Winter et al. 1998). However, torque-actuated inverted pendulum models demonstrate that the necessary ankle stiffness is higher than any experimental estimates of ankle stiffness during standing (Morasso and Schieppati 1999; Peterka 2002), providing evidence that some active control is necessary. Experimental studies of CoP displacement suggest that both feedforward and feedback mechanisms are at play. Competing concepts describe this as either a slow feedforward motion of the desired CoP location with faster feedback oscillations about that trajectory (Zatsiorsky and Duarte 1999), or alternately a feedforward mechanism that acts on short timescale with a slower feedback mechanism acting on longer timescales (Collins and De Luca 1993). Simple models of postural control on both sides of this debate can generate spontaneous body sway and reproduce experimentally observed patterns of CoP deviations. For example, a simple inverted pendulum model controlled using delayed position and velocity feedback on error that simulates sensory noise can reproduce a wide variety of common time and frequency domain CoP measures (Peterka 2000; Maurer and Peterka 2005). In contrast, sway-like motion has also been demonstrated using a predictive feedforward mechanism (Loram et al. 2005; Gawthrop et al. 2009). Because these models have only been used to reproduce observations, they only show that it is possible for either mechanism to control quiet stance. In order to provide strong evidence regarding the role of either mechanism in postural control, experiments must be carefully designed to dissociate the feedforward from feedback components. However, there are limitations in system identification techniques in conditions where sensory noise is dominant over external perturbations as in quiet stance where movement is minimal (van der Kooij et al. 2005).

In response to perturbations during standing balance, simple, torque-actuated inverted pendulum models have provided clear evidence that feedback involving the flexible integration of different sensory systems (e.g., proprioceptive, vestibular and visual information) is necessary for balance control. Modifiable multisensory integration mechanisms are important for providing accurate estimates of body orientation with respect to vertical in a range of biomechanical contexts. Conflicting sensory information can be generated experimentally through perturbations to either the visual system (rotating a visual surround) or the proprioceptive system (rotating the support surface). The reliance on different sensory modalities can be dissociated by fitting experimental sway data to a simple inverted pendulum model with variable weightings on different sensory channels. Whereas healthy subjects tend to rely on proprioceptive information and align themselves with small support-surface rotation amplitude and frequencies, they must increase their reliance on vestibular information to align with the vertical (Peterka 2002, and see Chap. 9 in this volume) for faster and larger components of perturbations. This simple model predicts that stability can no longer be maintained for larger perturbations in the absence of both visual and vestibular information, consistent with responses observed in individuals with bilateral vestibular loss. That the model can correctly predict how balance fails under sensory loss provides support for the proposed control scheme and demonstrates how computational models, when used to interpret carefully designed experiments, can provide evidence regarding underlying neural control and provide a mechanism for examining the cause and effect of neural impairments.

Evidence supporting the role of task-level feedback governing balance control has also come from inverted pendulum models incorporating muscles to produce the torque about the ankle. Examining muscle activity in response to perturbations has provided evidence that the transformation from sensory information to motor action is based on the estimation of the CoM, a task-level variable estimated through multisensory integration, rather than local joint-level variables. Whereas the initial burst of muscle activity was previously proposed to be due to feedforward processes, reflecting a stereotypical response to a perturbation (Diener et al. 1988), more recent studies demonstrate that the magnitude and timing of this initial burst is scaled according to the CoM acceleration signal at the onset of the perturbation prior to the observed muscle response (Lockhart and Ting 2007; Welch and Ting 2008, 2009). Moreover, the entire time course of muscle activity can be reproduced and used to stabilize a simple inverted pendulum model with delayed feedback on CoM acceleration, velocity, and displacement. Note that this model assumes that the CoM is accurately estimated by multisensory integration mechanisms (Peterka 2002, 2015). In combination with experiments in which accelerations were varied, the model demonstrated that acceleration information is necessary to accurately reproduce the measured muscle activity responses across a wide range of perturbation (Lockhart and Ting 2007; Welch and Ting 2009; Safavynia and Ting 2013a). The model also demonstrates how the delayed feedback on acceleration information generates peak muscle activity that occurs earlier than and appears to "predict" future peak CoM displacement (Welch and Ting 2009;

Insperger et al. 2013). Together, these models suggest that delayed acceleration feedback due to sensory noise could account for the appearance of a feedforward component in both perturbed and quiet standing, although as mentioned previously, challenges exist in system identification where sensory noise is dominant (van der Kooij et al. 2005).

Even when multisegmental movements are not well-described by inverted pendulum models, the response to perturbations can still be predicted based on task-level CoM kinematics, reflecting sensorimotor rather than biomechanical processes. For example, the sensorimotor integration model of Peterka (Peterka 2002, 2015) holds true even when hip movement occurs (personal communication). Muscle activity in response to perturbations can also be predicted based on measured CoM kinematic error even during a response predominated by hip motion (Welch and Ting 2008, 2009). This suggests that the delayed feedback sensorimotor transformation for reactive balance uses neutrally-computed CoM information to drive the excitation of multiple muscles independent of joint-level motions. This is further supported by evidence that CoM kinematics are a better predictor of muscle activity than joint level kinematics (Safavynia and Ting 2013a), and may drive the coordinated recruitment of multiple muscles throughout the body (Safavynia and Ting 2013b). Although a simplification of body dynamics appears inherent in the excitation of multiple muscles, more complex musculoskeletal models are necessary to demonstrate that such control signals are sufficient to provide postural stability in multisegmental models of balance.

### *7.2.7 Multisegmental Models of Balance*

Multisegmental models typically incorporate sagittal-plane ankle, hip, and sometimes knee joints to examine how multisensory information is integrated to control the multiple joints of the body. These models are able to include proprioception at joints other than the ankle, demonstrating more complicated integration of sensory information in controlling the body (e.g. information from hip proprioceptors may be important for controlling the ankle and vice-versa). Multisegmental models have also provided evidence that humans likely utilize some internal model to overcome noisy or conflicting signals, and have facilitated study of multi-joint coordination strategies that is not possible with simple single-inverted pendulum models.

Multisegmental biomechanical models reveal that biomechanical as well as neural influences govern the choice of multi-joint movement strategy in balance control. Different multi-joint coordination strategies can be used to achieve the same higher task-level outcome due to redundancy in joint-space. Typically, small perturbations to standing balance elicit motions predominantly about the ankle ("ankle" strategy) whereas larger perturbations that tend to place the CoM near the edge of the base of support, evoke flexion or extension at the hips ("hip" strategy) (Diener et al. 1988). Kuo and Zajac (1993) demonstrate the ankle strategy is insufficient to keep the feet on the ground for larger perturbations, necessitating a hip strategy response. Similarly, Alexandrov and colleagues (2005) also found that the hip strategy is more efficient at restoring the body at larger perturbations by comparing three eigenmovements that were primarily dominated by motion at one joint (e.g., ankle, knee and hip eigenmovements). However, in both cases there was no biomechanical reason for the selection of an ankle strategy over the hip strategy, thus there likely exist other reasons (such as minimizing motion head or increasing trunk stability) that lead to the preferred selection of the ankle strategy for small perturbations.

Models incorporating sensory dynamics further demonstrate that movement strategies may also depend on the ability of the nervous system to obtain reliable sensory information. Multisegmental models have shown that the healthy nervous system is very good at accounting for sensory noise and errors (van der Kooij et al. 1999, e.g., Kuo 2005). In combination with a feedback scheme to keep the body upright, such models also incorporate a state estimator to minimize the error associated with imperfect sensors and to represent an internal model of the body and sensor dynamics. Modifying the parameters of the model to simulate removing a sensor or increasing sensory noise can reproduce changes in postural responses strategies used in healthy individuals under altered sensory conditions using both two-link (hip and ankle, Kuo 2005) and three-link inverted pendulum models (hip, knee, and ankle, van der Kooij et al. 1999). They can also produce similar responses to those found experimentally in older adults and vestibular loss patients. Together, the ability of these models to fail in a similar way to the human system suggests that an internal model of sensorimotor dynamics is critical in controlling balance.

Optimal control theory applied to multisegmental models of standing balance provide insight into the modulation and variation of response strategies by the nervous system. Several models simulate postural response strategies based on local feedback at each joint where the torque produced is based on joint kinematics. However, a single set of feedback gains cannot generate the appropriate postural response across biomechanical contexts given biomechanical limitations, including differences in maximum muscle torque about each joint, different postural configurations, and different perturbation levels. Therefore the nervous system must use knowledge of the current biomechanical state to plan future responses. For example, Park and colleagues (Park et al. 2004) used a two degree-of-freedom model (hip and ankle) to demonstrate that feedback control gains are continuously scaled as perturbation levels increase. As the postural challenge increased, the gains at the hip increased while ankle gains decreased, consistent with increased hip strategy for larger perturbation. In this case, the optimal feedback gains were found by replicating experimentally-recorded data, rather than testing a specific neural hypothesis. However similar results were found in a data-free model by optimizing the criteria of minimizing CoM excursion while maintaining upright stance (Kuo 1995). Although promising, these types of studies can only provide evidence that it is *possible* that the nervous system selects strategies based on particular task-level goals, but do not rule out the possibility that controlling other variables may also produce similar results. A further challenge remains in identifying which neural systems produce such behaviors and how these idealized commands are translated into muscle recruitment and coordination signals.

### 7.2.8 Complex Models of Balance with Muscles

In order to examine how the different neural control strategies identified in simple models translate to multi-muscle coordination, detailed musculoskeletal models that incorporate realistic musculotendon elements are necessary. Such models face increased challenges for validation associated with neuromechanical redundancy in sensory systems, different joint-level strategies, and the control of multiple muscles crossing the joints. To date, there have been a few attempts at modeling standing balance control in the sagittal plane using detailed musculoskeletal models. For example, Jo and Massaquoi (2004) developed a planar muscle-driven model that was designed to demonstrate the possible cerebrocerebellar influence on postural control. Reinbolt and colleagues (Clark et al. 2011; Mansouri et al. 2012) developed a three-dimensional muscle-driven model that responds to perturbations to balance using local stretch-reflex mechanisms, and Nataraj and colleagues (Nataraj et al. 2010; 2012a, b) incorporated both local joint feedback and whole body center of mass feedback in a three-dimensional musculoskeletal model. Each of these models are capable of remaining upright in response to perturbations and reproduce a kinematic response that looks qualitatively like the experimentally-observed response. Similar to the neurally-driven models of walking discussed above, there are variety of different control mechanisms that can be used to achieve similar results, and thorough validation efforts must be performed to test any given neural control scheme. However, it is not clear that we have sufficient information about the neural control of balance to perform these validations.

## 7.3 Challenges in Developing Validated Neuromechanical Models

Substantial challenges remain in the development of neuromechanical models of locomotion and balance that can be used to understand and predict mechanisms of motor dysfunction and rehabilitation. Combining complex musculoskeletal and neural control models exponentially compounds the redundancy "problem" facing neuromechanical modelers. Yet it is exactly this redundancy–or flexibility–between mechanical and neural contributions to movement that we need to understand in order to identify how compensatory mechanisms may facilitate movement in neuromotor deficits, and how different people could find different solutions for performing the same task. Ultimately, predictive neuromechanical models would incorporate both the musculoskeletal complexity used in many biomechanical walking simulations as well as the robust feedforward and feedback control systems found in balance models and other complexities including multisensory integration models and parallel feedback mechanisms with different time delays and tunable passive mechanics (Ting et al. 2009).

The integration of neuromechanical modeling and experimental approaches will be critical in testing specific hypotheses concerning how and why neuromechanical flexibility is both exploited and constrained under various movement contexts. As neuromechanical redundancy increases as more elements are included, the more critical it will be to go beyond simply reproducing a single measured behavior. While optimization is an important tool for predicting movement and in resolving redundancy, evidence suggests that humans do not always use optimal solutions (Muller and Sternad 2004; Welch and Ting 2008; de Rugy et al. 2012; Loeb 2012). Further, optimal solutions are highly dependent on the structure of the model as well as the constraints and the costs specified in the problem formulation. Thus, identifying neural constraints are critical for identifying relevant optimal solutions (Ting et al. 2012). To support the principled addition of model complexity, studies demonstrating where extant models fail to reproduce experimental data will be important (McKay et al. 2007, e.g., McKay and Ting 2008, McKay and Ting 2012) as well as robustly reproducing multiple experimental conditions. Even better, specific experiments should be designed to support or refute explicit hypotheses about how redundancy is exploited or constrained. In fact, neuromechanical models may be *necessary* to help interpret data by dissociating the effect of various underlying mechanisms of movement. Identifying neural constraints on muscle activity will likely be a more effective and physiologically-relevant way to resolve redundancy than through optimization alone. Below, we give some examples of how such constraints and limits can be identified through investigations combining both experimental and computational analyses and how they aid in the advancement of neuromechanical models for balance and walking and can define a range of possible motor patterns for the same movement as well as variations in movement observed across individuals.

### 7.3.1 Neuromechanical Models Dissociate Neural Versus Mechanical Contributions to Movement

Any given movement results from interactions between neural and mechanical dynamics, including the passive mechanics of the body and muscles, changes in muscle properties with excitation level and muscle state, as well as parallel neural pathways with different delays and different information content (Ting et al. 2009; Roth et al. 2014). As passive mechanical models have the capability of exhibiting movements similar to those observed during behavior, it is important to dissociate the contributions that are due to neural control signals versus mechanical dynamics. Under various conditions and especially in impaired populations, the dependence upon neural versus mechanical elements could vary substantially. Neuromechanical models may be necessary to identify both the possible range of variability in neural signals as well as the degree of neural versus mechanical control for a given measured behavior.

One example of how a simple neuromechanical model can be used to define the necessary contributions and possible variations in neural control signals to balance control is a frontal plane model of balance control driven by delayed feedback (Scrivens et al. 2006; Bingham et al. 2011; Bingham and Ting 2013). We sought to understand why the dynamics of the CoM after a postural perturbation are similar when subjects stand at different stance widths even though the magnitude of muscle activity differs dramatically (Henry et al. 2001; Torres-Oviedo and Ting 2010). This suggests a variation in the neural and mechanical contributions to standing stability within an individual across biomechanical contexts. To understand and quantify these neuromechanical interactions, we developed a model of lateral balance control using a four-bar linkage model where the distance between the feet can be altered. A delayed feedback signal based on hip angle position and velocity drive the torque about that joint. This is reasonable as the acceleration signal in the muscle activation pattern is essentially low-passed filtered by muscle activation contraction dynamics in the production of force. When altering the stance width, we found that the same feedback gain values could no longer be used to stabilize the system and that similar variations in postural stability could be generated in the model based on changing only feedback gains or stance width, demonstrating neuromechanical redundancy in postural control (Scrivens et al. 2006). Indeed, large differences in torques are necessary to produce the same CoM motion when standing at wide vs. narrow stance such that that the set of possible delayed feedback gains varies dramatically with stance width (Fig. 7.1) (Bingham and Ting 2013). An analysis of stability in our model further revealed substantial variability in the magnitudes of the feedback gains that can be used within each stance, which is corroborated with the variations in feedback gains that we observe across individuals (Bingham et al. 2011).

While we can use the model to investigate the effects of non-delayed passive stiffness and damping due to tonic muscle activity prior to the perturbations, we were not able to accurately identify the passive versus active contributions to kinematics even in simulated data where those components were known. While there is a delay between the effects of passive versus active torque generation at the beginning of the perturbation, there was simply not enough information in the kinematic signals in the perturbations that we used to reliably dissociate their effects (Bingham et al. 2011). In healthy individuals the passive contributions are typically about 10% of the overall torque generating in perturbed balance control, but this value could increase substantially in individuals with neuromotor impairments (Dietz and Sinkjaer 2007). To allow the effects to be more apparent in the kinematics data, we further designed specific experiments that modified the effects of passive versus active dynamics by altering subject mass, passive joint stiffness, and delayed feedback gains and extended the duration of perturbation (Bingham 2013). Similar variations in the model were implemented to examine how feedback gains are altered. Across stance widths, we also found that the level of stability could vary across individuals, and that these differences were preserved across stance widths, suggesting a higher level goal driving the selection of redundant neuromechanical strategies that could differ across individuals and movement contexts.

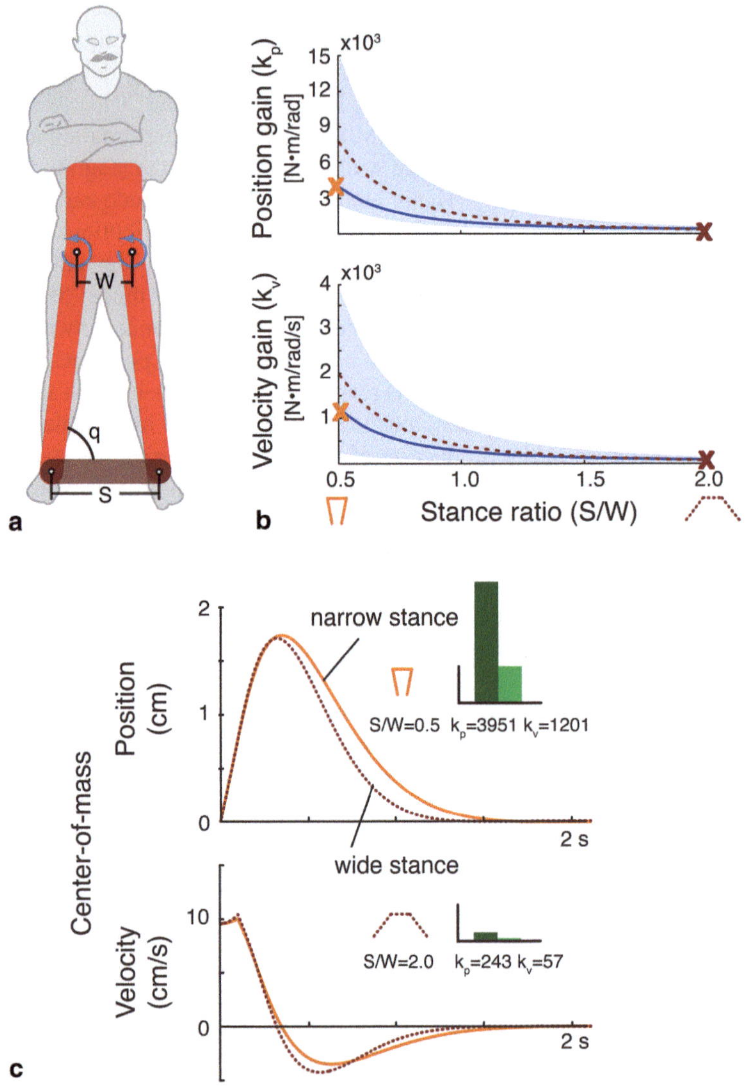


**Fig. 7.1** Frontal plane model of human mediolateral balance control. **a** Frontal plane motion of the body is modeled as a four-bar linkage. Two bars represent the legs, the third bar is the torso, and the fourth bar is the ground. Perturbations are applied as ground translations. Important parameters of the model are the hip width (W), stance width (S), hip torque ($T_H$) and ankle angle ($q_A$). **b** Stable feedback gains vary across stance widths. The dotted line indicates the feedback gain pairs that produced maximum stability across stance widths. The solid line indicates the feedback gain values that produced identical stability characteristics across stance widths. **c** Simulated CoM position across stance widths that have similar stability characteristics. Narrow (*top*) and wide (*bottom*) stance responses correspond to feedback gains from orange and red X's in **b**, respectively. Although feedback gain values differed substantially across stance widths, the resulting CoM motion produced in response to a change in the initial state of the system was similar in narrow (*solid*) and wide (*dotted*) stance widths when feedback gains with the same stability characteristics were used. (Reprinted with permission from Bingham and Ting 2013)

### 7.3.2 EMG Provides Better Information for Validating Neuromechanical Models

Using kinematic measures to validate model outputs is an obvious and necessary step, but may be a very blunt instrument for validating neuromechanical systems. Indeed, Bernstein's original formulation of the redundancy problem arose from the very idea that the internal forces that generated a given kinematic trajectory could not be uniquely identified (Bernstein 1967). This redundancy is due to the fact that forces are integrated twice to produce movements, computations that are also dependent upon the position, velocity, and inertial properties of the limb. Adding ground-reaction forces is helpful, but still leaves intersegmental dynamics at the level of kinematics. An example of this problem can be seen in the use of modern digital control in controlling motors; it is well known that a smooth trajectory can be generated either from a constant torque or a series of pulses as long as the area under the curves are equivalent. The differences in the resulting trajectories are too small to be discerned and cannot be used to back out the forces going in. Therefore, using kinematics alone is insufficient to distinguish different neural control strategies that result from different forces or patterns of muscle activation, potentially indicating different mechanisms of sensorimotor control.

Muscle activity as recorded through electromyography (EMG) can provide important information as it represents amplified motor neuron pool activity and is also related to muscle force (Milner-Brown and Stein 1975; Basmajian and De Luca 1985; Winter 2009). Not only can EMGs help validate neuromechanical models, but models may also be necessary to understand EMG activity. Although EMG provides only partial information about the outputs of the nervous system, the full information contained in such signals has yet to be fully exploited in the development and validation of neuromechanical models. While using EMG to drive models has been attempted e.g., (Lloyd and Besier 2003; Buchanan et al. 2004; Shao et al. 2009; Sartori et al. 2012), there are issues in appropriately converting EMG to muscle force and in many cases only the general on and off and mean amplitude of muscle activation patterns are used to validate optimal model excitations (Zajac et al. 2002; Thelen et al. 2003; Damsgaard et al. 2006). One reason could be that EMG patterns are highly variable compared to biomechanical measures making them difficult to analyze and interpret using statistical tests. Yet the relative consistency of motor outputs compared to the flexible neural strategies we use to generate them are at the crux of the questions that neuromechanical models can and should answer. The variability observed both within and across subjects can be used to identify the task-level goals that are controlled by the nervous system (Scholz and Schoner 1999; Todorov and Jordan 2003) and the underlying structure of the signals can be used to identify constraints on the organization of motor activity (Torres-Oviedo and Ting 2007; Welch and Ting 2008, 2009; Torres-Oviedo and Ting 2010; Safavynia and Ting 2013b), some of which could improve optimization and simulation of walking and balance (McKay and Ting 2012; Borzelli et al. 2013; Walter et al. 2014). Neuromechanical models become essential to understanding the impact of variations in muscle activity on biomechanical performance (e.g., Kutch and Valero-Cuevas 2011; Sohn et al. 2013; Simpson et al. 2015). While it may be considered ideal to access neural signals directly, there are substantial limitations in current invasive and noninvasive techniques and advances in understanding variability in muscle activity will provide insights for incorporating more complex neural models

As an example of how a simple neuromechanical model used to predict muscle activity can reveal fundamental neural mechanisms of balance, our sensorimotor feedback model demonstrated the importance of acceleration feedback in reactive balance responses. A simple delayed feedback loop using COM acceleration, velocity, and displacement can reproduce the entire time course of muscle activity during postural perturbations to standing balance in both humans and other animals (Fig. 7.2) (Lockhart and Ting 2007; Welch and Ting 2008). We found that the steep initial rise of muscle activity occurring at a long latency (~40 ms in cats, ~100 ms in humans) after a perturbation follows the initial acceleration induced by the perturbation. Feedback on the acceleration signal can account for what can appear to be a predictive, or feedforward, burst of muscle activity (Diener et al. 1988) where the peak in muscle activity precedes the peak in CoM displacement induced by the perturbation. Removing acceleration feedback from the model only reduced the goodness of fit to recorded data by about 2%, yet the initial burst was eliminated, altering the qualitative shape of the response (Lockhart and Ting 2007); adding jerk to the model did not improve fits. Using current quantitative metrics of similarity it would thus be possible to simulate muscle activation patterns that quantitatively account for the variance in the data, but which generate qualitatively different response patterns. Further, clues to the physiological basis of the acceleration information were obtained by demonstrating that the initial burst is lost after large-fiber peripheral neuropathy (Fig. 7.2) (Lockhart and Ting 2007). In this condition, the large diameter muscle spindle, Golgi tendon organ, and cutaneous afferents were damaged (Stapley et al. 2002). The selection of feedback gains could be predicted by the same optimization criteria with a further constraint on removing acceleration feedback, suggesting there was a common motor goal driving the selection of the motor pattern both before and after neuropathy. While qualitative changes in the muscle activity were found, the kinematics of the CoM were quite similar before and after neuropathy, differing only in magnitude (Fig. 7.3). This highlights the relative insensitivity of the kinematic signals to changes in the underlying neural control system.

Since qualitative differences may not be easily identified using typical quantitative statistical analyses, better exploratory statistical tools are necessary to facilitate better neuromechanical models that can identify subtle but potentially important differences between and across conditions as well as to compare simulated and experimental kinematic and muscle activity data. High variability across trials and subjects often hinder our ability to rigorously test the fidelity of neuromechanical models, leaving many to be validated "by eye". Often models are considered valid if they roughly lie within the broad 95% confidence limits of the data, which allow for

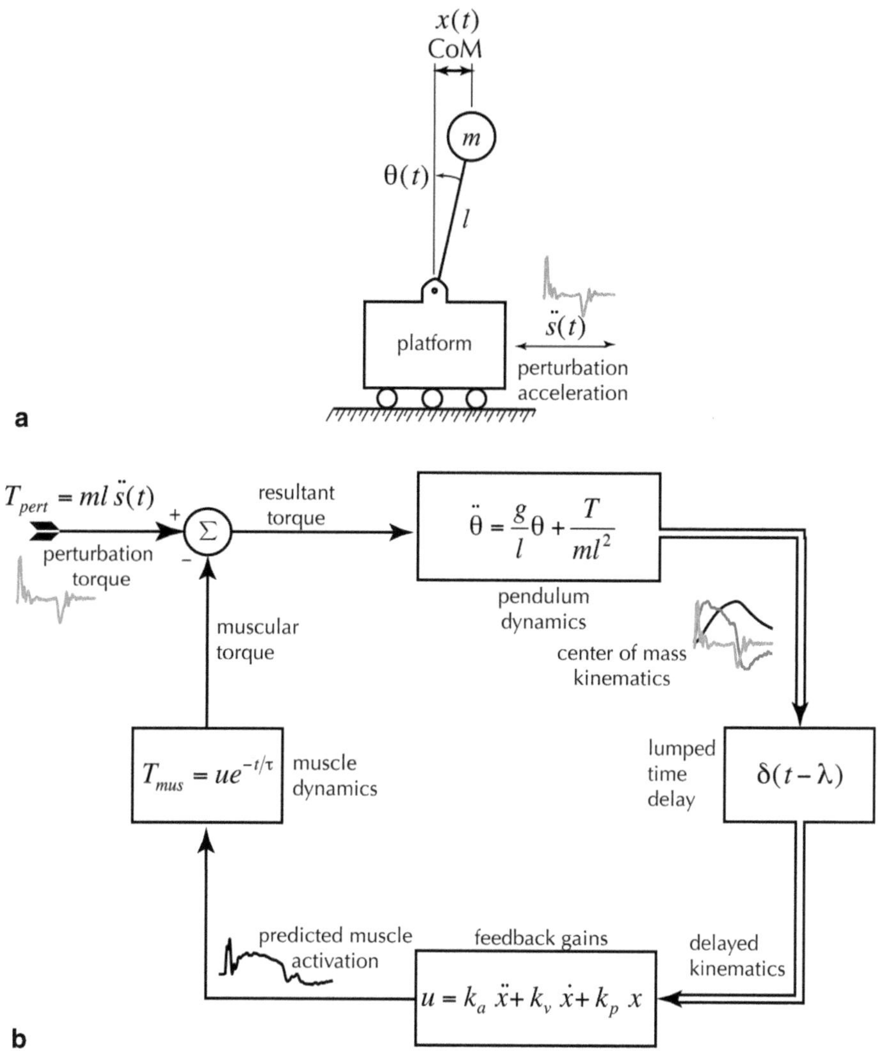


**Fig. 7.2** Simple feedback model of postural control used to predict muscle activity during reactive balance responses. **a** The mechanics of the body during balance is approximated as an inverted pendulum on a moving cart. Experimentally measured accelerations of the platform were applied to the cart so that realistic acceleration, velocity and displacement trajectories of the platform were modeled. **b** The perturbation acceleration generates a disturbance torque at the base of the pendulum. Delayed kinematics of the horizontal CoM were used in a simple feedback law to generate model muscle-activation patterns, which were compared with those measured experimentally. The modeled muscle activation then generated a stabilizing torque about the representative joint. (Reprinted with permission from Lockhart and Ting 2007)

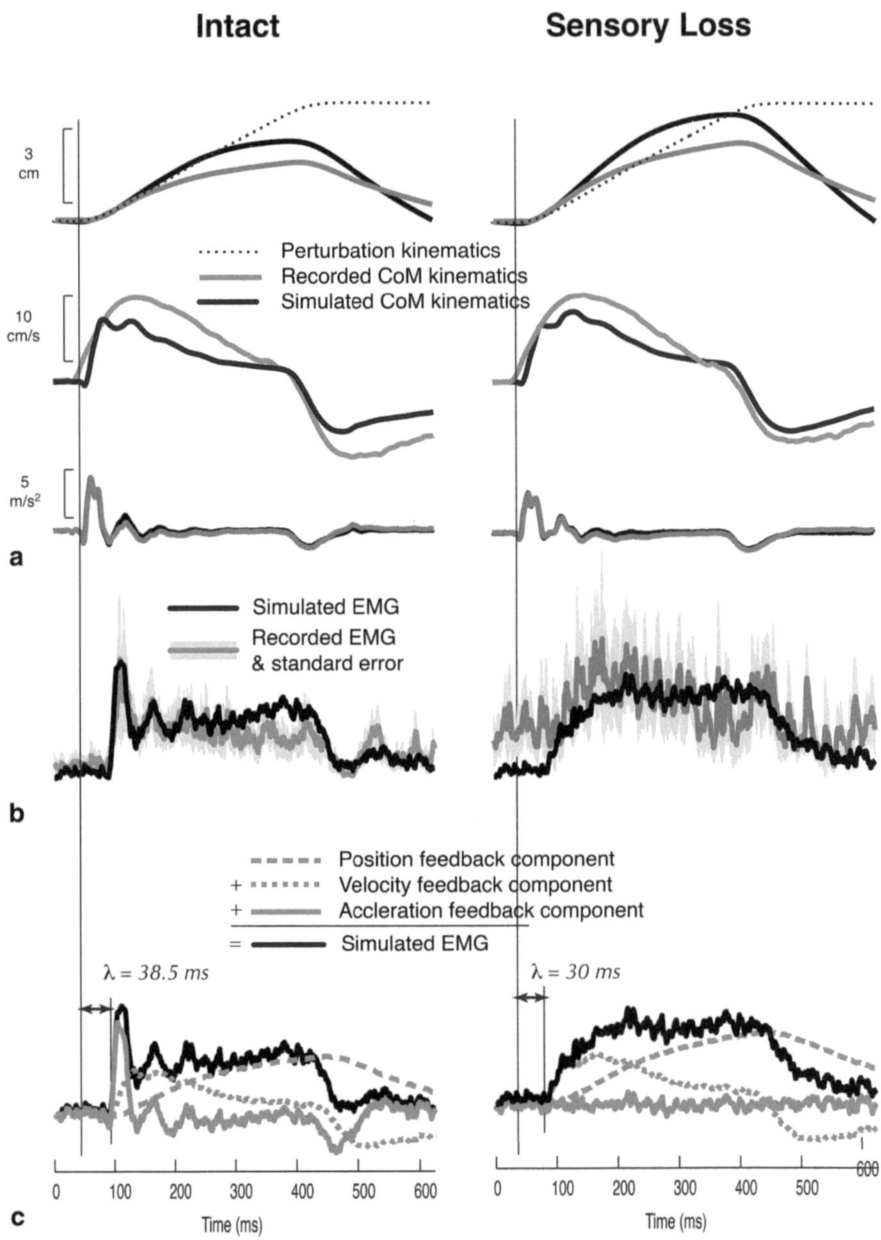


**Fig. 7.3** Comparison of recorded and simulated muscle activation and CoM kinematics before and after large-fiber peripheral neuropathy. **a** Recorded (*gray lines*) and simulated (*black lines*) CoM displacement, velocity and acceleration. **b** Recorded (*gray line*) muscle activity and simulated (*black line*) muscle activity from the best-match model parameters. **c** Decomposition of simulated muscle activity (*black line*) into components arising from CoM position feedback (*gray dashed line*), CoM velocity feedback (*gray dotted line*), and CoM acceleration feedback (*gray solid line*). Note that the initial burst in the intact condition (*left panel*) is due primarily to acceleration feedback and the absence of the initial burst and acceleration feedback in the sensory loss condition (*right panel*). (Reprinted with permission from Lockhart and Ting 2007)

qualitative differences in the traces, or only certain features such as peak amplitude and timing are used in statistical analysis. Toward the development of better tools to compare curves, we recently proposed a new method for functional analysis of variance (fANOVA) based on the wavelet representations of EMG signals (McKay et al. 2013). The statistically significant differences identified within the wavelet domain are then transformed back into the time domain rendering clearly interpretable difference curves between conditions. These analyses have the potential to reveal critical features that are typically hard to quantify such as inflection points and small bursts of activity and could be expanded to demonstrate differences between model outputs and experimental data with more power and better temporal resolution than traditional methods, ultimately enhancing our ability to develop and validate neuromechanical models.

### *7.3.3 Neuromechanical Constraints on Musculoskeletal Redundancy*

Developing validated neuromechanical models using muscles has the further challenge of dealing with musculoskeletal redundancy in a physiologically-relevant way as adding neural complexity only furthers the "problem" of redundancy and model validation (Prinz et al. 2004). It is well known that many different coordination patterns across multiple muscles can be used to generate a given set of joint torques at an instant in time. Some evidence suggests that the activity of multiple muscles during movements can be predicted based on optimality criteria typically minimizing muscle stress or force (Crowninshield and Brand 1981; Thelen et al. 2003; Kurtzer et al. 2006; Erdemir et al. 2007). However, these produce only a single solution for a given task, whereas great variability within and across subjects is often observed. Our work using a musculoskeletal model demonstrates that substantial variations in the amplitude of activity in a given muscle are possible to achieve a given biomechanical task (Sohn et al. 2013; Simpson et al. 2015). How thus do we interpret and quantify these deviations in muscle coordination? And to what degree do these need to be accounted for in neuromechanical models?

In order to develop a new framework for understanding, quantifying, and predicting muscle activation patterns we have performed a number of experimental studies to investigate constraint of the spatial activation of muscles and their relationship to biomechanical functions. Our work in combination with studies from upper extremity movements suggest that muscles are not activated independently, as assumed by optimization models, but are constrained to be active in modular units specifying fixed spatial patterns of muscle activity (d'Avella et al. 2003; Bizzi et al. 2008; Ting and Chvatal 2010). These motor modules, also referred to as muscle synergies, coordinate muscles across multiple joints and can produce consistent biomechanical outputs necessary to achieve a task (Ting and Macpherson 2005; Chvatal et al. 2011). Within this framework, movements are constructed by combining and varying the recruitment of motor modules to achieve a task (Fig. 7.4). Thus, motor

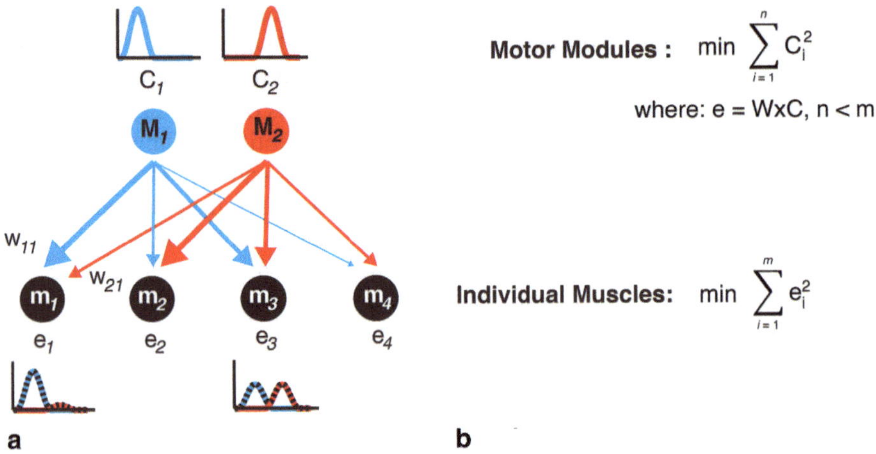


**Fig. 7.4** Motor module concept. **a** Each motor module ($M_i$) contributes to the activation ($e_i$) of a single muscle ($m_i$) with a fixed weight ($w_{ij}$, where i=1 to number of motor module and j=1 to number muscles) with an activation profile that can vary over time ($C_i$). In this example, the weight of the connection between each motor module and muscle is depicted by the width of the arrow. For example, the activity of muscle 1 ($e_1$) has a large contribution from module 1 (*blue*) and a much smaller contribution from module 2 (*red*). In contrast, the activity of muscle 3 ($e_3$) has approximately equal contributions from both modules. **b** This allows the search space of neural control inputs to satisfy the objective function be constrained to a smaller number

modules provide a framework for understanding variability observed both within and across subjects. For example, trial-by-trial variations in reactive balance and cycle-by-cycle variations in walking can be explained by different levels of recruitment of motor modules rather than muscles (Clark et al. 2010; Torres-Oviedo and Ting 2010). Further, we observe different number and structure of motor modules across individuals (Torres-Oviedo and Ting 2007). Moreover, these are preserved across different biomechanical contexts consistent with the idea that motor modules are neural constraints (Torres-Oviedo and Ting 2010; Chvatal et al. 2011) that could represent preferred motor patterns (de Rugy et al. 2012). The same motor modules are even shared across walking and balance tasks (Chvatal and Ting 2012; 2013). Finally, in the upper extremity, performance of novel force generation tasks that can be achieved using existing motor modules are much easier for subjects to learn than those that require muscle activation patterns that are not compatible with existing muscle synergies (Berger et al. 2013).

Incorporating motor modules into neuromechanical models is necessary to effectively understand motor function; in return they may also improve the predictive capabilities of model performance. Several studies have demonstrated that similar motor performance can be achieved through optimal control of individual muscle versus recruitment of muscle groups (Raasch and Zajac 1999; McKay and Ting 2012). In reactive balance responses, we demonstrated that minimizing mo-

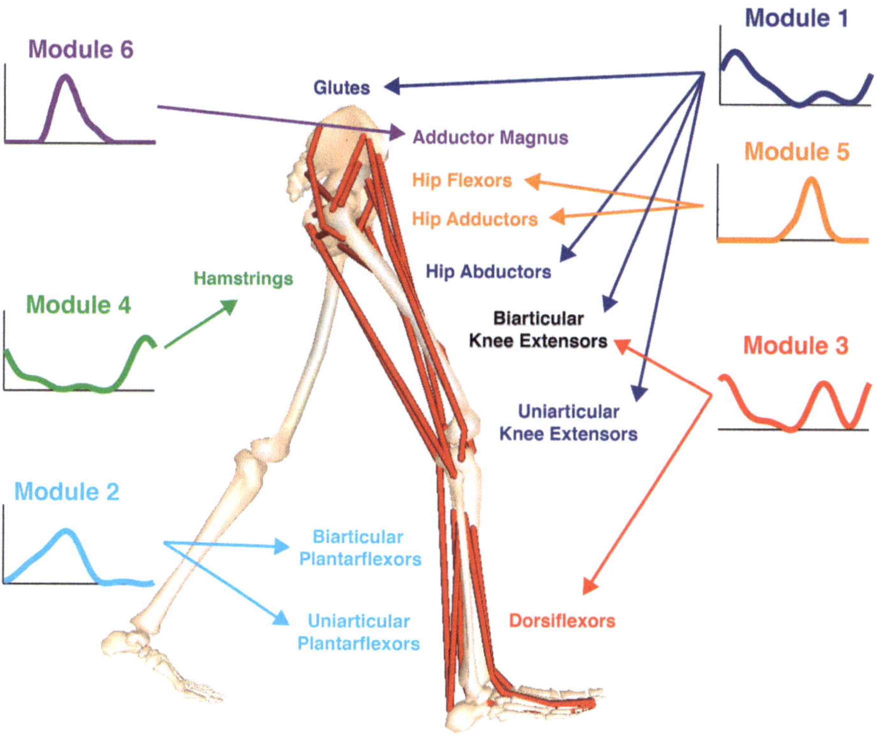


**Fig. 7.5** Detailed musculoskeletal models with motor modules instead of individual muscles as the control inputs can be used to successfully reproduce well-coordinated walking patterns. Allen and Neptune 2012 demonstrated that that the six motor modules shown here are sufficient to satisfy the biomechanical demands present during walking. Each motor module coordinates multiple muscles with a common temporal recruitment pattern

tor module recruitment better matched the force evoked during reactive balance responses compared to minimizing individual muscle activation (McKay and Ting 2012). While the total level of muscle activation was increased slightly compared to the minimum possible activity, the computation time required to find an optimal solution was improved. The feasibility of using motor modules in simulations of human walking has also been demonstrated (Neptune et al. 2009; Allen and Neptune 2012; Sartori et al. 2013) (Fig. 7.5), as well as the detrimental consequences of altered motor modules to walking function observed in stroke (Allen et al. 2013). Further, the internal joint loading force predictions are improved in a torque-based model of walking when motor module constraints are considered (Walter et al. 2014). Therefore, using motor modules constraints may be an important step to both validating motor modules as a mechanism for motor control as well as improving the predictive power of model on an individual basis.

## 7.4 Conclusion

In conclusion, neuromechanical models are becoming increasingly important in elucidating principles of movement in balance and locomotion. Advancing the integration of neuromechanical modeling and experimental approaches will require researchers to have a training in a diverse set of disciplines spanning neural and musculoskeletal systems, experimental design, as well as computational modeling techniques. As computational tools for analyzing and simulating both musculoskeletal models and their control improve (Cofer et al. 2010; Bunderson et al. 2012; Bunderson and Bingham 2015; Markin et al. 2015) our ability to develop models that can not only describe but also predict movement strategies and control will also improve. Focus on a principled increase in complexity of models, with explicit tests to demonstrate the contributions of each component will help to justify the necessity of each component to explain the robustness of human locomotor and balance behaviors.

## References

Ackermann M, van den Bogert AJ (2010) Optimality principles for model-based prediction of human gait. J Biomech 43:1055–1060

Alexandrov AV, Frolov AA, Horak FB, Carlson-Kuhta P, Park S (2005) Feedback equilibrium control during human standing. Biol Cybern 1–14

Allen JL, Neptune RR (2012) Three-dimensional modular control of human walking. J Biomech 45:2157–2163

Allen JL, Kautz SA, Neptune RR (2013) The influence of merged muscle excitation modules on post-stroke hemiparetic walking performance. Clin Biomech (Bristol Avon) 28:697–704

Aoi S, Ogihara N, Funato T, Sugimoto Y, Tsuchiya K (2010) Evaluating functional roles of phase resetting in generation of adaptive human bipedal walking with a physiologically based model of the spinal pattern generator. Biol Cybern 102:373–387

Aoi S (2015) Neuromusculoskeletal modeling for the adaptive control of posture during locomotion. In: Prilutsky BI, Edwards DH Jr (eds) Neuromechanical modeling of posture and locomotion. Springer, New York (in press)

Basmajian JV, De Luca C (1985) Muscles alive. Williams & Wilkins, Baltimore

Berger DJ, Gentner R, Edmunds T, Pai DK, d'Avella A (2013) Differences in adaptation rates after virtual surgeries provide direct evidence for modularity. J Neurosci 33:12384–12394

Bernstein N (1967) The coordination and regulation of movements. Pergamon Press, New York

Bingham JT (2013) The effects of delayed feedback and configuration on stable interactions between the neural and musculoskeletal systems[Thesis]. Type, Georgia Institute of Technology, Atlanta

Bingham JT, Ting LH (2013) Stability radius as a method for comparing the dynamics of neuromechanical systems. IEEE Trans Neural Syst Rehabil Eng 21:840–848

Bingham JT, Choi JT, Ting LH (2011) Stability in a frontal plane model of balance requires coupled changes to postural configuration and neural feedback control. J Neurophysiol 106:437–448

Bizzi E, Cheung VC, d'Avella A, Saltiel P, Tresch M (2008) Combining modules for movement. Brain Res Rev 57:125–133

Blickhan R (1989) The spring-mass model for running and hopping. J Biomech 22:1217–1227

Borzelli D, Berger DJ, Pai DK, d'Avella A (2013) Effort minimization and synergistic muscle recruitment for three-dimensional force generation. Front Comput Neurosci 7:186

Brown TG (1914) On the nature of the fundamental activity of the nervous centres; together with an analysis of the conditioning of rhythmic activity in progression, and a theory of the evolution of function in the nervous system. J Physiol 48:18–46

Buchanan TS, Lloyd DG, Manal K, Besier TF (2004) Neuromusculoskeletal modeling: estimation of muscle forces and joint moments and movements from measurements of neural command. J Appl Biomech 20:367–395

Bunderson N, Bingham J (2015) Better science through predictive modeling: numerical tools for understanding neuromechanical interactions. In: Prilutsky BI, Edwards DH Jr (eds) Neuromechanical modeling of posture and locomotion. Springer, New York (in press)

Bunderson NE, Bingham JT, Sohn MH, Ting LH, Burkholder TJ (2012) Neuromechanic: a computational platform for simulation and analysis of the neural control of movement. Int J Numer Method Biomed Eng 28:1015–1027

Cavagna GA, Saibene FP, Margaria R (1964) Mechanical work in running. J Appl Physiol 19:249–256

Cavagna GA, Heglund NC, Taylor CR (1977) Mechanical work in terrestrial locomotion: two basic mechanisms for minimizing energy expenditure. Am J Physiol 233:R243–R261

CDC (2008) Falls among older adults: an overview. Department of Health and Human Services, Centers for Disease Control

Chvatal SA, Ting LH (2012) Voluntary and reactive recruitment of locomotor muscle synergies during perturbed walking. J Neurosci 32:12237–12250

Chvatal SA, Ting LH (2013) Common muscle synergies for balance and walking. Front Comput Neurosci 7:48

Chvatal SA, Torres-Oviedo G, Safavynia SA, Ting LH (2011) Common muscle synergies for control of center of mass and force in non-stepping and stepping postural behaviors. J Neurophysiol 106:999–1015

Clark DJ, Ting LH, Zajac FE, Neptune RR, Kautz SA (2010) Merging of healthy motor modules predicts reduced locomotor performance and muscle coordination complexity post-stroke. J Neurophysiol 103:844–857

Clark AE, Seth A, Reinbolt JA (2011) Biarticular muscles influence postural responses: implications for treatment of stiff-knee gait. Proceedings of the XIII International Symposium on Computer Simulation in Biomechanics. Leuven, Belgium

Cofer D, Cymbalyuk G, Reid J, Zhu Y, Heitler WJ, Edwards DH (2010) AnimatLab: a 3D graphics environment for neuromechanical simulations. J Neurosci Methods 187:280–288

Collins JJ (1995) The redundant nature of locomotor optimization laws. J Biomech 28:251–267

Collins JJ, De Luca CJ (1993) Open-loop and closed-loop control of posture: a random-walk analysis of center-of-pressure trajectories. Exp Brain Res 95:308–318

Collins SH, Ruina A (2005) A bipedal walking robot with efficient and human-like gait. Proceedings of the 2005 IEEE International Conference on Robotics and Automation, pp. 1983–1988

Collins SH, Wisse M, Ruina A (2001) A three-dimensional passive-dynamic walking robot with two legs and knees. Int J Rob Res 20:607–615

Collins S, Ruina A, Tedrake R, Wisse M (2005) Efficient bipedal robots based on passive-dynamic walkers. Science 307:1082–1085

Crowninshield RD, Brand RA (1981) A physiologically based criterion of muscle force prediction in locomotion. J Biomech 14:793–801

Damsgaard M, Rasmussen J, Christensen ST, Surma E, de Zee M (2006) Analysis of musculoskeletal systems in the AnyBody modeling System. Simul Model Pract Th 14:1100–1111

d'Avella A, Saltiel P, Bizzi E (2003) Combinations of muscle synergies in the construction of a natural motor behavior. Nat Neurosci 6:300–308

de Rugy A, Loeb GE, Carroll TJ (2012) Muscle coordination is habitual rather than optimal. J Neurosci 32:7384–7391

Dickinson MH, Farley CT, Full RJ, Koehl MAR, Kram R, Lehman S (2000) How animals move: an integrative view. Sci New Ser 288:100–106

Diener HC, Horak FB, Nashner LM (1988) Influence of stimulus parameters on human postural responses. J Neurophysiol 59:1888–1905

Dietz V, Sinkjaer T (2007) Spastic movement disorder: impaired reflex function and altered muscle mechanics. Lancet Neurol 6:725–733

Erdemir A, McLean S, Herzog W, van den Bogert AJ (2007) Model-based estimation of muscle forces exerted during movements. Clin Biomech (Bristol Avon) 22:131–154

Full RJ, Koditschek DE (1999) Templates and anchors: neuromechanical hypotheses of legged locomotion on land. J Exp Biol 202(Pt 23):3325–32

Gawthrop P, Loram I, Lakie M (2009) Predictive feedback in human simulated pendulum balancing. Biol Cybern 101:131–146

Geyer H, Seyfarth A, Blickhan R (2006) Compliant leg behaviour explains basic dynamics of walking and running. Proc Biol Sci 273:2861–2867

Hall AL, Peterson CL, Kautz SA, Neptune RR (2011) Relationships between muscle contributions to walking subtasks and functional walking status in persons with post-stroke hemiparesis. Clin Biomech (Bristol Avon) 26:509–515

Henry SM, Fung J, Horak FB (2001) Effect of stance width on multidirectional postural responses. J Neurophysiol 85:559–570

Higginson JS, Zajac FE, Neptune RR, Kautz SA, Delp SL (2006) Muscle contributions to support during gait in an individual with post-stroke hemiparesis. J Biomech 39:1769–1777

Hoy MG, Zernicke RF (1986) The role of intersegmental dynamics during rapid limb oscillations. J Biomech 19:867–877

Insperger T, Milton J, Stepan G (2013) Acceleration feedback improves balancing against reflex delay. J R Soc Interface 10:20120763

Jo S, Massaquoi SG (2004) A model of cerebellum stabilized and scheduled hybrid long-loop control of upright balance. Biol Cybern 91:188–202

Jo S, Massaquoi SG (2007) A model of cerebrocerebello-spinomuscular interaction in the sagittal control of human walking. Biol Cybern 96:279–307

Kuo AD (1995) An optimal control model for analyzing human postural balance. IEEE Trans Biomed Eng 42:87–101

Kuo AD (2005) An optimal state estimation model of sensory integration in human postural balance. J Neural Eng 2:S235–S249

Kuo AD, Zajac FE (1993) Human standing posture: multi-joint movement strategies based on biomechanical constraints. Prog Brain Res 97:349–358

Kurtzer I, Pruszynski JA, Herter TM, Scott SH (2006) Primate upper limb muscles exhibit activity patterns that differ from their anatomical action during a postural task. J Neurophysiol 95:493–504

Kutch JJ, Valero-Cuevas FJ (2011) Muscle redundancy does not imply robustness to muscle dysfunction. J Biomech 44:1264–1270

Liu MQ, Anderson FC, Schwartz MH, Delp SL (2008) Muscle contributions to support and progression over a range of walking speeds. J Biomech 41:3243–3252

Lloyd DG, Besier TF (2003) An EMG-driven musculoskeletal model to estimate muscle forces and knee joint moments in vivo. J Biomech 36:765–776

Lockhart DB, Ting LH (2007) Optimal sensorimotor transformations for balance. Nat Neurosci 10:1329–1336

Loeb GE (2012) Optimal isn't good enough. Biol Cybern 106:757–765

Loram ID, Maganaris CN, Lakie M (2005) Human postural sway results from frequent, ballistic bias impulses by soleus and gastrocnemius. J Physiol 564:295–311

Mansouri M, Clark AE, Reinbolt JA (2012) The use of a platform for dynamic simulation of movement: application to balance recovery. Proceedings of the American Society of Biomechanics. Gainesville, FL

Markin SN, Klishko AN, Shevtsova NA, Lemay MA, Prilutsky BI, Rybak IA (2015) A neuromechanical model of spinal control of locomotion. In: Prilutsky BI, Edwards DH Jr (eds) Neuromechanical modeling of posture and locomotion. Springer, New York (in press)

Martelli S, Calvetti D, Somersalo E, Viceconti M, Taddei F (2013) Computational tools for calculating alternative muscle force patterns during motion: a comparison of possible solutions. J Biomech 46:2097–2100

Maurer C, Peterka RJ (2005) A new interpretation of spontaneous sway measures based on a simple model of human postural control. J Neurophysiol 93:189–200
McGeer T (1990a) Passive dynamic walking. Int J Rob Res 9:62–82
McGeer T (1990b) Passive walking with knees. Proceedings of the IEEE International Conference on Robotics and Automation, pp. 1640–1645, Cincinnati, OH
McKay JL, Ting LH (2008) Functional muscle synergies constrain force production during postural tasks. J Biomech 41:299–306
McKay JL, Ting LH (2012) Optimization of muscle activity for task-level goals predicts complex changes in limb forces across biomechanical contexts. PLoS Comput Biol 8:e1002465
McKay JL, Burkholder TJ, Ting LH (2007) Biomechanical capabilities influence postural control strategies in the cat hindlimb. J Biomech 40:2254–2260
McKay JL, Welch TD, Vidakovic B, Ting LH (2013) Statistically significant contrasts between EMG waveforms revealed using wavelet-based functional ANOVA. J Neurophysiol 109:591–602
McLean SG, Su A, van den Bogert AJ (2003) Development and validation of a 3-D model to predict knee joint loading during dynamic movement. J Biomech Eng 125:864–874
McMahon TA, Cheng GC (1990) The mechanics of running: how does stiffness couple with speed? J Biomech 23(Suppl 1):65–78
Milner-Brown H, Stein R (1975) The relation between the surface electromyogram and muscular force. J Physiol 246:549–569
Mochon S, McMahon TA (1980) Ballistic walking. J Biomech 13:49–57
Morasso PG, Schieppati M (1999) Can muscle stiffness alone stabilize upright standing? J Neurophysiol 82:1622–1626
Muller H, Sternad D (2004) Decomposition of variability in the execution of goal-oriented tasks: three components of skill improvement. J Exp Psychol Hum Percept Perform 30:212–233
Nataraj R, Audu ML, Kirsch RF, Triolo RJ (2010) Comprehensive joint feedback control for standing by functional neuromuscular stimulation-a simulation study. IEEE Trans Neural Syst Rehabil Eng 18:646–657
Nataraj R, Audu ML, Kirsch RF, Triolo RJ (2012a) Center of mass acceleration feedback control for standing by functional neuromuscular stimulation: a simulation study. J Rehabil Res Dev 49:279–296
Nataraj R, Audu ML, Triolo RJ (2012b) Comparing joint kinematics and center of mass acceleration as feedback for control of standing balance by functional neuromuscular stimulation. J Neuroeng Rehabil 9:25
Neptune RR, Kautz SA, Zajac FE (2001) Contributions of the individual ankle plantar flexors to support, forward progression and swing initiation during walking. J Biomech 34:1387–1398
Neptune RR, Clark DJ, Kautz S (2009) Modular control of human walking: a simulation study. J Biomech 42:1282–1287
Ogihara N, Yamazaki N (2001) Generation of human bipedal locomotion by a bio-mimetic neuro-musculo-skeletal model. Biol Cybern 84:1–11
Park S, Horak FB, Kuo AD (2004) Postural feedback responses scale with biomechanical constraints in human standing. Exp Brain Res 154:417–427
Paul C, Bellotti M, Jezernik S, Curt A (2005) Development of a human neuro-musculo-skeletal model for investigation of spinal cord injury. Biol Cybern 93:153–170
Perry J (1992) Gait analysis: normal and pathological function. Slack Incorporated, Thorofare, NJ
Peterka RJ (2000) Postural control model interpretation of stabilogram diffusion analysis. Biol Cybern 82:335–343
Peterka RJ (2002) Sensorimotor integration in human postural control. J Neurophysiol 88:1097–1118
Peterka RJ (2015) Model-based interpretations of experimental data related to the control of balance during stance and gait in humans. In: Prilutsky BI, Edwards DH Jr (eds) Neuromechanical modeling of posture and locomotion. Springer, New York (in press)
Peterson CL, Kautz SA, Neptune RR (2011) Muscle work is increased in pre-swing during hemiparetic walking. Clin Biomech (Bristol Avon) 26:859–866

Prinz AA, Bucher D, Marder E (2004) Similar network activity from disparate circuit parameters. Nat Neurosci 7:1345–1352
Raasch CC, Zajac FE (1999) Locomotor strategy for pedaling: muscle groups and biomechanical functions. J Neurophysiol 82:515–525
Roth E, Sponberg S, Cowan NJ (2014) A comparative approach to closed-loop computation. Curr Opin Neurobiol 25:54–62
Safavynia SA, Ting LH (2013a) Long-latency muscle activity reflects continuous, delayed sensorimotor feedback of task-level and not joint-level error. J Neurophysiol 110:1278–1290
Safavynia SA, Ting LH (2013b) Sensorimotor feedback based on task-relevant error robustly predicts temporal recruitment and multidirectional tuning of muscle synergies. J Neurophysiol 109:31–45
Sartori M, Reggiani M, Farina D, Lloyd DG (2012) EMG-driven forward-dynamic estimation of muscle force and joint moment about multiple degrees of freedom in the human lower extremity. PLoS One 7:e52618
Sartori M, Gizzi L, Lloyd DG, Farina D (2013) A musculoskeletal model of human locomotion driven by a low dimensional set of impulsive excitation primitives. Front Comput Neurosci 7:79
Scholz JP, Schoner G (1999) The uncontrolled manifold concept: identifying control variables for a functional task. Exp Brain Res 126:289–306
Scrivens JE, Ting LH, Deweerth SP (2006) Effects of stance width on control gain in standing balance. Conf Proc IEEE Eng Med Biol Soc 1:4055–4057
Shao Q, Bassett DN, Manal K, Buchanan TS (2009) An EMG-driven model to estimate muscle forces and joint moments in stroke patients. Comput Biol Med 39:1083–1088
Silverman AK, Neptune RR (2012) Muscle and prosthesis contributions to amputee walking mechanics: a modeling study. J Biomech 45:2271–2278
Simpson CS, Sohn MH, Allen JL, Ting LH (2015) Feasible muscle activation ranges based on inverse dynamics analyses of human walking. J Biomech 48:2990–2997
Sohn MH, McKay JL, Ting LH (2013) Defining feasible bounds on muscle activation in a redundant biomechanical task: practical implications of redundancy. J Biomech 46:1363–1368
Srinivasan M, Ruina A (2006) Computer optimization of a minimal biped model discovers walking and running. Nature 439:72–75
Stapley PJ, Ting LH, Hulliger M, Macpherson JM (2002) Automatic postural responses are delayed by pyridoxine-induced somatosensory loss. J Neurosci 22:5803–5807
Steele KM, van der Krogt MM, Schwartz MH, Delp SL (2012) How much muscle strength is required to walk in a crouch gait? J Biomech 45:2564–2569
Steele KM, Seth A, Hicks JL, Schwartz MH, Delp SL (2013) Muscle contributions to vertical and fore-aft accelerations are altered in subjects with crouch gait. Gait Posture 38:86–91
Sutherland DH (1984) Gait disorders in childhood and adolescence. Williams & Wilkins, Baltimore
Taga G (1995a) A model of the neuro-musculo-skeletal system for human locomotion. I. Emergence of basic gait. Biol Cybern 73:97–111
Taga G (1995b) A model of the neuro-musculo-skeletal system for human locomotion. II Realtime adaptability under various constraints. Biol Cybern 73:113–121
Thelen DG, Anderson FC, Delp SL (2003) Generating dynamic simulations of movement using computed muscle control. J Biomech 36:321–328
Ting LH, Chvatal SA (2010) Decomposing muscle activity in motor tasks: methods and interpretation. In: Danion F, Latash ML (eds) Motor control: theories, experiments, and applications. Oxford, pp. 102–138.
Ting LH, Macpherson JM (2005) A limited set of muscle synergies for force control during a postural task. J Neurophysiol 93:609–613
Ting LH, van Antwerp KW, Scrivens JE, McKay JL, Welch TD, Bingham JT et al (2009) Neuromechanical tuning of nonlinear postural control dynamics. Chaos 19:026111
Ting LH, Chvatal SA, Safavynia SA, McKay JL (2012) Review and perspective: neuromechanical considerations for predicting muscle activation patterns for movement. Int J Numer Method Biomed Eng 28:1003–1014

Todorov E, Jordan MI (2003) A minimal intervention principle for coordinated movement. Adv Neural Inf Process Syst 15:27–34
Torres-Oviedo G, Ting LH (2007) Muscle synergies characterizing human postural responses. J Neurophysiol 98:2144–2156
Torres-Oviedo G, Ting LH (2010) Subject-specific muscle synergies in human balance control are consistent across different biomechanical contexts. J Neurophysiol 103:3084–3098
van der Kooij H, Jacobs R, Koopman B, Grootenboer H (1999) A multisensory integration model of human stance control. Biol Cybern 80:299–308
van der Kooij H, van Asseldonk E, van der Helm FC (2005) Comparison of different methods to identify and quantify balance control. J Neurosci Methods 145:175–203
Walter J, Kinney AL, Banks SA, D'Lima D, Besier TF, Lloyd DG et al (2014) Muscle synergies may improve optimization prediction of knee contact forces during walking. J Biomech Eng 136:1054–1060
Welch TD, Ting LH (2008) A feedback model reproduces muscle activity during human postural responses to support-surface translations. J Neurophysiol 99:1032–1038
Welch TD, Ting LH (2009) A feedback model explains the differential scaling of human postural responses to perturbation acceleration and velocity. J Neurophysiol 101:3294–3309
Winter DA (2009) Biomechanics and motor control of human movement. Wiley, Hoboken, New Jersey
Winter DA, Patla AE, Prince F, Ishac M, Gielo-Perczak K (1998) Stiffness control of balance in quiet standing. J Neurophysiol 80:1211–1221
Wisse M, Feliksdal G, Van Frankkenhuyzen J, Moyer B (2007) Passive-based walking robot. IEEE Rob Autom Mag 14:52–62
Zajac FE, Neptune RR, Kautz SA (2002) Biomechanics and muscle coordination of human walking. Part I: introduction to concepts, power transfer, dynamics and simulations. Gait Posture 16:215–232
Zatsiorsky VM, Duarte M (1999) Instant equilibrium point and its migration in standing tasks: rambling and trembling components of the stabilogram. Motor Control 3:28–38

# Chapter 8
# Neuromusculoskeletal Modeling for the Adaptive Control of Posture During Locomotion

**Shinya Aoi**

**Abstract** People and animals produce adaptive locomotion in diverse environments by cooperatively and skillfully manipulating their complicated and redundant musculoskeletal systems. To establish such locomotion, the control of leg movement to transport the entire body against gravity and the control of posture to prevent falling are required. However, these controls affect one another for the posture of the body during locomotion because leg movement disturbs the posture. The underlying mechanism for stabilizing posture during locomotion remains unclear. In this chapter, simulation studies are presented to investigate the functional roles of the nervous system to maintain the posture of the body during locomotion by focusing on the adaptive walking of humans during disturbances and on obstacle avoidance during walking by the hind legs of rats. Neuromusculoskeletal models for humans and rats were constructed by integrating the musculoskeletal model using anatomical data and the nervous system model based on physiological findings. The leg movement control was modeled based on the physiological concepts of central pattern generators and muscle synergy and on sensory regulation by phase resetting and interlimb coordination. The posture control was also modeled to regulate the postural behavior using somatosensory information. We also examine how these controls contribute to stabilizing posture during locomotion.

**Keywords** Neuromusculoskeletal model · Human · Rat · Locomotion · Obstacle avoidance · Posture · Central pattern generator (CPG) · Muscle synergy · Phase resetting · Interlimb coordination

## 8.1 Introduction

People and animals produce adaptive locomotion in diverse environments by cooperatively and skillfully manipulating their complicated and redundant musculoskeletal systems. To put it simply, locomotion involves moving the entire body against

---

S. Aoi (✉)
Department of Aeronautics and Astronautics, Graduate School of Engineering, Kyoto University, Kyoto Daigaku-Katsura, Nishikyo-ku, Kyoto 615-8540, Japan
e-mail: shinya_aoi@kuaero.kyoto-u.ac.jp



B. I. Prilutsky, D. H. Edwards (eds.), *Neuromechanical Modeling of Posture and Locomotion,* Springer Series in Computational Neuroscience,
DOI 10.1007/978-1-4939-3267-2_8

gravitational force by using the legs. To establish locomotion, leg movement needs to be controlled to transport the entire body and posture needs to be controlled to prevent falling during locomotion. However, although posture is maintained via the posture control, the movement of the legs disturbs the posture. In other words, the controls of posture and movement affect one another during locomotion, making the adequate balancing of these controls very important. The underlying mechanism for stabilizing posture during locomotion remains unclear.

So far, the abilities of humans and other animals to generate adaptive movements have been investigated by examining the structure and activities of neural systems. For example, physiological studies on lampreys and cats have greatly contributed to the elucidation of locomotor mechanisms (Grillner 1975; Shik and Orlovsky 1976; Orlovsky et al. 1999). However, locomotion is a well-organized motion generated through dynamic interactions among the nervous system, the musculoskeletal system, and the environment. It is difficult to fully analyze locomotor mechanisms solely in terms of the nervous system. In addition to understanding the nervous system, it is crucial to elucidate the dynamic characteristics inherent in the musculoskeletal system. Integrative studies of the nervous and musculoskeletal systems are required to clarify locomotor mechanisms. Physiological and anatomical findings now enable the construction of reasonably realistic models of the nervous and musculoskeletal systems. Thus, to overcome the limitations of behavioral studies based only on the nervous system, simulation studies have recently investigated specific functional roles of the nervous system in locomotor behavior (Taga et al. 1991; Ogihara and Yamazaki 2001; Ivashko et al. 2003; Yamasaki et al. 2003; Yakovenko et al. 2004; Ekeberg and Pearson 2005; Pearson et al. 2006; Jo and Massaquoi 2007; Prochazka and Yakovenko 2007; Jo 2008; Nomura et al. 2009; Markin et al. 2010).

In an actual travel path, obstacles that must be stepped over to continue locomotion are often encountered. Stepping over obstacles to avoid tripping is an essential movement for safe, smooth locomotion. Such obstacle avoidance is a skillful, intentional movement, whereby humans and other animals must recognize the dimensions of an obstacle, and determine how to control their legs to avoid colliding with it while maintaining their posture. This task requires a highly coordinated control of the leg movements and posture, which highlights the relationship between movement and postural controls.

In this chapter, the functional roles of the nervous system in maintaining posture during locomotion are investigated by focusing on the adaptive walking of humans during applications of perturbing forces and sudden environmental variations (Aoi et al. 2010, 2012) and on obstacle avoidance by the hind legs of rats during walking (Aoi et al. 2013). Neuromusculoskeletal models for humans and rats were constructed by integrating the anatomically realistic musculoskeletal model and the physiologically based nervous system model. The leg movement control was modeled based on the physiological concepts of central pattern generators (CPGs) and muscle synergy and on sensory regulation by phase resetting and interlimb coordination. The posture control was modeled to regulate the postural behavior using

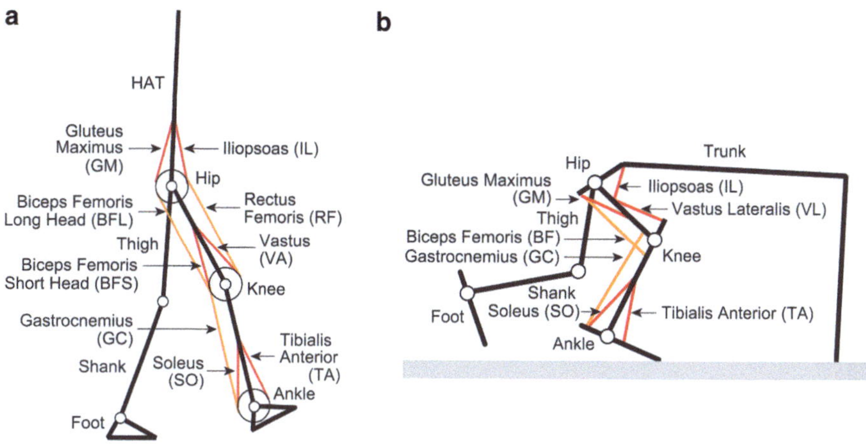


**Fig. 8.1** Musculoskeletal models for humans (**a**) and rats (**b**)

somatosensory information. We have examined how these controls contribute to stabilizing posture during locomotion.

## 8.2 Neuromusculoskeletal Model

### *8.2.1 Musculoskeletal Model*

Figure 8.1a and b show the musculoskeletal models for humans and rats, respectively. The human skeletal model consists of seven rigid links that represent the HAT (head, arms, and trunk), thighs, shanks, and feet. For the rat skeletal model, seven rigid links were used for the trunk and hind legs, where the front legs are fixed on the trunk and slide on the ground without friction, which is similar to the models of previous studies (Yakovenko et al. 2004; Ekeberg and Pearson 2005; Pearson et al. 2006; Prochazka and Yakovenko 2007). The current two models are two-dimensional and the walking behaviors are constrained in the sagittal plane. The contact between their feet and the ground was modeled using viscoelastic elements. Physical parameters of the skeletal models were determined from measured anatomical data.

The human model has nine principal muscles for each leg, including uniarticular and biarticular muscles. The rat model has seven principal muscles for each hind leg. Each muscle receives signals from the corresponding motoneuron and generates muscle tension depending on the force-length and force-velocity relationships and muscle activation. The muscle tension was modeled based on a contractile element and passive elements parallel to the contractile element. The muscle activation for the contractile element is given through a low-pass filter to motor commands of

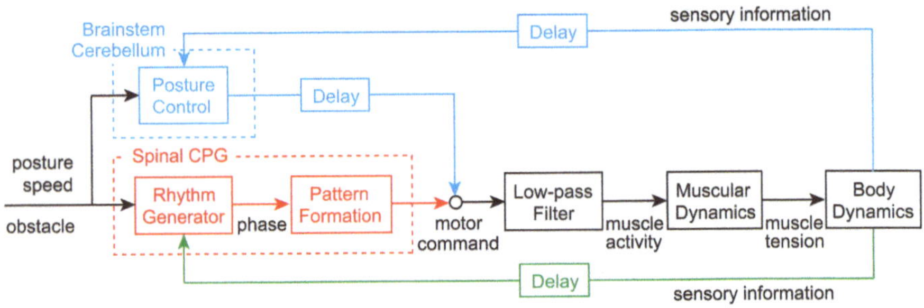


**Fig. 8.2** Nervous system model for the human and rat

the motoneuron determined in the nervous system model. The physical parameters of the muscle models were determined from the measured anatomical data.

## *8.2.2 Nervous System Model*

A nervous system model at the brainstem, cerebellar, and spinal cord levels was developed based on the physiological findings (Fig. 8.2), which were used for both the human and rat models.

### 8.2.2.1 Central Pattern Generators

Physiological studies suggest that CPGs in the spinal cord strongly contribute to rhythmic leg movements, such as locomotion (Grillner 1975; Shik and Orlovsky 1976; Orlovsky et al. 1999). Although the organization of CPGs remains unclear, physiological findings suggest that CPGs consist of hierarchical networks, including rhythm generator (RG) and pattern formation (PF) networks (Burke et al. 2001; Lafreniere-Roula and McCrea 2005; Rybak et al. 2006a, b). The RG network generates the basic rhythm and alters it by producing phase shifts and rhythm resetting based on sensory afferents and perturbations. The PF network shapes the rhythm into spatiotemporal patterns of activated motoneurons through interneurons. CPGs separately control the locomotor rhythm and pattern of motoneuron activation in the RG and PF networks, respectively.

Such a two-layered hierarchical network was created for the CPG model. For the RG model, two simple phase oscillators, each of which produces basic rhythm and phase information for the corresponding leg, were used. In the PF model, the motor commands for the motoneurons were determined to produce periodic leg movements using the oscillator phase based on the physiological concept of muscle synergy, which is explained in the next section.

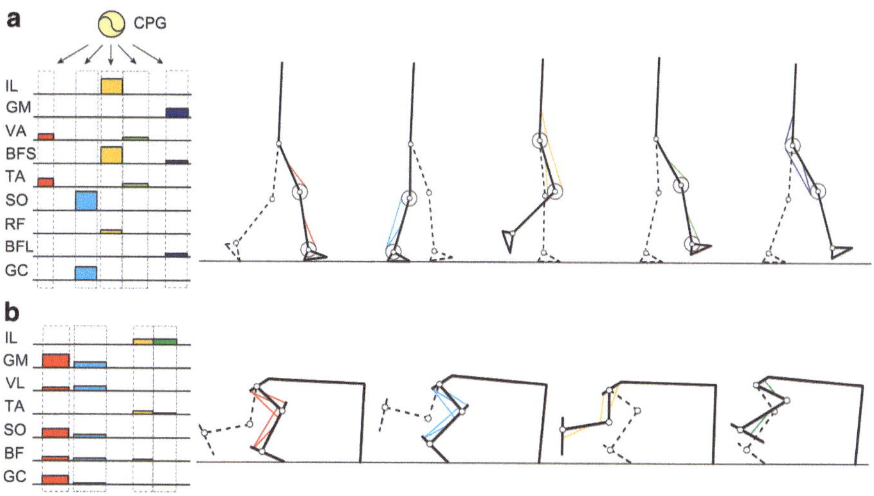


**Fig. 8.3** Rectangular pulses from the CPG model delivered to the muscles of the human model (**a**) and rat model (**b**). Five pulses were used for the human model and four pulses were used for the rat model to generate the basic patterns of locomotion

#### 8.2.2.2 Muscle Synergy

Physiological studies have suggested an important concept of muscle synergy, which explains the coordinated structure of muscle activity and is viewed as one way of coping with motor redundancy by decreasing the number of degrees of freedom (Todorov and Jordan 2002; d'Avella et al. 2003; d'Avella and Bizzi 2005; Ting and Macpherson 2005; Drew et al. 2008; Latash 2008). In regards to muscle synergy for locomotion, many studies have shown that although the recorded electromyography (EMG) data during locomotion are complex, they can be accounted for by the combination of only a small number of basic patterns (Ivanenko et al. 2004; Ivanenko et al. 2005; Cappellini et al. 2006; Ivanenko et al. 2006; Dominici et al. 2011).

For the PF model of the CPG model, five rectangular pulses for the human model and four rectangular pulses for the rat model were used for the basic patterns of the motor commands for walking (Fig. 8.3), which is similar to the models of a previous study (Jo and Massaquoi 2007). The timing of the initiation of bursting and the burst duration were determined in accordance with the oscillator phase of the RG model. The rectangular pulses were delivered to muscles by using weighting coefficients.

Muscle synergy analysis has also shown that the addition of another pattern to the basic patterns for walking explains the muscle activities for obstacle avoidance during walking (Ivanenko et al. 2005, 2006), which means that this additional pattern controls the intralimb (intersegmental) coordination of the leg movement to step over an obstacle. To establish the obstacle avoidance task in the rat model,

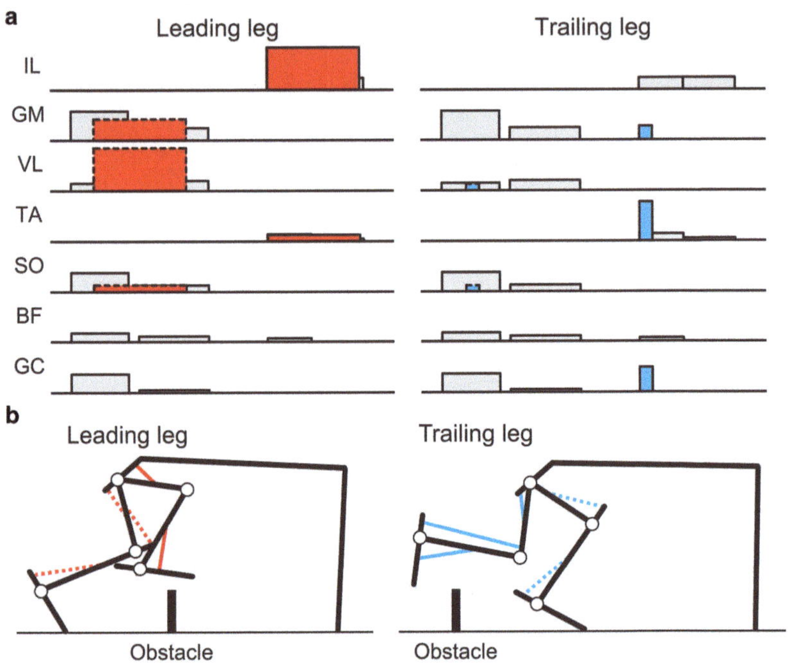


**Fig. 8.4** Additional rectangular pulses for obstacle avoidance (**a**) and muscles activated by the additional pulses (**b**). *Solid* and *dotted lines* correspond to the contributions to the swinging and supporting legs, respectively

another rectangular pulse, similar to that used in a previous study (Jo and Massaquoi 2007), was used. Because the leading and trailing legs have different roles during obstacle avoidance, different rectangular pulses for the leading and trailing legs were used (Fig. 8.4). To step over an obstacle, a rat must not only swing its legs more than usual, but also support its body with its contralateral legs. The additional rectangular pulses contribute to both the swinging and supporting legs.

#### 8.2.2.3 Phase Resetting

Because basic motor patterns for walking and obstacle avoidance are produced by rectangular pulses, adequate timing to generate these pulses is crucial. Although CPGs can produce oscillatory behaviors even in the absence of rhythmic input and proprioceptive feedback, CPGs must use sensory information to produce adaptive and effective locomotion. In particular, from the muscle synergy analysis, physiological findings suggest that CPGs manage the timing to produce the basic patterns based on events, such as foot contact, to achieve adaptive locomotion (Ivanenko et al. 2006).

The locomotor rhythm and phase have been shown to be modulated by producing phase shifts and rhythm resetting based on sensory afferents and perturbations (Duysens 1977; Conway et al. 1987; Guertin et al. 1995; Schomburg et al. 1998;

Lafreniere-Roula and McCrea 2005; Rybak et al. 2006b). As cutaneous afferents were observed to contribute to these phase shifts and rhythm resetting behaviors (Duysens 1977; Schomburg et al. 1998), they were modeled by resetting the oscillator phase in the RG model based on foot-contact information (phase resetting) for the sensory regulation model.

In cat locomotion, two types of sensory information are suggested to be used for the phase transition from stance to swing: force-sensitive afferents in the ankle extensor muscles (Duysens and Pearson 1980; Whelan et al. 1995) and position-sensitive afferents from the hip (Grillner and Rossignol 1978; Hiebert et al. 1996). When the force in the ankle extensor muscle is low (unloading rule) or when the hip joint is sufficiently extended (hip extension rule), the phase changes from the stance to the swing. However, it is unclear which rule has more contribution to the generation of robust walking (Ekeberg and Pearson 2005; Pearson et al. 2006). To investigate the sensory mechanism to regulate this transition for adaptive walking, the oscillator phase was reset not only based on foot-contact information but also on the unloading and hip extension rules in the human model.

#### 8.2.2.4 Control of Interlimb Coordination

Because locomotor behavior is produced by alternating leg movements between the left and right legs, interlimb coordination is an important factor. In the discussed above models, weak potential was used in the oscillator dynamics to stabilize the antiphase movement of the oscillators.

During obstacle avoidance, when the swinging leg steps over an obstacle, the contralateral leg must support the entire body to maintain its posture. Because obstacle avoidance will fail without this support, interlimb coordination is crucial for the success of this task. To satisfy this adequate interlimb coordination during obstacle avoidance, the oscillator phase was regulated to delay the additional rectangular pulse for stepping over an obstacle until the contralateral leg contacted the ground to support the body.

#### 8.2.2.5 Posture Control

At the brainstem and cerebellum levels, postural behavior is regulated based on the somatosensory information. For the walking motion of a human, it is crucial to maintain a vertical trunk pitch and move the center of mass (COM) forward at the desired velocity. For the walking motion of a rat, it is important to maintain a constant hip height and forward velocity. For posture control, these factors were focused on for simplicity and motor commands were produced in a feedback fashion using specific muscles to regulate posture during locomotion (Fig. 8.5). Because this posture control is managed at the brainstem and cerebellar levels, receiving the somatosensory information at the brainstem and cerebellar levels and sending the signal to the spinal cord level are delayed. Due to these delays, gain parameters of

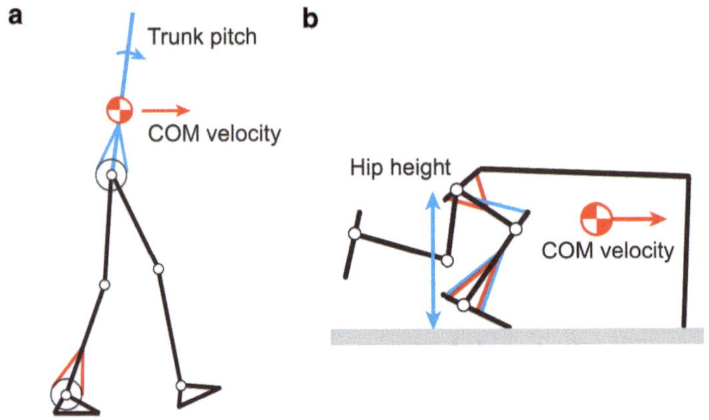


**Fig. 8.5** Posture control for the human model (**a**) and rat model (**b**)

the posture control were small, so that the contribution of the posture control was smaller than that of the muscle synergy.

The motor commands for motoneurons were given by the summation of the rectangular pulses based on muscle synergy and the motor commands given by the posture control (Fig. 8.2).

## 8.3 Results

The dynamic characteristics of our neuromusculoskeletal models for humans and rats that were obtained via forward dynamic simulations are described in this section.

### *8.3.1 Locomotion in the Human Model*

Here we verify our model by comparing the simulation results with the measured data obtained during human walking and investigate the functional roles of the nervous system.

#### 8.3.1.1 Generation of Walking

Stable walking was established by adequately determining the model parameters. To verify our model, we compared the simulation results with the measured data obtained during human walking (Fig. 8.6), where the recorded joint angles and ground reaction forces were taken from (Winter 2004) and the EMG data from (Inman 1953). The properties of the simulation results are similar to those of the measured data, and in particular the vertical reaction force has a double-peaked shape also seen in the human walking results.

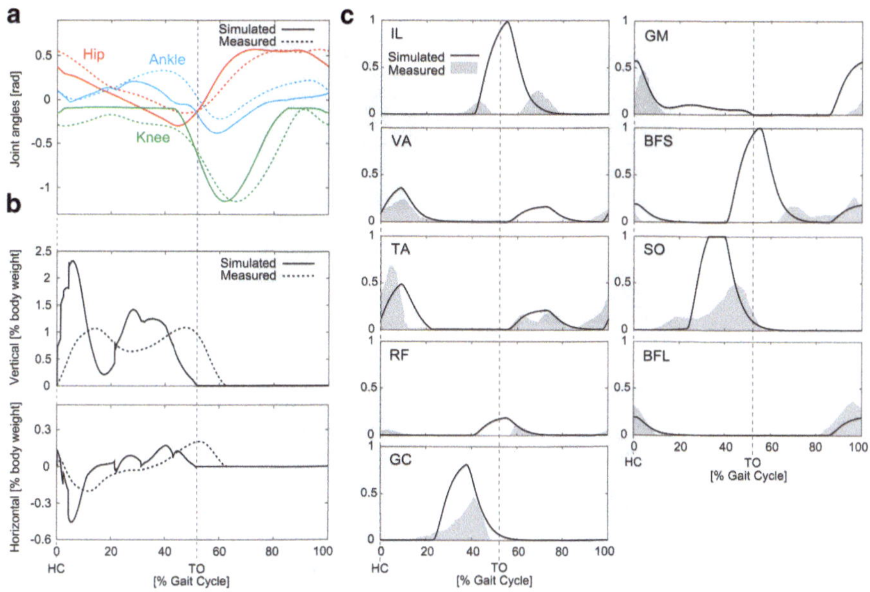


**Fig. 8.6** Comparison between the simulation results and the measured data that were obtained during human walking (**a**: joint angles, **b**: ground reaction forces, and **c**: muscle activities). *HC* and *TO* indicate heel contact and toe-off, respectively, for the simulation results

#### 8.3.1.2 Contribution of the Posture Control

While the motor commands are given by the summation of the rectangular pulses and posture control, the contribution of the posture control was only a few percent during steady walking. This contribution appears relatively small, but the human model easily fell when the posture control was eliminated from our nervous system model. This shows that the posture control plays an important role in the generation of walking.

#### 8.3.1.3 Contribution of Phase Resetting Based on Foot-Contact Information

To determine the contribution of sensory regulation by phase resetting to the generation of adaptive walking, we examined the ability of our model to adapt to the perturbing forces; that is, we determined if the human model could recover after being perturbed. More specifically, after steady walking was established in the model, a perturbing force was applied for 0.1 s to the COM of HAT in the horizontal direction (forward or backward) using various magnitudes and timings of the perturbation. In particular, four cases were compared: (1) without the use of phase resetting, (2) use of phase resetting only at foot off, (3) use of phase resetting only at foot contact, and (4) use of phase resetting at both foot off and foot contact. Figure 8.7 shows the results, where the white boxes indicate that the human

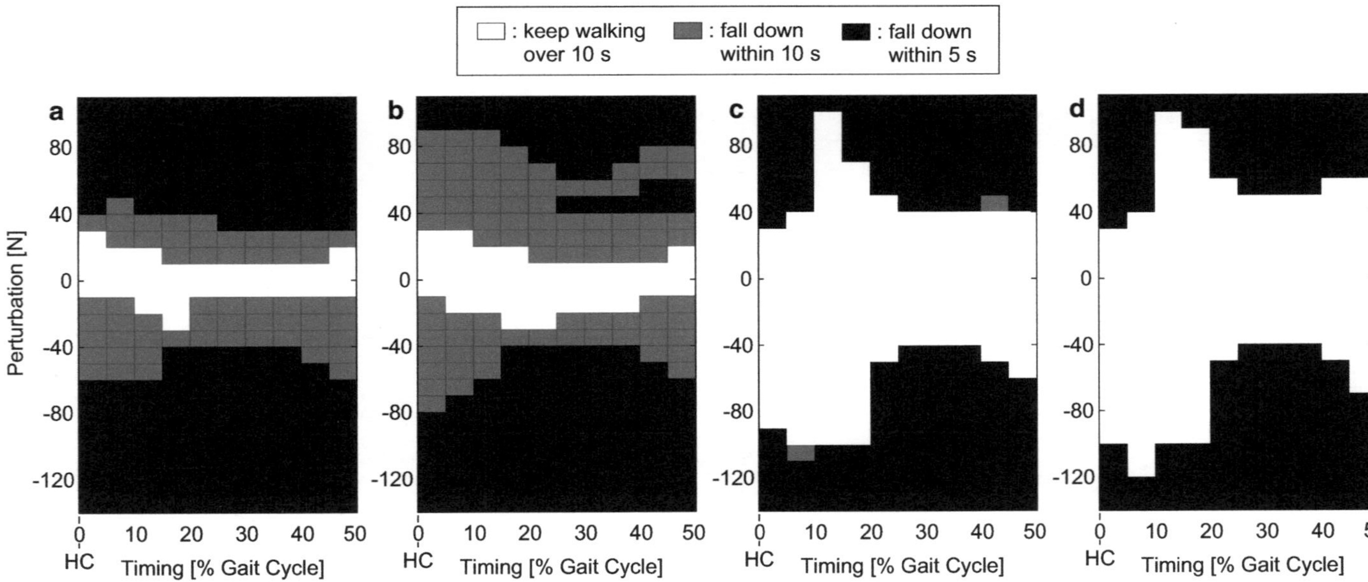


**Fig. 8.7** Tolerance to the perturbing forces. (**a**) without the use of phase resetting, (**b**) use of phase resetting only at foot off, (**c**) use of phase resetting only at foot contact, and (**d**) use of phase resetting at both foot off and foot contact

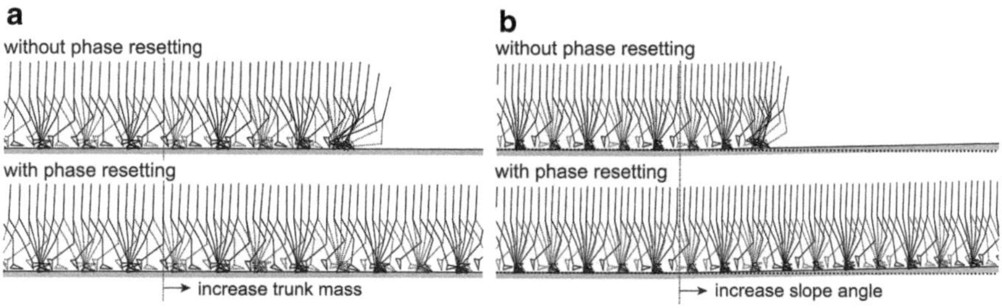


**Fig. 8.8** Simulated walking behavior for the sudden environmental changes, with and without the phase resetting. **a** Sudden increase in the trunk mass, and **b** sudden increase in the slope angle

model continued walking for over 10 s after the disturbance was applied, the gray boxes indicate that the human model fell within 10 s after the disturbance was applied, and the black boxes indicate that the human model fell within 5 s after the disturbance was applied. When phase resetting was not incorporated, the human model easily fell. The human model walked longer when phase resetting was applied. The number of white boxes suggests that the application of phase resetting at foot contact contributes more significantly to counteracting the perturbing forces than when phase resetting was applied at foot off. The use of phase resetting at both foot off and foot contact yielded the greatest degree of robustness among the four cases.

The ability of our model to adapt to sudden environmental changes was also investigated. To alter the environment, the trunk mass (Fig. 8.8a) and slope angle (Fig. 8.8b) were instantaneously increased for the cases with and without phase resetting. The human model without the phase resetting easily fell after the sudden environmental changes. In contrast, the human model with the phase resetting continued walking against the environmental variations. These environmental changes induced a decrease in walking speed and changes in the joint motions through the sensory regulation of the motor patterns.

#### 8.3.1.4 Unloading Rule vs. Hip Extension Rule

To investigate the roles of the unloading and hip extension rules that regulate the stance-to-swing transition during walking, the ability of our model to adapt to the perturbing forces was examined. The following three cases of model walking were compared: (1) without the use of phase resetting, (2) use of phase resetting based on the hip extension rule, and (3) use of phase resetting based on the unloading rule (Fig. 8.9). The human model for case 2 easily fell after being disturbed, compared to the model without phase resetting. However, for case 3, the human model walked longer, which indicates that the unloading rule increased the robustness of the responses, similar to the results of a previous modeling study of cat locomotion (Ekeberg and Pearson 2005; Pearson et al. 2006).

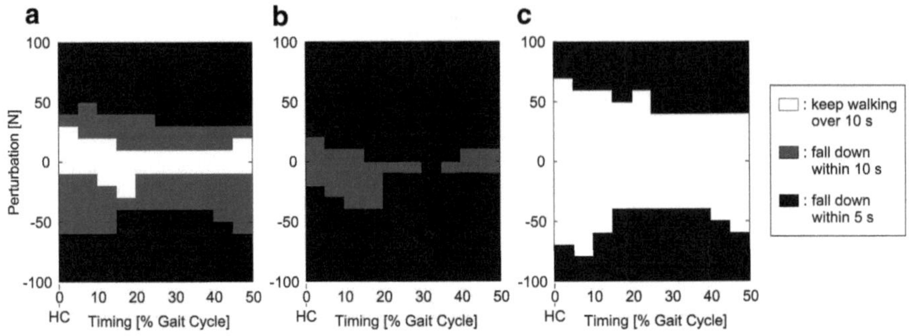


**Fig. 8.9** Tolerance to the perturbing forces. **a** Without the use of phase resetting, **b** use of phase resetting based on the hip extension rule, and **c** use of phase resetting based on the unloading rule

## *8.3.2 Locomotion in the Rat Model*

In this section, the neuromusculoskeletal model of rats is verified by comparing the simulation results with the measured data obtained during rat walking, and the functional roles of the nervous system during walking and obstacle avoidance are investigated.

### 8.3.2.1 Generation of Walking

An adequate determination of the model parameters produced stable walking of the rat. We verified our model by comparing the simulation results with the measured data obtained during rat walking (Aoi et al. 2013; Fig. 8.10). The properties of the simulation results are similar to those of the measured data.

### 8.3.2.2 Contribution of the Posture Control

In the rat model, the contribution of the posture control was only a few percent for the motor commands during steady walking. However, the rat model fell when the posture control was removed from the nervous system model. This indicates the importance of the posture control to generate walking, which was similarly shown in the human model.

### 8.3.2.3 Stepping Over an Obstacle

By using additional rectangular pulses, the rat model stepped over an obstacle and the walking behavior of the rat soon recovered after the obstacle avoidance process (Fig. 8.11c). Figure 8.11a and b show stick diagrams of the measured kinematics of the leading and trailing legs, respectively, during the obstacle avoidance process of the rat (Aoi et al. 2013) as a comparison with the simulation results. When the

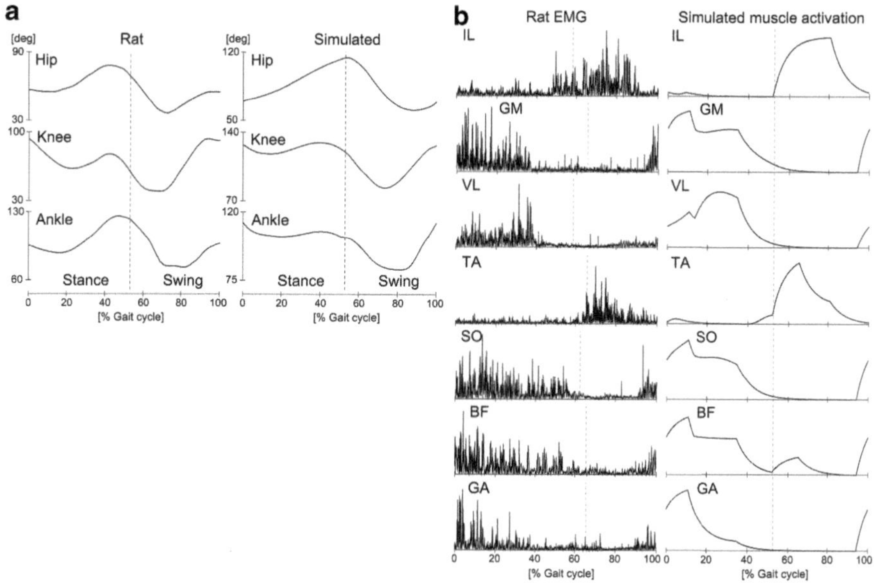


**Fig. 8.10** Comparison between the simulation results and the measured data obtained during the rat walking (**a**: joint angles and **b**: muscle activities)

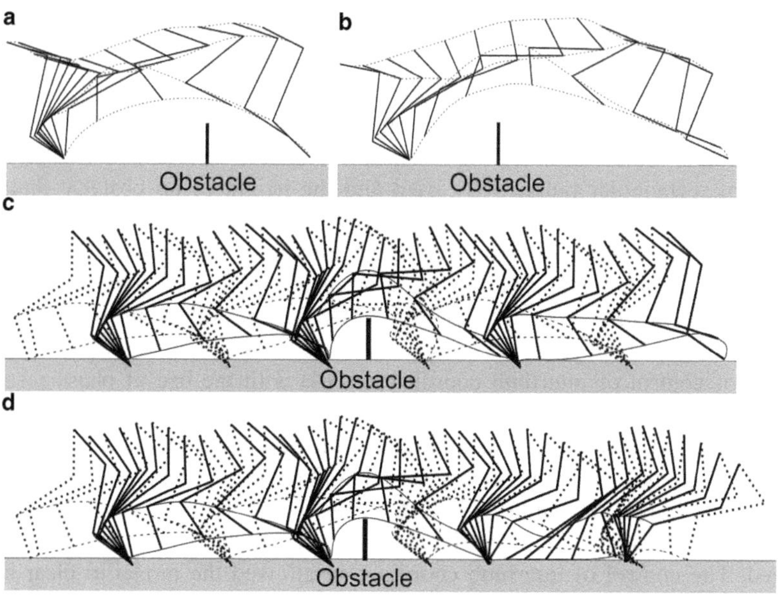


**Fig. 8.11** Stick diagrams during obstacle avoidance. **a** Measured kinematics of the leading leg, **b** measured kinematics of the trailing leg, **c** simulated obstacle avoidance behavior, and **d** falling after stepping over an obstacle due to the lack of phase resetting

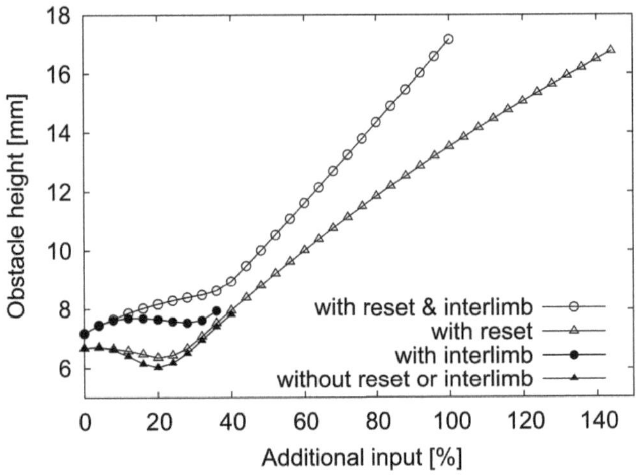


**Fig. 8.12** Comparison between obstacle height with and without phase resetting and control of interlimb coordination

phase resetting based on foot contact was not incorporated, the rat model fell after stepping over an obstacle (Fig. 8.11d).

#### 8.3.2.4 Contribution of Phase Resetting and Interlimb Coordination During Obstacle Avoidance

To investigate the contribution of the sensory regulation based on phase resetting and interlimb coordination during obstacle avoidance, various magnitudes of the additional rectangular pulses were used and the height of an obstacle that a rat could step over without falling was examined. The height of an obstacle that a rat could step over without collision was determined from the resultant simulated kinematics of the leading and trailing legs, in which the obstacle was assumed to be a zero-width bar in the sagittal plane. In particular, the following four cases were compared: (1) without phase resetting or control of interlimb coordination, (2) with the use of control of interlimb coordination, (3) with the use of phase resetting, and (4) with the use of both phase resetting and control of interlimb coordination. Figure 8.12, in which various magnitudes of the additional inputs are used, shows the height of an obstacle that the rat model could clear. When phase resetting was not used, the rat model stepped over an obstacle of 8 mm (40 % of additional inputs) at best. Phase resetting contributed to a quick recovery after the obstacle was cleared. The control of interlimb coordination allowed the model to clear higher obstacles. Although the rat model with phase resetting also stepped over higher obstacles, the magnitudes of the additional inputs for the model needed to be higher than those of the model with both phase resetting and control of interlimb coordination, which cleared high obstacles by using small additional inputs without falling after stepping over the obstacles.

## 8.4 Discussion

To investigate the functional roles of the nervous system for the stabilization of posture during locomotion, our models focus on the leg movement control, which is based on the physiological concepts of CPGs and muscle synergy and on sensory regulation by phase resetting and interlimb coordination, and the posture control to regulate the postural behavior via somatosensory information.

### *8.4.1 Control of the Leg Movement and Posture*

To establish locomotion, it is necessary to control the leg movement needed to move the entire body and the posture needed to prevent the body from falling during locomotion. In our models, the leg movement control uses basic motor command patterns that are based on the physiological concepts of CPGs and muscle synergy, which are regulated through sensory information based on phase resetting and interlimb coordination. In contrast, the posture control regulates the postural behavior in a feedback fashion by using somatosensory information.

When the sensory regulation in the leg movement control was eliminated, the human and rat models fell easily and the robustness against the force disturbances and sudden environmental variations decreased. When the posture control was eliminated from the nervous system model, the human and rat models fell easily. By adequately integrating these controls in the nervous system model, our models produced adaptive locomotor behaviors.

### *8.4.2 Leg Movement Control Based on Muscle Synergy*

Humans and other animals produce adaptive movements from a combination of various degrees of freedom, from which they must solve the motor redundancy problem. Physiological studies suggest the importance of muscle synergies for controlling movements (Todorov and Jordan 2002; d'Avella et al. 2003; Ivanenko et al. 2004; d'Avella and Bizzi 2005; Ivanenko et al. 2005; Ting and Macpherson 2005; Ivanenko et al. 2006; Drew et al. 2008; Latash 2008; Dominici et al. 2011), which is viewed as one solution to the redundancy problem. Muscle synergy is related to the co-variation of muscle activities. Humans and other animals share some basic patterns for producing muscle activation patterns among various movements (e.g., the jumping,, and walking patterns of frogs and the walking, obstacle avoidance, kicking motion, and running of humans) and produce these various movements with the addition of other patterns (Ivanenko et al. 2004; d'Avella and Bizzi 2005; Ivanenko et al. 2005; Cappellini et al. 2006; Ivanenko et al. 2006). This means that some degrees of freedom are functionally connected depending on the task, which reduces the number of degrees of freedom and solves the problem of motor redundancy.

CPGs are considered to produce such basic patterns in a feedforward fashion to create various movements, and by adding another pattern to the basic patterns for walking, the motor control of stepping over an obstacle is achieved (Ivanenko et al. 2005, 2006). In addition, the timing to produce the basic patterns is managed by CPGs based on events, such as foot contact (Ivanenko et al. 2006). Based on these physiological findings and hypotheses, we developed a simple rectangular pulse model for walking and obstacle avoidance and modulated the rectangular pulses by incorporating a sensory regulation model based on phase resetting and interlimb coordination.

For successful obstacle avoidance during locomotion, two factors are crucial; the leading and trailing legs must clear the obstacle without collision and the walking behavior must recover soon after stepping over the obstacle. As the obstacle height increases, the toe heights of the leading and trailing legs must also increase, which disturbs the posture and causes instability and falling. Therefore, the processes of stepping over a high obstacle and recovering soon after obstacle avoidance are not consistent. The simulation results of the rat model showed that the sensory regulation based on phase resetting, which was achieved by using the foot-contact information, contributed to a quick recovery after stepping over an obstacle and the sensory regulation based on interlimb coordination contributed to efficiently stepping over a high obstacle.

### 8.4.3 Sensory Regulation by Phase Resetting

For the sensory regulation model, phase resetting was used. Although physiological evidence showed that locomotor rhythm and phases are modulated by phase shifts and rhythm resetting that is produced based on sensory afferents and perturbations (Duysens 1977; Conway et al. 1987; Guertin et al. 1995; Schomburg et al. 1998; Rybak et al., 2006b), such rhythm and phase modulations have been investigated, for the most part, during fictive locomotion in cats, and their functional roles during actual locomotion remain unclear. However, spinal cats produce locomotor behaviors on treadmills and change their gaits depending on the belt speed (Forssberg and Grillner 1973; Orlovsky et al. 1999), suggesting that the tactile sensory information between their feet and the belt influences the locomotion phase and rhythm generated by the CPG (Duysens et al. 2000). In addition, cutaneous afferents were observed to contribute to phase resetting (Duysens 1977; Schomburg et al. 1998). Our sensory regulation model, in which phase resetting is utilized, is consistent with these observations. Furthermore, previous neuromechanical models have demonstrated that phase resetting contributes to the generation of adaptive walking (Yamasaki et al. 2003; Nomura et al. 2009).

Locomotor behavior can be determined from the spatiotemporal patterns of motor commands and phase resetting manages the temporal regulation based on foot-contact events. Even if the timing of the foot-contact event is affected by disturbances, the phase resetting regulates the timing to generate motor commands based on the event. Early foot contact induces a phase shift in the periodic motor commands to interrupt the locomotor rhythm, and delayed foot contact results in a phase shift in the periodic motor commands to prolong the locomotor rhythm. Phase resetting creates various phase profiles and locomotor rhythms depending on the situation, thus improving the stability and robustness of the locomotion.

People and animals integrate sensory information to produce motor commands. Different sensory information causes different dynamic characteristics in locomotor behavior. To produce adaptive and efficient movements, the type of sensory information they use and when and how they use the sensory information are crucial. Our simulation results showed that sensory regulation based on foot-contact information helps maintain posture during locomotion and the unloading rule related to the force information in the ankle extensor muscle increased the robustness of locomotion more than the hip extension rule that is related to the angle information of the hip joint. Computer simulations are useful to examine sensorimotor integration mechanisms during locomotion.

### *8.4.4 Sensory Regulation Based on Interlimb Coordination*

During obstacle avoidance, as the additional input for the leading leg increases, the toe height of the leading leg increases and its contact with the ground is delayed. When the delay is longer than the onset of the additional input for the trailing leg, the rat model begins to step over an obstacle without support from its contralateral leg. This reduces the performance of obstacle avoidance. The sensory regulation to produce adequate interlimb coordination to support the body by the contralateral leg allowed the rat model to clear a high obstacle with little additional input.

Although the sensory regulation based on this interlimb coordination increased the performance of obstacle avoidance, it shifted the relative phase of the rectangular pulses for the basic patterns of locomotion between the legs from an antiphase state. Because this shift causes instability and falling during walking, the relative phase should return to antiphase after stepping over an obstacle. Weak potential was used in the oscillator dynamics to stabilize the antiphase movement of the oscillators, which increased the robustness of locomotion. Adequate control of the interlimb coordination is required during walking and obstacle avoidance.

## 8.5 Conclusion

Simulation studies that were conducted by integrating musculoskeletal models based on anatomical and biomechanical findings and nervous system models based on physiological findings have become useful tools to elucidate the locomotor mechanisms in biological systems, which overcome the limitations of behavioral studies based only on the nervous system. In this chapter, we constructed the neuromusculoskeletal models of humans and rats to investigate the functional roles of the nervous system for stabilizing the posture of the body during locomotion. The physical structure of the musculoskeletal models is simple and is constrained in the two- dimensional sagittal plane. The nervous system is limited to the brainstem, cerebellar, and spinal cord levels and only focuses on the leg movement control that is based on CPGs, muscle synergy, and sensory regulation via phase resetting and interlimb coordination, and focuses on the posture control by using somatosensory information. To further elucidate adaptive functions in the locomotion dynamics of biological systems, we intend to employ a more sophisticated and plausible model in the future.

**Acknowledgements** This chapter is based on collaboration studies with Kazuo Tsuchiya, Naomichi Ogihara, Dai Yanagihara, Tetsuro Funato, Nozomi Tomita, Takahiro Kondo, Naohiro Hayashi, Sho Aoki, Hiroshi Yamaura, Yasuhiro Sugimoto, and Kei Senda. This chapter is supported in part by KAKENHI (23360111, 26120006) from MEXT, Japan, by JST, CREST, Japan, and by Uehara Memorial Foundation, Japan.

## References

Aoi S, Ogihara N, Funato T, Sugimoto Y, Tsuchiya K (2010) Evaluating functional roles of phase resetting in generation of adaptive human bipedal walking with a physiologically based model of the spinal pattern generator. Biol Cybern 102:373–387

Aoi S, Ogihara N, Funato T, Tsuchiya K (2012) Sensory regulation of stance-to-swing transition in generation of adaptive human walking: a simulation study. Robot Auton Syst 60:685–691

Aoi S, Kondo T, Hayashi N, Yanagihara D, Aoki S, Yamaura H et al (2013) Contributions of phase resetting and interlimb coordination to the adaptive control of hindlimb obstacle avoidance during locomotion in rats: a simulation study. Biol Cybern 107:201–216

Burke RE, Degtyarenko AM, Simon ES (2001) Patterns of locomotor drive to motoneurons and last-order interneurons: clues to the structure of the CPG. J Neurophysiol 86:447–462

Cappellini G, Ivanenko YP, Poppele RE, Lacquaniti F (2006) Motor patterns in human walking and running. J Neurophysiol 95:3426–3437

Conway BA, Hultborn H, Kiehn O (1987) Proprioceptive input resets central locomotor rhythm in the spinal cat. Exp Brain Res 68:643–656

d'Avella A, Bizzi E (2005) Shared and specific muscle synergies in natural motor behaviors. Proc Natl Acad Sci U S A 102:3076–3081

d'Avella A, Saltiel P, Bizzi E (2003) Combinations of muscle synergies in the construction of a natural motor behavior. Nat Neurosci 6:300–308

Dominici N, Ivanenko YP, Cappellini G, d'Avella A, Mondi V, Cicchese M et al (2011) Locomotor primitives in newborn babies and their development. Science 334:997–999

Drew T, Kalaska J, Krouchev N (2008) Muscle synergies during locomotion in the cat: a model for motor cortex control. J Physiol 586:1239–1245

Duysens J (1977) Fluctuations in sensitivity to rhythm resetting effects during the cat's step cycle. Brain Res 133:190–195

Duysens J, Pearson KG (1980) Inhibition of flexor burst generation by loading ankle extensor muscles in walking cats. Brain Res 187:321–332

Duysens J, Clarac F, Cruse H (2000) Load-regulating mechanisms in gait and posture: comparative aspects. Physiol Rev 80:83–133

Ekeberg O, Pearson K (2005) Computer simulation of stepping in the hind legs of the cat: an examination of mechanisms regulating the stance-to-swing transition. J Neurophysiol 94:4256–4268

Forssberg H, Grillner S (1973) The locomotion of the acute spinal cat injected with clonidine i.v. Brain Res 50:184–186
Grillner S (1975) Locomotion in vertebrates: central mechanisms and reflex interaction. Physiol Rev 55:247–304
Grillner S, Rossignol S (1978) On the initiation of the swing phase of locomotion in chronic spinal cats. Brain Res 146:269–277
Guertin P, Angel MJ, Perreault MC, McCrea DA (1995) Ankle extensor group I afferents excite extensors throughout the hindlimb during fictive locomotion in the cat. J Physiol 487(Pt 1):197–209
Hiebert GW, Whelan PJ, Prochazka A, Pearson KG (1996) Contribution of hind limb flexor muscle afferents to the timing of phase transitions in the cat step cycle. J Neurophysiol 75:1126–1137
Inman VT (1953) The pattern of muscular activity in the lower extremity during walking. In: Technical Report Series II, Issue 25, Prosthetic Devices Research Project, Institute of Engineering Research, University of California, Berkeley, CA
Ivanenko YP, Poppele RE, Lacquaniti F (2004) Five basic muscle activation patterns account for muscle activity during human locomotion. J Physiol 556:267–282
Ivanenko YP, Cappellini G, Dominici N, Poppele RE, Lacquaniti F (2005) Coordination of locomotion with voluntary movements in humans. J Neurosci 25:7238–7253
Ivanenko YP, Poppele RE, Lacquaniti F (2006) Motor control programs and walking. Neuroscientist 12:339–348
Ivashko DG, Prilutsky BI, Markin SN, Chapin JK, Rybak IA (2003) Modeling the spinal cord neural circuitry controlling cat hindlimb movement during locomotion. Neurocomputing 52–54:621–629
Jo S (2008) Hypothetical neural control of human bipedal walking with voluntary modulation. Med Biol Eng Comput 46:179–193
Jo S, Massaquoi SG (2007) A model of cerebrocerebello-spinomuscular interaction in the sagittal control of human walking. Biol Cybern 96:279–307
Lafreniere-Roula M, McCrea DA (2005) Deletions of rhythmic motoneuron activity during fictive locomotion and scratch provide clues to the organization of the mammalian central pattern generator. J Neurophysiol 94:1120–1132
Latash ML (2008) Synergy. Oxford University Press, New York
Markin SN, Klishko AN, Shevtsova NA, Lemay MA, Prilutsky BI, Rybak IA (2010) Afferent control of locomotor CPG: insights from a simple neuromechanical model. Ann N Y Acad Sci 1198:21–34
Nomura T, Kawa K, Suzuki Y, Nakanishi M, Yamasaki T (2009) Dynamic stability and phase resetting during biped gait. Chaos 19:026103
Ogihara N, Yamazaki N (2001) Generation of human bipedal locomotion by a bio-mimetic neuro-musculo-skeletal model. Biol Cybern 84:1–11
Orlovsky GN, Deliagina TG, Grillner S (1999) Neuronal control of locomotion: from mollusc to man. Oxford University Press, New York
Pearson K, Ekeberg O, Buschges A (2006) Assessing sensory function in locomotor systems using neuro-mechanical simulations. Trends Neurosci 29:625–631
Prochazka A, Yakovenko S (2007) Predictive and reactive tuning of the locomotor CPG. Integr Comp Biol 47:474–481
Rybak IA, Shevtsova NA, Lafreniere-Roula M, McCrea DA (2006a) Modelling spinal circuitry involved in locomotor pattern generation: insights from deletions during fictive locomotion. J Physiol 577:617–639
Rybak IA, Stecina K, Shevtsova NA, McCrea DA (2006b) Modelling spinal circuitry involved in locomotor pattern generation: insights from the effects of afferent stimulation. J Physiol 577:641–658
Schomburg ED, Petersen N, Barajon I, Hultborn H (1998) Flexor reflex afferents reset the step cycle during fictive locomotion in the cat. Exp Brain Res 122:339–350
Shik ML, Orlovsky GN (1976) Neurophysiology of locomotor automatism. Physiol Rev 56:465–501

Taga G, Yamaguchi Y, Shimizu H (1991) Self-organized control of bipedal locomotion by neural oscillators in unpredictable environment. Biol Cybern 65:147–159

Ting LH, Macpherson JM (2005) A limited set of muscle synergies for force control during a postural task. J Neurophysiol 93:609–613

Todorov E, Jordan MI (2002) Optimal feedback control as a theory of motor coordination. Nat Neurosci 5:1226–1235

Whelan PJ, Hiebert GW, Pearson KG (1995) Stimulation of the group I extensor afferents prolongs the stance phase in walking cats. Exp Brain Res 103:20–30

Winter DA (2004) Biomechanics and motor control of human movement (3rd edn.). Wiley, New York

Yakovenko S, Gritsenko V, Prochazka A (2004) Contribution of stretch reflexes to locomotor control: a modeling study. Biol Cybern 90:146–155

Yamasaki T, Nomura T, Sato S (2003) Possible functional roles of phase resetting during walking. Biol Cybern 88:468–496

# Chapter 9
# Model-Based Interpretations of Experimental Data Related to the Control of Balance During Stance and Gait in Humans

**Robert J. Peterka**

**Abstract** An important goal in developing a model is to explain experimental data from a physiological system in a manner that provides insight into the function of that system. We begin by using data from experiments that characterized the dynamic properties of the human balance control system that regulates body orientation during stance. The dynamic properties of stance control are expressed as frequency response functions derived from body sway evoked by pseudorandom stimuli that tilted the surface upon which subjects stood or the visual surround that they viewed. A feedback control model is developed in a step-wise manner in order to illustrate how different subsystems of the model combine to explain the features of the experimental data and to reveal (1) the contributions of feedback control based on sensory measures of body motion from proprioceptive, visual, and vestibular systems, (2) the regulation of the responsiveness to perturbations using sensory reweighting, (3) the contribution of positive torque feedback, and (4) the influence of passive dynamics of muscle/tendon systems. The insights obtained from this stance control model are then applied to aid in the interpretation of new results from experiments that investigate the control of body orientation during a gait-like task of stepping-in-place.

**Keywords** Modeling · Balance · Stance · Gait · Stepping · Sensory integration · Sensory reweighting · Human

## 9.1 Introduction

Our objectives in this chapter are (1) to demonstrate the steps involved in the development of a mathematical model that has sufficient complexity to provide insight into how the nervous system controls balance during quiet stance and (2) to illustrate how this model can provide insight into the mechanisms contributing to balance during gait.

---

R. J. Peterka (✉)
Department of Biomedical Engineering, Oregon Health & Science University,
3303 SW Bond Avenue, CH13B, Portland, OR 97239, USA
e-mail: peterkar@ohsu.edu



B. I. Prilutsky, D. H. Edwards (eds.), *Neuromechanical Modeling of Posture and Locomotion,* Springer Series in Computational Neuroscience,
DOI 10.1007/978-1-4939-3267-2_9

For the control of balance during quiet stance, it is well accepted that humans use orientation and motion information derived from sensory systems to generate corrective actions that resist the destabilizing effects of gravity, external perturbations, and internal perturbations in order to maintain a desired body orientation. These active, sensory-driven contributions to balance control act together with passive contributions that arise primarily from the intrinsic mechanical properties of muscle/tendon systems. The primary sensory systems contributing to balance control are the somatosensory/proprioceptive system (signaling forces applied to and within the body in addition to the orientation and motion of body segments relative to one another), visual system (signaling head orientation and motion relative to the visual environment), and vestibular system (signaling head orientation and motion in space) (Nashner 1981; Horak and Macpherson 1996).

In principle, the central nervous system could combine information from visual and vestibular sensors with multi-segmental proprioceptive cues in order to derive an estimate of body center-of-mass (CoM) orientation (Mergner 2004). Because of the importance of the concept of the CoM in describing and predicting the motion of mechanical systems in classical physics, it is common in discussions of balance control to assume that the nervous system can derive an estimate the body's CoM orientation from available sensory information and then generate corrective responses based on the deviation of the CoM from a desired position. We will use this assumption in the balance control model developed below because our goal is to lead the reader, who might not be familiar with physiological system modeling, through a logical progression of steps that illustrate how a model of balance control can be developed and how that development can enhance our understanding of the balance control system.

We will also make the assumption that the body mechanics are those of a single-segment inverted pendulum with body sway occurring about an ankle joint such that the ankle-joint angle also defines the CoM sway angle with respect to the stance surface. Because we are focusing on the development of a relatively simple model, we will not be considering studies that have used modeling to understand balance control in a more complex system that controls a multi-segmental body (van der Kooij et al. 1999; Park et al. 2004; Alexandrov et al. 2005; Fujisawa et al. 2005; Kiemel et al. 2008; Hettich et al. 2014; Kim et al. 2012; Li et al. 2012; Pasma et al. 2012; Boonstra et al. 2013).

In quasi-static conditions of quiet or mildly perturbed stance, physics dictates that stability is assured if motor actions can maintain body orientation such that the body CoM remains within the base of support defined by the area under and between the feet. Therefore, the task of balance control during stance is mainly a task of controlling body orientation. However, in the dynamic conditions that occur during a typical walking gait, there are periods in the gait cycle when the CoM is not within the base of support (Winter 1995). In these dynamic conditions, stability requires that natural body dynamics and/or motor actions constrain the CoM motion to follow a nominal, repeating trajectory and to return to that trajectory following a perturbation. Theoretical and experimental studies of dynamic stability during gait have been concerned mainly with characterizing the repeatability of trajectories during the gait cycle (Dingwell and Cusumano 2000; Terry et al. 2012; Mummolo et al. 2013) and with investigating mechanisms that influence gait variability (Bauby and Kuo 2000; Dean et al. 2007; Ahn and Hogan 2013). Less well investigated are the mechanisms that regulate body orientation during gait.

In this chapter we will develop and apply a model that gives insights into how humans control body orientation during stance and gait. We will first describe the methods used to obtain an extensive experimental data set that has sufficient complexity to guide and constrain the development of a model for the control of body orientation during stance. We will develop this stance control model in stages so that the reader can clearly see that different components of the model account for different features of the experimental data. Then the results from stance control experiments will be compared to results from similar experiments performed during gait. To interpret the results from gait experiments, knowledge gleaned from the model-based interpretation of stance data will be used to determine the extent to which balance mechanisms contributing to body orientation control during stance also contribute to body orientation control during gait.

## 9.2 An Experimental Data Set Worthy of Modeling

In order to develop a comprehensive model of the human stance control system, an extensive set of experimental data are required. If a data set is too limited then multiple models may be equally effective in accounting for the experimental data with no way to distinguish among them. A comprehensive data set provides the constraints necessary to capture functionally important characteristics of the control system and distinguishing between different models.

In our studies of balance control we have extensively used a particular type of pseudorandom stimulus. Specifically, our pseudorandom stimulus is derived from a "maximal length ternary sequence" of numbers (Davies 1970). The ternary sequence of 0's, 1's, and -1's are mapped into a set of velocity steps with amplitudes of 0, +v, and −v with each value in the sequence having a step duration of $\Delta t$ seconds. This velocity waveform has the property of a white noise stimulus where the amplitude spectrum of the stimulus is approximately constant out to a bandwidth of about $1/3\Delta t$ Hz. The velocity waveform is integrated to give a stimulus position waveform that we have used as stimuli to control, for example, the angular position of the support surface that subjects stand on or the visual surround that subjects view (Peterka 2002; Goodworth and Peterka 2010b).

Figure 9.1a shows an example of one complete cycle of a pseudorandom stimulus waveform (2° peak-to-peak amplitude) based on a maximal length ternary sequence and the corresponding evoked CoM body sway (averaged across five cycles). Data are from a subject with normal sensory function standing with eyes closed on the rotating surface. The subject's ankle joints were aligned with the rotation axis of the surface such that surface rotations evoked body sway in the sagittal plane.

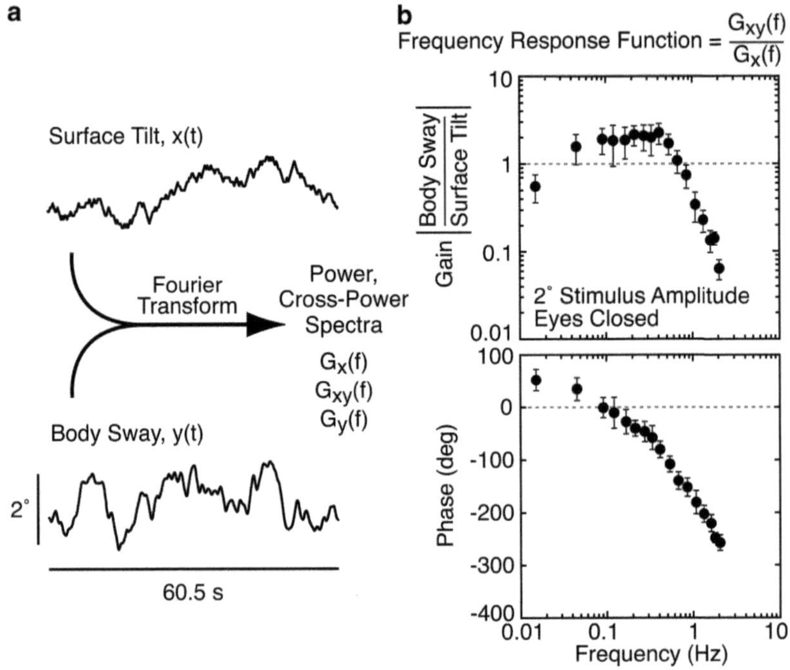


**Fig. 9.1** Method to determine dynamic characteristics of the human balance control system during stance using wide-bandwidth pseudorandom stimulation. **a** A subject's balance was perturbed using pseudorandom stimuli that evoked a sagittal plane body sway response consisting of an angular tilt of the subject's center-of-mass from a vertical orientation. One cycle of a pseudorandom stimulus is shown (*top* trace; peak-to-peak amplitude of 2°). The stimulus controlled the angular tilt of the surface upon which the subject stood with eyes closed. The average body sway response (*bottom* trace; averaged across five cycles) is shown. Discrete Fourier transforms were applied to the stimulus and response data, and the transformed data were used to calculate power spectra of the stimulus and response, and the cross-power spectrum between the stimulus and response. **b** The ratio of the cross-power spectrum to the stimulus power spectrum provides an estimate of the frequency response function that defines the dynamic characteristics of the balance control system. The frequency response function can be expressed as gain and phase functions that describe the relative amplitude and timing, respectively, of body sway evoked by the stimulus across a range of stimulus frequencies. Error bars show 95 % confidence intervals calculated following the procedures defined in Otnes and Enochson (1972).

We applied a standard frequency-domain analysis (defined in Otnes and Enochson (1972). Bendat and Piersol 2000; Pintelon and Schoukens 2012) using a discrete Fourier transform of the stimulus and CoM response waveforms to break down the time-domain waveforms into an equivalent set of sinusoidal components ranging from a frequency of 1/(cycle duration) = 0.0165 Hz in this example to an upper frequency of about 2 Hz. Results from the discrete Fourier transforms are used to calculate the power spectra of the stimulus and response, and the cross-power spectrum between stimulus and response for each stimulus cycle. The spectra are smoothed by averaging power spectra across the stimulus cycles and across selected ranges of frequency components in order to reduce variability.

The ratio of the cross power spectrum between stimulus and response to the stimulus power spectrum provides an estimate of the frequency response function (FRF). An FRF can be expressed as gain and phase functions, as shown in Fig. 9.1b, with the gain values indicating the ratio of CoM response amplitude to stimulus amplitude and the phase indicating the timing of the response relative to the stimulus. For a linear system, the FRF provides full knowledge of the system dynamics such that the response to any arbitrary stimulus can be predicted (Pintelon and Schoukens 2012).

Although body sway responses to a stimulus at any particular amplitude showed no evidence for strong nonlinearities, it was evident, from results of an earlier study using sinusoidal stimuli of varying amplitudes, that the response gain was not constant across all stimulus amplitudes (as expected for a linear system), but gain decreased with increasing amplitude (Peterka and Benolken 1995). Therefore, when we performed experiments using pseudorandom stimuli (Peterka 2002; Cenciarini and Peterka 2006; Goodworth and Peterka 2009, 2010a, b), a range of amplitudes was applied in each of the various test conditions. For example in our first study using pseudorandom stimuli (Peterka 2002) we applied stimuli with peak-to-peak amplitudes of 0.5°, 1°, 2°, 4°, and 8° in each of six test conditions. Examples of the family of FRFs for two of the test conditions, one for a surface stimulus with eyes closed and the other a visual stimulus during stance on a fixed and level surface, are shown in Fig. 9.2a and b, respectively.

The general shapes of the gain and phase data of all FRFs were quite similar across amplitude for both surface and visual stimuli. For each FRF, the gain was largest in the mid-frequency region of ~0.1–1 Hz and decreased with both decreasing and increasing frequency. The phase data showed phase leads at the lowest frequencies, increasing phase lags with increasing frequency, and a zero crossing at about 0.2 Hz. Although the general FRF shapes were similar for both surface and visual stimuli, there were some notable differences. For surface-tilt stimuli, the FRF gains from the various stimulus amplitudes tended to converge at about 1 Hz while this did not occur for FRF gains from the visual stimulus. The higher frequency phases from different stimulus amplitudes tended to diverge for surface-tilt stimuli with the phase from higher amplitude stimuli showing less phase lag. In contrast, the higher frequency phases from visual stimuli showed no change with stimulus amplitude.

The FRFs in Fig. 9.2a and b provide a representative data set that guides the model development described in the next section.

## 9.3 Development of a Stance Control Model

In the following subsections we will demonstrate the step-by-step development of a model that accounts for a wide variety of experimental FRF results. The stages of model development, from simple to more complex, correspond quite well with the thought processes that were applied in the original model development.

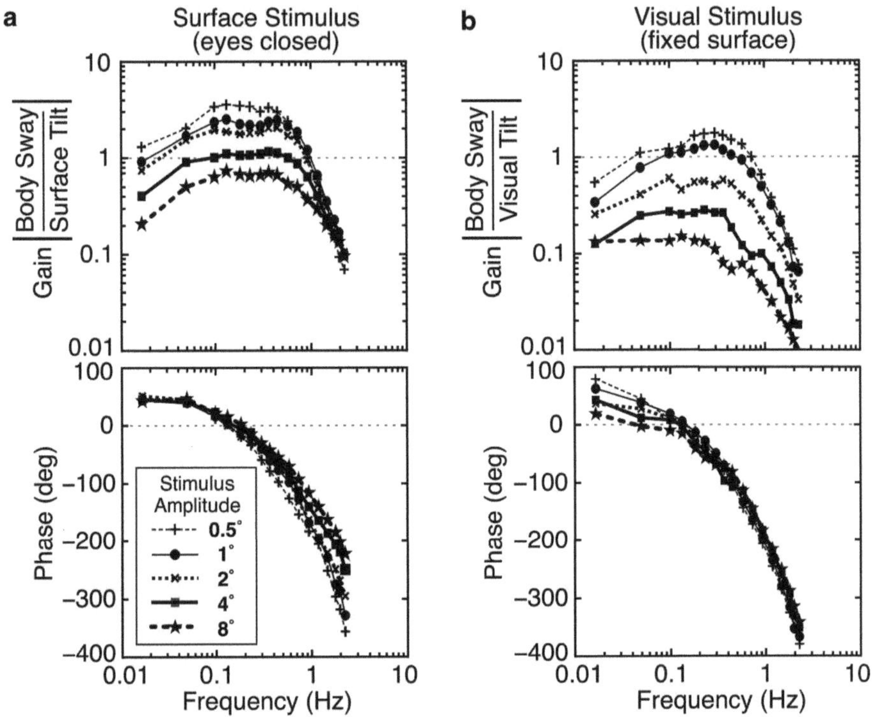


**Fig. 9.2** Sets of frequency response functions calculated using pseudorandom stimuli with five different amplitudes. Stimuli were either (**a**) support surface tilts during eyes closed stance or (**b**) visual surround tilts during stance on a fixed and level surface. (Data plots abstracted from Peterka (2002))

## *9.3.1 Basic Feedback Control Model of Stance*

It is widely recognized that balance control is organized as a feedback control system with the feedback provided by body orientation information obtained from sensory systems (Johansson et al. 1988; Horak and Macpherson 1996). Figure 9.3a shows a block diagram of a simple model that is capable of partially describing experimental FRFs and that includes the basic system components that we can reasonably assume to be present in the system. These include the body mechanics, a sensory system that detects body sway, a "neural controller" that converts the detected body sway into a corrective torque, $T_c$, applied about the ankle joints, and a time delay element representing all the delays in the system (i.e., sensory transduction, transmission of sensory information, processing of sensory and motor control information, transmission of motor control signals, and muscle activation delays).

The body mechanics are that of a single-segment inverted pendulum. The equation of motion for the inverted pendulum is:

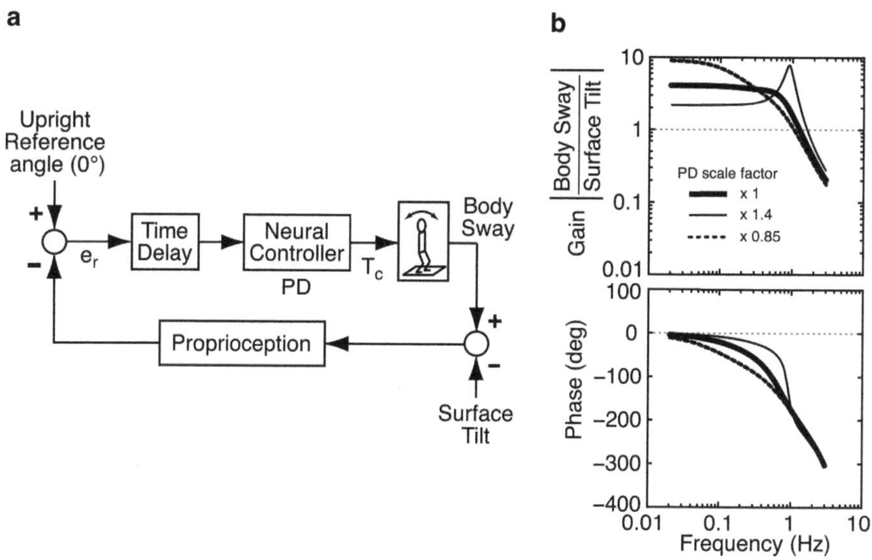


**Fig. 9.3** Simple feedback control model of the balance control system (**a**) and model predictions expressed in the form of frequency response functions (**b**). Model assumes a one-segment inverted pendulum body is controlled by sensory feedback from proprioception signaling body sway angle relative to the surface. The frequency response functions show the effect of changing the values of the PD parameters (that define proportional and derivative feedback gains) on the dynamic characteristics of the system. The "x 1" PD scale factor uses PD parameter values defined in Table 9.2 and produces gain and phase functions that most closely resemble experimental data. The body inverted pendulum parameters are given in Table 9.1

$$J\frac{d^2\theta_{bs}(t)}{dt^2} = mgh \cdot sin(\theta_{bs}(t)) + T_c(t) \tag{9.1}$$

where $\theta_{bs}(t)$ is the time course of body-in-space sway angle (i.e., body sway with respect to earth vertical), $J$ is the moment of inertia of the body about the ankle-joint axis, $m$ is body mass (not including mass of the feet), $h$ is height of the body CoM above the ankle joint, $g$ is the gravitational constant, and $T_c(t)$ is a torque applied at the ankle joint. This time-domain equation can be linearized for small angles of motion (i.e., $\theta_{bs}(t) \cong \sin(\theta_{bs}(t))$) and expressed in the Laplace domain:

$$\frac{\theta_{bs}(s)}{T_c(s)} = \frac{1}{Js^2 - mgh} \tag{9.2}$$

where $s$ is the Laplace variable. The Laplace representation is useful because (1) it allows for the algebraic manipulation of the differential equations that describe the input-output relations of each model component (i.e., each block in the block diagram) and (2) it allows for the calculation of FRFs by substitution of $s = j2\pi f$ with $j$ being the imaginary number $\sqrt{-1}$ and $f$ being frequency in Hz.

The sensory system provides information about body sway. In the Fig. 9.3a model we consider that only proprioception is available and that proprioception signals body orientation relative to the surface (i.e., it encodes the ankle-joint angle). We also assume that the proprioceptive sensory signal is a perfect encoder of the ankle joint angle such that there are no sensory dynamics involved. Specifically, the "Proprioception" block in Fig. 9.3a is set to a value of "1" to represent this perfect encoding with no dynamics. Effectively, we are assuming that the spinal and central nervous systems are capable of deriving an essentially perfect encoding of limb motion by processing the complex afferent signals available from the peripheral sensors such as stretch receptors. Justification for this assumption comes from recordings from the cerebellum and the dorsal spinocerebellar tract where the neural signals do not show responses typical of stretch receptors, but rather show activity representing kinematic variables of limb motion (Bosco et al. 2000; Casabona et al. 2004).

The sensory-derived body orientation is compared to an internal reference indicating the desire to remain in an upright orientation. Any deviation from the reference produces and error signal $e_r(t)=\theta_{ss}(t)-\theta_{bs}(t)$ where $\theta_{ss}(t)$ is the tilt angle of the support surface. The time delayed error signal, $e_r(t-t_d)$ where $t_d$ is length of the delay, is the input to the neural controller. The neural controller then generates a corrective torque in relation to the time-delayed error signal according to the time-domain equation:

$$T_c(t) = K_p \bullet e_r(t-t_d) + K_d \bullet \frac{de_r(t-t_d)}{dt} \qquad ((9.3)$$

The Laplace domain version is:

$$\frac{T_c(s)}{e_r(s)} = (K_p + K_d \bullet s) \bullet e^{-t_d s} \qquad (9.4)$$

where $K_p$ is the proportionality constant for the error angular position and $K_d$ is the proportionality constant for the error angular velocity. In engineering systems this type of controller is referred to as a PD (proportional-derivative) controller. With all the components of the model now defined we can write an equation in the Laplace domain and then solve for the ratio of the body sway response to the surface-tilt stimulus:

$$\theta_{bs}(s) = \frac{T_c(s)}{e_r(s)} \cdot \frac{\theta_{bs}(s)}{T_c(s)} \cdot e_r(s) = \frac{(K_p + K_d s) \cdot e^{-t_d s}}{Js^2 - mgh}(\theta_{ss}(s) - \theta_{bs}(s)) \qquad (9.5)$$

$$\frac{\theta_{bs}(s)}{\theta_{ss}(s)} = \frac{(K_p + K_d s) \bullet e^{-t_d s}}{Js^2 - mgh + (K_p + K_d s) \bullet e^{-t_d s}} \qquad (9.6)$$

**Table 9.1.** Model parameters for inverted pendulum body mechanics

| Parameter | Description | Value |
|---|---|---|
| $J$ | Moment of inertia about ankle joint | 81 kg $m^2$ |
| $m$ | Body mass (not including feet) | 83 kg |
| $h$ | Center-of-mass height above ankle joint | 0.9 m |
| $mgh$ | Gravity torque constant | 733 kg $m^2/s^2$ |

**Table 9.2** Model parameters for feedback control

| Parameter | Description | Value |
|---|---|---|
| $K_p$ | Neural controller proportional (P) gain constant | 968 (1060) Nm/rad |
| $K_d$ | Neural controller derivative (D) gain constant | 350 (286) Nms/rad |
| $K_t$ | Torque feedback gain constant | $8.7 \times 10^{-4}$ rad/Nm |
| $\tau_t$ | Torque feedback low-pass filter time constant | 15 s |
| $t_d$ | Time delay | 0.18 s |

$K_p$ and $K_d$ values in parentheses were used in the model shown in Fig. 9.6a that included stretch reflex and Hill muscle/tendon subsystem

After making the $s = j2\pi f$ substitution, FRF values expressed as real and imaginary numbers at each frequency, $f$, can be calculated and then represented as gain and phase values by:

$$\text{FRF gain} = \left| \frac{\theta_{bs}(j2\pi f)}{\theta_{ss}(j2\pi f)} \right| = \sqrt{Re(\bullet)^2 + Im(\bullet)^2} \tag{9.7}$$

$$\text{FRF phase} = \frac{180^\circ}{\pi} \bullet \text{atan}(Im(\bullet)/Re(\bullet)) \tag{9.8}$$

where $Re(\cdot)$ and $Im(\cdot)$ refer to the real and imaginary components, respectively, of the FRF values and phase is expressed in degrees.

Assuming fixed values for the $J$, $m$, and $h$ parameters for a give subject (see Table 9.1), the shapes of the gain and phase curves depend on the particular values of the parameters $K_p$, $K_d$, and $t_d$ (see Table 9.2). The rather large time delay value (0.18 s) accounts for a portion of the large and increasing phase lags seen at higher frequencies (consistent with experimental FRF phases). Three different gain and phase curves are shown in Fig. 9.3b to demonstrate the effect of varying $K_p$ and $K_d$ values. The curves labeled as having PD controller values scaled by a "x 1" scale factor result in FRF gain and phase curves that are closest to the shape of experimental FRFs shown in Fig. 9.2.

Consistent with most experimental FRF gains shown in Fig. 9.2, the FRF gains at low to mid-frequencies are greater than unity for all FRFs shown in Fig. 9.3b. These low-frequency gain values calculated for the Fig. 9.3a model are determined by the value of $K_p$ in relation to the value of the product $mgh$. Specifically, the low frequency gain is given by $K_p/(K_p - mgh)$ and this value is obviously always greater than one. The value *mgh* is a scale factor that specifies the amount of destabilizing torque due to gravity created per unit of body sway away from the upright orientation. The value of $K_p$ must be greater than *mgh* in order that enough corrective torque is generated by the neural controller to resist the destabilizing torque due to gravity. For the FRF shown in Fig. 9.3b that most closely resembles the shape of the experimental FRFs (the times one PD scale factor), the low frequency gain is approximately equal to 4 and thus $K_p$ has a value that is about 1/3rd greater than *mgh*.

This relatively low value of $K_p$ makes this simple control system quite sensitive to even small perturbations caused by a non-level support surface. With this simple feedback model, the only way to reduce the sensitivity to surface tilt would be to increase the value of $K_p$. However, increasing $K_p$ produces an FRF with a resonant peak at ~1 Hz. (Fig. 9.3b shows the results of increasing both $K_p$ and $K_d$ but similar results are seen when only $K_p$ is increased.) This necessitates a careful regulation of the neural controller PD values in order to avoid resonant behavior or even unstable operation. Furthermore, larger time delay values make the system shown in Fig. 9.3a more difficult to stabilize such that no PD values can be found that provide stability if the time delay is greater than about 0.35 s (van der Kooij and Peterka 2011).

In summary, the very simple model in Fig. 9.3a is successful in predicting the basic shape seen in experimental FRFs, but this model is clearly incomplete. The model cannot account for the stimulus amplitude-dependent variation in overall FRF gain magnitude while shapes of the FRF gain and the phase curves remain relatively unaffected. In this simple model, the gain can only be influenced by changes in controller parameters, but these changes also change the shapes of the gain and phase curves. The next subsection modifies the basic Fig. 9.3a model in order to account for stimulus-dependent gain changes.

### 9.3.2 *Accounting for Stimulus Amplitude-Dependent FRF Gain*

The addition to the Fig. 9.3a model that accounts for amplitude-dependent changes in overall FRF gain values is shown in Fig. 9.4a. This model is expanded to include feedback from visual and vestibular systems in addition to proprioception. The model includes a "sensory integration" subsystem that represents the process by which the orientation information from the three sensory systems is combined, by a weighted summation, to form an internal estimate of body orientation. This internal estimate is compared to an internal reference of the desired upright body orientation (as in the Fig. 9.3a model but not shown in Fig. 9.4a) with the difference between the estimate and the reference forming an orientation error signal, $e_r(t)$. This error signal is the input to the neural controller that produces the corrective torque.

The three sensory-system weights represent the relative contributions of the sensory systems to the overall internal estimate of body orientation. Mathematically this means that, for a given test condition, the sum of the weights of all sensory systems contributing to balance control is equal to unity. As in the Fig. 9.3a model,

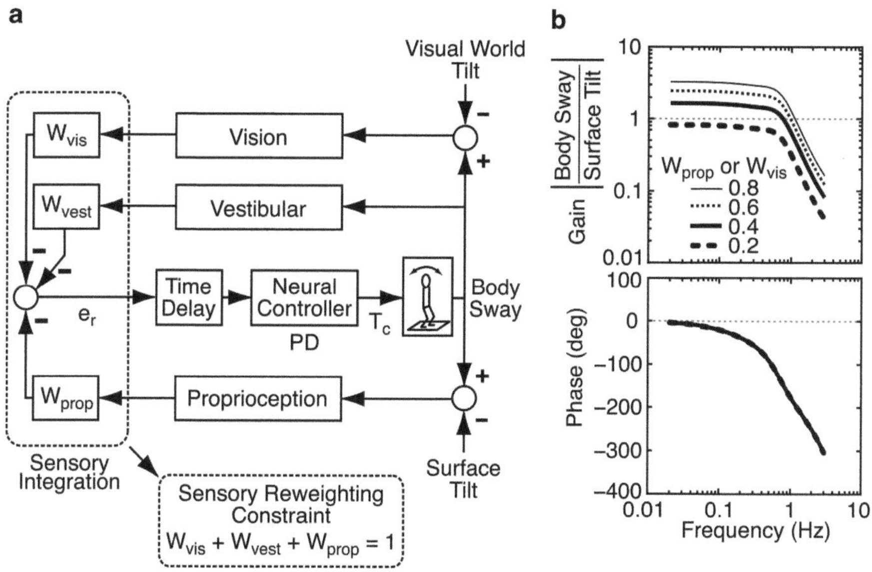


**Fig. 9.4** A feedback control model that includes sensory integration (**a**) and model predictions expressed in the form of frequency response functions (**b**). The model predictions demonstrate how changes in the proprioceptive weight (for surface-tilt stimulation) or in visual weight (for visual-tilt stimuli) affect the frequency response functions. The body mechanics parameters and feedback control parameters are defined in Tables 9.1 and 9.2, respectively

we continue to assume that ankle proprioception provides a perfect sensory signal encoding the ankle joint angle and therefore the block labeled "Proprioception" has a value of "1". We make the same assumption for the "Vision" and "Vestibular" blocks.

For the vestibular system the justification for the assumption of perfect encoding of head tilt with respect to earth vertical comes from experimental studies that show evidence that central processing of sensory signals from the semicircular canals (encoding angular head velocity) and otolith organs (encoding an ambiguous combination of head linear acceleration and acceleration due to gravity) can separate the transient component of linear acceleration from the component of linear acceleration due to gravity (Angelaki et al. 1999; Merfeld et al. 1999; Zupan et al. 2000; Angelaki et al. 2004; Peterka et al. 2004). For our purposes the conclusion is that nervous system has available to it an accurate, wide-bandwidth estimate of the orientation of the gravity vector with respect to the head and therefore a vestibular-derived estimate of body orientation with respect to earth vertical.

For the visual system, justification for the presence of wide bandwidth encoding of visual motion information comes from experiments showing that (1) eye movements in humans evoked by optokinetic stimuli have approximately constant FRF gains over a wide bandwidth (Peterka et al. 1990) and (2) neural recordings from Purkinje cells in the cerebellar flocculus of Java monkeys have high sensitivity to optokinetic stimuli out to 3 Hz (Markert et al. 1988).

The Laplace equation for the Fig. 9.4a model when a surface-tilt stimulus is applied is:

$$\frac{\theta_{bs}(s)}{\theta_{ss}(s)} = \frac{W_{prop} \bullet (K_p + K_d s) \bullet e^{-t_d s}}{Js^2 - mgh + (K_p + K_d s) \bullet e^{-t_d s}} \tag{9.9}$$

When a visual-tilt stimulus is applied the equation is:

$$\frac{\theta_{bs}(s)}{\theta_{vs}(s)} = \frac{W_{vis} \bullet (K_p + K_d s) \bullet e^{-t_d s}}{Js^2 - mgh + (K_p + K_d s) \bullet e^{-t_d s}} \tag{9.10}$$

Both of these equations are derived with the assumption that the sum of the sensory weights is always unity. These two equations differ from Eq. (9.6) only by having an additional multiplying factor of either $W_{prop}$ or $W_{vis}$ in the numerator. Thus if the $K_p$, $K_d$, and $t_d$ parameters of Eqs. (9.9) and (9.10) are set to the same values as used to calculate the FRFs shown in Fig. 9.3b, the FRF gain and phase curves will have the same shape as those shown in Fig. 9.3b. The only difference will be that the overall magnitude of the gain curves will depend on the value of $W_{prop}$ for test conditions where a surface-tilt stimulus was applied and on the value of $W_{vis}$ for conditions where a visual-tilt stimulus was applied.

The FRFs shown in Fig. 9.4b were calculated for 4 different $W_{prop}$ or $W_{vis}$ values ranging from 0.2 to 0.8 and with the remaining parameters set to values in Tables 9.1 and 9.2. The phase curves of the FRFs are not affected by changing $W_{prop}$ or $W_{vis}$ values, so the phase curves calculated for different sensory weights overlay one another in Fig. 9.4b. The gain curves calculated using Eqs. (9.9) and (9.10) are identical when $W_{prop}$ and $W_{vis}$ values are equal to one another. Changing the value of $W_{prop}$ or $W_{vis}$ simply shifts a gain curve up or down (when gain is plotted on a log-log scale).

The FRFs shown in Fig. 9.4b provide a better match to the experimental FRFs in Fig. 9.2 in that we now have an explanation that accounts for the decrease in FRF gains with increasing stimulus amplitude while the FRF phases are not affected by stimulus amplitude. Specifically, when applying a surface-tilt stimulus, the fact that the FRF gains are largest at the lowest stimulus amplitude and decrease with increasing stimulus amplitude implies that the proprioceptive contribution to balance control, quantified by $W_{prop}$, is greatest at the lowest stimulus amplitude and decreases with increasing amplitude. For eyes closed tests using surface-tilt stimuli, a decrease in $W_{prop}$ with increasing stimulus amplitude also implies that the vestibular contribution to balance control (represented by $W_{vest}$ in the model) increased with increasing stimulus amplitude. This increase in $W_{vest}$ is implied because of the constraint that sensory weights must sum to unity.

The interpretation of experimental results based on the Fig. 9.4a model allows us to recognize the contribution of a "sensory reweighting phenomenon" to the regulation of balance. Specifically, an amplitude-dependent sensory reweighting occurs such that the contribution to balance control from the sensory system encoding the primary perturbing stimulus (e.g., proprioception for surface movements) is decreased while the contribution from secondary sensory systems (e.g., the vestibular system in eyes closed tests) is increased. This model prediction of reciprocal reweighting has been tested and confirmed experimentally (Cenciarini and Peterka 2006).

However, the model-predicted FRFs in Fig. 9.4b are not yet entirely consistent with the experimental FRFs in Fig. 9.2. In particular, each of the experimental FRFs for both surface and visual stimuli show a decrease in gain and phase advance with decreasing frequency for frequencies below about 0.1 Hz. The next subsection will modify the Fig. 9.4a model to account for this low frequency behavior.

### *9.3.3 Accounting for Low Frequency FRF Gain Reduction and Phase Advance*

If balance control were governed entirely by feedback control with the properties of the Fig. 9.4a model, the regulation of body orientation would be very sensitive to the static conditions in the environment. For example, the Fig. 9.4a model predicts that if a subject was relying 80 % on proprioception for balance control and was standing on a surface that was tilted by only 1°, the subject's body would remain tilted from upright by almost 4°. That is, the Fig. 9.4a model predicts that equilibrium between the torque due to gravity and the corrective torque generated by the control system is achieved at a body angle of $W_{prop} \cdot K_p/(K_p - mgh)$ for a 1° surface-tilt angle. This equation shows that the only way to reduce the sensitivity to surface tilt would be to reduce $W_{prop}$ (and therefore increase $W_{vest}$ in this eyes-closed condition) or to greatly increase $K_p$. As shown in the Fig. 9.3b FRFs, it is not possible to greatly increase $K_p$. Furthermore, results based on optimal control concepts indicate that it is not desirable to rely heavily on vestibular information due to the high noise levels in the vestibular sensory system (van der Kooij and Peterka 2011). Therefore some other mechanism must be contributing to balance control to regulate low frequency or static body orientation.

An earlier sensory integration model attempted to account for the low frequency gain declines and phase advances by having a neural controller that includes integral control action in addition to PD control (i.e., PID control) (Peterka 2002). However, the PID controller did not fully account for the low frequency gain and phase data and motivated us to consider that the balance control system might exploit its ability to sense a sustained corrective torque in order to move the body toward a more upright orientation. The model originally proposed in Peterka (2002) was modified to include an additional feedback pathway based on sensory systems that encode the force exerted by muscle stretch and activation (e.g., by Golgi tendon organs) or by other somatosensory contributions such as sensing center-of-pressure (CoP) shifts on the feet (Peterka 2003; Cenciarini and Peterka 2006).

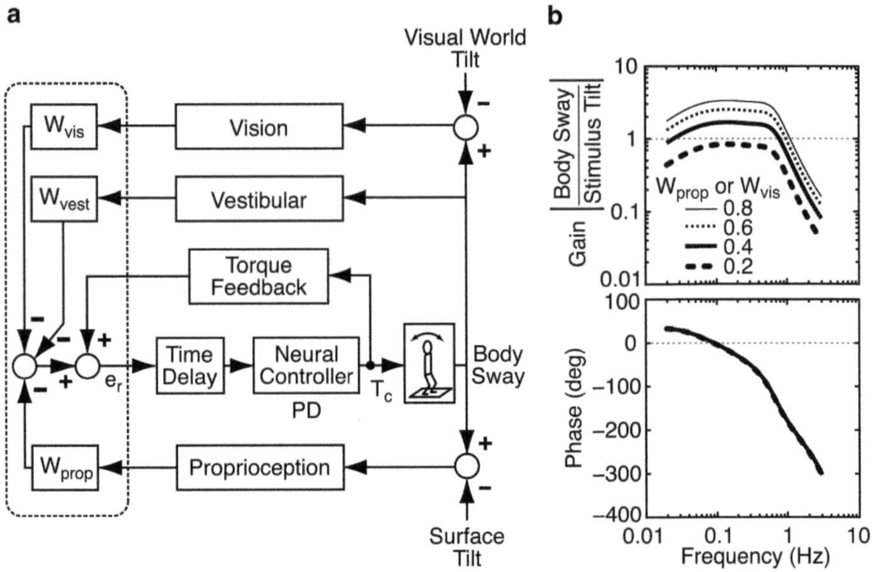


**Fig. 9.5** A feedback control model that includes sensory integration and torque feedback (**a**) and model predictions expressed in the form of frequency response functions (**b**). The model predictions demonstrate how changes in the proprioceptive weight (for surface-tilt stimulation) or in visual weight (for visual-tilt stimuli) affect the frequency response functions. The torque feedback loop provides positive feedback such that the magnitude of corrective torque, $T_c$, is increased if $T_c$ is sustained for longer periods of time. The increased $T_c$ will move the body toward an upright position. The effect of the positive torque feedback is to reduce the low frequency gain and advance the low frequency phase of the frequency response functions. The body mechanics parameters and feedback control parameters are defined in Tables 9.1 and 9.2, respectively

A model that incorporates torque feedback is shown in Fig. 9.5a along with FRFs that demonstrate how the dynamic characteristics of the system are affected by inclusion of a torque feedback pathway. Note that the torque feedback pathway provides positive feedback. Positive force feedback has previously been understood to contribute to motor control in specific circumstances such as load compensation (Prochazka et al. 1997; Duysens et al. 2000). Functionally, the Fig. 9.5a model predicts that when a subject is leaning forward and generating a sustained corrective torque, the positive torque feedback causes the error signal to increase in value. The increased error signal generates additional corrective torque that overcomes the torque due to gravity such that the body is moved toward an upright position. As the body-sway angle decreases, the magnitude of corrective torque decreases, and the contribution of the positive torque feedback pathway diminishes.

To account for the experimental FRFs we found it necessary to include an integrator, or to be more physiologically realistic, a leaky integrator (i.e., a low-pass filter with a long time constant) in the torque feedback pathway. When torque feedback is included in the model, the equation describing the FRF becomes much more complicated and it is more informative to present it as follows:

$$\frac{\theta_{bs}(s)}{\theta_{ss}(s)} = \frac{W_{prop} \bullet B \bullet NC \bullet TD}{1 - TF \bullet NC \bullet TD + B \bullet NC \bullet TD} \tag{9.11}$$

where *NC*, *B*, *TD*, and *TF* are the Laplace equations for the neural controller, body mechanics, time delay, and torque feedback, respectively, given by $NC = K_p + K_d \bullet s, B = 1/(J \bullet s^2 - mgh)$, $TD = e^{-t_d s}$, and $TF = K_t / (\tau_t \bullet s + 1)$ The parameters $K_t$ and $\tau_t$ are the gain and time constants, respectively, of the low-pass filter in the torque feedback pathway. As with the Fig. 9.4a model the assumption is that the sum of sensory weights is unity. The functional form of the FRF equation for a visual stimulus is the same except that $W_{prop}$ in Eq. (9.11) is replaced with $W_{vis}$. As with the Fig. 9.4a model, variation in $W_{prop}$ or $W_{vis}$ produces FRF gain curves with different amplitudes but with identical shapes (when plotted on log-log axes) and the FRF phase curves are the same for all values of $W_{prop}$ or $W_{vis}$. The model-predicted FRFs now show a gain decrease and phase advance at lower frequencies, consistent with the experimental FRFs shown in Fig. 9.2.

The FRFs predicted by the Fig. 9.5a model now closely resemble FRFs across the entire bandwidth of test frequencies from experiments using visual stimuli (Fig. 9.2b) but the resemblance is not quite as good for surface stimuli (Fig. 9.2a). Specifically, experimental FRFs obtained using different amplitudes of surface stimuli show a convergence of the gain curves in the 1–2 Hz region and a divergence of the phase curves with increasing frequency. The next subsection considers an extension of the Fig. 9.5a model that includes additional mechanisms.

### 9.3.4 Accounting for FRF Differences for Surface and Visual Stimuli

The differences between experimental FRFs from surface and visual stimuli occur mainly at higher frequencies. If a single model can account for both sets of FRFs then there must be some asymmetry in the balance control system that causes responses evoked by a surface-tilt stimuli to differ from those evoked by visual stimuli. There are two obvious oversights in the models considered so far. One is the absence of a contribution from passive muscle/tendon mechanics, and the second is the absence of a stretch reflex contribution to muscle activation.

The model shown in Fig. 9.6a adds these two components. The muscle/tendon properties were modeled using a linearized Hill-type model (McMahon 1984) that included an elastic element in series with a parallel combination of a contractile component (generating internal muscle force), and elastic and damping elements. Inclusion of this muscle/tendon system alters the balance control model in two ways. First, the passive mechanical properties of the muscle/tendon system generate an ankle torque, $T_{pas}$, as a function of ankle joint motion. This torque is generated without time delay. $T_{pas}$ sums with the torque generated by active muscle contractions, $T_{act}$, to produce that total corrective ankle torque, $T_c$. Second, the activation of

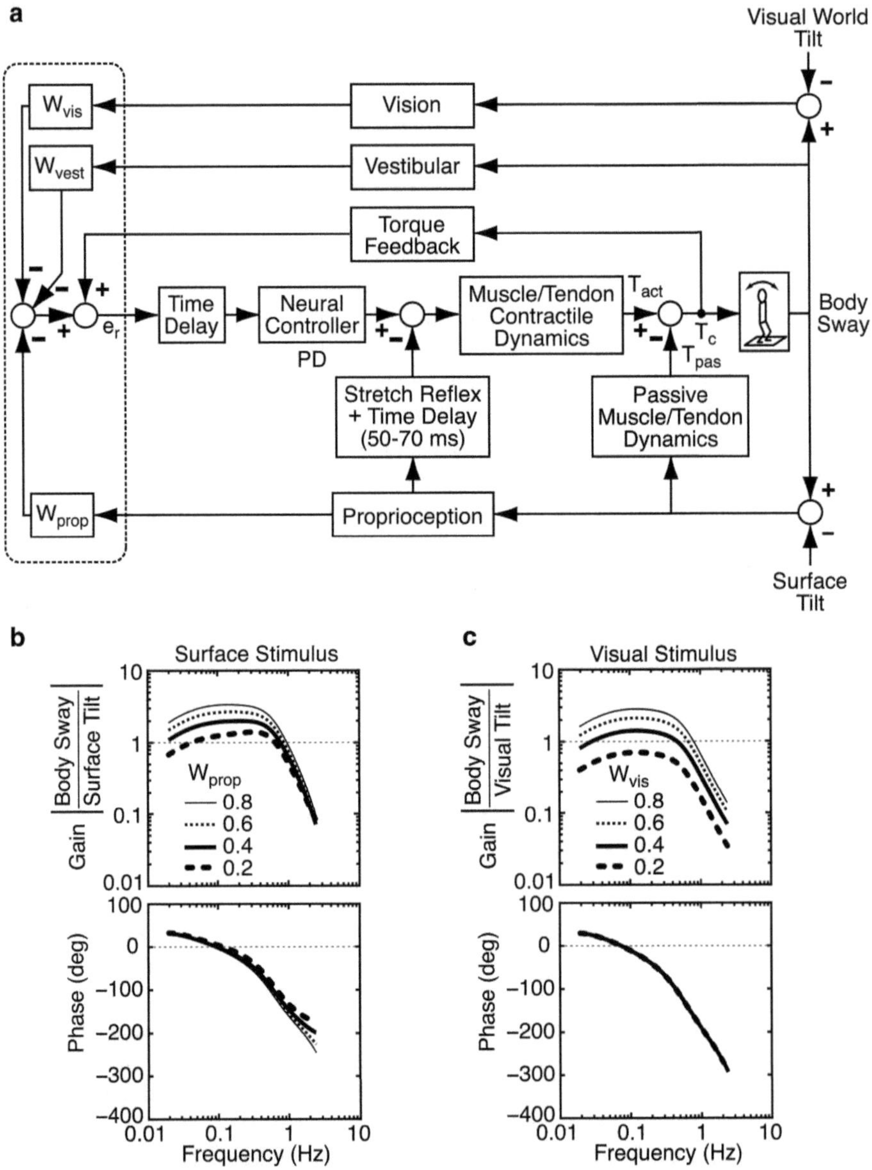


**Fig. 9.6.** A feedback control model that includes sensory integration, torque feedback, stretch reflex, muscle/tendon contractile dynamics, and passive muscle/tendon dynamics (**a**) and model predictions expressed in the form of frequency response functions (**b**, **c**). The stretch reflex is assumed to generate a muscle activation signal proportional to the velocity of ankle joint motion. The muscle/tendon contractile and passive dynamics are based on a linearized Hill-type muscle/tendon system. With the addition of stretch reflex and muscle/tendon components, the model-predicted frequency response functions for surface-tilt stimuli (**b**) and visual-tilt stimuli (**c**) now resemble experimental frequency response functions in Fig. 9.2. The body mechanics parameters, feedback control parameters, and stretch reflex and muscles/tendon parameters are defined in Tables 9.1, 9.2, and 9.3, respectively

the muscle contractile element generates an internal muscle force that acts through, and is effectively low-pass filtered by the muscle mechanics to produce $T_{act}$. The output of the neural controller is now considered to be a muscle activation signal, rather than an applied ankle torque.

The stretch reflex is represented as a subsystem whose output is a muscle activation signal that is a function of the angular velocity of the ankle joint motion. The stretch reflex output is summed with the activation signal from the neural controller. The stretch reflex includes a time delay that is assumed to be shorter than the delay associated with the sensory integration mechanism.

With the addition of muscle/tendon and stretch reflex components, the Laplace equations describing body sway responses to surface and visual stimuli now differ from one another. These equations are:

$$\frac{\theta_{bs}(s)}{\theta_{ss}(s)} = \frac{B \bullet (P+SR) + W_{prop} \bullet B \bullet NC \bullet M \bullet TD}{1 - TF \bullet NC \bullet M \bullet TD + B \bullet (P+SR) + B \bullet NC \bullet M \bullet TD} \tag{9.12}$$

$$\frac{\theta_{bs}(s)}{\theta_{vs}(s)} = \frac{W_{vis} \bullet B \bullet NC \bullet M \bullet TD}{1 - TF \bullet NC \bullet M \bullet TD + B \bullet (P+SR) + B \bullet NC \bullet M \bullet TD} \tag{9.13}$$

where $M$, $P$, and $SR$ are the Laplace equations for linearized Hill-type muscle/tendon contractile dynamics, passive muscle/tendon dynamics, and stretch reflex dynamics, respectively. The Laplace equations for these components are $M = K_{se} / (K_{ce} + K_{se} + B_{ce} \cdot s)$, $P = (K_{se} \cdot K_{ce} + K_{se} \cdot B_{ce} \cdot s) / (K_{ce} + K_{se} + B_{ce} \cdot s)$, and $SR = B_{sr} \bullet s \bullet e^{-t_{sr}s}$. The additional model parameters (values given in Table 9.3 include the linearized Hill-type muscle/tendon properties of series elastic stiffness, $K_{se}$, contractile element stiffness, $K_{ce}$, and contractile element damping, $B_{ce}$. The functional stretch reflex is assumed to contribute a muscle activation signal proportional to ankle velocity (proportionality constant $B_{sr}$) and with time delay $t_{sr}$. Note that the passive muscle/tendon dynamics have properties of a "lead/lag" system that generates more passive torque for higher compared to lower frequencies of ankle joint motion. The muscle/tendon contractile dynamics have the functional form of a low-pass filter. The static gain of this filter is $K_{se}/(K_{se}+K_{ce})$, which is less than unity. Therefore adjustments in system parameters are necessary to make up for this gain reduction such that the overall dynamic characteristics of the balance control system remain similar to the ones shown for the Fig. 9.5a model. The adjusted parameter values are shown in parentheses in Table 9.2.

**Table 9.3** Model parameters for stretch reflex and muscle/tendon subsystems

| Parameter | Description | Value |
|---|---|---|
| $B_{sr}$ | Stretch reflex velocity gain constant | 115 Nms/rad |
| $t_{sr}$ | Stretch reflex time delay | 0.07 s |
| $K_{se}$ | Hill muscle/tendon subsystem series element stiffness | 660 Nm/rad |
| $K_{ce}$ | Hill muscle/tendon subsystem contractile element stiffness | 147 Nm/rad |
| $B_{ce}$ | Hill muscle/tendon subsystem contractile element damping | 22 Nms/rad |

With a proper selection of parameters, Eqs. (9.12) and (9.13) now predict a set of FRFs that differ for surface and visual stimuli (Fig. 9.6b and c) and that display characteristics that approximately match those of experimental FRFs. The set of visual FRFs shown in Fig. 9.6c are very similar to the ones predicted by the Fig. 9.5a model that did not include the stretch reflex or muscle/tendon mechanics. This similarity occurs because a visual stimulus only indirectly evokes body sway causing ankle-joint motion. That is, the dynamics of the balance control system effectively filters out the higher frequency components of the visual stimulus leaving only lower frequency component of the ankle-joint motion. This limits the higher frequency, velocity-related contributions to corrective torque arising from the stretch reflex and passive muscle/tendon mechanics. In contrast, the same stimulus applied to the surface directly activates the stretch reflex and passive muscle/tendon pathways such that the higher frequency components of the stimulus contribute additional corrective torque at higher frequencies (as compared to the visual stimulus), thus affecting the higher frequency dynamics of the balance control system (Fig. 9.6b). Furthermore, as the surface-tilt amplitude increases, the proprioceptive weight, $W_{prop}$, decreases. With decreasing $W_{prop}$ the influence of the stretch reflex contribution and passive muscle/tendon mechanics on overall response dynamics increases in comparison to the influence of the proprioceptive contribution from the sensory integration subsystem. Thus, the FRF gain curves from surface stimuli change in shape as well as magnitude, and the FRF phase curves change shape as $W_{prop}$ changes.

### 9.3.5 Limitations and Extensions of the Stance Control Model

The predictions from the Fig. 9.6a model suggest that we can now account for the major features of experimental FRFs. It seems like it should be possible to obtain reliable measures of the various model parameters by applying optimal estimation methods to fit the model FRF equation to the experimental FRF data (see details of fitting methods in (Peterka 2002; van der Kooij and Peterka 2011)). Assuming that we use estimates of parameters associated with body mechanics ($J$, $m$, $h$) obtained from anthropometric measures, there are 11 additional free parameters in Eqs. (9.12) and (9.13).

Unfortunately, limitations in our ability to reliably identify the free parameters are quickly revealed from the results of optimal estimation procedures to determine model parameters. A major difficulty is that there is considerable redundancy in that, for some combinations of parameters, more than one parameter can account for particular features of the FRFs. To overcome this problem of parameter redundancy, our approach in previous studies (Peterka 2002, 2003; Cenciarini and Peterka 2006; van der Kooij and Peterka 2011) has been to simplify the model to reduce redundancies in parameters. The simplified model in these previous studies did not include the stretch reflex subsystem and muscle/tendon contractile dynamics subsystem, and the passive muscle/tendon dynamics were simplified by representing this subsystem as a summation of stiffness and damping elements (i.e., $P = K_{pas} + B_{pas} \cdot s$) rather than the more complex definition of $P$ used in Eqs. (9.12) and (9.13). With these simplifications, the number of free parameters is reduced to eight and the optimal fits to the experimental FRFs typically yield reliable parameters in the sense that the variance of parameters across subjects is relatively low. However, the tradeoff for this reduced variability is the recognition that the model is deviating from reality to some extent.

Another approach to enhancing our ability to identify parameters of more realistic models is to gather more complex data sets. More complex data sets could be obtained from experiments that include stimuli with wider bandwidths (Goodworth and Peterka 2009, 2010b, 2012), stimuli that include multiple types of perturbations that are presented simultaneously (e.g., combinations of visual-tilt stimuli, surface-tilt stimuli, surface translations, galvanic vestibular stimulation, and/or applications of external force; see (Cenciarini and Peterka 2006; Pasma et al. 2012; Boonstra et al. 2013)), and recordings of additional "in-the-feedback-loop" signals such as muscle activation signals recorded using electromyography (Kiemel et al. 2008) or muscle motion recorded using ultrasound techniques (Loram et al. 2005). The idea is that the added richness of the experimental data will provide sufficient additional information to allow for the reliable identification of more realistic models that include the added parameters that are necessary to represent the various subsystems contributing to balance control.

Since a particular balance control model represents a quantitative hypothesis about the organization and function of the system, the model can be used to make predictions that motivate new experiments to test these predictions and thereby test the hypothesis represented by the current model. Failure of the model to account for some aspects of new experimental data should then motivate refinements of the existing model or the consideration of alternative model structures (for example see (Mergner et al. 2002; Mergner 2010) for the description of a model structure that is quite different from the one discussed here). The ongoing cycle of model development, followed by experimental challenges, followed by model refinement simply represents the productive application of the scientific method. The general trend will be for the models to become more complex as they become better able to explain the complex nature of balance control.

However, it is also worth noting that simpler models retain some value. As a specific example, the Fig. 9.5a model does account for the main dynamic characteristics of the stance control system. Parameters determined from a fit of this model to FRFs of individual subjects provide quantitative measures with physiological relevance, such as sensory weights. Tests that quantify the function of the balance control system using a model-based interpretation of stimulus-response behavior could be used clinically to diagnose neurological disorders, track changes over time, or monitor the effectiveness of therapy.

## 9.4 Investigation of Balance Control During Gait

Of interest to us is whether or not the sensory integration principles identified for stance control also apply to the control of body orientation during gait. If so, then we would expect the mean body orientation during gait would be significantly influenced by environmental conditions such as walking on a sloped surface.

As with our studies of stance control, we want to have an experimental data set that is rich enough to promote the development of models that help us to understand the mechanisms contributing to the control of body orientation and dynamic stability during gait. Our prior use of pseudorandom stimuli and FRF analysis methods to study stance control motivated us to apply similar methods to study gait. The experimental results reported here focus on the control of body orientation in the frontal plane.

In order to apply pseudorandom stimuli during gait, we adopted a stepping-in-place (SiP) protocol as a surrogate for walking gait. The roll motions of the body CoM during SiP (Fig. 9.7a) closely resemble the frontal plane motions of the CoM during forward walking (Brenière 1996; Hof et al. 2005). Specifically, the CoM oscillates right and left with a sinusoidal motion, and the peak lateral displacement of the CoM is typically located medial of the stance foot. The CoP oscillates right and left with plateaus occurring when only 1 ft is on the surface. The CoP shows a ramp trajectory during the double leg support phase as weight is shifted from 1 ft to the other. From analysis of the CoP trajectory, the timing of the gait-cycle events can be measured.

All SiP tests were performed with eyes closed. To maintain an approximately constant location of the subject during SiP on the surface of our balance test device, soft foam guides were taped to the surface in a T-shaped configuration. Subjects were instructed to make small corrections in stepping location when they made contact with the T.

Balance in the frontal plane was perturbed using pseudorandom surface-tilt stimuli (six cycles per test trial, peak-to-peak amplitudes of 0° (control trial), 1°, 2°, and 4°) while subjects performed SiP. Metronome clicks were used to pace the stepping such that subjects performed 38 step cycles in each 48.4-s cycle of the pseudorandom stimulus (i.e., stepping frequency of ~ 0.8 Hz). Measurements included subject CoM sway angle in the frontal plane, lateral displacement of the CoP, and lateral displacements of right and left heels. The same stimuli were used to evoke frontal-plane sway during stance. Stance trials were performed eyes closed with the feet separated by 8 cm (intermalleolar distance). CoM sway in response the surface-tilt stimulus was analyzed to calculate FRFs from SiP trials and stance trials. Step width and step timing were also measured in order to investigate mechanisms contributing to the control of dynamic balance (Kuo 1999; Hausdorff 2005; Maufroy et al. 2010), but only results related to the control of body orientation are discussed below.

### 9.4.1 Control of Body Orientation During Gait

Do the principles of sensory integration and sensory reweighting identified for stance control also apply to the control of body orientation during gait? The results from sway measured during eyes closed SiP suggest there are similarities. Figure 9.7b shows one subject's frontal-plane CoM sway during a SiP test while balance was perturbed by a continuous pseudorandom tilt of the surface upon which the subject was stepping. Results are shown for three different stimulus amplitudes

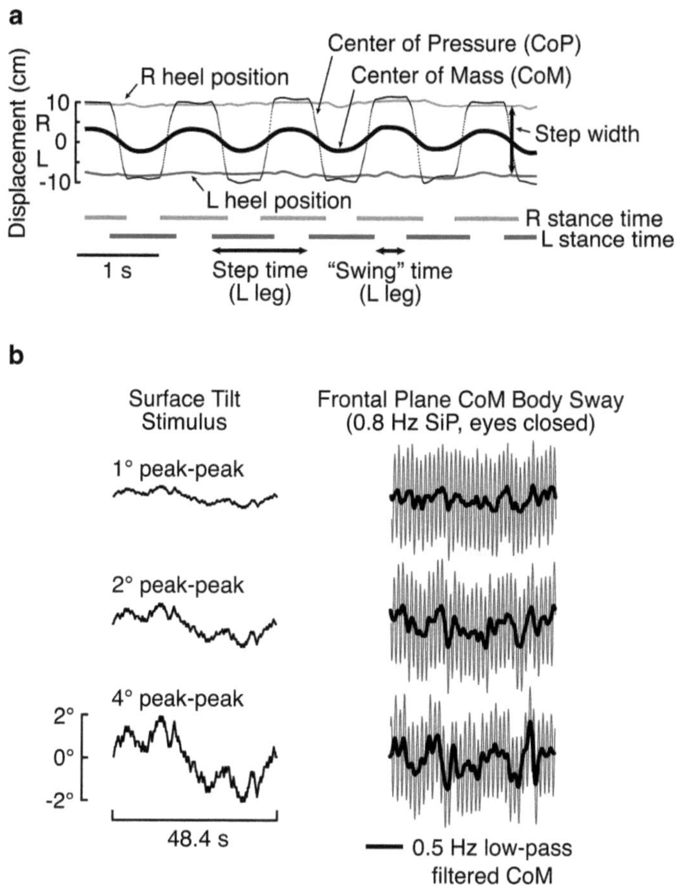


**Fig. 9.7** Results from stepping-in-place (*SiP*) experiments. **a** Example data recorded during SiP showing frontal-plane displacements of center-of-pressure (*CoP*), center-of-mass (*CoM*), and *right* and *left* heel position from which step-by-step measures of step width and step timing (stance times and swing times of *right* and *left* legs) were taken. **b** Example CoM sway recorded from an individual subject during eyes-closed SiP on a surface that was rotated according to a pseudorandom waveform at three different amplitudes. CoM sway data, averaged across five cycles of the pseudorandom stimuli, shows that the oscillating pattern of CoM in the frontal plane (*gray* traces) is disturbed by the surface-tilt stimulus. Application of a 0.5 Hz low-pass filter to the CoM eliminates the oscillatory component at the 0.8 Hz SiP frequency and reveals the deviations of the CoM orientation across the stimulus cycle (thick *black* traces)

and the CoM sway data were averaged across the last five cycles of the pseudorandom stimulus. The regularity of oscillations occurring at the 0.8 Hz stepping frequency become more disrupted at larger stimulus amplitudes due to changes in body orientation evoked by the stimulus. The time course of the change in body orientation is made more evident by low-pass filtering (0.5 Hz cutoff frequency) the CoM sway data to remove the 0.8-Hz oscillation at the stepping frequency. The low-pass filtered CoM traces (thick traces) tend to increase with increasing stimulus amplitude and show some resemblance to the time course or the stimulus waveform.

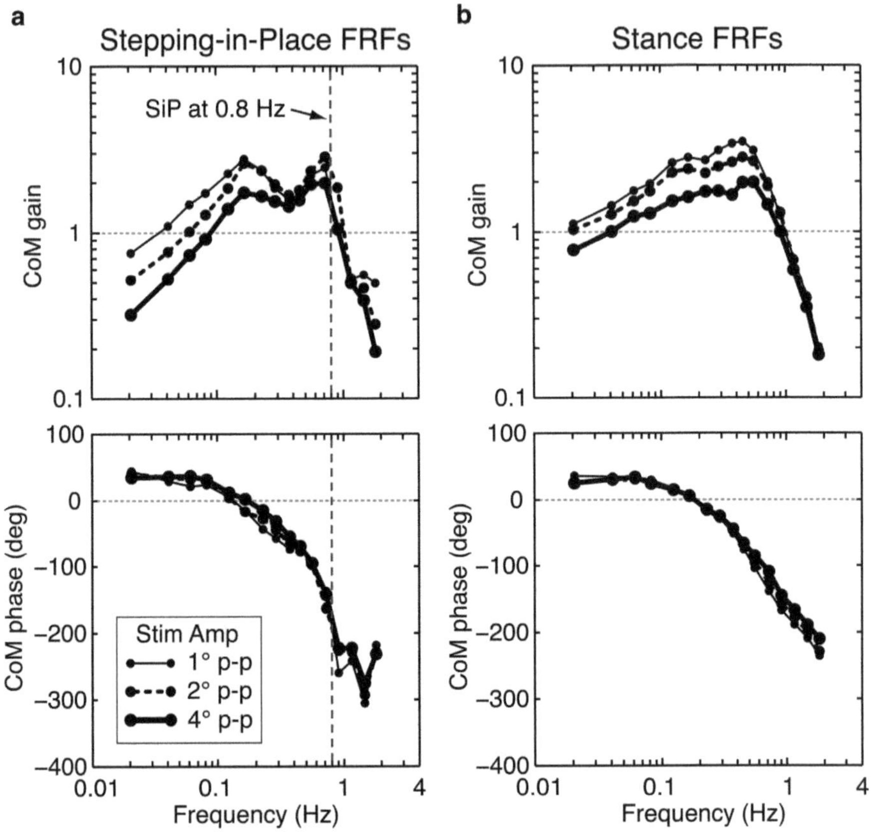


**Fig. 9.8** Frequency response functions (*FRFs*) from SiP experiments (**a**) and stance experiments (**b**). Pseudorandom surface-tilt stimuli at three different amplitudes were used to evoke CoM sway in the frontal plane. FRF results were averaged across the data from six subjects with normal sensory function who performed both SiP and stance experiments with eyes closed

FRFs calculated from the CoM sway evoked by surface-tilt stimulation during SiP are shown in Fig. 9.8a. For comparison, FRFs from stance trials of the same subjects are also shown in Fig. 9.8b. Each set of FRFs show results for the three different stimulus amplitudes and the FRFs are average results from six young adult subjects with normal sensory function. There were similarities and differences between FRFs from stance and SiP trials. Both showed the largest gains in the mid-frequency region, decreasing gains with decreasing frequencies below about 0.1 Hz, and decreasing gains with increasing frequencies above about 0.8 Hz. Mid-frequencies gains were greater than unity for both SiP and stance indicating the high sensitivity to surface-tilt perturbations in this mid-frequency range. Results from SiP and stance both showed decreasing FRF gains with increasing stimulus amplitude. For the SiP FRFs, this gain decrease with increasing amplitude was evident at frequencies below about 0.2 Hz while for stance FRFs, the amplitude-dependent gain decrease was evident at frequencies below about 0.8 Hz. Both SiP and stance FRFs showed phase leads at frequencies below about 0.2 Hz, and increasing phase lags at higher frequencies. SiP and stance phase data was minimally influenced by stimulus amplitude.

Differences between SiP and stance FRFs relate to the detailed shapes of the gain curves. The SiP gain curves show a notched decrease in gain at about 0.4 Hz and a peak near the stepping frequency while the stance gain curves are more uniform in this frequency region. It is perhaps notable that a previous study, that perturbed balance during treadmill walking using a sinusoidal visual stimulus at various frequencies, observed a similar enhancement of lateral sway amplitude when the stimulus frequency was close to the walking frequency (Kay and Warren 2001).

To the extent that we are confident that an amplitude-dependent change in FRF gains in the absence of phase changes is indicative of a sensory reweighting phenomenon, the pattern of amplitude-dependent changes seen in the SiP FRFs implies that the same or a very similar sensory reweighting phenomenon is contributing to the regulation of body orientation during SiP. Because the frontal-plane body motion during SiP and actual walking gait are similar, we anticipate that the balance mechanisms regulating the control of quiet stance are also influencing frontal-plane balance control during walking. As discussed previously, the balance mechanisms regulating body orientation during stance do a relatively poor job of maintaining an upright body orientation in conditions where the surface is not level or the visual system is not providing accurate orientation information. Therefore when walking on a tilted surface, for example, the mechanisms that control body orientation will evoke a leaning posture, and this leaning posture will impose a mechanical asymmetry that would be expected to complicate the task of maintaining dynamic stability during gait.

The FRF gain values at the lowest frequencies are notably smaller for the SiP trials than the stance trials, particularly in the results from the largest stimulus amplitude where the SiP gains are about half the value of the stance trials. The gain decrease and phase advance at lower frequencies are consistent with a torque feedback mechanism contributing to the control of body orientation at these lower frequencies. The lower low-frequency gains during SiP compared to stance suggest that this mechanism made a stronger contribution to orientation control during SiP. Assuming that SiP results apply to actual walking, a low FRF gain at low frequencies predicts that the mean body orientation would be close to zero (i.e., vertical) if a subject were walking for an extended period of time across a surface with a constant slope. However, if the slope was changing, as it would be on uneven terrain, the mean body orientation will be affected to a much greater extent and thus have a greater impact on the task of maintaining both static and dynamic stability, thus perhaps increasing the likelihood of falling.

**Acknowledgements** All experiments were performed according to protocols approved by the IRB of Oregon Health & Science University. Work was supported by NIH grants R01AG17960 and R01DC010779.

## References

Ahn J, Hogan N (2013) Long-range correlations in stride intervals may emerge from non-chaotic walking dynamics. PLoS One 8:e73239
Alexandrov AV, Frolov AA, Horak FB, Carlson-Kuhta P, Park S (2005) Feedback equilibrium control during human standing. Biol Cybern 93:309–322
Angelaki DE, McHenry MQ, Dickman JD, Newlands SD, Hess BJ (1999) Computation of inertial motion: neural strategies to resolve ambiguous otolith information. J Neurosci 19:316–327
Angelaki DE, Shaikh AG, Green AM, Dickman JD (2004) Neurons compute internal models of the physical laws of motion. Nature 430:560–564
Bauby CE, Kuo AD (2000) Active control of lateral balance in human walking. J Biomech 33:1433–1440
Bendat JS, Piersol AG (2000) Random data: analysis and measurement procedures, 3rd edn. Wiley, New York
Boonstra TA, Schouten AC, van der Kooij H (2013) Identification of the contribution of the ankle and hip joints to multi-segmental balance control. J Neuroeng Rehabil 10:23
Bosco G, Poppele RE, Eian J (2000) Reference frames for spinal proprioception: limb endpoint based or joint-level based? J Neurophysiol 83:2931–2945
Brenière Y (1996) Why we walk the way we do. J Mot Behav 28:291–298
Casabona A, Stella Valle M, Bosco G, Perciavalle V (2004) Cerebellar encoding of limb position. Cerebellum 3:172–177
Cenciarini M, Peterka RJ (2006) Stimulus-dependent changes in the vestibular contribution to human postural control. J Neurophysiol 95:2733–2750
Davies WDT (1970) System identification for self-adaptive control. Wiley-Interscience, London
Dean JC, Alexander NB, Kuo AD (2007) The effect of lateral stabilization on walking in young and old adults. IEEE Trans Biomed Eng 54:1919–1926
Dingwell JB, Cusumano JP (2000) Nonlinear time series analysis of normal and pathological human walking. Chaos 10:848–863
Duysens J, Clarac F, Cruse H (2000) Load-regulating mechanisms in gait and posture: comparative aspects. Physiol Rev 80:83–133
Fujisawa N, Masuda T, Inaoka H, Fukuoka Y, Ishida A, Minamitani H (2005) Human standing posture control system depending on adopted strategies. Med Biol Eng Comput 43:107–114
Goodworth AD, Peterka RJ (2009) Contribution of sensorimotor integration to spinal stabilization in humans. J Neurophysiol 102:496–512
Goodworth AD, Peterka RJ (2010a) Influence of bilateral vestibular loss on spinal stabilization in humans. J Neurophysiol 103:1978–1987
Goodworth AD, Peterka RJ (2010b) Influence of stance width on frontal plane postural dynamics and coordination in human balance control. J Neurophysiol 104:1103–1118
Goodworth AD, Peterka RJ (2012) Sensorimotor integration for multisegmental frontal plane balance control in humans. J Neurophysiol 107:12–28
Hausdorff JM (2005) Gait variability: methods, modeling and meaning. J Neuroeng Rehabil 2:19
Hettich G, Assländer L, Gollhofer A, Mergner T (2014) Human hip-ankle coordination emerging from multisensory feedback control. Human Movement Science 37:123–146
Hof AL, Gazendam MGJ, Sinke WE (2005) The condition for dynamic stability. J Biomech 38:1–8
Horak FB, Macpherson JM (1996) Postural orientation and equilibrium. In: Rowell LB, Shepherd JT (eds) Handbook of physiology: section 12: exercise: regulation and integration of multiple systems. Oxford University Press, New York, pp 255–292
Johansson R, Magnusson M, Akesson M (1988) Identification of human postural dynamics. IEEE Trans Biomed Eng 35:858–869
Kay BA, Warren WH (2001) Coupling of posture and gait: mode locking and parametric excitation. Biol Cybern 85:89–106
Kiemel T, Elahi AJ, Jeka JJ (2008) Identification of the plant for upright stance in humans: multiple movement patterns from a single neural strategy. J Neurophysiol 100:3394–3406.

Kim S, Atkeson CG, Park S (2012) Perturbation-dependent selection of postural feedback gain and its scaling. J Biomech 45:1379–1386
Kuo AD (1999) Stabilization of lateral motion in passive dynamic walking. Int J Robot Res 18:917–930
Li Y, Levine WS, Loeb GE (2012) A two-joint human posture control model with realistic neural delays. IEEE Trans Neural Syst Rehabil Eng 20:738–748
Loram ID, Maganaris CN, Lakie M (2005) Non-invasive tracking of contractile length. J Appl Physiol 100:1311–1323
Markert G, Büttner U, Straube A, Boyle R (1988) Neuronal activity in the flocculus of the alert monkey during sinusoidal optokinetic stimulation. Exp Brain Res 70:134–144
Maufroy C, Kimura H, Takase K (2010) Integration of posture and rhythmic motion controls in quadrupedal dynamic walking using phase modulations based on leg loading/unloading. Auton Robot 28:331–333
McMahon TA (1984) Muscles, reflexes, and locomotion. Princeton University Press, Princeton
Merfeld DM, Zupan L, Peterka RJ (1999) Humans use internal models to estimate gravity and linear acceleration. Nature 398:615–618
Mergner T (2004) Meta level concept versus classic reflex concept for the control of posture and movement. Arch Ital Biol 142:175–198
Mergner T (2010) A neurological view on reactive human stance control. Anu Rev Control 34:177–198
Mergner T, Maurer C, Peterka RJ (2002) Sensory contributions to the control of stance: a posture control model. Adv Exp Med Biol 508:147–152
Mummolo C, Mangialardi L, Kim JH (2013) Quantifying dynamic characteristics of human walking for comprehensive gait cycle. J Biomech Eng 135:91006
Nashner LM (1981) Analysis of stance posture in humans. In: Towe AL, Luschei ES (eds) Handbook of behavioral neurobiology, vol 5. Plenum Publishing Corporation, New York, pp 527–565
Otnes RK, Enochson L (1972) Digital time series analysis. Wiley, New York
Park S, Horak FB, Kuo AD (2004) Postural feedback responses scale with biomechanical constraints in human standing. Exp Brain Res 154:417–127
Pasma JH, Boonstra TA, Campfens SF, Schouten AC, van der Kooij H (2012) Sensory reweighting of proprioceptive information on the left and right leg during human balance control. J Neurophysiol 108:1138–1148
Peterka RJ (2002) Sensorimotor integration in human postural control. J Neurophysiol 88:1097–1118
Peterka RJ (2003) Simplifying the complexities of maintaining balance. IEEE Eng Med Biol Mag 22:63–68
Peterka RJ, Benolken MS (1995) Role of somatosensory and vestibular cues in attenuating visually induced human postural sway. Exp Brain Res 105:101–110
Peterka RJ, Black FO, Schoenhoff MB (1990) Age-related changes in human vestibulo-ocular and optokinetic reflexes: Pseudorandom rotation tests. J Vestib Res 1:67–71
Peterka RJ, Gianna-Poulin CC, Zupan LH, Merfeld DM (2004) Origin of orientation-dependent asymmetries in vestibulo-ocular reflexes evoked by caloric stimulation. J Neurophysiol 92:2333–2345
Pintelon R, Schoukens J (2012) System identification: a frequency domain approach. IEEE Press, New York
Prochazka A, Gillard D, Bennett DJ (1997) Positive force feedback control of muscles. J Neurophysiol 77:3226–3236
Terry K, Sinitski EH, Dingwell JB, Wilken JM (2012) Amplitude effects of medio-lateral mechanical and visual perturbations. J Biomech 45:1979–1986
van der Kooij H, Peterka RJ (2011) Non-linear stimulus-response behavior of the human stance control system is predicted by optimization of a system with sensory and motor noise. J Comput Neurosci 30:759–778

van der Kooij H, Jacobs R, Koopman B, Grootenboer H (1999) A multisensory integration model of human stance control. Biol Cybern 80:299–308

Winter DA (1995) Human balance and posture control during standing and walking. Gait Posture 3:193–214

Zupan LH, Peterka RJ, Merfeld DM (2000) Neural processing of gravito-inertial cues in humans. I. Influence of the semicircular canals following post-rotatory tilt. J Neurophysiol 84:2001–2015

# Part IV
# Neuromechanical Modeling of Locomotion

# Chapter 10
# Computing Motion Dependent Afferent Activity During Cat Locomotion Using a Forward Dynamics Musculoskeletal Model

**Boris I. Prilutsky, Alexander N. Klishko, Douglas J. Weber and Michel A. Lemay**

**Abstract** The structure and function of mammalian locomotor central pattern generators (CPGs) and their control by afferent feedback *in vivo* are not completely understood. The aim of this study was to develop a forward dynamics model of cat hindlimbs that using neural or muscle activity as input generates realistic locomotion mechanics and motion-dependent afferent activity. This model can be combined with CPG models to study the spinal control of locomotion using a comprehensive closed-loop neuromechanical model. The developed planar, 10-DOF model of two cat hindlimbs with 18 Hill-type muscle actuators generated realistic walking mechanics and firing rates of muscle type Ia, Ib, II and paw pad cutaneous afferents matching experimental results. The afferent activities were obtained from computed muscle fiber length and velocity, tendon force and simplified relationships transforming these mechanical variables to the afferent firing rates. The computed afferent signals were consistent with their suggested role in triggering locomotor phase transitions.

**Keywords** Locomotion · Afferent activity · Forward dynamics · Computational modeling · Cat

B. I. Prilutsky (✉) · A. N. Klishko
School of Applied Physiology, Center for Human Movement Studies, Georgia Institute of Technology, 555 14th Street NW, Atlanta, GA 30332, USA
e-mail: boris.prlutsky@ap.gatech.edu

D. J. Weber
Department of Physical Medicine and Rehabilitation, Department of Bioengineering, University of Pittsburgh, Pittsburgh, PA, USA

M. A. Lemay
Department of Bioengineering, Temple University, 1947 N 12th St, Philadelphia, PA 19122, USA



B. I. Prilutsky, D. H. Edwards (eds.), *Neuromechanical Modeling of Posture and Locomotion*, Springer Series in Computational Neuroscience,
DOI 10.1007/978-1-4939-3267-2_10

## 10.1 Introduction

Mammalian locomotion is a complex motor behavior which involves sensorimotor integration at different levels of the central nervous system (Grillner 1981; Orlovsky et al. 1999; Rossignol 2006). A network of neural circuits in the lumbar region of the mammalian spinal cord, called central pattern generators (CPGs), is involved in generating hindlimb locomotor activity in quadrupedal mammals. The CPG can produce locomotor-like activity of hindlimb motoneurons and flexor and extensor muscles in the absence of descending rhythmic input (spinal cord transaction at thoracic level; (Grillner 1981)) or proprioceptive feedback (deafferentation) in the cat (Brown 1914). In intact animals, CPG rhythmic activity is modulated by both descending commands (Shik et al. 1969; Grillner et al. 1999) and motion-dependent proprioceptive feedback (McCrea 2001; Pearson 2008).

The motion-dependent feedback from limb muscles and foot skin has been proposed to adjust CPG activity to the external environment and to regulate swing-stance and stance-swing transitions (McCrea 2001; Pearson 2008). Evidence for afferent regulation of mammalian CPG activity has been obtained in experiments performed in fictive locomotion preparations and on walking animals. In fictive locomotion preparations, neuromuscular transmission is blocked pharmacologically and CPG locomotor rhythm, recorded in muscle nerves and individual motoneurons and interneurons, can be elicited by electrical stimulation of the midbrain locomotor regions (Shik et al. 1969; Jordan 1998). During fictive locomotion, electrical stimulation of peripheral nerves exciting the large group I afferents (Ia, muscle length and velocity sensitive, and Ib, muscle force sensitive) of the ankle extensors enhances extensor activity in most extensor hindlimb muscle nerves if delivered during the extensor phase and resets the phase to extensor if delivered during the flexor phase (Conway et al. 1987; Guertin et al. 1995). Similar stimulations in walking animals during the stance phase enhances extensor activity (Pearson and Collins 1993; Whelan et al. 1995) as does stimulation of the cutaneous afferents innervating plantar surface of the foot (Duysens and Loeb 1980).

Although it is clear that afferent regulation of CPG activity during mammalian locomotion takes place, the detailed mechanisms of such regulation are still elusive. It is because the structure of the mammalian CPG networks and its elements receiving afferent feedback and supraspinal inputs has not been identified (McCrea and Rybak 2008; Kiehn 2011). In addition, it is impossible to distinguish effects of electrical stimulation of muscle length-velocity sensitive Ia and muscle force sensitive Ib afferents on the CPG activity because these afferents have similar excitation thresholds. Recent studies in transgenic mice without functioning spindle afferents or Golgi tendon organs revealed differential roles of these afferents during mammalian locomotion (Akay et al. 2014; Takeoka et al. 2014). Since in the above studies proprioceptive feedback in the transgenic mice was removed from all muscles, the contribution of spindle and Golgi tendon organ afferents from specific muscles to regulation of CPG activity is still unknown.

One approach to addressing this and other issues of the CPG control of mammalian locomotion is neuromechanical modeling and computer simulations. Several models of terrestrial locomotor CPGs integrated with a biomechanical system through motion-dependent feedback have been developed and analyzed (He et al. 1991; Ivashko et al. 2003; Yakovenko et al. 2004; Ekeberg and Pearson 2005; Markin et al. 2010; Aoi et al. 2013; see also Chapters by Aoi 2015; Bondy et al. 2015; Markin et al. 2015 in this book). In most of these studies, however, either the CPG model did not reproduce the relevant experimental observations (e.g., changes in rhythmic motoneuronal activity during fictive locomotion as a result of afferent stimulation (Rybak et al. 2006b) or the musculoskeletal models did not include biomechanical details important for determining motion-dependent afferent input (e.g., the muscle series elastic element that affects muscle fascicle length changes during walking (Hoffer et al. 1989; Maas et al. 2009). For example, the model of cat hindlimb locomotion by (Ivashko et al. 2003) included a CPG controlling nine groups of motoneurons in each hindlimb that excited the corresponding Hill-type muscle actuators. The CPG activity was modulated by motion-dependent feedback signals from muscles and paw skin receptors. Although this model demonstrated stable locomotion with patterns of muscle activity, kinematics and ground reaction forces somewhat similar to real cat locomotion, it had a number of limitations. Specifically, the CPG model was not based on and did not reproduce experimental observations obtained in fictive locomotion (Rybak et al. 2006a, b). The musculoskeletal parameters of the cat hindlimb model were not optimized within physiologically reasonable ranges. As a result, the model could not quantitatively reproduce muscle responses to imposed muscle length changes, or joint kinematics, ground reaction forces and joint moments of the walking cat when recorded muscle activity was used as input to the model. Furthermore, computed motion-dependent afferent signals were assumed proportional to muscle length, velocity, and force and did not reproduce nor were validated against the afferent signals recorded during locomotion. It is expected that in muscles with a relatively long tendon a substantial part of the muscle-tendon unit stretch can be taken up by the stretched tendon reducing elongation of muscle fibers and thus the output from length and stretch velocity sensitive spindle afferents (Hoffer et al. 1989; Maas et al. 2009). Therefore to accurately predict length- and velocity-dependent sensory feedback, the tendon elasticity must be incorporated in the model. Accurate tendon force estimation in individual muscle-tendon units during locomotion is also necessary for realistic predictions of sensory feedback signals from Golgi tendon organs.

Our long-term goal of modeling the spinal locomotor control has been to develop a comprehensive neuromusculoskeletal model of mammalian locomotor system that integrates the CPG model reproducing fictive locomotion experiments (Rybak et al. 2006a, b) and a detailed musculoskeletal model of the cat hindlimbs reproducing muscle force responses, walking mechanics and motion-dependent afferent signals. Such an integrated neuromusculoskeletal model will serve as a common computational framework for studying neural control of locomotion in the intact and spinalized animals and the effects of afferent feedback on restoring spinal locomotion (for details see (Markin et al. 2015)).

The aim of this work was four-fold: (1) Develop a forward dynamics musculoskeletal model of the cat hindlimbs that generates the activity of spindle (groups Ia, II), tendon organ (group Ib), and paw pad cutaneous afferents using neural or muscle EMG activity as input and computes muscle fiber length and velocity and tendon forces; (2) Validate simulated muscle force responses to imposed muscle length changes as well as mechanics of locomotion against the experimental joint angles, ground reaction forces, and joint moments; (3) Compare the computed muscle length- and force-dependent afferent signals with activity of selected afferents recorded during cat locomotion (Loeb and Duysens 1979; Loeb et al. 1985; Prochazka and Gorassini 1998a, b; Weber et al. 2007); and (4) Examine if computed proprioceptive signals were consistent with their suggested role in triggering locomotor phase transitions and modulating CPG activity. Our main hypothesis was that patterns of afferent activity during walking computed from the simulated muscle length and velocity, tendon force and ground reaction force applied to the paw would be similar to the patterns of afferent activity recorded during cat walking and reported in the literature.

Preliminary results of this work were published in an abstract form (Prilutsky and Klishko 2007; Prilutsky et al. 2013).

## 10.2 Model Development

The developed musculoskeletal model, although similar in some respects to several previously published models of cat hindlimbs (He et al. 1991; Ivashko et al. 2003; Yakovenko et al. 2004; Ekeberg and Pearson 2005; Markin et al. 2010; Bunderson et al. 2012; see also Chap. 1 by Bunderson and Bingham (2015) in this book) is unique because it incorporates all relevant model features that previously were not integrated in a single model. Specifically, the model uses as input electromyographic (EMG) activity recorded intramuscularly in relevant hindlimb muscles. The equations of muscle dynamics incorporate such muscle properties as muscle mass, the force-length-velocity relations of the contractile element and the force-length relations of the tendon and the muscle parallel elastic elements. Many parameters of the musculoskeletal model (i.e., positions of origin and insertion of each muscle) were directly measured; then these and other parameters were tuned to obtain a close match between the experimental and simulated locomotion. In addition, model output included motion-dependent proprioceptive signals that are computed using regression equations relating afferent firing rates with muscle length and velocity, tendon force and muscle activation. These regression equations were developed based on *in vivo* recordings in walking cats (Prochazka and Gorassini 1998b; Prochazka 1999; Weber et al. 2007).

### *10.2.1 Experimental Data for Model Development*

To ensure realistic simulations of cat hindlimb locomotor movements, parameters of the model were tuned to match as close as possible simulated and experimentally obtained hindlimb mechanics. The developed model was tested by comparing computed activity patterns of muscle and paw pad afferents with those recorded experimentally during unrestrained walking in cats.

#### 10.2.1.1 Recordings and Analysis of Muscle Activity and Walking Mechanics

To optimize tendon and muscle model parameters, we recorded EMG activity of 9 hindlimb muscles, muscle fascicle length and mechanics of walking in 5 adult cats (*Felis catus*) (mass 3.24 ± 0.40 kg) using experimental procedures consistent with US Public Health Service Policy on Humane Care and Use of Laboratory Animals and approved by the Institutional Animal Care and Use Committee of Georgia Institute of Technology (for details see (Gregor et al. 2006; Maas et al. 2009; Prilutsky et al. 2011). Briefly, the animals were trained for 3–5 weeks to walk on a Plexiglas enclosed walkway (3.0 m × 0.4 m) with 3 embedded force plates (0.16 m × 0.11 m and 0.11 m × 0.07 m; Bertec, USA) using food rewards. Mechanics of walking at self-selected speed were recorded using Vicon (UK) motion capture system and small reflective markers (6–9 mm in diameter) placed on hindlimbs' joints (Fig. 10.1a). After initial data collection, 9 pairs of thin Teflon-insulated multi-stranded stainless-steel wires (CW5402, Cooner Wire, USA) were implanted into 9 muscles of right hindlimb (Fig. 10.1c) under aseptic conditions and isoflurane anesthesia. The wires were attached to two multi-pin Amphenol connectors fixed to the skull by stainless steel screws and acrylic cement. After recovery (10–14 days), walking mechanics and muscle activity were collected for several weeks. No deviations from pre surgery walking patterns were noticed. Sampling rates were 120, 360, and 3000 Hz for kinematics, ground reaction forces and EMG, respectively. Marker displacements were low-pass filtered (5–6 Hz cutoff frequency, 4-order Butterworth zero-lag filter). Raw EMG signals were band-pass filtered (30–1500 Hz) and rectified. EMG bursts were identified by selecting a threshold EMG value (2 SD of the mean between-burst baseline activity) above which the muscle was considered active (Fig. 10.2a, SO); subsequently the baseline activity was set to zero. The modified rectified signals were low-pass filtered at 30 Hz, normalized to the EMG linear envelope maximum value across all cycles within the muscle and cat, and then averaged for each percent of the cycle time across all walking cycles of each cat and across 5 cats (Fig. 10.2b). Recorded or computed ground reaction forces, joint angles and joint moments (for details see (Gregor et al. 2006; Prilutsky et al. 2011)) were also averaged for each percent of the cycle time. After completion of data collection, the animal was euthanized with an overdose of pentobarbital sodium (120–180 mg/kg, intravenous administration). Immediately after euthanasia, both hindlimbs were dissected, placements of EMG electrodes were verified, and segment lengths and locations of origin and attachment midpoints of each muscle were measured using a caliper.

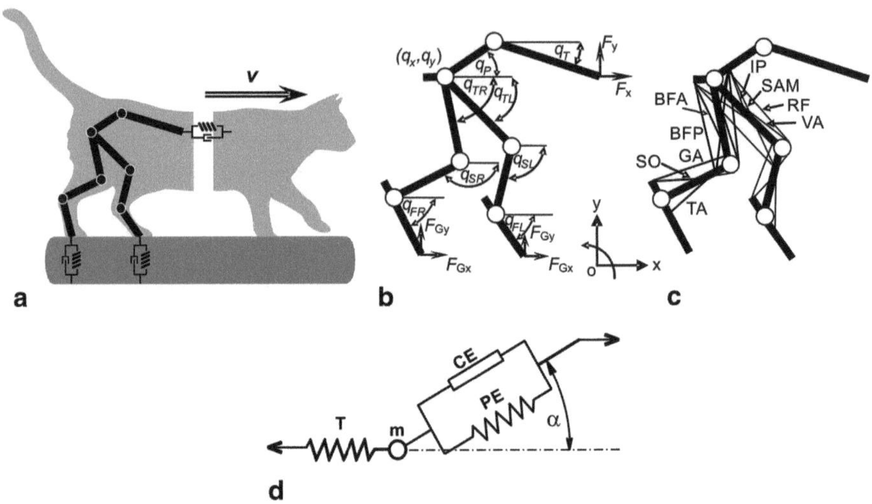


**Fig. 10.1** Schematic representation of the musculoskeletal model of the cat hindlimbs. **a** The hindlimbs, pelvis and posterior trunk interacting with the ground and the anterior trunk, head and forelimbs. These interactions are modeled as viscoelastic forces. **b** A 10-DOF planar system of rigid segments with frictionless revolute joints representing two hindlimbs, pelvis and posterior trunk. Each hindlimb consists of the foot, shank, and thigh. Ten generalized coordinates describe kinematics of the system: hip horizontal and vertical coordinates and hindlimb segment angles with respect to the horizontal axis. Interactions of the feet and the trunk with external environment are simulated by viscoelastic horizontal and vertical forces ($F_{GX}$, $F_{GY}$) (see text for details). **c** Schematic representation of muscles of the model: *IP* iliopsoas, *BFA* biceps femoris anterior, *RF* rectus femoris, *BFP* biceps femoris posterior, *SAM* sartorius medial, *VA* vastii, *GA* gastrocnemii, *TA* tibialis anterior, and *SO* soleus. **d** A schematic representation of a Hill-type model of the muscle-tendon unit (MTU). *T* tendon, *m* muscle mass located at the muscle-tendon junction, *CE* the contractile element, *PE* the parallel elastic element and *α* angle of pennation

#### 10.2.1.2 Recording and Analysis of Activity from Paw Cutaneous Afferents

The cutaneous afferent recordings used in this study were obtained on one day of recordings from an adult cat implanted chronically with a microelectrode array in the L7 dorsal root ganglia (DRG). The methods for implanting electrodes and recording in the DRG have been described in detail in (Weber et al. 2007). A summary of these methods follows. All procedures were approved by the University of Alberta Animal Care and Use Committee. All surgical procedures were conducted with the use of isoflurane anesthesia. A laminectomy was performed to expose the DRG on one side of the lower lumbar spine. A single Utah Electrode Array (UEA, Blackrock Microsystems, Inc.) comprising 36 electrodes arranged in a 9 × 4 configuration was implanted in the L7 DRG. The lead wires were anchored to the L5 spinous process and terminated in a percutaneous connector attached to an implanted stainless steel saddle affixed to the iliac crests.

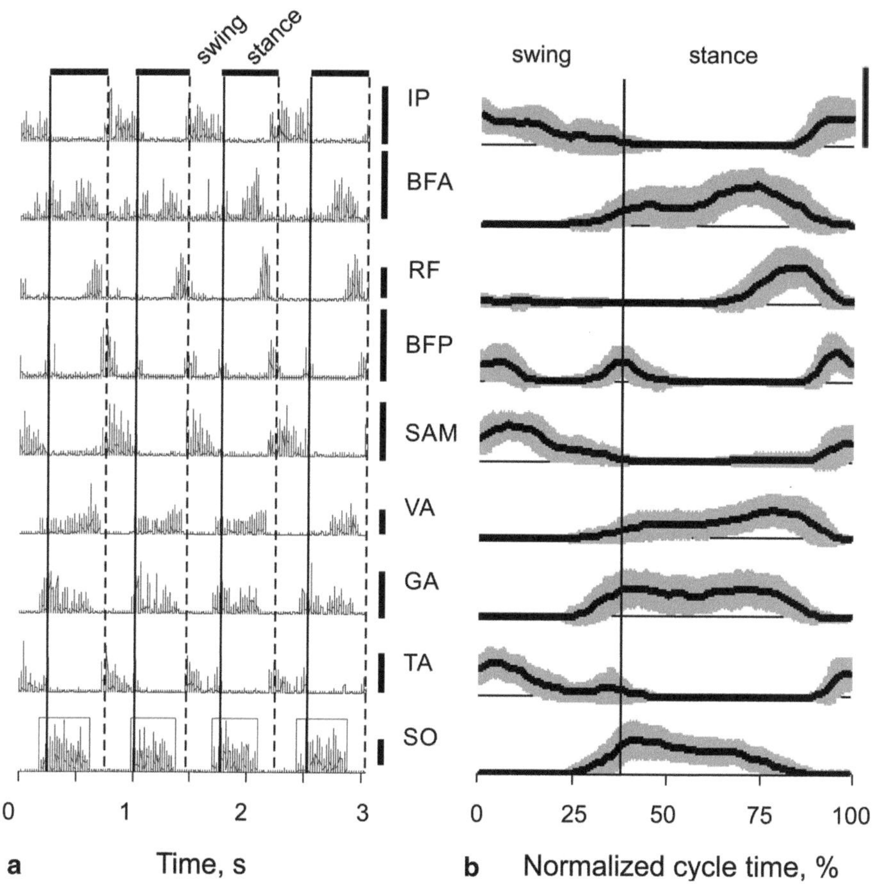


**Fig. 10.2** EMG activity of 9 cat hindlimb muscles during walking. Muscle abbreviations are the same as in Fig. 10.1c. **a** Example of rectified EMG signals recorded during 4 cycles of walking in one representative cat. Horizontal bars on the top and corresponding vertical dashed lines indicate stance phases identified using force plates or kinematics. *Vertical bars* on the *right* indicate scale of 0.5 mV. The *dashed-line rectangles* in the *bottom plot* (SO) demonstrate an example of identified EMG bursts; the activity between the bursts was set to zero before low-pass filtering was applied to the rectified signal. **b** Normalized mean ± SD rectified and low-pass filtered EMG patterns of multiple walking cycles and cats (total number of cycles analyzed for each muscle were between 84 and 169; data for SAM were obtained from 2 cats, for IP from 3 cats, for BFP from 4 cats, and for BFA, RF, VA, GA, TA, and SO from 5 cats). The *vertical line* separates the swing and stance phase. The *vertical bar* on *top right* corresponds to the maximal muscle recruitment magnitude. During walking, peak EMG activities for each muscle were set to be a certain proportion of the maximum: IP = 0.2, BFA = 0.5, RF = 0.5, BFP = 0.2, SAM = 0.5, VA = 0.3, GA = 0.215, TA = 0.2, and SO = 0.9; adopted and modified from (He et al. 1991)

Neural recordings and hindlimb kinematics were recorded as the cat walked on a treadmill at an average speed of 0.4 m/s. Hindlimb kinematics were recorded using a high-speed video camera and reflective markers stuck to the skin over the iliac crest and centers of the hip, knee, ankle, and metatarsophalangeal (MTP) joints. Heel-strike and toe liftoff times were identified manually in the video records and marked for segmentation of individual step cycles. During recording sessions, a 100-channel pre-amplifier was connected to the UEA connector and the amplified signals were sampled at 30,000 samples/s. Thresholds for spike detection were set for each channel and a 1–1.6 ms segment of the spike was recorded each time the signal exceeded the threshold. Spike-sorting for each electrode was performed off-line using custom Matlab (Mathworks, Inc) software as described in (Shoham et al. 2003). The sorted spike event times were used to calculate a time series of instantaneous firing rates for each sensory neuron (see below).

A series of tests were performed before or after the treadmill experiments to identify each receptor type. During the unit identification tests, the cat was anesthetized with isoflurane gas to permit a thorough examination of the limb. The identification process began with the localization of the receptive field of each receptor, identified by palpating the leg. Then, its response to (1) joint motion, (2) pressure, (3) blowing and (4) vibration was studied. Units that responded to blowing and very light brushing of the fur were identified as hair receptors (located in the skin at the base of the hair follicle). Units that responded reliably to light pressure or localized skin stretch were also classified as cutaneous afferents.

For the purpose of this study, activity of four paw pad cutaneous afferents continuously recorded in 11 walking cycles were selected for further analysis. Instantaneous firing rates of each cell were computed at 25 ms intervals (Weber et al. 2007) and then interpolated for each percent of a walking cycle using a cubic spline function. The ensemble activity of paw pad cutaneous afferents was obtained by averaging the interpolated firing rates for each percent of the cycle across all recorded cycles and four cells (Fig. 10.7d, thin line). To compare the mean afferent activity with the mean vertical ground reaction force recorded separately in five different animals as described above, the mean duty factor (stance time/cycle time) of the cycles collected for afferent activity was made equal to the mean duty factor obtained in locomotor mechanics experiments. Subsequently, the mean afferent activity was normalized to the new stance and swing durations separately.

## *10.2.2 Musculoskeletal Model*

### 10.2.2.1 Limb Dynamics

The cat hindlimbs with the pelvis and trunk were modeled as a 10 degrees-of-freedom sagittal plane system of rigid segments interconnected by frictionless revolute joints (Fig. 10.1a, b). Interactions of the hindlimbs with the ground and the trunk with the fore-legs, neck and head were modeled by linear springs and dampers (Fig. 10.1a and Table 10.1; e.g., (van den Bogert et al. 1989)). The ground reaction viscoelastic forces were computed as a function of velocity and displacement of the leg endpoint during stance:

**Table 10.1** Parameters of the limb model

| Parameters | Segments | | | | |
|---|---|---|---|---|---|
| | Trunk | Pelvis | Thigh | Shank | Foot[a] |
| Center of mass location[b], mm | 122.5 | 27.5 | 43.1 | 43.4 | 33.5 |
| Mass, g | 1448.8 | 324.6 | 149.5 | 63.5 | 21.7 |
| Moment of inertia, $g \cdot mm^2$ | 7249930 | 81524 | 140063 | 59201 | 9147 |
| Length, mm | 245.0 | 54.9 | 97.3 | 102.5 | 68.9 |
| Parameters of linear springs and dampers at model contact sites | | | | | |
| | Foot-ground | Joints | | Distal trunk | |
| Coefficient of stiffness | 1.25 N/mm | 3000 N/rad | | 6 N/mm | |
| Coefficient of viscosity | 28.5 N · ms/mm | 18000 N · ms/rad | | 35 N · ms/mm | |

Body segment parameters were calculated from body mass and segment length using the regression equations from (Hoy and Zernicke 1985)

[a] Foot included tarsals and digits

[b] Distance from the proximal joint

$$\begin{aligned} F_{Gx} &= -k_g \Delta x - b_g \dot{x} \\ F_{Gy} &= -k_g \Delta y - b_g \dot{y}, \end{aligned} \tag{10.1}$$

where $F_{Gx}$ and $F_{Gy}$ are the horizontal and vertical components of the ground reaction force vector, constants $k_g$ and $b_g$ ($b_g$=0 for $\dot{y} > 0$) are stiffness and viscosity coefficients (Table 10.1) found using optimization (see below); $\Delta x$ and $\Delta y$ are foot endpoint displacements during the stance phase from the initial point of contact at the ground level; dots indicate time derivatives. Similarly, the viscoelastic force applied at the distal trunk was a function of the trunk's tip deviation from preset horizontal and vertical positions and velocities. Mass, moment of inertia, and the location of the center of mass of each body segment (Table 10.1) were calculated based on animal mass and measured segment lengths using the regression equations from (Hoy and Zernicke 1985).

The dynamic equations of system motion can be written in a vector form:

$$\ddot{\mathbf{q}} = \mathbf{I}^{-1}(\mathbf{q}) \left[\mathbf{C}(\mathbf{q},\dot{\mathbf{q}}) + \mathbf{G}(\mathbf{q},\dot{\mathbf{q}}) + \mathbf{S}(\mathbf{q},\dot{\mathbf{q}}) + \mathbf{M}(\mathbf{q},\dot{\mathbf{q}},\mathbf{t})\right] \tag{10.2}$$

where $\mathbf{q}$, $\dot{\mathbf{q}}$ and $\ddot{\mathbf{q}}$ are vectors of the generalized displacements, velocities and accelerations, **I** is the system inertia matrix, **C** is the vector of Coriolis and centrifugal forces, **G** is the vector of external generalized forces, **S** is the vector of viscoelastic generalized forces at the joints (passive joint moments) and **M** is the vector of muscle generalized forces (muscle moments). Viscoelastic passive moments at the joints were engaged only if computed joint angles exceed the nominal locomotor range of joint motion: 127° (fixed angle), 50°–180°, 45°–160°, and 45°–160° for pelvis-trunk articulation, hip, knee and ankle, respectively. These generalized passive forces at *j*-th joint were computed as $S_j = -k_p \Delta q_j - b_p \dot{q}_j$, where $\Delta q_j$ is the deviation of joint angle from the nominal range, $\dot{q}_j$ is angular velocity, and $k_p$ and $b_p$ coefficients of stiffness and viscosity given in Table 10.1.

#### 10.2.2.2 Muscle Dynamics

The hindlimb model was driven by 18 Hill-type muscle actuators (9 per hindlimb) that represented major hindlimb muscles (Fig. 10.1c). The location of each muscle-tendon unit (MTU) with respect to the joints was described geometrically using the measured origin and insertion points of the muscle and pulleys representing bone protuberances over which the muscle wraps (Table 10.2). MTU lengths and moment arms were computed using muscle and joint parameters from Table 10.2 as functions of joint angles.

*Equations of Muscle Dynamics and Properties of Tendon and Muscle* The contraction dynamics of the MTU were described by a Hill-type model (Fig. 10.1d, e.g., (Zajac 1989; He et al. 1991; Brown et al. 1996b; Sandercock and Heckman 1997) taking into account muscle mass, angle of pennation, the force-length-velocity relations of the contractile element and the force-length relation of the serial (tendon) and parallel elastic elements. Muscle contractile dynamics were described by the following differential equation:

$$\dot{V}_T = \left[F_T - F_M \cos\alpha\right]/m, \tag{10.3}$$

where

$$F_T = F_T^{Max}\left[F_{TN}\left(L_T\right) + b_T V_T\right]; \tag{10.4}$$

$$F_M = F_M^{Max}\left[F_{CE}\left(L_M\right)F_{CE}\left(V_M\right)k_u^{\max}A + F_{PEN}\left(L_M\right) + b_M V_M\right]; \tag{10.5}$$

$F_T$, $L_T$, and $V_T$ are tendon force, length and velocity; $F_M$, $L_M$, and $V_M$ are muscle fiber force, length and velocity; $\alpha$ is the pennation angle; $b_T$ and $b_M$ are coefficients of viscosity for the tendon and muscle; $m$ is muscle mass; $F_{TN}(L_T)$, $F_{PEN}(L_M)$ and $F_{CE}(L_M)$ are the normalized force-length relations for the tendon, muscle parallel elastic element, and the muscle contractile element (see equations for these relationships below and in Table 10.3); $F_{CE}(V_M)$ is the normalized force-velocity relation for the muscle contractile element (see below and Table 10.3); $F_T^{Max}$ and $F_M^{Max}$ are the maximal isometric tendon and muscle force at optimum length of the contractile element and $F_T^{Max} = F_M^{Max}\cos(\alpha_0)$ ($\alpha_0$ is pennation angle at the optimal contractile element length); $k_u^{\max}$ is muscle maximal activation during walking ($0 \le k_u^{\max} \le 1$, modified from (He et al. 1991), see Table 10.2); and $A$ is time dependent muscle activation obtained from the first-order differential equation describing the muscle excitation (EMG)-activation (active state) dynamics (Zajac 1989; He et al. 1991):

$$\frac{dA(t)}{dt} + \left\{\frac{\tau_{act}}{\tau_{deact}} + \left[1 - \frac{\tau_{act}}{\tau_{deact}}\right]u(t)\right\}\frac{A(t)}{\tau_{act}} = \frac{u(t)}{\tau_{act}} \tag{10.6}$$

**Table 10.2** Geometric and physiological parameters of hindlimb muscles

| Parameter \ Muscle | IP | BFA | RF | BFP | SAM | VA | GA | TA | SO |
|---|---|---|---|---|---|---|---|---|---|
| $m$, g | 17.6 | 28 | 9.5 | 41.3 | 8.5 | 27.5 | 24.8 | 5.5 | 3.5 |
| $F_M^{max}$, N | 64.3 | 82.9 | 77.3 | 147.5 | 35.2 | 435 | 54.8 | 36.2 | 36.4 |
| $\alpha$, deg | 0 | 14.0 | 7.0 | 7.0 | 0.0 | 13.7 | 19.0 | 7.0 | 7.0 |
| $L_{T0}$, mm | 15.5 | 17.5 | 80.0 | 14.0 | 17.5 | 35.5 | 80.8 | 25.0 | 20.9 |
| $L_{T0}/L_0$ | 0.5 | 0.5 | 4.0 | 0.1 | 0.1 | 2.0 | 6.0 | 1.0 | 1.0 |
| $L_{PE0}/L_0$ | 1.0 | 0.9 | 1.0 | 1.0 | 0.7 | 0.9 | 0.85 | 1.0 | 0.9 |
| $R_1$, mm | 5.0 | 5.0 | 5.0 | 5.0 | 5.0 | 7.1 | 5.0 | 5.0 | 5.0 |
| $R_2$, mm | – | – | 7.1 | 5.0 | 5.0 | – | 5.0 | – | – |
| $a_1$, mm | 36.8 | 8.0 | 14.6 | 8.0 | 32.0 | 55.1 | 5.0 | 6.2 | 36.1 |
| $a_2$, mm | 5.0 | 40.8 | 7.1 | 24.9 | 20.1 | 7.1 | 15.0 | 7.3 | 15.0 |
| $\phi_1$, deg | 0 | 141 | 0 | 141 | 0 | 0 | 0 | 0 | 0 |
| $\phi_2$, deg | 0 | 0 | 145 | 0 | 0 | 145 | −174 | 0 | −174 |
| $k_u^{max}$ | 0.2 | 0.6 | 0.59 | 0.22 | 0.6 | 0.28 | 0.17 | 0.24 | 1 |
| $R_{Ia_0}$, imp/s | 12 | 16 | 34 | 18 | 31 | 58 | 70 | 128 | 62 |
| $R_{II_0}$, imp/s | 32 | 62 | 16 | 16 | 36 | 85 | 108 | 166 | 33 |

*IP* iliopsoas, *BFA* biceps femoris anterior, *RF* rectus femoris, *BFP* biceps femoris posterior, *SAM* sartorius medial, *VA* vastus, *GA* gastrocnemius, *TA* tibialis anterior, *SO* soleus. $m$, muscle mass; $F_M^{max}$, maximal isometric muscle force at optimal fiber length $L_{Mo}$; $\alpha$, pennation angle at $L_{Mo}$, estimated from (Sacks and Roy 1982); $L_{T0}$, tendon rest (slack) length; $L_{T0}/L_{M0}$, tendon rest length–fiber optimal length ratio, estimated from (Sacks and Roy 1982; Zajac 1989; Prilutsky et al. 1996); $L_{PE0}/L_{M0}$, resting (slack) length of the parallel elastic element normalized to optimal fiber length (estimated from (Gareis et al. 1992; Brown et al. 1996a)); $R_1$, joint radius; $R_2$, radius of the other (distal joint) crossed by a two-joint muscle; $a_1$, muscle-tendon unit (MTU) origin-to-joint center distance; $a_2$, MTU attachment-to-joint center distance; $\phi_1$ angle between segment axis and position vector from joint center to muscle origin, $\phi_2$, angle between segment axis and position vector from joint center to muscle attachment; $k_u^{max}$, peak of normalized activation during walking, obtained using estimates from (He et al. 1991); $R_{Ia_0}$ and $R_{II_0}$, mean firing frequencies of Ia and II afferents, respectively, set based on experimental recordings (see Fig. 10.7). Initial values for muscle mass $m$ and maximal force $F_M^{max}$ before optimization were calculated as $m = PCSA \cdot L_M \cdot \rho_m$ and $F_M^{max} = \sigma \cdot PCSA$, respectively, where $PCSA$ and $L_M$ are the muscle physiological cross-sectional area and fiber length taken from (Sacks and Roy 1982), $\rho_m$ and $\sigma$ are muscle density (1.0564 g/cm$^3$) and maximum muscle stress (2.25 N/cm$^{-2}$) (Spector et al. 1980). MTU length was calculated as the length of straight line from muscle origin to attachment except when this line touched a joint pulley in which case the part of muscle length was calculated assuming that the muscle wraps around the joint surface represented by a pulley with a given diameter

**Table 10.3** Parameters of tendon and muscle model

| Parameter | | | Equation |
|---|---|---|---|
| Tendon $F_T(L_T)$ relation | | | $F_T(L_T)=\frac{k_{T1}}{k_{T2}}\{\exp(k_{T2}[L_T/L_{T0}-1])-1\}$ |
| $k_{T1}$ | | 0.1 | |
| $k_{T2}$ | | 90 | |
| Parallel element $F_{PE}(L_{PE})$ relation | | | $F_{PE}(L_M)=\frac{k_{PE1}}{k_{PE2}}\{\exp(k_{PE2}[L_M/L_0-1])-1\}$ |
| $k_{PE1}$ | | 0.0075 | |
| $k_{PE2}$ | | 11.6 | |
| Muscle $F_{CE}(V_M)$ relation | | | $F_{CE}(V_M)=\begin{cases}\frac{b_0+b_1V_M}{b_0+V_M}(V>0)\\ \frac{V_M^{max}+V_M}{V_M^{max}-a_VV_M}(V<0)\end{cases}$ |
| Muscle | $a_V$ | $V_{max}$ (m/s) | |
| IP | 3.70 | 0.578 | |
| BFA | 9.50 | 0.643 | |
| VA | 2.40 | 0.266 | $b_0=V_M^{max}\frac{0.8}{a_V+1}$, |
| TA | 5.90 | 0.813 | |
| SO | 1.10 | 0.176 | |
| RF | 5.00 | 0.288 | $b_1=\frac{1.8(V_M^{max}+b_0)-b_0}{V_M^{max}}$ |
| BFP | 9.50 | 0.624 | |
| SAM | 3.70 | 1.491 | |
| GA | 2.28 | 0.259 | |
| Muscle $F_{CE}(L_M)$ relation | | | $F_{CE}(L_M)=\exp\left(-\left\|\frac{(L_M/L_M^0)^\beta-1}{\omega}\right\|^\rho\right)$ |
| $\omega$ | | 0.55 | |
| $\rho$ *for* $L_{T0}/L_0 \le 1$ | | 6 | |
| $\rho$ *for* $L_{T0}/L_0 > 1$ | | 3 | |
| $\beta$ | | 1.55 | |
| Muscle and tendon viscosity coefficients | | | |
| $b_T$ (N·s/m) | | 0.02 | See Eq. (10.4) |
| $b_M$ (N·s/m) | | 0.02 | See Eq. (10.5) |

**Table 10.3** (continued)

| Parameter | | | Equation |
|---|---|---|---|
| Muscle activation dynamics parameters | | | |
| Muscle | $\tau_{act}$, ms | $\tau_{act}/\tau_{deact}$ | See Eq. (10.6) |
| IP | 10.0 | 0.8 | |
| BFA | 10.0 | 0.5 | |
| VA | 20.2 | 0.5 | |
| TA | 24.6 | 0.5 | |
| SO | 42.6 | 0.6 | |
| RF | 10.0 | 0.5 | |
| BFP | 24.8 | 0.6 | |
| SAM | 10.0 | 0.5 | |
| GA | 20.7 | 1.0 | |

$L_T$ is tendon length; $L_M$ and $V_M$ are muscle fiber length and velocity; $V_{max}$ is muscle maximum shortening velocity (taken from (Spector et al. 1980) for SO and GA and estimated for the other muscles based on the assumption that $V_{max}$ is proportional to the expression $L_{Mo} \cdot (1 - S/100)$, where $L_{Mo}$ is optimal muscle fiber length (Sacks and Roy 1982) and $S$ is the percentage of slow-twitch fibers (Ariano et al. 1973); $a_V$ is coefficient in the equation for $F_{CE}(V_M)$, it was calculated from equation $a_V = 0.00915 \cdot S - 0.00467$, $r=0.974$; (Baratta et al. 1995), where $S$ is the percentage of slow-twitch fibers; $b_T$ and $b_M$ are coefficients of viscosity for tendon and muscle contractile element, respectively; $F_{CE}(L_M)$ and $F_{CE}(V_M)$ are normalized force-length and force-velocity relationships for contractile element; $F_T(L_T)$ and $F_{PE}(L_M)$ are tendon and parallel element force-length relations; $\tau_{act}$ and $\tau_{deact}$ are muscle activation and deactivation time constants; $A$ is muscle activation (see Eq. 10.6). Coefficient $\rho$ was set to 6 for muscles with short tendon (LT0/L0≤1): IP, BFA, BFP, SAM, TA, SO; $\rho$ was set to 3 for muscles with long tendon (LT0/L0>1): RF, VA, GA (see text for further explanations) where $u(t)$ is rectified, low-pass filtered and normalized to its peak EMG activity ($0 \le u(t) \le 1$, see Fig. 10.2b); $\tau_{act}$ and $\tau_{deact}$ are the activation and deactivation time constants. Definitions, values, and sources for the above muscle-tendon-activation parameters and relations are given in Tables 10.2 and 10.3.

*Tendon Force-Length Relation* Two general forms of the normalized force-length relation for the tendon were examined (see Eq. 10.7 and Table 10.3). The first form of the normalized relation $F_{TN}(L_T)$ (see Eq. 10.4) was adopted from (Siebert et al. 2008) and slightly modified (see Fig. 10.3a):

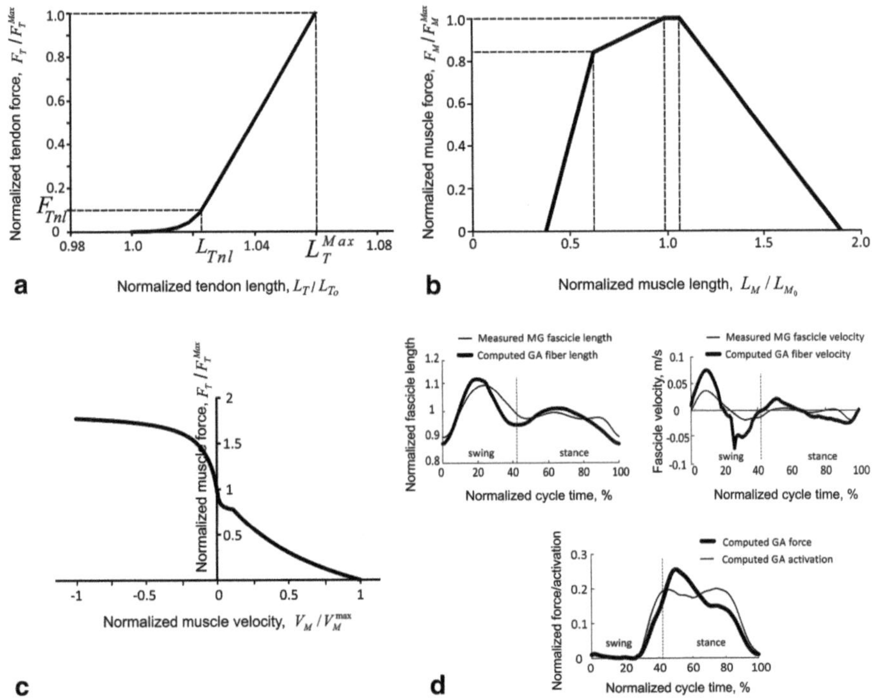


**Fig. 10.3** Modeling the tendon force-length and contractile element force-length and force-velocity properties. **a** Normalized tendon force-length relation adopted from (Siebert et al. 2008). Parameters of the relation (see Eq. 10.7) were found by minimizing the difference between simulated and measured muscle fascicle length during the cycle of cat walking (see text and panel **d**). **b** Normalized force-length relation for the contractile element adopted from (Gordon et al. 1966) and (Siebert et al. 2008) (see text). **c** Normalized force-velocity relation for the contractile element adopted from (Edman 1988) (see Eq. 10.8 and the corresponding text). **d** Fascicle length and velocity during the walking cycle (*top* panels) computed for gastrocnemius (GA) after minimizing cost function $Z_1$ (Eq. 10.10) and measured in medial gastrocnemius (MG) in (Maas et al. 2009). *Bottom* panel shows computed MG force and activation after minimizing cost function $Z_1$ (see text)

$$F_T = \begin{cases} F_T^{Max} \cdot F_{Tnl} \dfrac{\exp\{k_{Tnl} \cdot [(L_T / L_{To}) / L_{Tnl}]\} - 1}{\exp(k_{Tnl}) - 1}, & 1 < L_T / L_{To} < L_{Tnl} \\ F_T^{Max} \cdot [F_{nl} + k_{Tl} \cdot (L_T / L_{To} - L_{Tnl})], & L_T / L_{To} \geq L_{Tnl} \end{cases}, \quad (10.7)$$

where $F_T$ and $L_T$ are tendon force and length, $L_{To}$ is the tendon rest (slack) length in mm, $L_{Tnl}$ and $F_{Tnl}$ are the tendon length normalized to $L_{To}$ and force corresponding to the point of separation between the nonlinear and linear parts of the tendon force-length relation, $k_{Tnl}$ and $k_{Tl}$ are the stiffness constants for the nonlinear and linear parts of the relation, $L_T^{Max}$ is the tendon length corresponding to $F_T^{Max}$ (see Eq. 10.4).

The second form of the relation $F_{TN}(L_T)$ and its parameters are presented in Table 10.3. This form was used in computations of afferent feedback signals (see below).

*Isometric Force-Length Relation for the Contractile Element* Again, two general forms of the normalized isometric force-length relation $F_{CE}(L_M)$ were examined (see Fig. 10.3b and Table 10.3). The first form (Fig. 10.3b) was derived from the normalized force-length relation of the frog sarcomere (Gordon et al. 1966) and adapted to the cat muscle (Siebert et al. 2008). The relation is described by 4 straight lines connecting the following points in the $L_M / L_{M_0} - F_M / F_M^{Max}$ plane: (0.39, 0), (0.64, 0.84), (1, 1), (1.08, 1) and (1.9, 0). The second form of the relation was adopted from (Markin et al. 2010); the equation and its parameters are included in (Table 10.3).

*Force-Velocity Relation for the Contractile Element* Initially we adopted the force-velocity relation of the contractile element, $F_{CE}(V_M)$, in the double-hyperbolic form proposed by (Edman 1988) (Fig. 10.3c):

$$F_{CE} = \begin{cases} \dfrac{V_M^{\max} + V_M}{V_M^{\max} - aV_M}, & V_M > V_{Mtr} \\ k_{FV} + \dfrac{g}{V_M - h}, & V_{Mtr} > V \end{cases}, \quad (10.8)$$

where $V_{Mtr}$ corresponds to velocity at which the Hill force-velocity equation (Hill 1938) starts to deviate from the Edman's equation. Parameters $k_{FV}$, $g$, and $h$ describe the hyperbolic equation in the velocity range $0 < V_M < V_{Mtr}$ and were derived based on the following three conditions:

1. The derivative of the normalized force over the normalized velocity at $V_M / V_M^{\max} = 0$ is $k_d = 100$ (based on Fig. 2c in Edman 1988).
2. $F_{CE}^{\max} = 1$ at $V_M = 0$.
3. Normalized muscle force at the transition from the Hill equation to Edman equation (at velocity $V_M = V_{Mtr} = 0.11V_M^{\max}$) is $F_{CE} = 0.78$ (Edman 1988).

Condition 1 leads to

$$g = -k_d h^2, \tag{10.9a}$$

Condition 2 combined with Eq. (10.9a) gives

$$k_{FV} = F_M^{\max} - k_d h, \tag{10.9b}$$

and condition 3 combined with Eqs. (10.9a) and (10.9b) gives

$$h = \frac{V_{Mtr}\left(F_{CE}^{\max} - F_{CEtr}\right)}{V_{Mtr} \cdot k_d - \left(F_{CE}^{\max} - F_{CEtr}\right)}. \tag{10.9c}$$

Parameters $k_{FV}$, $g$, and $h$ for the hyperbolic equation at muscle stretch velocities $V_M < 0$ were derived from the following conditions:

1a. The derivative of the normalized force over the normalized velocity at $V_M / V_M^{\max} = 0$ is $k_d = 50$ (based on Fig. 7 in Edman 1988).
2a. $F_{CE}^{\max} = 1$ at $V_M = 0$.
3a. $F_{CE}$ approaches an asymptote $k_{FV} = 1.85$ at muscle elongation velocity $V_M$ approaching minus infinity.

From condition 1a follows Eq. (10.9a); condition 2a and Eq. (10.9a) leads to

$$h = \frac{F_{CE}^{\max} - k_{FV}}{k_d} \tag{10.9d}$$

Thus the parameters for the force-velocity relation are (see Fig. 10.3c):

$$k_{FV} = \begin{cases} F_{CE}^{\max} - k_d h, & 0 < V_M / V_M^{\max} < V_{Mtr} \\ 1.85, & V_M / V_M^{\max} < 0 \end{cases},$$

$$h = \begin{cases} \dfrac{V_{Mtr}\left(F_{CE}^{\max} - F_{CEtr}\right)}{V_{Mtr} \cdot k_d - \left(F_{CE}^{\max} - F_{CEtr}\right)}, & 0 < V_M / V_M^{\max} < V_{Mtr} \\ \dfrac{F_{CE}^{\max} - k_{FV}}{k_d}, & V_M / V_M^{\max} < 0 \end{cases},$$

$$k_d = \begin{cases} 100, & 0 < V_M / V_M^{\max} < V_{Mtr} \\ 50, & V_M / V_M^{\max} < 0 \end{cases}.$$

The other form of the force-velocity relation and its parameters are in Table 10.3 (see below).

#### 10.2.2.3 Tuning Muscle Model Parameters and Validation of Muscle-Tendon Model

Since the output of muscle models (typically muscle force) is sensitive to errors in model parameters (Brown et al. 1996b; Scovil and Ronsky 2006; De Groote et al. 2010), it is important to accurately measure the parameters that have the greatest effect on the model performance (typically tendon properties as well as muscle maximum force) when possible or tune them so that the computed muscle force and simulated locomotor mechanics match the experimental data as close as possible. In this study, parameters of the equations of muscle dynamics (Eqs. 10.3–10.8) were optimized within the physiological ranges in several stages.

*Stage 1: Tuning Tendon Model Parameters* In stage 1, parameters of the tendon $L_{T_0}$, $k_{Tl}$, $L_{Tnl}$ and $L_T^{\max}$ (see Eq. 10.7) and optimal muscle fiber (contractile element) length $L_{M_0}$ were tuned by minimizing the difference (Eq. 10.10) between computed and measured during cat level walking muscle fascicle length and velocity separately for muscle-tendon actuators with a relatively long tendon, e.g., medial gastrocnemius (MG, $L_{T0} / L_{M0} > 1$, where $L_{T0}$ and $L_{M_0}$ are the tendon rest and muscle fiber optimum length, respectively) and a short tendon, e.g., soleus (SO, $L_{T0} / L_{M0} \le 1$). Inputs for these calculations were EMG activity of these muscles and hindlimb joint angles recorded during walking in 5 cats as described above. It was assumed that other hindlimb muscles with relatively long and short tendons had the same normalized tendon properties as those of MG and SO, respectively. The muscle actuators with relatively long tendons included gastrocnemius (GA), vastii (VA) and rectus femoris (RF); the actuators with relatively short tendons consisted of soleus (SO), tibialis anterior (TA), sartorius medial (SAM), biceps femoris posterior (BFP), biceps femoris anterior (BFA) and iliopsoas (IP). This muscle classification was based on values of $L_{T0}$ and $L_{M_0}$ reported by (Sacks and Roy 1982; Zajac 1989; Prilutsky et al. 1996). Tendon model parameters $L_{T_0}$, $k_{Tl}$, $L_{Tnl}$ and $L_T^{\max}$ and optimal muscle fiber length $L_{M_0}$ for MG and SO were found using a numerical minimization of the cost function $Z_1$ for each muscle separately:

$$Z_1 = \frac{\sum_{t=1}^{N}\left(L_{M_t} - L_{M_t}^e\right)^2}{R_{L_M}^2} + \frac{\sum_{t=1}^{N}\left(V_{M_t} - V_{M_t}^e\right)^2}{R_{V_M}^2} + \frac{\sum_{t=1}^{N}\left(F_{M_t} - A_t\right)^2}{A_0^2}, \tag{10.10}$$

where $L_{M_t}$ and $L^e_{M_t}$ are computed and measured during cat walking mean muscle fascicle length at the time instant $t$ ($t$=1,2,…,$N$), respectively; $R_{L_M}$ is the range of measured muscle fascicle length changes in the walking cycle; $V_{M_t}$ and $V^e_{M_t}$ are computed and measured muscle fascicle velocity, respectively; $R_{V_M}$ is the range of measured muscle fascicle velocity changes in the walking cycle; $F_{M_t}$ is computed muscle force; $A_t$ is muscle activation computed from recorded muscle activity (low-pass filtered EMG, Fig. 10.2b) using Eq. (10.6) in which muscle activation and deactivation constants $\tau_{act}$ and $\tau_{deact}$ were assumed to be 25 and 50 ms, respectively, for all muscles for the purpose of this optimization. The third term in Eq. (10.10) was introduced to ensure that in the optimal solution the muscle does not produce noticeable passive force when the muscle is not activated.

The fascicle length and fascicle velocity of SO and MG muscles during the walking cycle were measured in 4 cats in a previous study (Maas et al. 2009) and averaged across walking cycles and cats (Fig. 10.3d, thin lines). Muscle-tendon unit (MTU) length trajectories of SO and MG were computed from recorded joint angles using a hindlimb geometric model.

Dynamic optimization problem of minimizing cost function $Z_1$ (Eq. 10.10) was solved under the constraints described by equations of muscle dynamics (10.3–10.6) and Eq. (10.11):

$$L_{M_1} = L^e_{M_1};\quad V_{M_1} = V^e_{M_1};\quad L_{M_t}\cdot\cos\alpha + L_{T_t} = L_{MTU_t}\quad (t = 1,2,\ldots,N), \qquad (10.11)$$

where $L_{M_1}$ and $L^e_{M_1}$ are initial values of computed and experimental muscle fascicle length; $V_{M_1}$ and $V^e_{M_1}$ are initial values of computed and experimental muscle fascicle velocity; $L_{MTU_t}$ is MTU length at time instant $t$ computed from recorded joint angles and the geometric model of SO and MG. Additional inequality constraints kept the optimized model parameters within the physiological ranges reported in the literature (Spector et al. 1980; Sacks and Roy 1982; Siebert et al. 2008).

The optimization problem was solved using a parallel simulated annealing algorithm (Corana et al. 1987) and Open Multi-Processing Interface for SO and GA separately. The equations of muscle dynamics (Eqs. 10.3–10.8) were integrated numerically using a custom C++ program on a PC with Intel Xeon Quad Core processors. The equations were integrated over a complete walking cycle by a second order Runge-Kutta method with a constant 0.075-ms time step.

The found optimal tendon model parameters provided the minimal possible difference between computed and measured muscle fascicle length and velocity for both SO and MG muscles. An example of this comparison for MG during a cycle of walking is shown in Fig. 10.3d, along with the comparison between computed MG force and activation. The tuned tendon model parameters for SO and MG were used in the subsequent calculations for the muscles with relatively short and long tendons, respectively. The optimal parameters for SO were $L_{M_0} = 42.00$ mm, $L_{T_0} = 56.03$ mm, $k_{Tnl} = 4.70$, $L_{Tnl} = 0.024$, $L_T^{Max} = 0.060$; for MG they were $L_{M_0} = 27.00$ mm, $L_{T_0} = 75.30$ mm, $k_{Tnl} = 8.00$, $L_{Tnl} = 0.01$, $L_T^{Max} = 0.120$.

To validate the muscle model and its parameters, we compared simulated force time-histories of fully activated SO during imposed isovelocity stretch and shortening computed using Eqs. (10.3–10.8) with experimental recordings from a cat SO muscle obtained in (Sandercock and Heckman 1997). In those experiments, cat SO was fully activated by 60-Hz tetanic stimulation trains and when the muscle reached its maximum isometric force, MTU length isovelocity ramps were imposed by a muscle puller; shortening and lengthening speeds ranged between −0.2 m/s and 0.08 m/s (Fig. 10.4). For the simulations, the optimum fascicle length and maximum

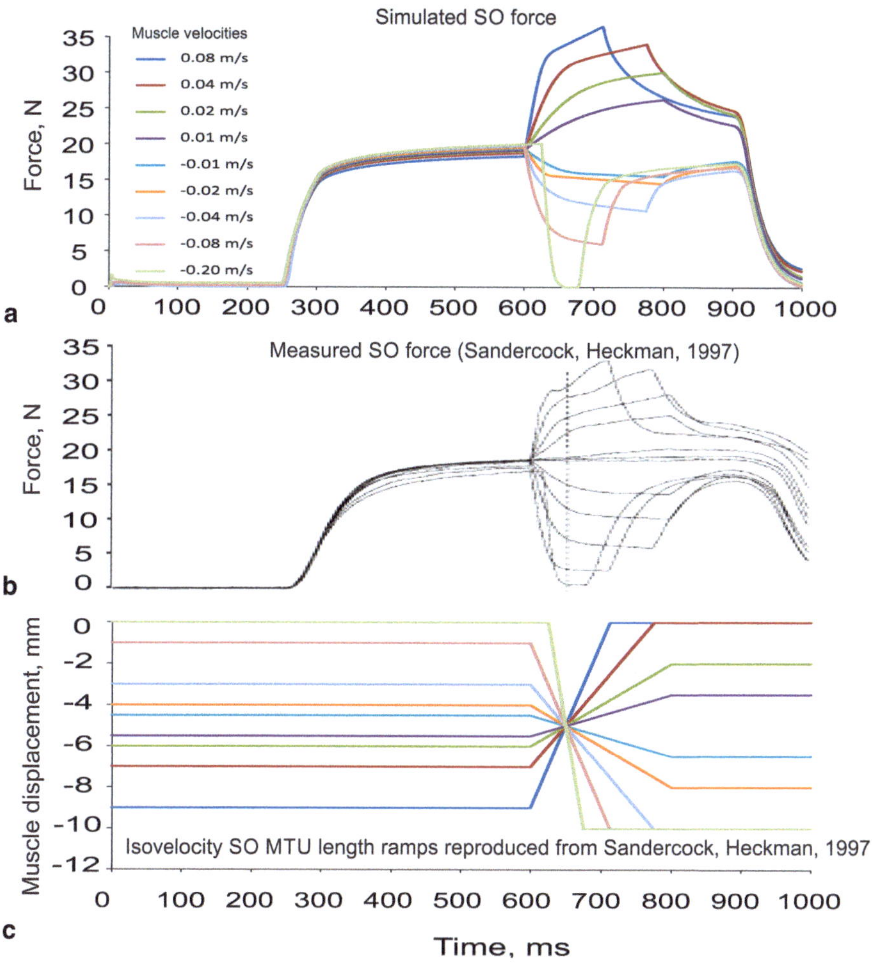


**Fig. 10.4** Simulated **a** and experimental **b** force traces of fully activated cat soleus (SO) in response to isovelocity changes in muscle-tendon unit length **c**. Experimental force traces were adopted with permission from Fig. 2 of (Sandercock and Heckman 1997), American Physiological Society. Parameters of SO tendon and muscle contractile element used in these simulations were: $F_M^{\max} = 20.8$ N, $L_{M_0} = 42.00$ mm, $L_{T_0} = 56.03$ mm, $k_{Tnl} = 4.70$, $L_{Tnl} = 0.024$, $L_T^{Max} = 0.060$ (see Eq. 10.7)

SO force at the optimum muscle length were selected to match values reported in (Sandercock and Heckman 1997). Simulated SO forces were in good agreement with the measured ones in terms of peak values and patterns.

*Stage 2: Tuning Muscle Model Parameters* In stage 2 of tuning model parameters, the found optimal tendon parameters were fixed and muscle specific parameters $F_M^{Max}$, $\tau_{act}$ and $\tau_{deact}$ (see Eqs. 10.4–10.6), and distances of muscle origin and attachments from the joint centers $a_1$ and $a_2$ were tuned for each muscle. The values of these muscle parameters were found by minimizing the difference between computed and experimental resultant muscle moments at hindlimb joints and between computed normalized muscle force and computed normalized activation (cost function $Z_2$, Eq. 10.12) using as input to computations the recorded EMG activity of each muscle (Fig. 10.2b) and experimental angles at the ankle, knee and hip joints (Fig. 10.5a):

$$Z_2 = \frac{w_M}{3N}\sum_{j=1}^{3}\sum_{t=1}^{N}\left[\left(M_{jt} - M_{jt}^e\right)/ SD_j^M\right]^2 + \frac{w_F}{9N}\sum_{i=1}^{9}\sum_{t=1}^{N}\left[\left(F_{Mit} - A_{it}\right)/ SD_i^u\right]^2, \quad (10.12)$$

where $M_{jt}$ and $M_{jt}^e$ are computed and experimental joint moments at the $j$-th joint of the right hindlimb at the $t$-th time frame; $F_{Mit}$ is computed force of the $i$-th muscle; $A_{it}$ is muscle activation obtained from recorded EMG (see Eq. 10.6) of the $i$-th muscle; $SD_j^M$ and $SD_i^u$ are standard deviations of the $j$-th joint moment and the $i$-th muscle normalized EMG averaged over a walking cycle; $N$ is number of time frames; $w_M = 5$ and $w_F = 1$; subscripts $j$ and $i$ designate 3 joints and 9 muscles of the right hindlimb, respectively. Equations of muscle dynamics (10.3–10.8) and MTU length and velocity of each muscle during walking cycle (computed from recorded joint angles) were used as constraints for the optimization problem. In addition, the optimal solution was constrained by the physiological range of values reported in the literature for the parameters $F_M^{Max}$, $\tau_{act}$ and $\tau_{deact}$ (Sacks and Roy 1982; Zajac 1989; He et al. 1991; Gareis et al. 1992; Baratta et al. 1995; Scott and Loeb 1995; Brown et al. 1996a, b). Minimization of cost function $Z_2$ gave a close match between the computed and experimental resultant joint moment (Fig. 10.5c, blue dotted lines).

*Stage 3: Tuning Parameters of Viscoelastic Interactions with the External Environment* In stage 3, the found optimal parameters of the tendon and muscle models were fixed (except constants $\tau_{act}$ and $\tau_{deact}$) and parameters of viscoelastic interactions of the hindlimb model with the external environment (coefficients of stiffness and viscosity, Table 10.1), constant forward velocity of the trunk distal endpoint (Fig. 10.1a), and activation and deactivation constants $\tau_{act}$ and $\tau_{deact}$ were tuned by minimizing the differences between simulated and recorded joint angles, ground reaction forces and joint moments. The constants $\tau_{act}$ and $\tau_{deact}$ obtained in stage 2 optimization were re-optimized for each muscle to ensure the best possible match between simulated and experimental walking mechanics.

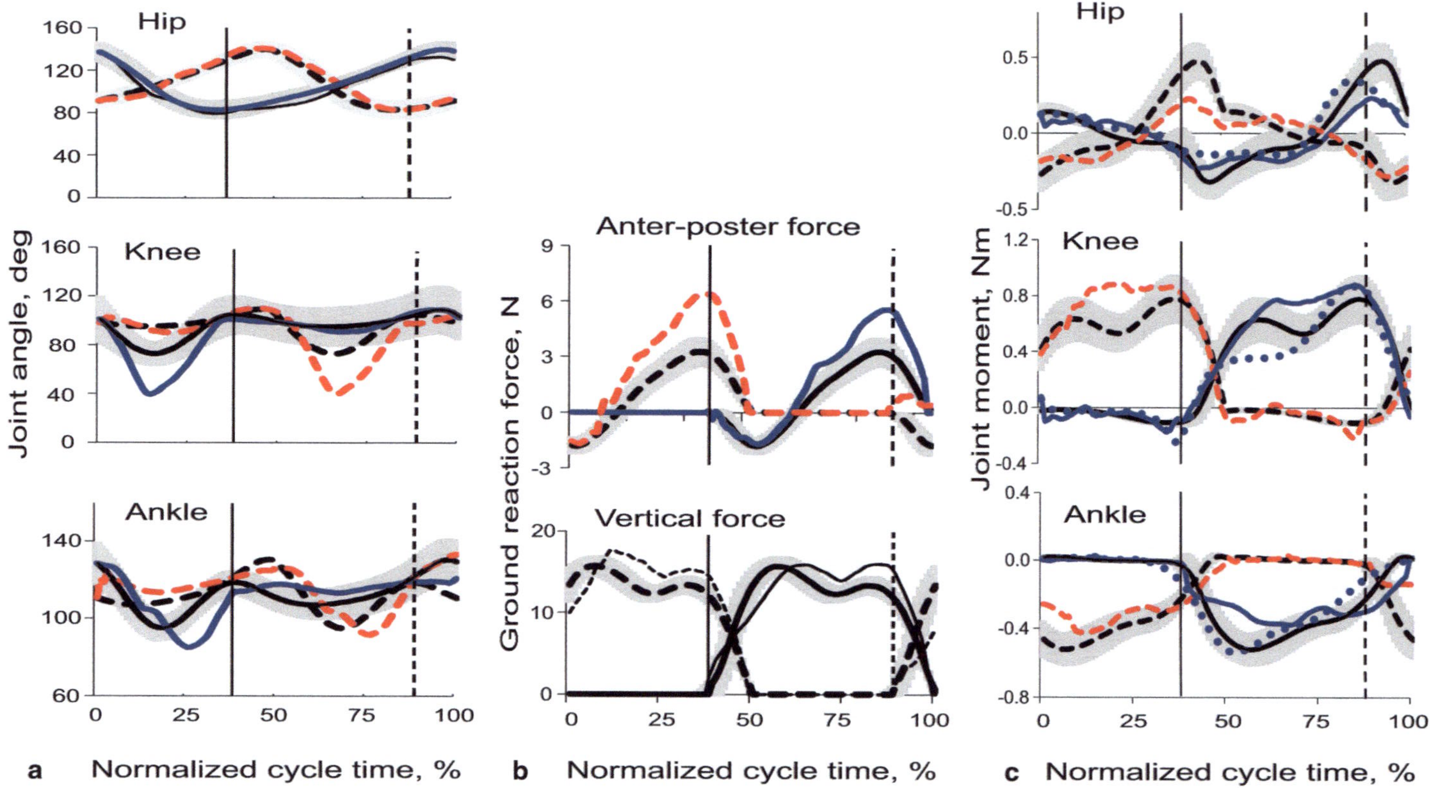


**Fig. 10.5** Recorded mean (*black lines*) ± SD (*gray shadow*) and computed (*blue* and *red lines*) joint angles **a**, ground reaction forces **b** and joint moments **c** for two hindlimbs during a cycle of walking. Simulations were based on the tendon and muscle properties presented in Fig. 10.3. The *solid lines* correspond to the left hindlimb and the *dashed lines* correspond to the right hindlimb. The *dotted blue lines* in **c** indicate joint moments computed in the first stage optimization using as input experimental joint angles and EMG activity (see Eq. 10.10). The *solid* and *dashed vertical lines* indicate paw contact with the ground by the left and right hindlimbs, respectively. Positive moments are flexor for the ankle and hip and extensor for the knee. The experimental mean and SD values for the mechanical variables were obtained from 5 cats (total number of cycles 61; cat 1 = 13, cat 2 = 10, cat 3 = 19, cat 4 = 11, cat 5 = 8)

The following cost function $Z_3$ was used for this optimization:

$$Z_3 = \frac{w_M}{6N}\sum_{j=1}^{6}\sum_{t=1}^{N}\left[\left(M_{jt} - M_{jt}^e\right)/SD_j^M\right]^2 + \frac{w_F}{18N}\sum_{i=1}^{18}\sum_{t=1}^{N}\left[\left(F_{Mit} - A_{it}\right)/SD_j^u\right]^2 +$$

$$\frac{w_a}{6N}\sum_{j=1}^{6}\sum_{t=1}^{N}\left[\left(\theta_{jt} - \theta_{jt}^e\right)/SD_j^a\right]^2 + \frac{w_{FG}}{4N}\sum_{c=1}^{4}\sum_{t=1}^{N}\left[\left(F_{Gct} - F_{Gct}^e\right)/SD_c^F\right]^2, \quad (10.13)$$

where $M_{jt}$, $M_{jt}^e$, $F_{it}$, $A_{it}$, $SD_j^M$ and $SD_j^u$ are the same as in Eq. (10.12) except here these variables were obtained for the two hindlimbs; subscripts $j=1,2,\ldots,6$ in $M_{jt}$ and $M_{jt}^e$ and subscripts $i=1,2,\ldots,18$ in $F_{it}$ and $A_{it}$ designate joints and muscles, respectively, in two hindlimbs; $\theta_{jt}$ and $\theta_{jt}^e$ are computed and experimental joint angles at the $j$-th joint of two hindlimbs at the $t$-th time frame, subscripts $j=1,2\ldots,6$ designate hip, knee and ankle joints of two hindlimbs; $F_{Gct}$ and $F_{Gct}^e$ are computed and experimental ground reaction forces, subscripts $c=1,2,\ldots,4$ designate anterior-posterior and vertical components of ground reaction forces applied to two hindlimbs; $SD_j^\theta$ and $SD_c^F$ are the experimental standard deviations of $j$-th joint angle and $c$-th ground reaction force component; $w_M=1$; $w_F=1$; $w_a=10$; and $w_{FG}=10$.

EMG linear envelopes obtained for 9 muscles of one hindlimb were used as input to each muscle of the model assuming symmetry between the two hindlimbs. The initial conditions for integration of equations of hindlimb and muscle dynamics—the generalized segment positions and velocities as well as MTU lengths and velocities at swing onset of the left hindlimb—were determined experimentally as described above. The equations of hindlimb and muscle dynamics also served as constraints for the optimization problem.

The found optimal solution allowed a relatively close match between computed and experimental hip joint angle, two components of the ground reaction forces and resultant muscle moments at the joints (Fig. 10.5). However, there were rather large discrepancies in the computed knee joint angles during the swing phase (up to 40°) and in the ankle joint angles during stance, Fig. 10.5a). Since our goal was to evaluate length and velocity dependent afferent feedback from muscles based on muscle fascicle length and velocity, computed in turn from joint kinematics, such large errors were considered unacceptable.

Although we were not able to determine the exact reasons for the inability of our model to fit the experimental data more closely, the examination of previous publications that demonstrated better fits (e.g. (Neptune et al. 2009)) suggested that possible inaccuracies in our computations could result from Eqs. (10.7) and (10.8) describing the tendon force-length and contractile element force-velocity relations, as well as the force-length relation for the contractile element consisting of a set of straight lines (see above). The description of the tendon and muscle properties in our model contained discontinuities in the derivatives of force over length and velocity that appeared to be the major distinct feature of our model. These discontinuities did not cause visible problems during simulations of force production during isovelocity MTU ramps of the fully activated SO (Fig. 10.4), but could potentially cause the observed large deviations in knee and ankle angles when MTU lengths changed with variable velocity at alternating muscle activation.

We therefore substituted the force-length relations for the tendon and parallel elastic element, as well as the force-length and force-velocity relations for the muscle contractile element with more conventional ones that are typically used in musculoskeletal modeling and are smooth (e.g., (Markin et al. 2010)). These equations are presented in Table 10.3. Parameters of these equations $F_M^{\max}$, $L_{T0}$, $a_1$ and $a_2$ were again optimized for each muscle, whereas parameters $k_{T2}$ and $\rho$ were optimized for two groups of muscles and considered the same within each group—muscles with a relatively short ($L_{T0}/L_{M0} \le 1$) and long ($L_{T0}/L_{M0} > 1$) relative tendon length (as identified in the literature, (Sacks and Roy 1982), Tables 10.2 and 10.3). These parameters were tuned by minimizing the cost function $Z_2$ (Eq. 10.12), i.e. minimizing the difference between simulated and experimental resultant joint moments and between simulated normalized muscle forces and activation using the experimentally obtained mean muscle activity (Fig. 10.2b) and joint angles (Fig. 10.5a) as input. The tuned muscle parameters (Table 10.3) produced a close match with the experimental joint moments during the walking cycle (Fig. 10.6c, blue dotted lines).

The parameters of viscoelastic interactions with the environment, constant forward velocity of the distal trunk endpoint and constants $\tau_{act}$ and $\tau_{deact}$ were tuned by minimizing cost function $Z_3$ (Eq. 10.13) while all other muscle parameters were fixed. The results of this optimization demonstrated a much better fit between simulated and experimental joint angles, ground reaction forces and joint moments (Fig. 10.6)—in most cases the simulated variables were within one standard deviation of the experimental ones.

#### 10.2.2.4 Computation of Motion-Dependent Afferent Feedback

Computed muscle fiber length and velocity, and tendon force for each muscle during walking were used to estimate the firing rates of the corresponding muscle afferents. Muscle length-dependent afferent signals, the firing rates of muscle Ia and II spindle afferents, were calculated as functions of muscle fascicle length, velocity and muscle activity using the modified regression equations developed by (Prochazka and Gorassini 1998b; Prochazka 1999). These authors developed the equations based on their measurements of the firing rates of group Ia and II spindle afferents and MTU lengths in walking cats and on earlier similar models (e.g. (Houk et al. 1981). To compute the firing rate of Ib Golgi tendon afferents, we assumed the Ib activity being proportional to muscle force during the mid-range of forces (see for example (Houk et al. 1981; Crago et al. 1982). The Ia, II and Ib activity was computed as follows:

$$R_{Ia} = 4.3 V_M^{\ 0.6} + 2\Delta L_M + 100 k_u^{\max} u + R_{Ia_0}, \tag{10.14}$$

$$R_{II} = 13.5 \Delta L_M + 20u + R_{II_0}, \tag{10.15}$$

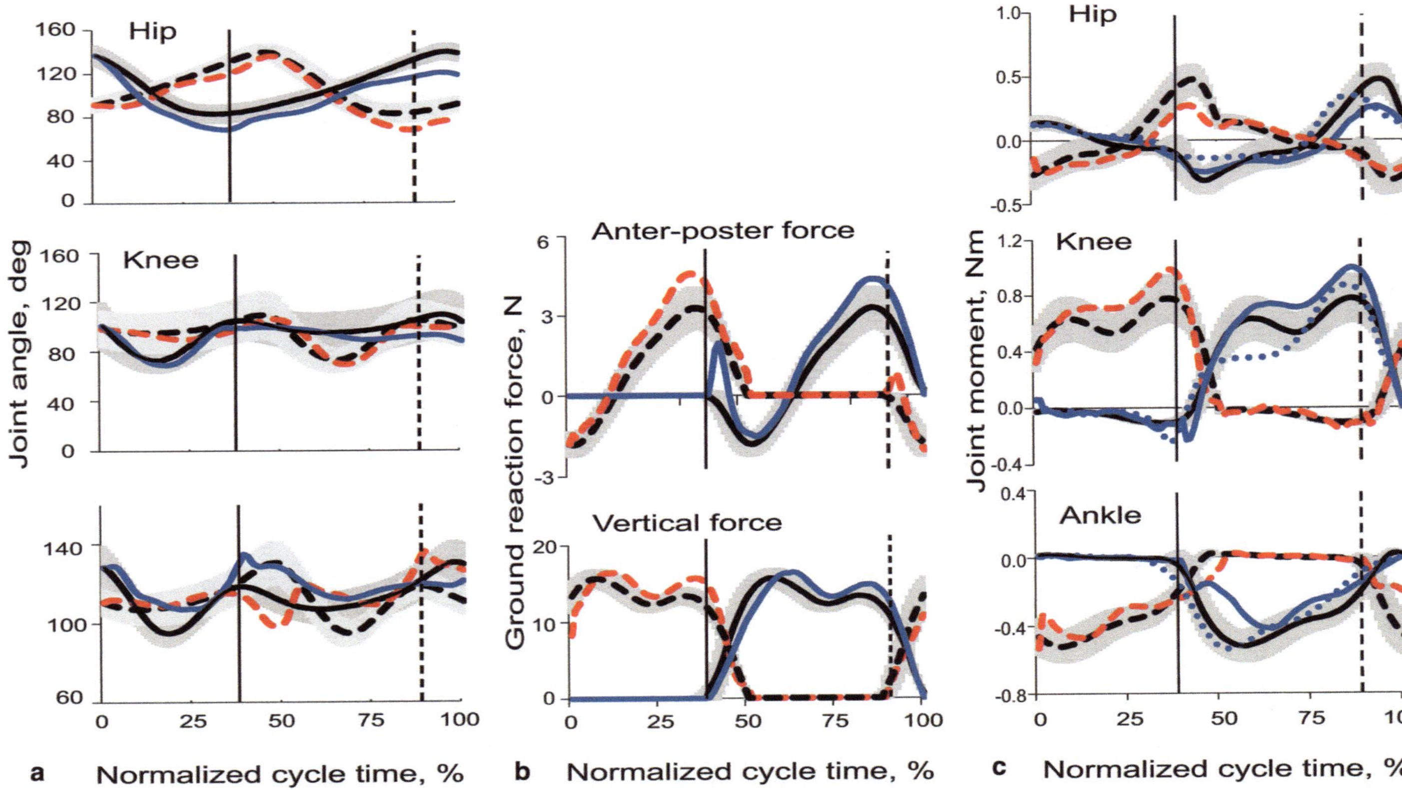


**Fig. 10.6** Recorded mean (*black lines*) ± SD (*gray shadow*) and computed (*blue* and *red lines*) joint angles **a**, ground reaction forces **b** and joint moments **c** for two hindlimbs during a cycle of walking. Simulations were based on the tendon and muscle properties presented in Tables 10.2 and 10.3. For other information see the legend for Fig. 10.5

$$R_{Ib} = k_{Ib}F_M \,/\, F_M^{Max}, \tag{10.16}$$

where $R_{Ia}$, $R_{II}$ and $R_{Ib}$ are the firing rates of spindle primary and secondary afferents and Golgi tendon organ afferents (imp/s), respectively; $V_M$ is muscle (fascicle) velocity (mm/s); $\Delta L_M$ is muscle length change from the mean value in a cycle (mm); $F_M \,/\, F_M^{Max}$ is normalized muscle force; $u$, normalized rectified and low-pass filtered EMG activity (Fig. 10.2b); $k_{Ib}$ is the firing rate of Ib afferents per unit of normalized muscle force ($k_{Ib}$=333 imp/s, (Prochazka 1999)); $R_{Ia_0}$ and $R_{II_0}$ are the mean firing rates of Ia and II afferents in a walking cycle (see Table 10.2).

In addition, the firing rate of cutaneous afferents from the plantar surface of the paw was calculated as a function of the vertical ground reaction force and the positive rate of its change during stance:

$$R_c = k_{c1}\left(F_{Gy} + k_{c2}\dot{F}_{Gy}\right), \tag{10.17}$$

where $R_c$ is the firing rate of cutaneous afferents from paw pad (imp/s), $k_{cl}=1$, and $k_{c2} = \begin{cases} 160, & \text{if } \dot{F}_y > 0 \\ 0, & \text{if } \dot{F}_y \le 0 \end{cases}$. The equation was derived based on recordings of cutaneous afferent activity and simulated ground reaction forces (see above).

## 10.3 Comparison of Simulated and Experimental Walking Mechanics and Afferent Activity

### *10.3.1 Input to the Simulation Model—EMG Patterns*

Patterns of EMG activity of major hindlimb muscles obtained in this study for individual animals (Fig. 10.2a) and EMG envelopes averaged across 5 cats (Fig. 10.2b) were consistent with previous reports (e.g., (Smith et al. 1998; Krouchev et al. 2006; Markin et al. 2012)). Specifically, major hindlimb extensors (SO, GA, VA, BFA) were active primarily during stance, whereas flexors (TA, SAM, IP) were active mostly during swing. Two-joint thigh muscles with flexion and extension function at two joints (RF, BFP) were active close to or at the swing-stance and stance-swing transitions.

Averaged EMG pattern of each muscle normalized to a cycle time (Fig. 10.2b) was locally fitted by a cubic spline function and EMG values were calculated at each 0.075-ms time step. These re-sampled EMG patterns (with the initial positions and velocities of the generalized coordinates determined experimentally) were used as input to the model. Muscle activation was computed first (Eq. 10.6), followed by muscle and tendon forces (Eqs. 10.4–10.5), and tendon acceleration of each muscle (Eq. 10.3); finally muscle and limb dynamics were computed by integrating Eqs. (10.3) and (10.2).

### *10.3.2 Walking Kinematics and Kinetics*

The mean patterns of joint angles, ground reaction forces and joint moments during level overground walking in the cat obtained experimentally here (Figs. 10.5 and 10.6, thin black lines) were consistent with previously published results (Manter 1938; Lavoie et al. 1995; Prilutsky et al. 2005; Gregor et al. 2006).

Identified model parameters (see above) and parameters taken from the literature or measured in this study (Tables 10.1–10.3) allowed for a close match between simulated walking mechanics and the corresponding mean mechanical variables obtained experimentally—the computed variables were typically within on standard deviation from the experimental ones (Fig. 10.6). It should be pointed out that in the obtained forward dynamics solution the passive joint moments at the hip, knee and ankle (vector **S** in Eq. 10.2) were not engaged because computed joint angles were within normal locomotor ranges (see Sect. 10.2.2.1).

The average speed of simulated walking was 0.505 m/s, whereas the average experimental speed of 5 animals was 0.646 m/s. Among three joint angles, the computed hip and knee angles corresponded to the mean patterns of the 5 cats with slight deviation from them at the transition from swing to stance or in early stance, but were within or close to one SD of the mean experimental values (Fig. 10.6a). The computed ankle angle was typically within or close to one SD of the mean recorded angle during swing and most of stance, but deviated substantially from the mean pattern in early and terminal stance (Fig. 10.6a). This difference could have resulted from modeling the foot as a rigid segment, whereas it consists of the tarsals and digits (paw). Overall, however, the computed joint angles were essentially within one SD from the mean experimental values and thus could be considered satisfactory.

The computed joint moments generally reproduced the mean experimental patterns (Fig. 10.6c). For example, during stance, the computed ankle and knee moments were extensor and reproduced well general experimental patterns, although the computed ankle moments deviated from the mean experimental values in early stance by more than one SD; the computed knee moments in mid stance were slightly higher than the mean + SD of experimental moments. The hip moments were generally within one SD from the mean experimental moments, being extensor in first half of stance and flexor in the second half, although were slightly lower than the mean-SD in terminal stance. Generally similar results were obtained during the stage 2 optimization performed to identify muscle model parameters using as input experimentally recorded joint angles and muscle activity (see Eq. 10.12 and Fig. 10.6c, blue dotted line).

During optimization (Eqs. 10.12 and 10.13), it was assumed that muscle active state $A$ provides the greatest contribution to muscle force $F_M$ (Eq. 10.5) and thus the force-length $F_{CE}(L_M)$ and force-velocity $F_{CE}(V_M)$ properties of the contractile element do not substantially affect muscle force during walking. This assumption is supported by a very high correlation ($r^2>0.9$) between the muscle force and the low-pass filtered EMG, shifted in time to account for the excitation-activation coupling, synchronously recorded during cat walking from the same muscle—soleus (Norman et al. 1988; Liu et al. 1999) and plantaris (Herzog et al. 1998). Another evidence for a relatively small contribution of the CE force-length-velocity properties to the muscle force during gait was obtained by (Davy and Audu 1987; Anderson and Pandy 2001) who compared computed muscle forces during human walking with and without inclusion of the contractile muscle properties in their models—both calculations gave similar results.

The anterior-posterior and vertical ground reaction forces computed using the model likewise closely matched the mean experimental patterns except short phases at foot contact for the horizontal forces; Fig. 10.6b).

### *10.3.3 Computed Activity of Muscle and Cutaneous Afferents During Walking*

The computed firing rates of selected afferents could be compared with the corresponding rates recorded *in vivo* in walking cats (Loeb and Duysens 1979; Loeb 1981; Loeb et al. 1985; Prochazka and Gorassini 1998a, b; Weber et al. 2007). This comparison can be considered a model validation test because the model parameters were not tuned to match experimental patterns of afferent activity. It should be noted that the regression equations for computing activity of spindle Ia and II afferents (Eqs. 10.14 and 10.15) were based on measurements of MTU length as an independent variable (Prochazka and Gorassini 1998a, b). Therefore, derivations of the firing rates of spindle afferents from MTU lengths for muscles with a very large ratio $L_{T0}/L_0$, as in GA (Table 10.3), may be less accurate than those derived for muscles with a small $L_{T0}/L_0$ ratio, as in BFP or SAM (Table 10.3) because a substantial part of MTU length changes in muscles like GA can be taken up by the long tendon (Hoffer et al. 1989; Maas et al. 2009) rather than by muscle fascicles and thus the spindles. The maximum correlation coefficients computed between measured and predicted firing rates of group Ia and II afferents from cat triceps surae ($r^2=0.62-0.73$) were in fact lower than those for hamstrings ($r^2=0.942-0.80$) (Prochazka and Gorassini 1998a, b).

The comparison of computed activity of muscle length-dependent type Ia and II afferents with the corresponding measured activity for BFP, VA, RF and SAM afferents showed reasonable qualitative agreement (Fig. 10.7a, b). Specifically, the computed and recorded type Ia and II BFP afferents increase their activity from mid swing reaching a peak near the paw contact; the activity subsided during stance with a slight rising of Ia activity in terminal stance. Type Ia VA afferent activity had two peaks—in late stance–early swing and in mid stance—and low or

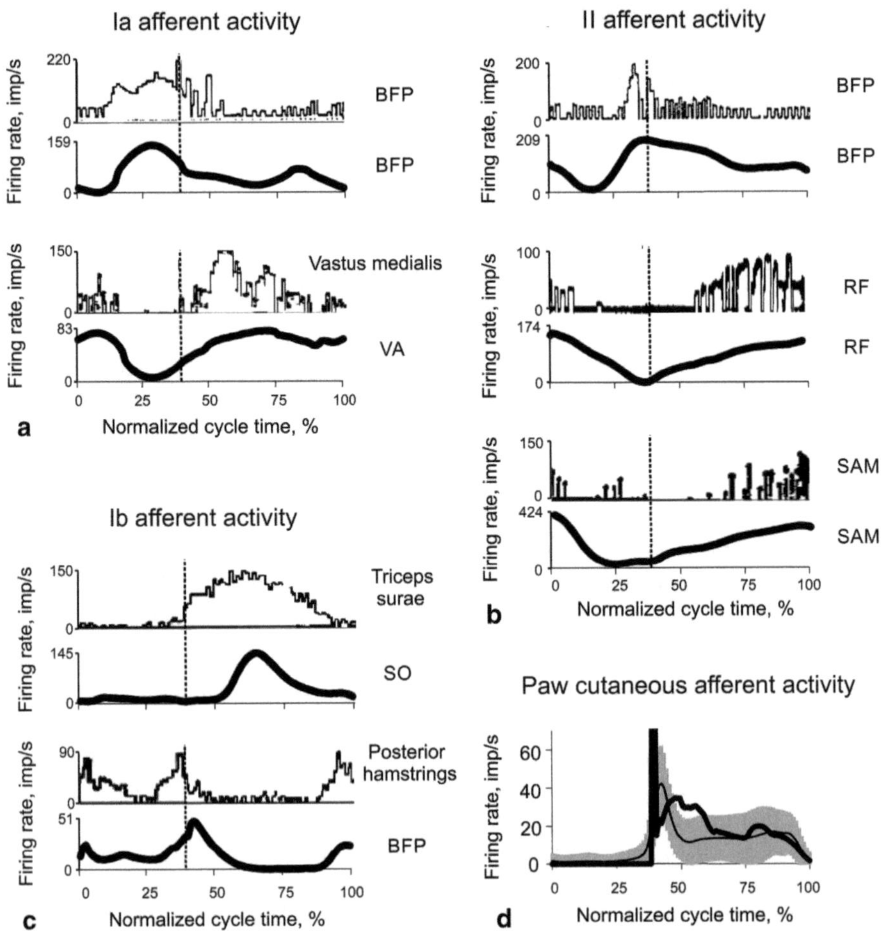


**Fig. 10.7** Comparison of computed (*thick lines*) and recorded *in vivo* (*thin lines*) firing rates of group Ia **a**, II **b** and Ib **c** afferents from selected muscles and cutaneous paw pad afferents **d** during the walking cycle (see text for explanations). The *vertical dotted lines* separate the swing and stance phases. Muscle abbreviations are the same as in Fig. 10.1c. *In vivo* activities of muscle afferents are reproduced with permission from: Ia biceps femoris posterior (BFP) and II BFP—Fig. 2 from (Loeb and Duysens 1979), American Physiological Society; Ia vastus medialis—Fig. 3 from (Loeb et al. 1985), American Physiological Society; spindle afferent from rectus femoris (RF)—Fig. 5 from (Loeb 1981), © Canadian Science Publishing, assumed to be spindle secondary afferent; sartorius medial (SAM)—Fig. 6 from (Loeb et al. 1985), American Physiological Society; Ib triceps surae and Ib posterior hamstrings—Fig. 6 (Prochazka and Gorassini 1998a), John Wiley and Sons. The mean (*thin line*) ± SD (*gray shadow*) activity of paw afferents in **d** is obtained from recorded 4 afferents of one animal collected during 11 walking cycles

zero activity during the second half of swing. The absolute peak values of computed firing rates of these afferents deferred substantially—up to 43 % (Fig. 10.7a, Ia VA). The computed patterns of group II afferents from RF and SAM were also in qualitative agreement with the measured activity, both increasing from early stance to the peak activity at terminal stance and early swing (Fig. 10.7b). The peak values of the firing rates, however, differed substantially between calculations and measurements.

The computed firing rate of Ib afferents from SO and BFP also demonstrated patterns qualitatively similar to those of recorded afferents (Fig. 10.7c). For instance, Ib SO afferents were mostly active during stance with very little activity during swing, whereas Ib BFP activity occurred at the stance-swing and swing-stance transitions. The difference in absolute peak values between the computed and recorded Ib activity was small for SO and substantial for BFP Ib afferents.

The instantaneous firing rate of cutaneous paw pad afferents computed as a function of the vertical ground reaction force and its time derivative (Eq. 10.17) demonstrated a sharp peak at paw contact and the moderate magnitude during the rest of stance, similar to the recorded activity of mechanoreceptors from paw pad (Fig. 10.7d). Constants $k_{c1}$ and $k_{c2}$ in Eq. 10.17 were selected to match the mean recorded firing rate values.

Given the reasonable qualitative agreement between patterns of the computed and recorded activity of type Ia, Ib and II afferents from selected muscles, as well as of paw pad afferents, the developed model can be integrated with a CPG model (Rybak et al. 2006a, b) and used for a closed-loop simulations and computational studies of spinal locomotion (see chapter by (Markin et al. 2015)). These simulations have the potential to provide additional more detailed information compared to the previous similar simulations (Ivashko et al. 2003; Markin et al. 2010) because the CPG model and the musculoskeletal model reasonably reproduce the activity patterns in fictive locomotion (Rybak et al. 2006a, b) and mechanics (Fig. 10.6) and afferent activity (Fig. 10.7) during real cat walking.

In addition, the developed musculoskeletal model provides patterns of type Ia, Ib and II muscle and paw pad cutaneous afferent activities during cat locomotion (Figs. 10.7 and 10.8). This information has not been available in such detail and can be used for investigating the role of different types of afferents in modulating muscle activity and controlling phase transitions. The results presented in Figs. 10.7 and 10.8 suggest that type II afferents from hip flexors IP, RF and SAM may trigger the extensor-flexor phase transition (or stance-swing transition) as their maximum activity is reached at terminal stance, while load sensitive afferents (type Ib from extensor muscles—SO, GA, VA and BFA and cutaneous paw receptors) approach their minimum activity at that time. This conclusion is consistent with earlier similar suggestions (Ekeberg and Pearson 2005; Pearson 2008). Type Ia and II afferents from BFP and TA as well as type Ia afferents from IP, RF and SAM could participate in controlling the flexor-extensor phase transition (or swing-stance transition), since these afferents reach their maximum activity at terminal swing (Fig. 10.8).

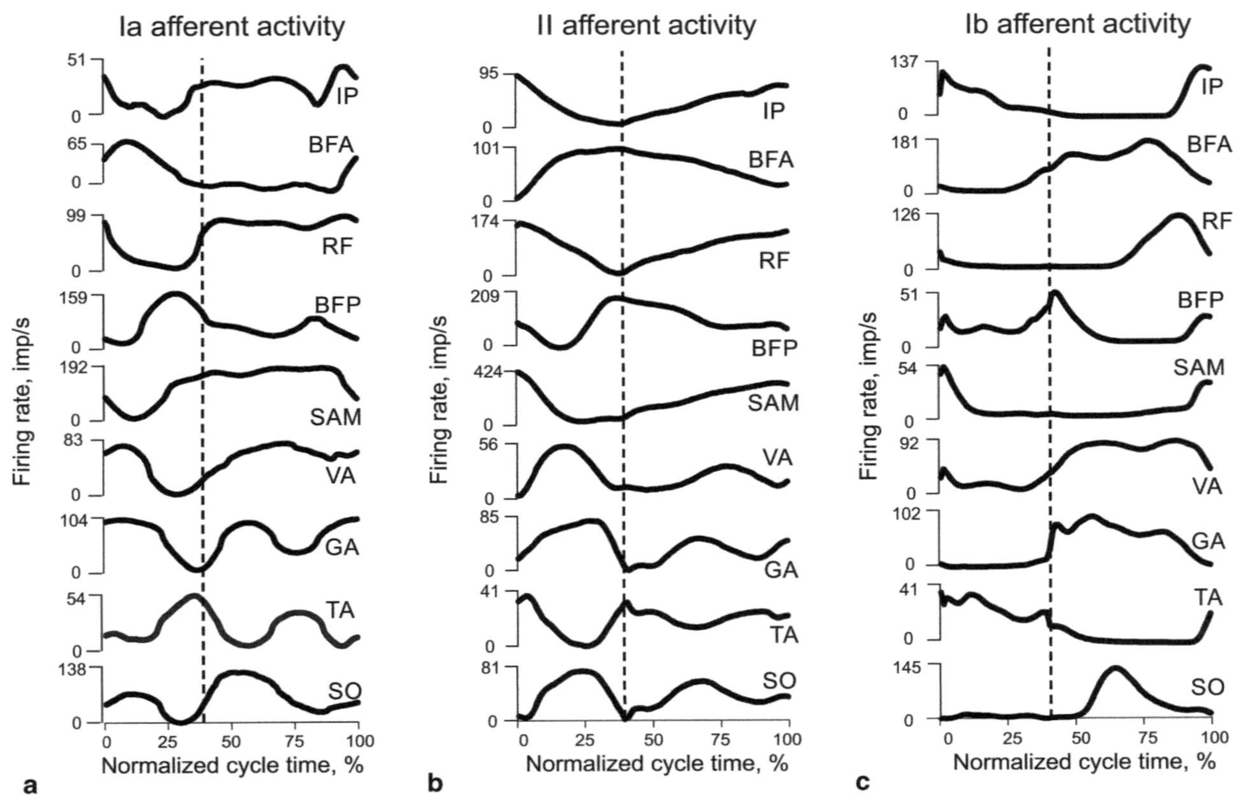


**Fig. 10.8** Computed type Ia **a**, II **b**, and Ib **c** afferent activities from 9 hindlimb muscles in the walking cycle. The *vertical dashed lines* separate the swing and stance phases. For muscle abbreviations see legends in Fig. 10.1c

## 10.4 Limitations of the Model

Although the forward dynamics model developed in this study reproduces reasonably well cat walking mechanics (Fig. 10.6) and activity of selected afferents (Fig. 10.7) using recorded EMG patterns of hindlimb muscles as input, several limitations of the model should be pointed out. First, it was assumed that the instantaneous muscle fiber velocity, length and force (and EMG magnitude in case of Ia afferents) uniquely determine the afferent activity. This assumption is difficult to verify. Simultaneous recordings from primary and secondary afferents and MTU length and force information during cat locomotion have been conducted in limited studies and in some instances revealed relatively low correlations between ensemble afferent activity and the corresponding mechanical variables (Loeb et al. 1985; Prochazka and Gorassini 1998a) especially for muscles with short fascicles and long tendons (i.e. gastrocnemius) in which fascicle (and thus spindle) length changes may differ from the recorded MTU length changes (Hoffer et al. 1989; Maas et al. 2009). The potential impact of compliant MTUs on the computed length-dependent afferent activity was reduced in the present model by accounting for tendon deformation.

Secondly, the employed models for computing muscle afferent signals (Eqs. 10.14–10.17) have been maximally simplified to enable online computations of afferent signals in the combined model of the CPG and musculoskeletal system (e.g. (Ivashko et al. 2003; Markin et al. 2010, 2015)). More accurate transfer functions relating muscle mechanical state with afferent activity (Houk et al. 1981; Prochazka 1999) or highly detailed models of the spindle and Golgi tendon receptors (e.g. (Mileusnic et al. 2006; Mileusnic and Loeb 2006)) would be difficult to implement in the online computations of afferent input.

Furthermore, the joint mechanoreceptor activities were not included in the current model, although they may affect the functional organization of spinal circuits (Rudomin et al. 2007). However, these afferents are mostly active at extreme joint positions (Ferrell 1980) which were not reached in the obtained forward dynamics simulation (Fig. 10.6a). This type of afferent information could be incorporated into the neuromechanical model of spinal control of locomotion in the future.

Finally, the model seems able to reproduce only pattern of changes in afferent activity rather than the absolute firing rate values. The impact of this limitation may be reduced by, for example, finding appropriate gains of the feedback signals in a combined closed-loop neuromechanical model that would permit a proper transformation of fictive locomotor activity patterns generated by the CPG model (Rybak et al. 2006b) to the EMG patterns of normal walking (Fig. 10.2); see (Markin et al. 2015).

**Aknowledgements** This work was supported by NIH grants HD-032571, NS-048844, and EB012855 and by the Center for Human Movement Studies at Georgia Institute of Technology. We would like to thank Dr. Guayhaur Shoe for his technical assistance.

## References

Akay T, Tourtellotte WG, Arber S, Jessell TM (2014) Degradation of mouse locomotor pattern in the absence of proprioceptive sensory feedback. Proc Natl Acad Sci U S A 111:16877–16882

Anderson FC, Pandy MG (2001) Static and dynamic optimization solutions for gait are practically equivalent. J Biomech 34:153–161

Aoi S (2015) Neuromusculoskeletal modeling for the adaptive control of posture during locomotion. In: Prilutsky BI, Edwards DH Jr (eds) Neuromechanical modeling of posture and locomotion. Springer, New York (in press)

Aoi S, Kondo T, Hayashi N, Yanagihara D, Aoki S, Yamaura H et al (2013) Contributions of phase resetting and interlimb coordination to the adaptive control of hindlimb obstacle avoidance during locomotion in rats: a simulation study. Biol Cybern 107:201–216

Ariano MA, Armstrong RB, Edgerton VR (1973) Hindlimb muscle fiber populations of five mammals. J Histochem Cytochem 21:51–55

Baratta RV, Solomonow M, Best R, Zembo M, D'Ambrosia R (1995) Architecture-based force-velocity models of load-moving skeletal muscles. Clin Biomech (Bristol Avon) 10:149–155

Bondy B, Klishko AN, Prilutsky BI, Cymbalyuk G (2015) Control of cat walking and paw-shake by a multifunctional central pattern generator. In: Prilutsky BI, Edwards DH (eds) Neuromechanical modeling of posture and locomotion. Springer, New York (in press)

Brown TG (1914) On the nature of the fundamental activity of the nervous centres; together with an analysis of the conditioning of rhythmic activity in progression, and a theory of the evolution of function in the nervous system. J Physiol 48:18–46

Brown IE, Liinamaa TL, Loeb GE (1996a) Relationships between range of motion, lo, and passive force in five strap-like muscles of the feline hind limb. J Morphol 230:69–77

Brown IE, Scott SH, Loeb GE (1996b) Mechanics of feline soleus: II. Design and validation of a mathematical model. J Muscle Res Cell Motil 17:221–233

Bunderson N, Bingham J (2015) Better science through predictive modeling: numerical tools for understanding neuromechanical interactions. In: Prilutsky BI, Edwards DH Jr (eds) Neuromechanical modeling of posture and locomotion. Springer, New York (in press)

Bunderson NE, Bingham JT, Sohn MH, Ting LH, Burkholder TJ (2012) Neuromechanic: a computational platform for simulation and analysis of the neural control of movement. Int J Numer Method Biomed Eng 28:1015–1027

Conway BA, Hultborn H, Kiehn O (1987) Proprioceptive input resets central locomotor rhythm in the spinal cat. Exp Brain Res 68:643–656

Corana A, Marchesi M, Martini C, Ridella S (1987) Minimizing multimodal functions of continuous variables with the "simulated annealing" algorithm. ACM Trans Math Softw 13:263–280

Crago PE, Houk JC, Rymer WZ (1982) Sampling of total muscle force by tendon organs. J Neurophysiol 47:1069–1083

Davy DT, Audu ML (1987) A dynamic optimization technique for predicting muscle forces in the swing phase of gait. J Biomech 20:187–201

De Groote F, Van Campen A, Jonkers I, De Schutter J (2010) Sensitivity of dynamic simulations of gait and dynamometer experiments to hill muscle model parameters of knee flexors and extensors. J Biomech 43:1876–1883

Duysens J, Loeb GE (1980) Modulation of ipsi- and contralateral reflex responses in unrestrained walking cats. J Neurophysiol 44:1024–1037

Edman KA (1988) Double-hyperbolic force-velocity relation in frog muscle fibres. J Physiol 404:301–321

Ekeberg O, Pearson K (2005) Computer simulation of stepping in the hind legs of the cat: an examination of mechanisms regulating the stance-to-swing transition. J Neurophysiol 94:4256–4268

Ferrell WR (1980) The adequacy of stretch receptors in the cat knee joint for signalling joint angle throughout a full range of movement. J Physiol 299:85–99

Gareis H, Solomonow M, Baratta R, Best R, D'Ambrosia R (1992) The isometric length-force models of nine different skeletal muscles. J Biomech 25:903–916

Gordon AM, Huxley AF, Julian FJ (1966) The variation in isometric tension with sarcomere length in vertebrate muscle fibres. J Physiol 184:170–192
Gregor RJ, Smith DW, Prilutsky BI (2006) Mechanics of slope walking in the cat: quantification of muscle load, length change, and ankle extensor EMG patterns. J Neurophysiol 95:1397–1409
Grillner S (1981) Control of locomotion in bipeds, tetrapods, and fish. In: Brooks V (ed) Handbook of physiology. Section I. The nervous system, vol II. American Physiological Society, Bethesda, pp 1179–1236
Grillner S, Georgopoulos AP, Jordan LM (1999) Selection and initiation of motor behavior. In: Stein PSG, Grillner S, Selverston AI, Stuart DG (eds) Neurons, networks, and motor behavior. MIT, Boston, pp 3–19
Guertin P, Angel MJ, Perreault MC, McCrea DA (1995) Ankle extensor group I afferents excite extensors throughout the hindlimb during fictive locomotion in the cat. J Physiol 487(Pt 1):197–209
He J, Levine WS, Loeb GE (1991) Feedback gains for correcting small perturbations to standing posture. IEEE Trans Autom Control 36:322–332
Herzog W, Sokolosky J, Zhang YT, Guimaraes AC (1998) EMG-force relation in dynamically contracting cat plantaris muscle. J Electromyogr Kinesiol 8:147–155
Hill AV (1938) The heat of shortening and the dynamic constants of muscle. Proc Royal Soc (Lond) B 126:136–95
Hoffer JA, Caputi AA, Pose IE, Griffiths RI (1989) Roles of muscle activity and load on the relationship between muscle spindle length and whole muscle length in the freely walking cat. Prog Brain Res 80:75–85; discussion 57–60
Houk JC, Rymer WZ, Crago PE (1981) Dependence of dynamic response of spindle receptors on muscle length and velocity. J Neurophysiol 46:143–166
Hoy MG, Zernicke RF (1985) Modulation of limb dynamics in the swing phase of locomotion. J Biomech 18:49–60
Ivashko DG, Prilutsky BI, Markin SN, Chapin JK, Rybak IA (2003) Modeling the spinal cord neural circuitry controlling cat hindlimb movement during locomotion. Neurocomputing 52–54:621–629
Jordan LM (1998) Initiation of locomotion in mammals. Ann N Y Acad Sci 860:83–93
Kiehn O (2011) Development and functional organization of spinal locomotor circuits. Curr Opin Neurobiol 21:100–109
Krouchev N, Kalaska JF, Drew T (2006) Sequential activation of muscle synergies during locomotion in the intact cat as revealed by cluster analysis and direct decomposition. J Neurophysiol 96:1991–2010
Lavoie S, McFadyen B, Drew T (1995) A kinematic and kinetic analysis of locomotion during voluntary gait modification in the cat. Exp Brain Res 106:39–56
Liu MM, Herzog W, Savelberg HH (1999) Dynamic muscle force predictions from EMG: an artificial neural network approach. J Electromyogr Kinesiol 9:391–400
Loeb GE (1981) Somatosensory unit input to the spinal cord during normal walking. Can J Physiol Pharmacol 59:627–635
Loeb GE, Duysens J (1979) Activity patterns in individual hindlimb primary and secondary muscle spindle afferents during normal movements in unrestrained cats. J Neurophysiol 42:420–440
Loeb GE, Hoffer JA, Pratt CA (1985) Activity of spindle afferents from cat anterior thigh muscles. I. Identification and patterns during normal locomotion. J Neurophysiol 54:549–564
Maas H, Gregor RJ, Hodson-Tole EF, Farrell BJ, Prilutsky BI (2009) Distinct muscle fascicle length changes in feline medial gastrocnemius and soleus muscles during slope walking. J Appl Physiol 106:1169–1180
Manter JT (1938) The dynamics of quadrupedal walking. J Exp Biol 15:522–540
Markin SN, Klishko AN, Shevtsova NA, Lemay MA, Prilutsky BI, Rybak IA (2010) Afferent control of locomotor CPG: insights from a simple neuromechanical model. Ann N Y Acad Sci 1198:21–34

Markin SN, Lemay MA, Prilutsky BI, Rybak IA (2012) Motoneuronal and muscle synergies involved in cat hindlimb control during fictive and real locomotion: a comparison study. J Neurophysiol 107:2057–2071

Markin SN, Klishko AN, Shevtsova NA, Lemay MA, Prilutsky BI, Rybak IA (2015) A neuromechanical model of spinal control of locomotion. In: Prilutsky BI, Edwards DH Jr (eds) Neuromechanical modeling of posture and locomotion. Springer, New York (in press)

McCrea DA (2001) Spinal circuitry of sensorimotor control of locomotion. J Physiol 533:41–50

McCrea DA, Rybak IA (2008) Organization of mammalian locomotor rhythm and pattern generation. Brain Res Rev 57:134–146

Mileusnic MP, Loeb GE (2006) Mathematical models of proprioceptors. II. Structure and function of the Golgi tendon organ. J Neurophysiol 96:1789–1802

Mileusnic MP, Brown IE, Lan N, Loeb GE (2006) Mathematical models of proprioceptors. I. Control and transduction in the muscle spindle. J Neurophysiol 96:1772–1788

Neptune RR, McGowan CP, Kautz SA (2009) Forward dynamics simulations provide insight into muscle mechanical work during human locomotion. Exerc Sport Sci Rev 37:203–210

Norman R, Gregor RJ, Dowling J (1988) The prediction of cat tendon force from EMG in dynamic muscular contractions. In: Cotton CE, Lamontagne M, Robertson DGE et al (eds) Biennial Conference of Canadian Society for Biomechanics. SPODYN, London, Ontario, pp 120–121

Orlovsky GN, Deliagina TG, Grillner S (1999) Neuronal control of locomotion: from mollusc to man. Oxford University Press, New York

Pearson KG (2008) Role of sensory feedback in the control of stance duration in walking cats. Brain Res Rev 57:222–227

Pearson KG, Collins DF (1993) Reversal of the influence of group Ib afferents from plantaris on activity in medial gastrocnemius muscle during locomotor activity. J Neurophysiol 70:1009–1017

Prilutsky BI, Klishko AN (2007) Activity of muscle spindle and tendon organ afferents during cat locomotion computed using a musculoskeletal model. In: Society for Neuroscience Meeting. Program No. 189.14. 2007 Neuroscience Meeting Planner. Online, San Diego, CA

Prilutsky BI, Herzog W, Leonard TR, Allinger TL (1996) Role of the muscle belly and tendon of soleus, gastrocnemius, and plantaris in mechanical energy absorption and generation during cat locomotion. J Biomech 29:417–434

Prilutsky BI, Sirota MG, Gregor RJ, Beloozerova IN (2005) Quantification of motor cortex activity and full-body biomechanics during unconstrained locomotion. J Neurophysiol 94:2959–2969

Prilutsky BI, Maas H, Bulgakova M, Hodson-Tole EF, Gregor RJ (2011) Short-term motor compensations to denervation of feline soleus and lateral gastrocnemius result in preservation of ankle mechanical output during locomotion. Cells Tissues Organs 193:310–324

Prilutsky BI, Klishko AN, Weber DJ, Lemay MA (2013) Activity of muscle and paw-skin afferents during cat locomotion computed using a forward dynamics neuromechanical model. In: XXIV Congress of International Society of Biomechanics. 2013. Natal, Brazil

Prochazka A (1999) Quantifying proprioception. Prog Brain Res 123:133–142

Prochazka A, Gorassini M (1998a) Ensemble firing of muscle afferents recorded during normal locomotion in cats. J Physiol 507(Pt 1):293–304

Prochazka A, Gorassini M (1998b) Models of ensemble firing of muscle spindle afferents recorded during normal locomotion in cats. J Physiol 507(Pt 1):277–291

Rossignol S (2006) Plasticity of connections underlying locomotor recovery after central and/or peripheral lesions in the adult mammals. Philos Trans R Soc Lond B Biol Sci 361:1647–1671

Rudomin P, Hernandez E, Lomeli J (2007) Tonic and phasic differential GABAergic inhibition of synaptic actions of joint afferents in the cat. Exp Brain Res 176:98–118

Rybak IA, Shevtsova NA, Lafreniere-Roula M, McCrea DA (2006a) Modelling spinal circuitry involved in locomotor pattern generation: insights from deletions during fictive locomotion. J Physiol 577:617–639

Rybak IA, Stecina K, Shevtsova NA, McCrea DA (2006b) Modelling spinal circuitry involved in locomotor pattern generation: insights from the effects of afferent stimulation. J Physiol 577:641–658

Sacks RD, Roy RR (1982) Architecture of the hind limb muscles of cats: functional significance. J Morphol 173:185–195

Sandercock TG, Heckman CJ (1997) Force from cat soleus muscle during imposed locomotor-like movements: experimental data versus Hill-type model predictions. J Neurophysiol 77:1538–1552

Scott SH, Loeb GE (1995) Mechanical properties of aponeurosis and tendon of the cat soleus muscle during whole-muscle isometric contractions. J Morphol 224:73–86

Scovil CY, Ronsky JL (2006) Sensitivity of a Hill-based muscle model to perturbations in model parameters. J Biomech 39:2055–2063

Shik ML, Severin FV, Orlovsky GN (1969) Control of walking and running by means of electrical stimulation of the mesencephalon. Electroencephalogr Clin Neurophysiol 26:549

Shoham S, Fellows MR, Normann RA (2003) Robust, automatic spike sorting using mixtures of multivariate t-distributions. J Neurosci Methods 127:111–122

Siebert T, Rode C, Herzog W, Till O, Blickhan R (2008) Nonlinearities make a difference: comparison of two common Hill-type models with real muscle. Biol Cybern 98:133–143

Smith JL, Carlson-Kuhta P, Trank TV (1998) Forms of forward quadrupedal locomotion. III. A comparison of posture, hindlimb kinematics, and motor patterns for downslope and level walking. J Neurophysiol 79:1702–1716

Spector SA, Gardiner PF, Zernicke RF, Roy RR, Edgerton VR (1980) Muscle architecture and force-velocity characteristics of cat soleus and medial gastrocnemius: implications for motor control. J Neurophysiol 44:951–960

Takeoka A, Vollenweider I, Courtine G, Arber S (2014) Muscle spindle feedback directs locomotor recovery and circuit reorganization after spinal cord injury. Cell 159:1626–1639

van den Bogert AJ, Schamhardt HC, Crowe A (1989) Simulation of quadrupedal locomotion using a rigid body model. J Biomech 22:33–41

Weber DJ, Stein RB, Everaert DG, Prochazka A (2007) Limb-state feedback from ensembles of simultaneously recorded dorsal root ganglion neurons. J Neural Eng 4:S168–S180

Whelan PJ, Hiebert GW, Pearson KG (1995) Stimulation of the group I extensor afferents prolongs the stance phase in walking cats. Exp Brain Res 103:20–30

Yakovenko S, Gritsenko V, Prochazka A (2004) Contribution of stretch reflexes to locomotor control: a modeling study. Biol Cybern 90:146–155

Zajac FE (1989) Muscle and tendon: properties, models, scaling, and application to biomechanics and motor control. Crit Rev Biomed Eng 17:359–411

# Chapter 11
# Modeling and Optimality Analysis of Pectoral Fin Locomotion

**Xinmin Liu, Frank Fish, R. Scott Russo, Silvia S. Blemker and Tetsuya Iwasaki**

**Abstract** During animal locomotion movements of the body and appendages generate propulsive force through dynamic interactions with the surrounding environment. Fundamental questions in understanding locomotion mechanisms include whether animals optimize the oscillation pattern (or gait) to achieve "efficient" locomotion, and if so, what is the optimality criterion. We address these questions in the context of swimming batoids, which propel the body by flapping their large pectoral fins. The flapping pattern varies from one species to another; batoids with round-shaped fins tend to send traveling waves across the fins and are called undulators, while batoids with triangular-shaped fins tend to flap the fins up and down and are called oscillators. We develop mathematical models of *Dasyatis sabina* and *Rhinoptera bonasus* as representatives of undulators and oscillators, respectively, to analyze optimal gaits with respect to various cost functions, and compare with observed kinematic data. The analysis shows that both gaits result from minimization of a metabolic cost, and a body resonance is exploited to reduce cost associated with negative work.

T. Iwasaki (✉) · X. Liu
Mechanical and Aerospace Engineering, University of California, 420 Westwood Plaza, Los Angeles, CA 90095, USA
e-mail: tiwasaki@ucla.edu

X. Liu
e-mail: xinmin.liu@gmail.com

F. Fish
Department of Biology, Liquid Life Laboratory, West Chester University, 750 S. Church St, West Chester, PA 19383-2112, USA
e-mail: FFish@wcupa.edu

R. S. Russo
193 Paoli Pike, Malvern, PA 19355, USA
e-mail: rsr4q@virginia.edu

S. S. Blemker
Biomedical Engineering, University of Virginia, MR5 Building, Room 2231, PO Box 800759, Charlottesville, VA 22903, USA
e-mail: ssblemker@virginia.edu

 
B. I. Prilutsky, D. H. Edwards (eds.), *Neuromechanical Modeling of Posture and Locomotion,* Springer Series in Computational Neuroscience,
DOI 10.1007/978-1-4939-3267-2_11

**Keywords** Swimming · Flapping propulsion · Optimal gait · Energy efficiency · Dynamic model

## 11.1 Introduction

Mechanisms underlying animal locomotion, such as walking, flying, and swimming, involve dynamic interactions between periodic local motion of the body and surrounding environment to generate thrust, maintain speed, and travel through a global distance. The pattern of body oscillations (or gait) has a strong influence on efficiency of locomotion, and animals appear to optimize their gaits to exploit natural dynamics of the body and environment (Cavagna et al. 1977; Iwasaki and Zheng 2006). Understanding of gait selection principles may provide a stepping stone toward comprehension of neuronal control circuits that process environmental information from sensory feedback and determine muscle activation. Moreover, such understanding would have a direct impact on engineering of robotic locomotion systems. For instance, bio-inspired underwater vehicles, such as those inspired by batoid fishes (skates and rays), see e.g, Fish et al. 2011, are being developed as an alternative to conventional screw-propelled submarines in recent years. These bio-inspired vehicles hold the promise of enhanced efficiency, speed, maneuverability and stealth (Moored et al. 2011; Parson et al. 2011).

In this chapter, we study gait selection principles underlying batoid swimming. Batoid fishes have dorso-ventrally flattened bodies with large wing-like pectoral fins. The fins are boneless and flexible, having a cartilaginous skeleton that is held together by elastic ligaments. In steady swimming, batoids periodically move the pectoral fins to generate thrust by passing waves down the fin. The number of waves traveling along the wing-like fins differ among the species of batoids. The wave pattern generated by the movements of the expanded pectoral fin vary for species along a undulatory-oscillatory continuum (Rosenberger 2001). Along this continuum, the locomotor pattern or gait is described as undulatory if the number of waves within the fin is greater than 1, oscillatory (flapping) if the number of waves is less than 0.5, and as semi-oscillatory otherwise; see Schaefer and Summers 2005 and references therein.

Fundamental questions addressed in this chapter are the following: Do batoids choose the fin oscillation patterns (or gaits) through an optimization of certain locomotion performance, and if so, what is the optimality criterion? To answer the questions, Atlantic stingray (*Dasyatis sabina*), and cownose ray (*Rhinoptera bonasus*) are selected for our analysis. They are at opposite ends of the swimming mode continuum and thus demonstrate extreme modes of swimming that can be compared to understand the underlying physics that is regulated by the biology of the system. The Atlantic stingray is an undulator with round-shaped fins, while the cownose ray is an oscillator with triangular-shaped fins (Rosenberger 2001). For each species, we will develop a dynamical model (a set of ordinary differential equations) to describe how body movements (velocity, orientation, and shape in 3-dimensional space) result from muscle contractions under the influence of hydrodynamic forces.

The model is used to compute optimal gaits with respect to various criteria involving energy, muscle tension, and fin curvature. The optimal gaits are then compared with observed biological gaits to determine which criterion is used by the batoids for gait selection.

## 11.2 Modeling of Batoid Swimming

### *11.2.1 The Propulsive Model and Motion Variables*

The skeleton of the batoid pectoral fins (hereto referred to as wings) is formed by long, tapering fin radial elements (Schaefer and Summers 2005). Each radial comprises a cylindrical cartilage that is connected to other radials in a chain with some bifurcations, and multiple chains are arranged on a plane to form the wing shape. The radial structures for the Atlantic stingray and cownose ray were measured through computed tomography (CT) scans and the locations of the cartilage joints were quantified by Russo 2013, as shown in Figs. 11.1 and 11.2 (second column from left). The skeleton is actuated by long thin muscles that insert on each of the radials. The numerous interradial joints are mobile to collectively provide flexibility in the wing, but the radials are stiff enough to transmit the muscle force. The range of the elongation of muscles between any two adjacent radials is small. Especially for undulatory batoids, the large number the radials have a fan-like arrangement that creates a broadly expanded pectoral fin from anterior to posterior of the animal (Fig. 11.1).

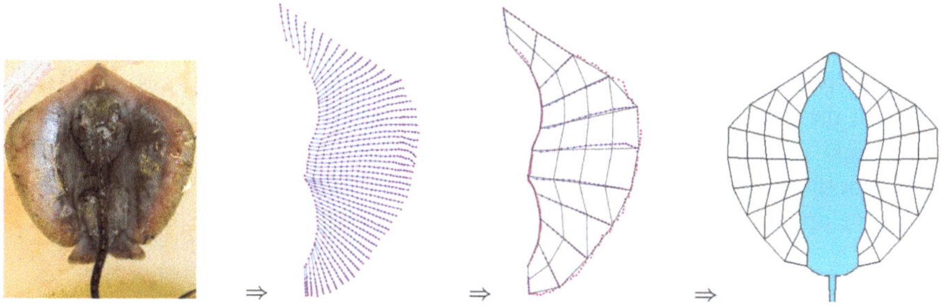

**Fig. 11.1** *D. sabina* modeling. (From cartilage structure to panel grid)

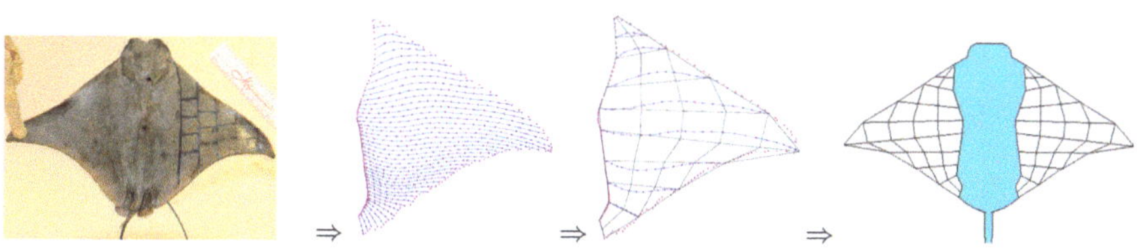

**Fig. 11.2** *R. bonasus* modeling. (From cartilage structure to panel grid)

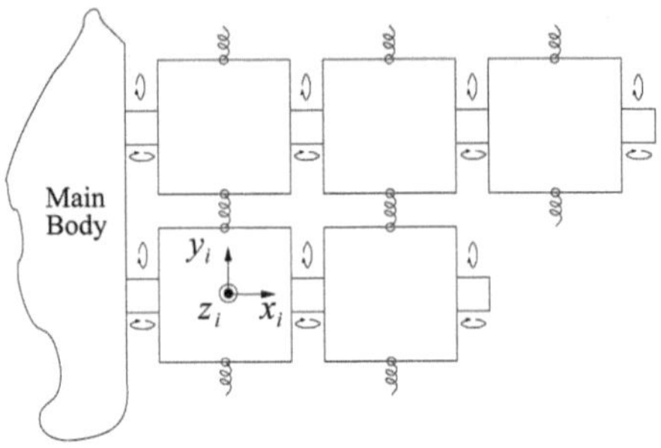


**Fig. 11.3** Panels in the wing are connected through joints and springs

The dynamical model approximates the body geometry and internal cartilage structure by a finite number of panels (Figs. 11.1 and 11.2). The whole body is represented by a main body and two wings. Each wing is gridded into multiple panels according to the directions of radials and locations of the radial joints. The adjacent panels in the spanwise direction are hinged together by flexible joints, through which each panel can rotate in both roll and pitch directions. On the other hand, the adjacent panels in the chordwise direction are connected by linear springs. Thus, two neighboring panels in the spanwise direction always share a joining point, while those in the chordwise direction can become apart during flapping motion (Fig. 11.3). To simplify the coding for numerical simulations, we grid the wing such that (i) each panel is a quadrilateral or triangle, (ii) roll and pitch axes of each panel are perpendicular to each other, and (iii) pitch axes of panels in each spanwise chain are aligned when the wing is flat and hence move within a common plane fixed to the main body (Fig. 11.4).

The motion of the batoid model in 3-dimensional space is described by 4n+6 variables, where n is the number of panels in each wing. In particular, we consider a body frame fixed to the main body, and describe the orientation of the $i$th panel by pitch angle $\beta_i$ and roll angle $\gamma_i$ defined with respect to the body frame. The body shape is thus captured by $\phi := (\beta, \gamma) \in R^{4n}$. The position and orientation of the body are described, respectively, by the Cartesian coordinates $w \in R^3$ of the center of mass (CM) of the whole system in the inertial frame, and the Euler angles $\psi \in R^3$ of the main body with respect to the inertial frame. Overall, the generalized coordinates for the body shape, orientation, and position are given by $q := (w, \psi, \phi) \in R^{4n+6}$. We will develop equations of motion in terms of $w$ and $\theta := (\psi, \phi)$.

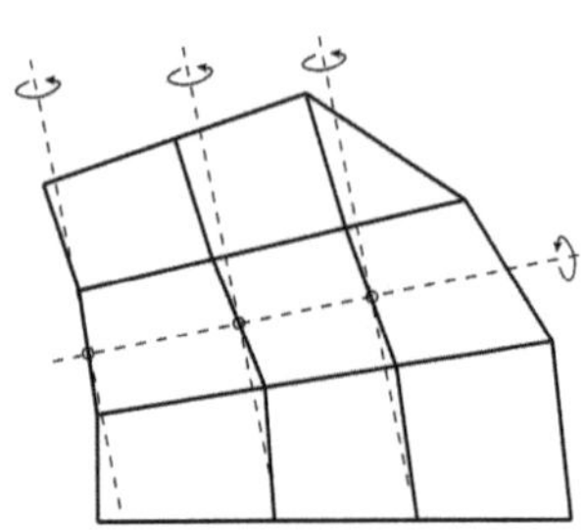

**Fig. 11.4** The arrangements of the pitch and roll axes for wing panels

### 11.2.2 Euler-Lagrange Equation

We develop a nonlinear model for batoid swimming in terms of the Euler-Lagrange equation, which requires descriptions of the kinetic and potential energies as functions of the generalized coordinates and their derivatives.

The total kinetic energy of the main body and all wing panels is given by

$$T = \frac{1}{2} m_b \dot{r}_b^T \dot{r}_b + \frac{1}{2} \omega_b^T J_b \omega_b + \sum_{i=1}^{2n} \left( \frac{1}{2} m_i \dot{r}_i^T \dot{r}_i + \frac{1}{2} \omega_i^T J_i \omega_i \right)$$

where $m_b$, $J_b$, $r_b$, and $\omega_b$ are the mass, moment of inertia, CM position, and angular velocity of the main body, and $m_i$, $J_i$, $r_i$, and $\omega_i$ are those of the $i$th panel. In this expression, the terms of the form $(1/2)m\dot{r}^T\dot{r}$ and $(1/2)\omega^T J\omega$ represent the translational and rotational kinetic energies, respectively, and the subscripts $b$ and $i$ stand for the main body and the $i$th wing panel. The potential energy $V$ consists of the elastic energies stored in the flexible joints and linear springs between adjacent wing panels:

$$V = \sum_{(i,j)\in A_s} \frac{1}{2} k_{ij}^r (\beta_i - \beta_j)^2 + \sum_{(i,j)\in A_c} \frac{1}{2} k_{ij}^p (\gamma_i - \gamma_j)^2 + \sum_{(i,j)\in A_c} \frac{1}{2} k_{ij}^S \left\| s_{ij} \right\|^2 \quad (11.1)$$

where $A_s$ and $A_c$ are the sets of index pairs such that $(i, j) \in A_s$ and $(i, j) \in A_c$ mean panels $i$ and $j$ (with $i > j$) are adjacent in the spanwise and chordwise directions, respectively, $k_{ij}^p$, $k_{ij}^r$, and $k_{ij}^s$ are the body stiffnesses for pitch bending, roll bending, and chordwise stretch between panels $i$ and $j$, and $s_{ij}$ is the distance between panels $i$ and $j$ as measured by the stretch of the linear spring connecting them. In the summations over $A_s$, the index for the main body is also included, where the roll and pitch angles of the main body are both zero by definition. Expressing the kinematic variables in terms of the generalized coordinates, the kinetic and potential energies can be described in the following forms:

$$T(q,\dot{q}) = \frac{1}{2} m \left\| \dot{w} \right\|^2 + \frac{1}{2} \dot{\theta}^T J(\theta) \dot{\theta}, \quad V = V(\phi),$$

where $m$ is the total mass, and $J(\theta)$ is the moment of inertia matrix at a particular shape $\theta$. The Euler-Lagrange equation for the system is now given by

$$\frac{d}{dt}\left( \frac{\partial L}{\partial \dot{q}} \right) - \frac{\partial L}{\partial q} = h, \quad (11.2)$$

where $L(q,\dot{q}) = T(q,\dot{q}) - V(\phi)$ is the Lagrangian, and $h$ is the generalized force resulting from muscle contraction and hydrodynamic forces acting on the body, as described in the next section.

### 11.2.3 Generalized Forces

The hydrodynamic forces acting on the moving batoid body are modeled as resistive forces that are functions of the relative velocity. The resistance experienced by a flat fin is larger in the normal direction than in the tangential direction, and the difference is essential for thrust generation in fish swimming (Saito et al. 2002; Chen et al. 2011). As in the classical resistive force theory by Taylor 1952, we model the force components in the two directions separately, but with linear approximations of nonlinear (quadratic) form drag for simplicity.

For the purpose of fluid force modeling, we ignore the thickness of the batoid body and regard the main body and wing panels in the model as thin plates of zero thickness. We consider an infinitesimal segment with area $dA$ of the batoid body and assume that the normal and tangential fluid forces acting on it are given by $df_n = -c_n v_n dA$ and $df_t = -c_t v_t (2dA)$, where $c_n$ and $c_t$ are the drag coefficients with unit Ns/m$^3$, and $v_n$ and $v_t$ are the respective component vectors of the velocity of the segment (Fig. 11.5). The factor of two in the tangential force appears since the force is assumed proportional to the wetted area, counting both dorsal and ventral sides of the segment. Thus the total force on the segment is given by

$$df = -(c_n v_n + 2c_t v_t)dA.$$

Let $s := (x, y)$ be the coordinates of the infinitesimal segment in the main body frame when the wings are flat, i.e., $s$ does not vary with time and uniquely specifies the location of the segment on the body. For instance, integrating $df$ over the area of a panel with $dA = dxdy$ would give the net force acting on it. The position $r$ and velocity $v$ of the segment in the inertial frame can be expressed as

$$r = w + \rho(\theta, s), \qquad v = \dot{r} = \dot{w} + \left(\frac{\partial \rho}{\partial \theta}\right)\dot{\theta},$$

for some function $\rho$. Let $n(\theta, s)$ be the unit vector normal to the body at the segment (which would be the normal vector of the panel or main body on which the segment lies). Then the normal and tangential velocity vectors are given by

$$v_n = nn^{\mathrm{T}} v, \quad v_t = n \times (v \times n),$$

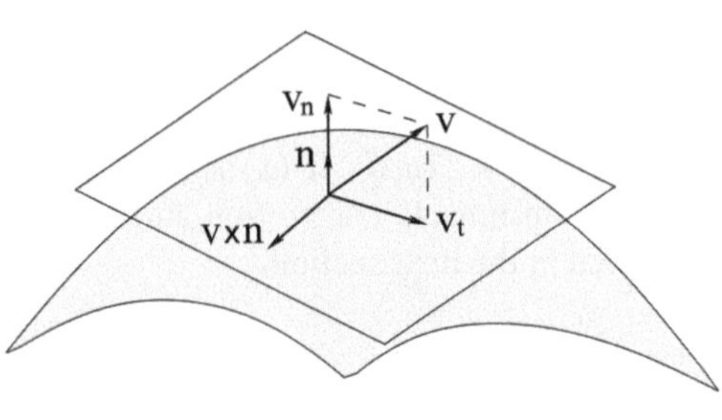


**Fig. 11.5** Normal and tangential projections of the velocity

where $\times$ denotes the cross product. The total virtual work done by the hydrodynamic forces can then be obtained by integrating the individual virtual work over the whole body:

$$\partial W_f = \int (\delta r)^T df = (\delta q)^T F(\theta)\dot{q}$$

where $F(\theta)$ is a symmetric positive definite matrix function of $\theta$, appropriately defined in terms of an integral involving $\partial p/\partial\theta$ and $n$. Thus, the generalized hydrodynamic force $F(\theta)\dot{q}$ appears as a term within $h$ in the Euler-Lagrange Eq. (11.2). This term can be viewed as the sum of the forces and torques acting on the CMs of the main body and wing panels.

The bodies of batoids have a slightly higher density than the sea water. But the difference is considered to have little effect on gait analyses, and hence we assume neutral buoyancy and neglect the effects of gravity and buoyancy in our model.

Muscle fibers run along the fin rays in the spanwise direction and project onto the cartilage structure. Differential contractions of dorsal and ventral muscles generate roll bending moments at the cartilage joints. This is modeled as torque inputs $u_i$ at the joints between spanwise adjacent panels $(i, j) \in A_s$ to induce roll motions of the panels about the $y_i$-axis and $y_j$-axis (see Fig. 11.3). The virtual work done by the muscle torque $u := (u_1, \ldots u_{2n})$ is given by

$$\delta W_m = \sum_{(i,j)\in A_s} (\delta\beta_i - \delta\beta_j)u_i = (\delta\theta)^{\mathrm{T}} Bu,$$

where $B$ is a constant coefficient matrix of dimension $(2n+3)\times 2n$ with elements 0, 1 or $-1$. Since $q := (w, \theta)$, the generalized force resulting from the muscle bending moments is given by a vector whose first three entries (corresponding to $w$) are zero and the remaining $2n+3$ entries are $Bu$. This vector will appear as a term in the overall generalized force $h$ in (11.2).

## 11.2.4 Equations of Motion

In the previous sections, we have developed the Euler-Lagrange Eq. (11.2) and expressions for the kinetic energy $T$, potential energy $V$, and generalized force $h$. Calculating the partial derivatives of $L := T - V$, we obtain the equations of motion for the batoid model:

$$m\ddot{w} + C(\theta)\dot{w} + E(\theta)^T\dot{\theta} = 0,$$

$$J(\theta)\ddot{\theta} + G(\theta,\dot{\theta})\dot{\theta} + k(\theta) + E(\theta)\dot{w} + D(\theta)\dot{\theta} = Bu, \qquad (11.3)$$

where

$$G(\theta,\dot{\theta}) := \left(\frac{\partial J(\theta)\dot{\theta}}{\partial\theta}\right) - \frac{1}{2}\left(\frac{\partial J(\theta)\dot{\theta}}{\partial\theta}\right)^T,$$

$$k(\theta) := \left( \frac{\partial V(\varphi)}{\partial \theta} \right), \quad \begin{bmatrix} C(\theta) & E(\theta)^T \\ E(\theta) & D(\theta) \end{bmatrix} := F(\theta).$$

Simulations show that the term $G(\theta,\dot{\theta})\dot{\theta}$ is small and has little effect on the steady swimming behavior. Since the presence of the term significantly increases the computational time, we chose to neglect the term when (11.3) was simulated in all the analyses reported here.

For our gait analysis, we further simplify the equations of motion in (11.3) by considering steady swimming along the *y*-axis of the main body (Fig. 11.3) with small-amplitude oscillations of the flapping wings. Assuming small $\theta$, Taylor series expansion and truncation of nonlinear terms yield

$$J(\theta) \approx J(0) =: J, \; D(\theta) \approx D(0) =: \mathrm{D}, \; k(\theta) \approx k(0) + K\theta, \; K := \frac{\partial k}{\partial \theta}(0),$$

$$e^T C(\theta) e \approx a + b^T \theta + \theta^T C \theta, \quad E(\theta) e \approx d + \Lambda \theta,$$

where $e := (0, 1, 0) \in \mathrm{R}^3$ projects the velocity vector $\dot{w}$ to the swim speed $v := e^T \dot{w}$. Since the batoid body experiences no restoring force due to elasticity when it is flat, we have $k(0) = 0$. Also, calculations show that $b = 0$ and $d = 0$ in the batoid model. Further assuming that the lateral and vertical velocities of the CM are small and can be ignored, the fully nonlinear equations of motion (11.3) can be approximated by

$$\begin{aligned} m\dot{v} + (a + \theta^T C \theta) v + \theta^T \Lambda^T \dot{\theta} &= 0, \\ \mathrm{J}\ddot{\theta} + \mathrm{K}\theta + v\Lambda\theta + \mathrm{D}\dot{\theta} &= Bu. \end{aligned} \tag{11.4}$$

We refer to the simplified equations of motion as the bilinear model since the term $\theta^T \Lambda^T \dot{\theta}$ captures the most essential dynamics underlying thrust generation by periodic body movements (Blair and Iwasaki 2011). The bilinear model (11.4) will be used for calculating the optimal gaits.

## 11.3 Model Parameters

The model parameters for the geometry, mass, elasticity, and hydrodynamic forces are fixed using experimental data of an Atlantic stingray and a cownose ray.

### *11.3.1 Body Shape/Structure and Mass Distribution*

A cownose ray and an Atlantic stingray were used to obtain typical body parameters. The lengths and masses of the two batoids are given in Table 11.1. The batoid bodies in the flat rest position were CT-scanned to determine the thickness

**Table 11.1** Size and mass of rays

| | Disc width [mm] | Mass [kg] | Each fin [kg] |
|---|---|---|---|
| *D. sabina* | 213 | 0.213 | 0.025 |
| *R. bonasus* | 372 | 0.637 | 0.086 |

and cartilage structure (Russo 2013). The thicknesses of wing cross-sections were measured at 15 grid points on the body indicated in Fig. 11.6. The thickness data is summarized in Table 11.2. The wing is tapered toward the wing tip, and each cross-section of a chordwise cut is streamlined like an aircraft wing with tapering toward the trailing edge.

The partition of the wing into multiple panels in the model is not the same as the partition shown in Fig. 11.6 but is in accordance with the cartilage structure as in Figs. 11.1 and 11.2. The thickness is assumed uniform over each panel, and the panel thickness $d_i$ is estimated by interpolating the 15 data points. The width (spanwise) $b_i$ of a panel is calculated as the distance between the two joints on the edges of the panel, and the length (chordwise) $l_i$ is calculated as the average distance, along the roll axis direction, of the other two edges (Fig. 11.4). We computed the mass of each panel and the main body, assuming a uniform density everywhere.

## 11.3.2 Wing Stiffness

We assume that the Young's modulus is uniform in each of the spanwise and chordwise directions, and denote them by $\kappa_s$ and $\kappa_c$. The stiffness coefficients in (11.1)

**Fig. 11.6** Section order in thickness measurement

**Table 11.2** Wing thickness [mm]

| Section | *D. sabina* | *R. bonasus* |
|---|---|---|
| 1 | 1.08 | 1.36 |
| 2 | 1.32 | 3.06 |
| 3 | 1.00 | 1.45 |
| 4 | 2.13 | 5.70 |
| 5 | 1.21 | 4.98 |
| 6 | 1.04 | 1.80 |
| 7 | 2.33 | 8.67 |
| 8 | 3.58 | 8.83 |
| 9 | 2.81 | 7.21 |
| 10 | 1.65 | 3.01 |
| 11 | 3.51 | 12.04 |
| 12 | 4.93 | 15.33 |
| 13 | 4.58 | 13.98 |
| 14 | 3.50 | 9.96 |
| 15 | 1.57 | 5.30 |

are related to the Young's moduli through the dimensions of the segments connecting the panels. The fictitious segments between panels $i$ and $j$ are assumed to have spanwise width $\bar{b}_{ij} := (b_i + b_j)/2$, chordwise length $\bar{l}_{ij} := (l_i + l_j)/2$, and thickness $\bar{d}_{ij}$, the definition of which is given in the next paragraph. Then the stiffness coefficients are given by

$$k_{ij}^s = \kappa_c \frac{\bar{b}_{ij}\bar{d}_{ij}}{\bar{l}_{ij}}, \quad k_{ij}^p = \kappa_c \frac{\bar{b}_{ij}\bar{d}_{ij}^3}{16\bar{l}_{ij}}, \quad k_{ij}^r = \kappa_s \frac{\bar{l}_{ij}\bar{d}_{ij}^3}{16\bar{b}_{ij}}. \tag{11.5}$$

These formulas are based on the following reasoning. The chordwise stretch stiffness $k_{ij}^s$ is proportional to the cross-sectional area $\bar{b}_{ij}\bar{d}_{ij}$ and inversely proportional to the chordwise length $l_{ij}$. The roll stiffness $k_{ij}^r$ arises as a result of the lever arm effect due to the difference in the dorsal and ventral tensions in the spanwise direction. The spanwise stretch stiffness is proportional to the cross-sectional area $\bar{l}_{ij}\bar{d}_{ij}$ and the inverse of the spanwise width $\bar{b}_{ij}$. The conversion of the stretch stiffness to the rotational stiffness involves the arm-length factor $\bar{d}_{ij}/4$ between the joint angle and stretch, and another factor $\bar{d}_{ij}/4$ between the contractive force and bending moment. A similar argument applies to the pitch stiffness $k_{ij}^s$. The stiffness coefficients can thus be calculated once the Young's moduli $\kappa_s$ and $\kappa_c$ are known.

We assume that the chordwise stiffness $\kappa_c$ is dependent on the muscle and body wall, and hence the thickness of the connecting segment in the chordwise direction is estimated using the panel thickness as $\bar{d}_{ij} = (d_i + d_j)/2$, which is used for calculating $k_{ij}^s$ and $k_{ij}^p$ in (11.5). On the other hand, the spanwise stiffness $k_s$ would be dominated by the property of the mineralized cartilage, and its thickness is used for $\bar{d}_{ij}$ in $k_{ij}^r$. The thickness values $\bar{d}_{ij}$ were determined by scaling the measurements from the images of mineralized cartilage in Schaefer and Summers 2005. For *Myliobatis californica* of disc width 170 mm, which has triangular-shaped wings

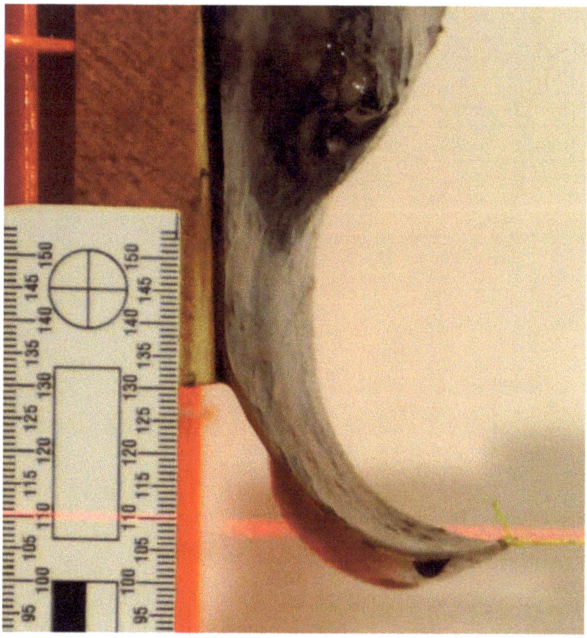

**Fig. 11.7** Stiffness experiment on *D. sabina*

similar to *R. bonasus*, the thickness of the mineralized cartilage is about 0.21 mm at the wing tip and 0.86 mm at the wing root. For *Urobatis halleri*, which has round-shaped wings similar to *D. sabina*, the cartilage thickness in the wing tip and root are about 0.35 and 1.43 mm, respectively. In both rays, the ratio of the root thickness to the tip thickness is about 4. This ratio with linear interpolation was used to set $\bar{d}_i$ for $k^r_{ij}$.

Experiments were conducted on a *D. sabina* and a *R. bonasus* to determine the wing stiffness of each species. A ray was clamped to a vertical support and the fin tip was pulled horizontally as shown in Fig. 11.7. The force and the tip displacement were then measured at different levels of the applied force, as summarized in Table 11.3, where the wing tip displacement is measured from the bottom of the body. The stiffness parameters $\kappa_c$ and $\kappa_s$ were determined as follows. Run the same experiments on the ray model numerically by applying the force $f_i$ at the wing tip and simulating for each grid point on the $(\kappa_c, \kappa_s)$ plane to determine the corresponding wing tip displacement $\delta(f_i)$. Plot the contours of the worst-case error between the experimental data and the model prediction:

$$e(\kappa_c, \kappa_s) := \max_{i \in \{1,2,3\}} \frac{|\delta(f_i) - \delta_i|}{\delta_i}$$

and then choose the stiffness parameters that minimize the error. The gravity effect was found significant during the experiments and hence was taken into account when simulating the ray models.

**Table 11.3** Applied force and wing tip displacement

| Measurement index $i$ | 1 | 2 | 3 |
|---|---|---|---|
| Applied Force $f_i$ [N] | 0.03 | 0.06 | 0.09 |
| Displacement $\delta_i$ [mm] *D. sabina* | 32.5 | 42.0 | 49.5 |
| Displacement $\delta_i$ [mm] *R. bonasus* | 60.0 | 72.5 | 80.0 |

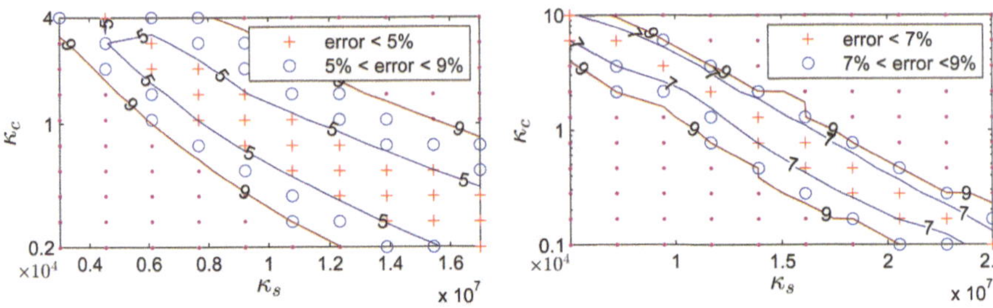


**Fig. 11.8** Contours of errors in wing tip displacements. *D. sabina* (*left*), *R. bonasus* (*right*)

For *D. sabina*, the error contour is given in Fig. 11.8 (left). The optimum with the minimum error is along a curve, rather than a point. Three example points ($\kappa_r$, $\kappa_s$) inside the 5% error contour are given in Table 11.4 (left). The force-displacement data alone cannot determine which point on the curve gives the most reasonable estimate for the stiffness. As an additional aspect for estimating the stiffness, the model-simulated wing shapes at the three points, together with the measured data, are shown in Fig. 11.9 (left). Comparing the horizontal distance of the wing along the line passing through the wing tip, we determine that Case 2 provides the wing shape closest to the data. In fact, the points on the following line segment have small displacement errors and good fits of the wing shape:

$$\log_{10}(\kappa_c) = -10^{-7}\kappa_s + 5, \quad 8\times10^6 \le \kappa_s \le 12\times10^6.$$

Based on these results, we chose $\kappa_s = 10^7$ and $\kappa_c = 10^4$ N/m$^2$.

Similarly, for *R. bonasus*, the error contour is given in Fig. 11.8 (right). Three points ($\kappa_s$, $\kappa_c$) inside the 7% error contour are given in Table 11.4 (right), and the corresponding wing shapes are shown in Fig. 11.9 (right). The points ($\kappa_s$, $\kappa_c$) with small displacement errors and good fit of wing curves are

$$\log_{10}(\kappa_c) = -10^{-7}\kappa_s + 5.6, \quad 1.5\times10^7 \le \kappa_s \le 2\times10^7.$$

**Table 11.4** Model-simulated wing tip displacements with various stiffness values under 0.06 N

| *D. sabina case* | 1 | 2 | 3 | *R. bonasus case* | 1 | 2 | 3 |
|---|---|---|---|---|---|---|---|
| $\kappa_s$ ($\times 10^6$) [N/m$^2$] | 5 | 10 | 17 | $\kappa_s$ ($\times 10^6$) [N/m$^2$] | 10 | 20 | 25 |
| $\kappa_c$ ($\times 10^3$) [N/m$^2$] | 30 | 10 | 2 | $\kappa_c$ ($\times 10^3$) [N/m$^2$] | 40 | 3 | 1 |
| $\delta$ (0.06) [mm] | 43.3 | 41.3 | 42.9 | $\delta$ (0.06) [mm] | 74.4 | 72.7 | 74.1 |

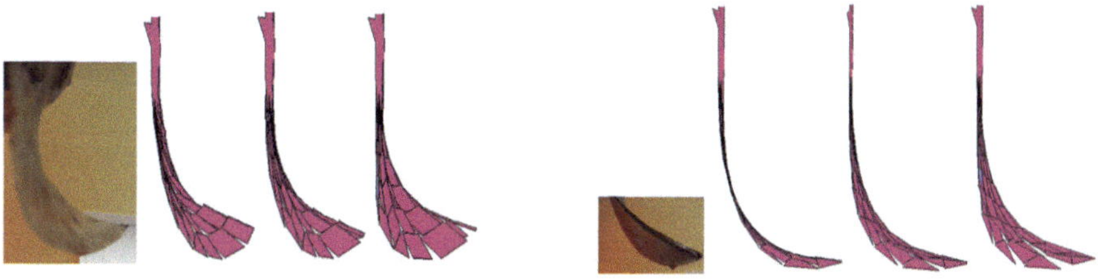

**Fig. 11.9** Wing curves with different ($\kappa_s$, $\kappa_c$) in Table 11.4. *D. sabina* (*left*) and *R. bonasus* (*right*). Cases 1, 2, and 3 from left to right for each species

Based on these results, we chose $\kappa_s = 2 \times 10^7$ and $\kappa_c = 3 \times 10^3$ N/m$^2$.

### *11.3.3 Fluid Drag Coefficients*

Consider a flat panel of area $A_c$ placed in a fluid flow of velocity $v_n$, perpendicular to the panel. The standard form drag model for the normal force experienced by the panel is given by

$$f_n = -\frac{1}{2} c_N \rho A_C \|v_n\| v_n, \tag{11.6}$$

where $\|v_n\|$ is the norm (magnitude) of the three-dimensional velocity vector $v_n$, and $\rho$ is the fluid density. The drag coefficient $c_N$ depends on the ratio of panel width to height (*b/h*) and the Reynolds number (*Re*). For instance, theoretical values are $c_N = 1.05$ when $b/h = 1$, and $c_N = 2.05$ when $b/h = \infty$ (Fox et al. 2003), provided $Re > 10^3$, which is the case for batoid swimming. For *D. sabina* and *R. bonasus*, the Reynolds number is about $10^4$–$10^5$ (midspan fin chord 0.2 m and swimming speed 0.2 m/s), and is about $10^6$ for Manta rays (Moored 2011). We use a linear approximation of this equation as a model for the normal drag $df_n$ on a small segment $dA$ on the body:

$$df_n = -c_n v_n dA, \quad c_n = \frac{1}{2} c_N \rho v_z,$$

where $v_z$ is a constant parameter approximating the time-varying signal $\|v_n\|$. We set $v_z$ to be a half of the average wing-tip speed. If the vertical wing tip displacement during a level swimming is given by $a \sin\omega t$, then the tip speed is $v(t) = a\omega \cos\omega t$, and a half of the average speed is calculated as

$$v_z = (1/2) \int_0^T v^2 dt \Big/ \int_0^T v dt = \pi a \omega / 8.$$

The values of $a$ and $\omega$ are set from the observed data summarized below.

Next consider a streamlined object placed in a fluid flow of velocity $v_t$. The tangential drag is dominated by surface friction and is given by

$$f_t = -\frac{1}{2} c_T \rho A_S \|v_t\| v_t, \tag{11.7}$$

where $\|v_t\|$ is the magnitude of the tangential velocity vector, and $A_s$ is the total surface area in contact with the fluid (wetted area) of the object. The drag coefficients for streamlined objects are rather small; $c_T$=0.0044 for a penguin with $Re \sim 10^6$, $c_T$=0.004-0.015 for fish with $Re \sim 5 \times 10^4$-$5 \times 10^5$ (Vogel 1996), and $c_T$=0.0058-0.009 for a fish-like robot (Barrett et al. 1999). A linear approximation yields the tangential force model for a small wing segment $dA$

$$df_t = -c_t v_t (2dA), \quad c_t = \frac{1}{2} c_T \rho v_y,$$

where $2dA$ is the wetted area and $v_y$ is set as the average of observed swimming speeds.

The drag coefficients $c_N$ and $c_T$, and associated velocity parameters $v_z$ and $v_y$, are determined from video data of swimming batoids. Seven episodes of *D. sabina* were filmed at 150 frames/s in our lab, and ten episodes of *R. bonasus* were filmed at 30 frames/s at Sea World in Orlando, FL, where the videos were taken from the side as batoids swam nearly horizontally. From a video sequence for each species, 20 frames with uniform temporal spacing were extracted to span roughly one flapping cycle. Vertical displacements of the wing edge were measured at several positions along the chord of the wing. Three dimensional wing deformation over the cycle were quantified by interpolating the skeletal joint positions across the wing (Russo 2013), and calculating the pitch angles $\beta$ of wing panels in the model accordingly. Figures 11.10 and 11.11 show samples of the video snapshots and the corresponding model motions generated from the quantified data.

Data for some characteristic kinematic variables are summarized in Table 11.5 with the values for average and standard deviation. The stride length is the swimming speed expressed as the distance traveled by a ray in one flapping cycle. As in Rosenberger 2001, the fin amplitude was calculated as half the dorsoventral displacement of the wing tip (i.e. widest portion of the fin). Both stride lengths and fin

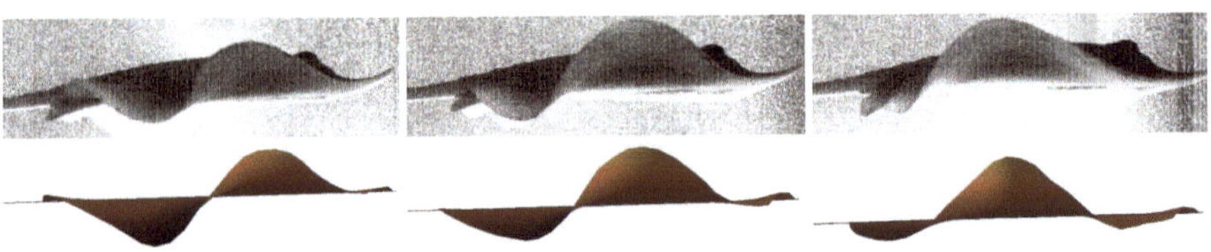

**Fig. 11.10** Snapshots of a swimming *D. sabina* (*above*) and the swimming model with quantified kinematic data (*below*), taken at $t$=0, $T$/6, $T$/3 from *left* to *right*, where $T$ is the cycle period

**Fig. 11.11** Snapshots of a swimming *R. bonasus* (*above*) and the swimming model with quantified kinematic data (*below*), taken at t=0, T/6, T/3 from *right* to *left*, where T is the cycle period

**Table 11.5** Kinematic parameters of observed batoid swimming

| | Episodes | Stride length (BL) | Frequency (Hz) | Fin amplitude (BL) |
|---|---|---|---|---|
| *D. sabina* | 7 | 0.59±0.07 | 1.22±0.16 | 0.17±0.02 |
| *R. bonasus* | 10 | 1.47±0.27 | 0.64±0.13 | 0.52±0.08 |

amplitudes are normalized by body lengths (BL). The body lengths in the video are about 0.26 m (*D. sabina*) and 0.23 m (*R. bonasus*). The model parameter $v_z = \pi a\omega/8$ was set using the average values of fin amplitude $a$ and frequency $\omega$. According to Table 11.5, the statistical swimming speeds are about 0.18 m/s (*D. sabina*) and 0.22 m/s (*R. bonasus*), and these values are used for $v_y$.

To determine the drag coefficients, the observed gait (pitch angles $\beta(t)$) was imposed on the model (11.3) for each point $(c_T, c_N)$ in the gridded $(c_T, c_N)$ plane. The stride length was computed via model simulation and its contours are plotted on the $(c_T, c_N)$ plane (Fig. 11.12). For *D. sabina*, the observed stride length 0.59 BL/cycle is reproduced when $c_N \approx 230 c_T$. On the other hand, for *R. bonasus*, the model achieves the observed value 1.47 BL/cycle when $c_N \approx 360 c_T$. Based on the tangential drag coefficients for streamlined aquatic bodies, we choose $c_T = 0.007$ and $c_N = 1.6$ for *D. sabina* and $c_T = 0.007$ and $c_N = 2.5$ for *R. bonasus*.

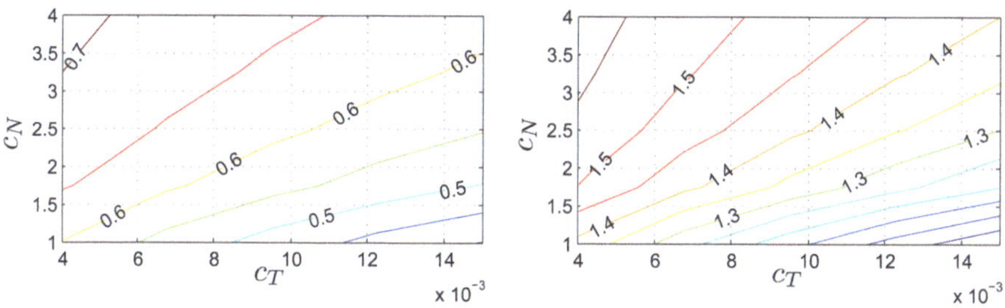


**Fig. 11.12** Contours of simulated stride length for *D. sabina* (*left*) and *R. bonasus* (*right*)

## 11.4 Gait Analysis

### *11.4.1 Objective and Methods*

Batoids employ different gaits for cruising through the water. *D. sabina* undulates its round-shaped fins by sending traveling waves down the chord, while *R. bonasus* flaps its triangular-shaped wings up and down with coordinated pitching motion. A fundamental question is why the batoids choose the particular fin movements (gaits). We attempt to answer the question in terms of optimality under the given wing shape and structure. In particular, we consider several cost functions capturing various energies, muscle tension, and strain on morphing wings, determine the optimal wing oscillation for each cost function, and compare the optimal gait with the observed gait. If the two gaits are found close to each other, then we conclude that the batoid may choose the observed gait as a result of optimizing the corresponding cost function.

The optimal gait problem we consider is to determine the periodic wing motion that minimizes a cost function while maintaining a given (observed) swim speed on average over the flapping cycle. The simplified equations of motion (11.4) are used as the dynamic constraints during the optimization because the simplicity allows us to compute the globally optimal solution in a systematic and numerically tractable manner (Blair and Iwasaki 2011). The optimal gait is given as sinusoidal oscillations of the wing shape variables, roll angles $\beta_i$ and pitch angles $\gamma_i$, where their amplitudes and phases are specified by the optimization. To verify that the model simplification did not introduce significant errors in the analysis, the original nonlinear model (11.3) was simulated by imposing the optimal gait through feedback control, and the motion was animated for visual inspection.

In general, the simulated steady swimming speed for the bilinear model (11.4) was found 5–10% larger than that for the nonlinear model (11.3) when the same parameter values are set and the same observed gaits are imposed on the models. Therefore, we set a larger target speed for the optimization than the observed value so that the nonlinear model swims near the observed speed when simulated. For example, we set the target speed to 0.63 BL/cycle, which is 7% larger than the observed value 0.59 BL/cycle, in an optimization for the *D. sabina* model. When the resulting optimal gait is imposed, the simulated bilinear model converges to the steady state swimming at average speed 0.63 BL/cycle as it should, and the simulated original nonlinear model yields 0.59 BL/cycle as desired.

### *11.4.2 Cost Functions*

The energetic cost of swimming is supplied by muscle, converted to kinetic and elastic energies of the body, and eventually lost into water through resistive drag and kinetic energy of accelerated fluid (the latter of which is not taken into account by our model). The instantaneous mechanical power supplied by muscle is given by the sum of the products of the roll bending moment $u_i$ and the corresponding roll angular velocity $\eta_i := \dot{\beta}_i - \dot{\beta}_j$ where $(i, j) \in A_S$ specifies two wing panels adjacent in the spanwise direction, and the summation is over all roll joints in the model; that is, $\dot{\eta}_1 u_1 + \ldots + \dot{\eta}_{2n} u_{2n} = \dot{\eta}^T u$. Hence the average power over a cycle of body oscillation is

$$E := \frac{1}{T} \int_0^T \dot{\eta}^T u dt$$

where $T$ is the cycle period. When $u$ and $\eta$ are sinusoids, the power $E$ admits a simple characterization. In particular, consider $u(t) = \Re[\hat{u} e^{j\omega t}]$ and $\eta(t) = \Re[\hat{\eta} e^{j\omega t}]$ where $\Re[\cdot]$ denotes the real part, $\omega := 2\pi/T$ is the oscillation frequency, and $\hat{\mu}$ and $\hat{\eta}$ are the phasors, i.e., complex vectors specifying the phases and amplitudes. Then $E = \Re[(j\omega\hat{\eta})^* \hat{u}]/2$ where $(\cdot)^*$ denotes the complex conjugate transpose. This simple formula allows us to find an analytical solution to the optimal gait problem of minimizing the power $E$.

During steady swimming, the average power consumption $E$ should always be a positive constant, but the instantaneous power $\dot{\eta}^T u$ oscillates with period $T/2$ and its value can be negative for a portion of the cycle. This is because the muscle torque $ui$ may act in the direction opposite to the joint angular velocity $\dot{\eta}_i$ and decelerate the motion, in which case the muscle does negative work. Thus energetic cost $E$ can be split into the positive power $P > 0$ and negative power $N < 0$:

$$E = P + N, \quad P := \frac{1}{T} \int_0^T (\dot{\eta}^T u)_+ dt, \quad N := \frac{1}{T} \int_0^T (\dot{\eta}^T u)_- dt,$$

where $(x)_+$ is the larger of $x$ and 0, and $(x)_-$ is the smaller (more negative) of $x$ and 0.

The negative power in the model means that the power is returned from the body back to muscle, reducing the total power consumption from $P$ to $E$. In general, some portion of the returned energy may be stored as potential energy in elastic elements (e.g. tendons) and used in the next cycle, but other portion may just be lost. In fact, negative work requires muscular efforts and there is an associated metabolic cost up to 20% that of positive work (Zarrugh 1981). Hence, we may consider the quantity proportional to the total metabolic cost:

$$M := P - N / 5.$$

The cost $M$ does not admit a simple mathematical characterization useful for optimization, and therefore we consider an approximation of the negative power $N$ by the reactive power $R$:

$$N \approx \sigma R, \quad R := \frac{1}{T} \int_0^T \omega \eta^T u dt,$$

where $\sigma = 0.138$ is chosen so that the average values of $N$ and $\sigma R$ over $0 < \phi < 90°$ are equal when $u = \sin(\omega t)$ and $\dot{\eta} = \sin(\omega t - \phi)$. We can then minimize the approximate metabolic cost

$$M = P + N - (6/5)N \approx E - (6/5)\sigma R,$$

which admits a simple formula since $R = \Re[\omega\hat{\eta}^* \hat{u}]/2$ in the phasor notation.

In addition to the power cost $E$ and metabolic cost $M$, we consider the following cost functions:

$$C = \frac{1}{T}\int_0^T \|\eta\|^2 dt = \frac{\|\hat{\eta}\|^2}{2},$$

$$B = \frac{1}{T}\int_0^T \|u\|^2 dt = \frac{\|\hat{u}\|^2}{2}.$$

The cost $C$ is associated with the curvature of the wings in the roll motion. Minimization of the curvature cost $C$ penalizes a large amplitude of wing tip oscillation, and has the tendency to keep the wings flat. The cost $B$ is associated with the magnitude of the muscle bending moment (or tension) and a batoid would minimize $B$ if it wants to swim with a minimal load on muscle.

### *11.4.3 What Cost Is Minimized by Batoids?*

We determine whether the observed gaits of batoid swimming can be explained by optimality with respect to one of the four cost functions: power $E$, metabolic cost $M$, curvature $C$, and bending moment $B$. The cycle frequency is fixed to the observed value (1.22 Hz for the *D. sabina* and 0.64 Hz for the *R. bonasus*), and the amplitudes and phases of the periodic wing movements are optimized by the method in Blair and Iwasaki 2011, so that the observed swim speed (0.59 BL/cycle for the *D. sabina* and 1.47 BL/cycle for the *R. bonasus*) is achieved with a minimum value of the cost function.

The optimal gaits for *D. sabina* are summarized in Fig. 11.13 in comparison with the observed gait quantified from the video recording data. For each wing panel, the displacement of its CM in the vertical direction of the main body frame oscillates sinusoidally. The amplitudes and phases of the displacement oscillations are indicated by different colors in the figure. Snapshots of the simulated batoid model are also shown. *D. sabina* is a representative undulator that swims by sending traveling waves posteriorly through the wings. The phase lag from the leading edge to trailing edge for the biological gait is about 470°. The number of traveling waves expressed by the wings is roughly equal to the phase lag divided by 360° in general. Hence, the biological gait has wave number 1.3.

For the minimum curvature gait, the tip amplitude is about 77% of that of the biological gait, and the phase lag is about 200°, which corresponds to the wave number 0.56. The body movements appear as an oscillation (flapping) gait rather than an undulation gait, and thus do not coincide with the biological observation. The optimal gait minimizing the bending moment cost has a phase lag close to the biological value,

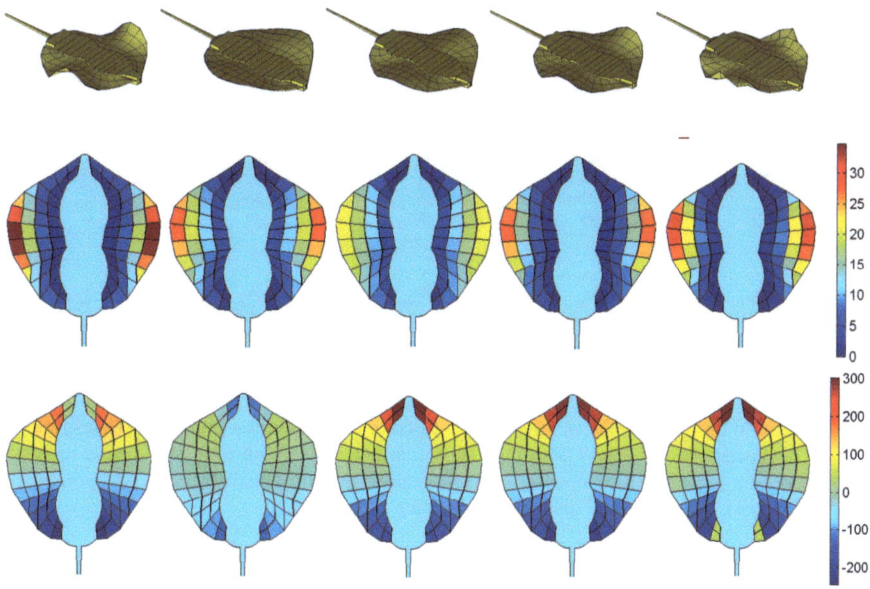


**Fig. 11.13** Optimal and observed gaits for *D. sabina*. The *top* row shows snapshots of the gaits. The second and third rows respectively indicate by color the amplitudes [mm] and phases [deg] of vertical displacements of panel CMs in the body frame. Each column from *left* to *right* indicates an observed biological gait, C-optimal gait (curvature), E-optimal gait (power), M-optimal gait (metabolic cost), and B-optimal gait (muscle bending moment)

but the panels near the wing tip oscillate with pitching motion (not quite captured by Fig. 11.13) that appear unnatural and far from the biological gait. For the power-optimal gait, the phase lag is close to that of the biological gait, but the tip amplitude is small and only 62 % of biological value. Finally, the optimal gait for the metabolic cost is the closest, among the four, to the biological gait in both amplitudes and phases. The wing tip amplitude is 79 % of the biological one, and the phase lag 490° is close to the biological value, indicating traveling waves with wave number 1.4.

The optimal gaits for *R. bonasus* are shown in Fig. 11.14. The results are essentially analogous to the case of *D. sabina*. The optimal gait that minimizes the bending moment does not appear similar to the observed gait, flapping only the wing tip portion up and down with almost no phase lag. The curvature-optimal gait has the phase lag much smaller than that of the observed gait, and the power-optimal gait has the phase lag comparable to the observed but the amplitude is much smaller. The optimal gait minimizing the metabolic cost is the closest to the observed gait. The oscillation amplitude is not quite as large as the observed, but unlike the power-optimal gait, the amplitude becomes larger toward the wing tip in a quadratic (rather than linear) manner, resulting in a curled-up wing shape as in the biological gait. The phase decreases toward the tail, with the phase lag from head to tail less than 180°. The resulting flapping motion is an oscillation rather than an undulation; traveling waves still exist but their wavelength is more than twice the body length, and hence the body expresses less than a half of a wave at each time instant.

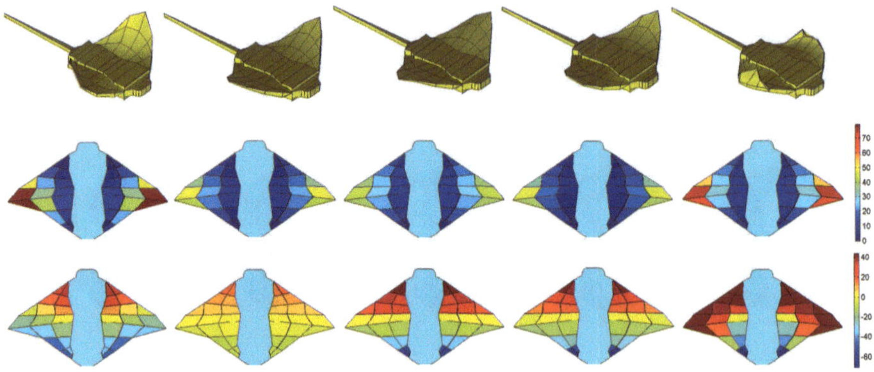


**Fig. 11.14** Optimal and observed gaits for *R. bonasus*. The top row shows snapshots of the gaits. The second and third rows respectively indicate by color the amplitudes [mm] and phases [deg] of vertical displacements of panel CMs in the body frame. Each column from *left* to *right* indicates an observed biological gait, C-optimal gait (curvature), E-optimal gait (power), M-optimal gait (metabolic cost), and B-optimal gait (muscle bending moment)

For both *D. sabina* and *R. bonasus*, the optimal gait minimizing the metabolic cost is found closest to the observed gait among those resulting from the four optimality criteria considered here. The optimality with respect to the metabolic cost can explain the two basic flapping gaits observed for various batoids—undulation and oscillation. Each gait results from the optimization of the same cost function for the particular wing shape and internal musculo-skeletal structure. For both cases, the amplitude is roughly constant along each chordwise chain of panels, and increases toward the wing tips. The phase is roughly constant along each fin ray, indicating no traveling waves in the spanwise direction. Waves travel down in the chordwise direction perpendicular to the fin rays. For *D. sabina*, this results in waves propagating along the circular perimeter of the wings. These observations suggest that the wing shape and internal structure are important for generation of the observed gaits. If the cartilage were arranged differently, the waves would travel along a different path. The particular shape and structure of batoids may have been optimized through evolution with respect to certain criteria (e.g. hydrodynamic efficiency), but the issue is beyond the scope of our analysis.

### *11.4.4 How Is the Flapping Frequency Chosen?*

The optimal gait analysis in the previous section did not optimize the flapping frequency but fixed it to the observed biological values. This section examines whether the observed frequency is also optimal with respect to the metabolic cost. To this end, the optimal gaits were computed for a range of frequency and the minimum value of the cost was plotted as a function of frequency. The optimal frequency that minimizes the cost was determined and compared with the observed frequency. The

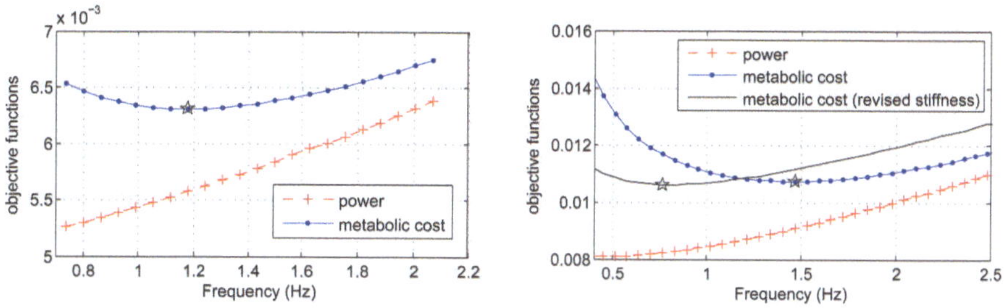


**Fig. 11.15** Minimum costs versus frequency. *Left*: *D. sabina*. *Right*: *R. bonasus*

process was repeated for both *D. sabina* and *R. bonasus* using the metabolic cost as well as the power cost for comparison, and the results are shown in Fig. 11.15.

For *D. sabina*, the cycle frequency of periodic body motion is 1.18 Hz for the optimal gait with the metabolic cost, while the observed frequency is 1.22 Hz; the two frequencies are fairly close. In contrast, the optimal frequency with respect to the power cost is not found in the frequency range since the power increases monotonically with the frequency. This means that the batoid would choose a much lower flapping frequency than observed if it attempts to minimize the power cost. Thus we conclude that the batoid may choose the flapping frequency by minimizing the metabolic cost rather than the power cost. The difference between the two cost functions is the penalty on negative work done by muscle when inertia and/or hydrodynamic effects overrule contractive tension to make it stretch. The power cost assumes that the negative work is fully returned to muscle for storage and re-use, which is not exactly true. The metabolic cost assumes an additional energetic cost associated with negative work, and thus batoids appear to choose the flapping frequency to avoid or reduce negative work.

Negative work would be zero if muscle generates tension only when it is shortening. This is analogous to pushing a child on a swing only when it is going forward. We do so because if we push when the swing is coming toward us, then the kinetic energy is lost and in addition our effort (metabolic energy) is wasted. The strategy for zero negative work leads to oscillation of the swing at its natural frequency. Thus, the strategy exploits natural dynamics to sustain oscillations with a minimal effort. Now, if batoids choose the flapping frequency based on this principle, we should expect that it is close to a natural frequency. Indeed, this turned out to be the case: the frequencies of the natural modes for the batoid body model (without hydrodynamic effects) are 1.25 Hz (1st mode) and 1.74 Hz (2nd mode). Thus, *D. sabina* appears to exploit the 1st mode of natural oscillation arising from the body inertia and elasticity with the particular mass and stiffness distributions over the wing.

For *R. bonasus*, the result is qualitatively identical to that for *D. sabina*, and we may reach the same conclusion regarding the exploitation of the natural dynamics. However, the optimal frequency for the metabolic cost is not precisely matched with the observed frequency; the former is 1.45 Hz, and the latter 0.64 Hz. We suspected that the difference was caused by a higher natural frequency than the flapping frequency. As expected, the frequencies of the natural modes are 1.77 Hz (1st mode) and 1.89 Hz (2nd mode). If the wing is softer than the model predicts, the natural frequencies would be lower and the optimal frequency could then be closer to the observed frequency. The logic has lead us to revisit the modeling of wing stiffness. We had chosen the Young's moduli ($\kappa_c$, $\kappa_s$) based on the model-error contours in Fig. 11.8 (right). However, small errors can be achieved along a line and hence we could have chosen different values. For instance, we may choose $\kappa_s = 2\times10^7 \times(0.64/1.45)^2 = 3.9\times10^6$ N/m$^2$ to reduce the natural frequency associated with flapping (roll) mode of oscillation, and set $\kappa_c = 2.5\times10^5$ N/m$^2$ so that the new point ($\kappa_c$, $\kappa_s$) is still on the line. In this case, the optimal gait analysis result remains essentially unchanged except for the optimal frequency, which is now reduced to 0.76 Hz. Thus, the revised stiffness values may be closer to reality and if so the optimal frequency with respect to the metabolic cost explains the observed flapping frequency with a quantitative precision.

## 11.5 Conclusion

We developed dynamical models for swimming of batoids, undulator *D. sabina* and oscillator *R. bonasus*. The body was represented by multiple panels connected by flexible joints and springs. The hydrodynamic effects are modeled by resistive forces proportional to the relative velocity in each of the normal and tangential directions, and the drag coefficients were determined to match the kinematic data from video recording. The stiffness distribution over the wings is derived in accordance with the measured thickness distribution, assuming uniform Young's modulus in each of the spanwise and chordwise directions. The Young's moduli were determined from experimental force-displacement data, and it is found that (a) the wing is more flexible chordwise than spanwise for both batoids, and (b) undulator/oscillator has more chordwise/spanwise flexibility than oscillaor/undulator.

For each batoid model, the optimal wing flapping pattern was calculated to minimize a cost function while maintaining a steady swimming speed. We considered four cost functions: the power output from muscle (or equivalently power lost into water), metabolic cost containing a fraction of negative power in addition to positive power, magnitude of muscle bending moment, and body curvature. Among the four, the metabolic cost was found to yield optimal gaits closest to observed biological gaits for both batoids, and thus the single criterion based on metabolic optimality explains the two distinctive gaits of undulator and oscillator. The optimal flapping frequencies were also found close to the biological data, as well as to the first mode of natural oscillation of the body. Thus, batoids appear to exploit natural dynamics of the body to minimize a metabolic (energetic) cost associated with both mechanical power output and negative work.

**Acknowledgements** This work was supported by ONR MURI Grant N00014-08-1-0642.

# Appendix: Model Variables and Parameters (Tables 11.6 and 11.7)

**Table 11.6** Model variables

| Variable | Dimension | Definition |
|---|---|---|
| $\psi$ | 3 | Euler angles specifying the orientation of the main body frame with respect to the inertial frame |
| $\omega_b$ | 3 | Angular velocity of the main body with respect to the inertial frame, expressed in the main body frame |
| $r_b$ | 3 | Cartesian coordinates of the CM of the main body in the inertial frame |
| $\omega_i$ | 3 | Angular velocity of the $i^{th}$ panel with respect to the inertial frame, expressed in the panel frame |
| $r_i$ | 3 | Cartesian coordinates of the CM of the $i^{th}$ panel in the inertial frame |
| $\beta$ | n | Column vector $(\beta_1,\ldots,\beta_n)$ where $\beta_i$ is the roll angle of the $i^{th}$ panel with respect to the main body |
| $\gamma$ | n | Column vector $(\gamma_1,\ldots,\gamma_n)$ where $\gamma_i$ is the pitch angle of the $i^{th}$ panel with respect to the main body |
| $\phi$ | 2n | Column vector $(\beta, \gamma)$ representing the wing shape |
| $\theta$ | 2n+3 | column vector $(\psi, \phi)$ representing shape and orientation of the whole body |
| $w$ | 3 | Cartesian coordinates of the CM of the whole body in the inertial frame |
| $q$ | 2n+6 | Generalized coordinates to describe the batoid motion |
| $v$ | 1 | Swim speed (the second entry of $\dot{w}$) |
| $u$ | n | Column vector $(u_1,\ldots,u_n)$, where $u_i$ is the muscle bending moment applied at joint $i$ in the roll direction |

**Table 11.7** Model parameters

| Parameter | Dimension | Definition |
|---|---|---|
| $J_b$ | $3\times3$ | Moment of inertia for the main body |
| $m_b$ | 1 | Mass of the main body |
| $J_i$ | $3\times3$ | Moment of inertia for the $i^{th}$ panel |
| $m_i$ | 1 | Mass of the $i^{th}$ panel |
| $b_i$ | 1 | Spanwise width of the $i^{th}$ panel |
| $l_i$ | 1 | Chordwise length of the $i^{th}$ panel |
| $d_i$ | 1 | Thickness of the $i^{th}$ panel |
| $n$ | 1 | Number of panels representing each wing |
| $m$ | 1 | Total mass of the whole body |
| $\kappa_c$ | 1 | Young's modulus of the wing in the chordwise direction |
| $\kappa_s$ | 1 | Young's modulus of the wing in the spanwise direction |
| $c_N$ | 1 | Fluid drag coefficient in the normal direction |
| $c_T$ | 1 | Fluid drag coefficient in the tangential direction |
| $\rho$ | 1 | Fluid density |

## References

Barrett D, Triantafyllou M, Yue D, Grosenbaugh M, Wolfgang M (1999) Drag reduction in fish-like locomotion. J Fluid Mech 392:183–212
Blair J, Iwasaki T (2011) Optimal gaits for mechanical rectifier systems. IEEE Trans Auto Control 56:59–71
Cavagna G, Heglund N, Taylor C (1977) Mechanical work in terrestrial locomotion: two basic mechanisms for minimizing energy expenditure. Am J Physiol 233:R243–R261
Chen J, Friesen W, Iwasaki T (2011) Mechanisms underlying rhythmic locomotion: body-fluid interaction in undulatory swimming. J Exp Biol 214:561–574
Fish F, Haj-Hariri H, Smits A, Bart-Smith H, Iwasaki T (2011) Biomimetic swimmer inspired by the manta ray. In: Bar-Cohen Y (ed) Biomimetics: nature based innovation. CRC Press, Boca Raton, pp 495–523
Fox R, McDonald A, Pritchard P (2003) Introduction to fluid mechanics. Wiley, New York
Iwasaki T, Zheng M (2006) Sensory feedback mechanism underlying entrainment of central pattern generator to mechanical resonance. Biol Cybern 94:245–261
Moored III K (2011) The design of a novel tensegrity-based synthetic pectoral fin for bio-inspired propulsion. Ph.D Dissertation, University of Virginia
Moored III K, Fish F, Kemp T, Bart-Smith H (2011) Batoid fishes: inspiration for the next generation of underwater robots. Marine Technol Soc J 45:99–109
Parson J, Fish F, Nicastro A (2011) Turning performance in batoid rays: limitations of a rigid body. J Exp Marine Biol Ecol 402:12–18
Rosenberger L (2001) Pectoral fin locomotion in batoid fishes: undulation versus oscillation. J Exp Biol 204:379–394
Russo R (2013) Biomechanical modeling of ray pectral fins to inform the design of AUV propulsion systems. Ph.D Dissertation, University of Virginia
Saito M, Fukaya M, Iwasaki T (2002) Serpentine locomotion with robotic snake. IEEE Control Syst Mag 22:64–81
Schaefer J, Summers A (2005) Batoid wing skeletal structure: Novel morphologies, mechanical implications, and phylogenetic patterns. J Morphol 264:298–313
Taylor G (1952) Analysis of the swimming of long and narrow animals. Proc Roy Soc Lond Ser A 214:158–183
Vogel S (1996) Life in moving fluids: the physical biology of flow. Princeton University Press, New Jersey
Zarrugh M (1981) Power requirements and mechanical efficiency of treadmill walking. J Biomech 14:157–165

# Chapter 12
# Control of Cat Walking and Paw-Shake by a Multifunctional Central Pattern Generator

**Brian Bondy, Alexander N. Klishko, Donald H. Edwards, Boris I. Prilutsky and Gennady Cymbalyuk**

**Abstract** Central pattern generators (CPGs) are oscillatory neuronal networks controlling rhythmic motor behaviors such as swimming, walking, and breathing. Multifunctional CPGs are capable of producing multiple patterns of rhythmic activity with different periods. Here, we investigate whether two cat rhythmic motor behaviors, walking and paw-shaking, could be controlled by a single multifunctional CPG. To do this, we have created a parsimonious model of a half-center oscillator composed of two mutually inhibitory neurons. Two basic activity regimes coexist in this model: fast 10 Hz paw-shake regime and a slow 2 Hz walking regime. It is possible to switch from paw-shaking to walking with a short pulse of conductance in one neuron, and it is possible to switch from walking to paw-shaking with a longer pulse of excitatory conductance in both neurons. The paw-shake and walking rhythms generated by the CPG model were used as input to a neuromechanical model of the cat hindlimbs to simulate the corresponding rhythmic behaviors. Simulation results demonstrated that the multifunctional half-center locomotor CPG could produce movement mechanics and muscle activity patterns typical for cat walking or paw-shake responses if synaptic weights in selected spinal circuits were altered during each behavior. We propose that the selection of CPG regimes and spinal circuitry is triggered by sensory input from paw skin afferents.

**Keywords** Multifunctional multistability · Paw-shaking · Walking · Neuromechanical modeling · Cat

---

B. Bondy (✉) · D. H. Edwards · G. Cymbalyuk
Neuroscience Institute, Georgia State University, Petit Science Center 818, 100 Piedmont Ave SE, Atlanta, GA 30303, USA
e-mail: bbondy2@student.gsu.edu

D. H. Edwards
e-mail: dedwards@gsu.edu

G. Cymbalyuk
e-mail: gcymbalyuk@gsu.edu

A. N. Klishko · B. I. Prilutsky
School of Applied Physiology, Center for Human Movement Studies, Georgia Institute of Technology, 555 14th Street NW, Atlanta, GA 30332, USA
e-mail: aklishko3@gatech.edu



B. I. Prilutsky, D. H. Edwards (eds.), *Neuromechanical Modeling of Posture and Locomotion,* Springer Series in Computational Neuroscience,
DOI 10.1007/978-1-4939-3267-2_12

## 12.1 Introduction

Rhythmic motor behaviors, such as walking, breathing, and chewing are controlled by specialized neuronal circuits known as central pattern generators (CPGs). CPGs can produce behaviors even when deprived of sensory feedback and/or input from the brain (Marder and Calabrese 1996; Marder and Bucher 2001; Grillner 2006; McCrea and Rybak 2008). The half-center oscillator (HCO) is a ubiquitous motif in CPG connectivity. It is a simple mechanism explaining how non-rhythmic neurons can produce rhythmic output. The simplest HCO is composed of two mutually inhibitory neurons which produce a pattern of alternating single spikes or bursts of spikes (Brown 1911; Wang and Rinzel 1992; Friesen 1994; Marder and Calabrese 1996). The rhythmic activity generated by the HCO drives pattern formation in downstream interneurons which in turn activate muscles via motoneurons (McCrea and Rybak 2008).

If a CPG can produce more than one functional behavior, it is referred to as a multifunctional CPG. A multifunctional CPG can be composed of two circuits that share neurons (Getting 1989; Briggman and Kristan 2008). In both vertebrates and invertebrates, individual interneurons can contribute to multiple motor rhythms (Ritzmann et al. 1980; Lockery and Kristan 1990; Syed and Winlow 1991; Lieske et al. 2000; Popescu and Frost 2002; Berkowitz et al. 2010). There are two main mechanisms for motor program selection for multifunctional CPGs. First, *the network may be transformed* through some extrinsic process that functionally configures the CPG to produce one rhythm or another. Alternatively, the *network may demonstrate multistability*, where it is *intrinsically* capable of producing multiple rhythms and can switch between them in response to appropriate transient inputs.

Transformation can be accomplished through neuromodulation. The presence of different neuromodulators or combinations of neuromodulators can functionally reconfigure a network (by changing electrical properties of neurons and synaptic strengths) and/or change the pool of recruited neurons to produce different behaviors (Hooper and Moulins 1989; Marder et al. 1997; Crisp and Mesce 2004; Doi and Ramirez 2008). In a broad sense, to produce the correct behavior, the CPG having the correct neuromodulatory tone is *necessary and sufficient*. Tonic sensory input that does not involve neuromodulators can also transform a network by activating or suppressing certain neurons (Soffe 1997; Liao and Fetcho 2008). In this paradigm, at any given time the CPG can only produce one activity regime, and that is determined by what input it is receiving.

It is also possible that multiple behaviors could be produced by one set of neurons without external influence in a *multistable* CPG. As with any neural system, a multistable CPG would still require a certain neuromodulatory tone (Hounsgaard and Kiehn 1989; Lechner et al. 1996), but it would be able to switch between multiple activity regimes without any changes in that neuromodulatory tone. Some input would be necessary to cause a transition between regimes, but this input can be transient.

Multistability of activity regimes in neurons and networks is a common phenomenon. Bistability between spiking and silence was predicted from the original Hodgkin-Huxley model of the squid giant axon (Hodgkin and Huxley 1952; Rinzel 1978) and then demonstrated experimentally (Guttman et al. 1980). Multistability allows normally silent motoneurons to fire at high frequencies persistently after a short stimulation (Hounsgaard et al. 1984). In the case of motoneurons and in some other situations where bistability has been observed, slow or non-inactivating currents, such as the persistent sodium current, calcium currents, or the A-type potassium current, have been implicated as the underlying mechanism for multistability (Turrigiano et al. 1996; Hughes et al. 1999; Carlin et al. 2000; Egorov et al. 2002; Fuentealba et al. 2004). There has been a great deal of work modeling multistability of firing modes and investigating its mechanisms (Canavier et al. 1993; Butera 1998; Hahn and Durand 2001; Cymbalyuk et al. 2002; Cymbalyuk and Shilnikov 2005; Shilnikov et al. 2005; Fröhlich and Bazhenov 2006; Newman and Butera 2010; Malashchenko et al. 2011).

Models of multifunctional CPGs have been produced where multistability between fast and slow rhythms or phase relationships is the result of slow variables in synaptic dynamics (Bose and Kunec 2001; Manor and Nadim 2001; Chuan et al. 2004). The possibility that multistability in a CPG could arise from cellular dynamics has not been explored. It is not clear whether cellular dynamics can provide multistability in a network of neurons. Cellular dynamics are an alternative mechanism for multistability in a CPG.

Specifically, we investigated whether a pair of well-studied cat behaviors, walking and paw-shaking, could be produced by a single HCO. Both behaviors involve many of the same muscles, and can be elicited in a spinalized cat, even when motion-dependent sensory feedback has been removed, indicating both regimes are produced to a large extent by CPG (Carter and Smith 1986; Koshland and Smith 1989; Pearson and Rossignol 1991). The paw-shake response is initiated when a stimulus, such as an adhesive tape or water, becomes stuck to cat's paw. It consists of rapid (about 10 Hz) rhythmic movements of the paw. Walking, commonly referred to as stepping or locomotion, is a slower regime, usually around 1–2 Hz.

## 12.2 Model of Multifunctional Central Pattern Generator

The neurons were modeled using the Hodgkin and Huxley formalism (Hodgkin and Huxley 1952). Each neuron contains six ionic currents: a fast sodium current ($I_{NaF}$), a slow sodium current ($I_{NaS}$), a potassium current ($I_K$), a slow calcium current ($I_{CaS}$), a synaptic current ($I_{syn}$), and a leak current ($I_{leak}$). Parameters for $I_{Na}$, $I_K$, and $I_{leak}$ were taken from (Rybak et al. 2006). Conductances were modified to make the model produce paw-shaking-like and walking-like bursting activities. Additionally, a seventh current, $I_{Mod}$, which could be expressed as either $I_I$ or $I_E$ (inhibitory or excitatory modulatory current) based on the equilibrium potential, could be turned on and off in a time dependent and cell specific manner. There are a total of seven state variables in each cell: voltage, inactivation of $I_{NaF}$, $I_{NaS}$, and $I_{CaS}$, activation of $I_{NaS}$, $I_K$, $I_{CaS}$ and $I_{Syn}$. Activations of $I_{NaF}$, $I_{Mod}$, and $I_{Leak}$ are considered instantaneous and are simply functions of the membrane potential. The membrane potential is governed by the current conservation equation:

$$CV' = -[I_{NaF} + I_{NaS} + I_K + I_{CaS} + I_h + I_{Mod} + I_{leak} + I_{syn}]$$

| | | |
|---|---|---|
| $I_{NaF} = \bar{g}_{NaF} m^3_{NaF,\infty} h_{NaF} [V - E_{Na}]$ | $E_{Na} = 65\ mV$ | $\bar{g}_{NaF} = 40$ nS |
| $I_{NaS} = \bar{g}_{NaS} m_{NaS} h_{NaS} [V - E_{Na}]$ | | $\bar{g}_{NaS} = 3.9$ nS |
| $I_K = \bar{g}_K m_K{}^4 [V - E_K]$ | $E_K = -70$ mV | $\bar{g}_K = 40$ nS |
| $I_{CaS} = \bar{g}_{CaS} m^3_{CaS} h_{CaS} [V - E_{Ca}]$ | $E_{Ca} = 160$ mV | $\bar{g}_{CaS} = 12$ nS |
| $I_{Mod} = \bar{g}_{Mod} [V - E_{Mod}]$ | $E_I = -75$ mV | $E_E = 70$ mV |
| $I_{leak} = \bar{g}_{leak} [V - E_{leak}]$ | $E_{leak} = -54$ mV | $\bar{g}_{leak} = 3$ nS |
| $I_{Syn} = \bar{g}_{Syn} m_{Syn} [V_{pre} - E_{Syn}]$ | $E_{Syn} = -75$ mV | $\bar{g}_{Syn} = 5.5$ nS |

$$m_{NaF,\infty} = \frac{1}{1 + exp\left(\frac{V + 18.5}{-7.8}\right)},$$

where $\bar{g}_x$ and $E_x$ are the maximal conductance and reversal potential for current $I_x$, respectively.

Membrane potential of the neuron in the above equations is indicated by $V$ while $V_{pre}$ (seen in last, synapse equation) indicates membrane potential of the other neuron. $m$ refers to activation variables, $h$ refers to inactivation variables. All equations are identical in both neurons except with respect to $I_{Mod}$, which has a term that can make it active in one, both, or neither neurons. The conductance of $I_{Mod}$ was zero except during specified time interval. When $I_{Mod}$ is excitatory ($I_E$), $E_{Mod}=E_E=70$ mV. When $I_{Mod}$ is inhibitory ($I_I$), $E_{Mod}=E_I=-75$ mV. In the text, $I_{Mod}$ will be referred to as either $I_E$ or $I_I$, and its conductance as either $g_E$ or $g_I$. All voltages are in mV, all conductances are in nS, and all currents are in pA. In the CPG model, time is in seconds, whereas in the text, time is in milliseconds. The parameters were not changed at any point during simulations presented in this article.

### 12.2.1 *Computer Simulations*

The differential equations were solved in Matlab (The MathWorks, Inc., Natick, MA, USA) using the ode15s differential equation solver. An absolute and relative integration tolerance of $1 \cdot 10^{-10}$ was used, as well as an initial step of 0.0001 s and a max step of 0.005 s. A time step size of at least 0.0001 s was used for simulations that were part of any analysis. Sections of simulations where the neurons had not reached a steady state were discarded.

Burst duration is calculated as the time between the first and last spike of a burst. Inter-burst interval is the time between the last spike of a burst and first spike of the next burst, period is the sum of inter-burst interval and burst duration, and duty cycle is burst duration divided by period. Frequency is the inverse of the inter-spike interval, and mean frequency is the mean frequency of all the spikes in a burst and is calculated as the number of spikes per burst divided by burst duration. Maximal frequency and minimum frequency were calculated as the inverse of the smallest and largest inter-spike intervals in a burst, respectively. Coefficient of variation is the standard deviation as a percent of the sample mean. Charge was calculated as the integral of current with respect to time for each burst period.

## *12.2.2 Simulation Results*

### 12.2.2.1 Model Activity

The developed half-center oscillator model can exhibit walking-like and paw shake-like regimes of rhythmic activity. This HCO contains two mutually inhibitory model neurons that are not endogenously bursting. When coupled, the neurons burst in antiphase.

The isolated neuron ($\bar{g}_{Syn} = 0\,\text{nS}$) spikes tonically (Fig. 12.1). During baseline spiking activity, $I_{CaS}$ is close to zero ($I_{CaS}$ trace in Fig. 12.1, before conductance pulse) because $I_{CaS}$ is almost completely inactivated; the inactivation variable for $I_{CaS}$ ($h_{CaS}$) was never higher than $5 \cdot 10^{-6}$. The rhythmic spiking activity is predominantly driven by $I_{NaS}$. However, if the neuron is strongly hyperpolarized by an inhibitory current, $I_I$ (Fig. 12.1, $g_I$=1 nS, $E_I$=−75 mV), $I_{CaS}$ and $I_{NaS}$ will deinactivate and produce transient rebound bursting. The conductance pulse (100 ms duration) produced a rebound burst that is similar to the bursts in the walking regime (Fig. 12.2b). The first burst has duration of 290 ms, a maximal frequency of 641 Hz, a minimum frequency of 51 Hz, and 82 spikes per burst (Fig. 12.1). Over the course of subsequent bursts, burst duration, spikes per burst, and spike frequency all decrease along with $h_{CaS}$ until the neuron returns to tonic spiking.

We developed an HCO based on the rebound bursting properties seen in the isolated neuron. The introduction of inhibitory synaptic coupling with the same equilibrium potential (−75 mV) as $I_I$, and a conductance of 5.5 nS results in antiphase bursting. The synapse is activated during spiking, and higher frequency spiking in the postsynaptic neuron results in increased summation of $I_{syn}$ in the postsynaptic neuron. The model exhibited two basic modes of activity, which we refer to as paw-shake and walking (Fig. 12.2). These regimes differ both in burst characteristics and dynamics of bursting. Formally, the model exhibits two paw-shake regimes which are very similar in every measure. They differ in number of spikes per burst; one has 5 spikes per burst and the other has 6 spikes per burst. Here, we refer to them as paw-shake regime. Paw-shake is a fast regime. It has a period of about 100 ms and a duty cycle of 32 % (Fig. 12.2a). The bursts display a maximum spike frequency of

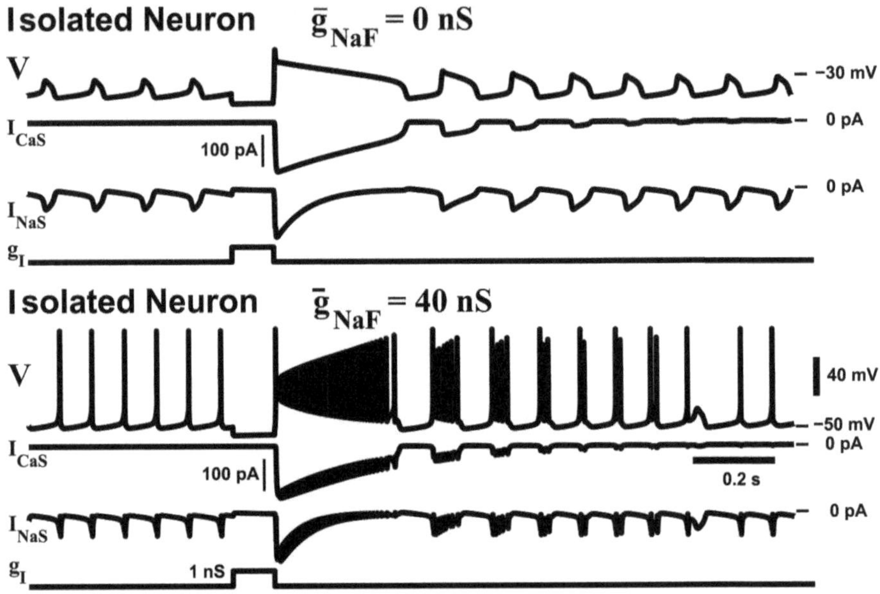


**Fig. 12.1** Activity in an isolated model neuron. Traces show spontaneous activity of the isolated neuron and response to a pulse of inhibitory conductance ( $g_I$). V is membrane potential of the isolated neuron. The neuron is isolated by setting all of the conductances in Neuron 2 to zero. The conductance for the inhibitory modulatory current ($I_I$) is shown as the bottom trace. It is zero, except for a 100 ms window, where it is equal to 1 nS. $I_I$ has an equilibrium potential of −75 mV

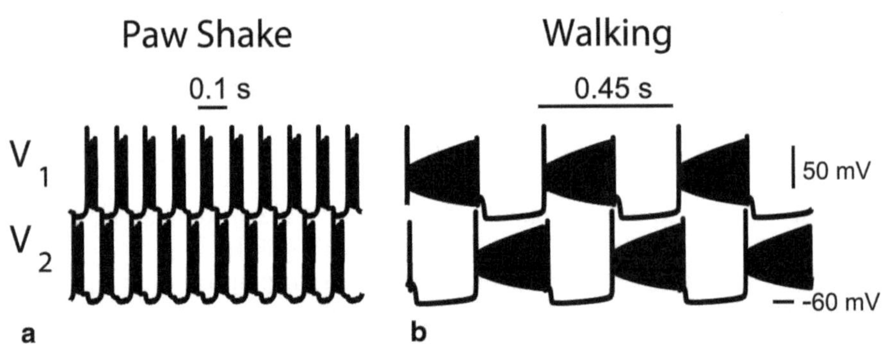


**Fig. 12.2** Coexistence of paw shake and walking regimes in the model. All simulations were given at least 1000 s to reach steady state

196 Hz at the beginning of a burst, a minimum frequency of 100 Hz at the end of a burst, and a mean frequency of 152 Hz.

The walking regime is the slowest of the exhibited regimes. It has a period of about 450 ms and a duty cycle of 52 %. The bursts display a maximum spike frequency of 620 Hz near the beginning of a burst, a minimum frequency of 100 Hz at the end of a burst, and a mean frequency of 399 Hz (Fig. 12.2b).

#### 12.2.2.2 Comparisons Between Walking and Paw-Shake Regimes

There are several major differences between the regimes. First is that there are very different levels of $I_{CaS}$ contribution. The peak value for $I_{CaS}$ in each burst is 40 times larger in the walking regime compared to paw-shake. Second is the magnitude of $I_{Syn}$. The mean value for $I_{Syn}$ in each burst is twice as larger in the walking regime compared to paw-shake. Finally, there is a large difference in maximal frequency between regimes. Paw-shake has a peak frequency of 196 Hz and walking has a peak frequency of 620 Hz. These three measures are interdependent. Increases in $I_{CaS}$ increases spike frequency, which in turn increases $I_{Syn}$ through summation of inhibitory postsynaptic potentials. All of this worked to change the period of bursting activity.

There are also many similarities between these bursting regimes. Burst initiation in all regimes is the result of $I_{NaS}$ activation/deinactivation. Likewise, burst termination is the result of inactivation of the slow inward currents. In the case of walking, escape of the postsynaptic neuron from inhibition also plays a role in burst termination, but this escape occurs when the presynaptic neuron's spike frequency falls close to 100 Hz.

#### 12.2.2.3 Switching Regimes

An important feature of multistable systems such as the developed HCO is that transient perturbations can cause a switch between regimes. Studying how our HCO model switches between the walking and paw-shake regimes can help us make testable predictions about these behaviors in the cat. In a CPG, perturbations can come from sensory neurons or brain areas. Theoretically, an object attached to a cat's paw could activate cutaneous afferents that synapse with interneurons comprising CPG circuitries and induce a switch between walking and paw-shake regimes. We used pulses of conductance (either excitatory or inhibitory) to mimic such perturbations with the goal of studying CPG transitions between walking and paw-shake regimes.

*Switch from Walking to Paw-Shake* Switching from walking to paw-shake CPG regimes requires inactivating $I_{CaS}$. This is possible by introducing an excitatory modulatory current $I_E$ with an equilibrium potential of +70 mV. The conductance ($g_E$) was set so that both neurons would be spiking during stimulation (Fig. 12.3, during $g_E$ pulse). As long as the neurons spiking tonically, $I_{CaS}$ will inactivate (as in Fig. 12.1). The threshold (in terms of the values for the inactivation variables) for switching between regimes is closer to paw-shake than walking. This means $I_{CaS}$ must be almost completely inactivated for the model to switch to paw-shake. Therefore, the length of stimulation time (pulse duration) depends on the time constant of inactivation for $I_{CaS}$, and thus a long pulse duration is needed for the switch from walking to paw-shake.

*Switch from Paw-Shake to Walking* A brief hyperpolarization to an isolated model neuron can cause a substantial deinactivation of $I_{CaS}$ and a large rebound burst

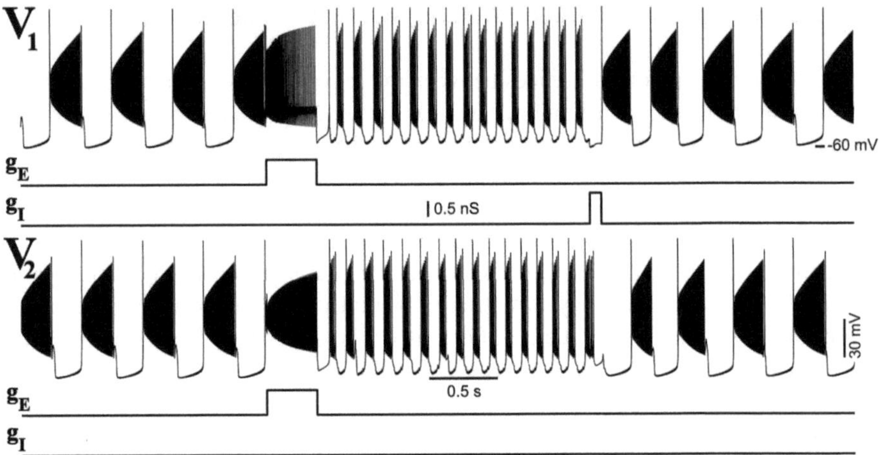


**Fig. 12.3** Switching back and forth between rhythms. $V_1$ and $V_2$ represent Neuron 1 and Neuron 2, respectively. The neurons begin in the walking rhythm. An optimally timed pulse of excitatory conductance ( $g_E$ for $I_E$, $g_E = 0.8$ nS, $I_E$ equilibrium potential = 70 mV, pulse duration = 370 ms) delivered to both neurons causes a switch to paw-shaking. During the pulse, cells are not firing in antiphase. A pulse of inhibitory conductance ( $g_I$ for $I_I$, $g_I$ = 0.85 nS, $I_I$ equilibrium potential = −75 mV, pulse duration = 90 ms) is delivered to neuron 1 only, causing a rebound burst and a switch back to walking

(Fig. 12.1). This is equally true in the coupled neurons. A brief hyperpolarization of sufficient magnitude could produce a rebound burst just as long as a burst in the walking regime (Fig. 12.1). In the isolated neuron, the resulting bursting would be transient. However, if a hyperpolarization was given to a neuron in the HCO while in the paw-shake rhythm, it could cause a switch to the walking regime.

Delivering a hyperpolarizing pulse to a neuron in the HCO while in paw-shake can have two outcomes, depending on the strength of the hyperpolarization. A hyperpolarization will still produce a rebound burst with a longer period and a higher spike frequency than a normal paw-shake burst. The rebound burst will inhibit the non-stimulated neuron more strongly and deinactivate $I_{CaS}$. If the original hyperpolarization is strong enough, each subsequent burst will have a longer period and higher spike frequency until the HCO reaches the steady state walking rhythm. If the original hyperpolarization is too small, each subsequent burst will be smaller, until the HCO returns to the steady state paw-shake regime.

External inhibitory command signal was simulated using an inhibitory modulatory current $I_I$, with an equilibrium potential of −75 mV. By having $I_I$ active in one neuron for as little as 45 ms it was possible to induce a switch from paw-shake to walking. Longer pulses were needed for a rapid transition (Fig. 12.3, 90 ms inhibitory conductance pulse), although shorter pulse durations could have the same effect if $g_I$ was larger. Depending on the pulse duration and the conductance of $I_I$, the transition could occur very rapidly, with the walking regime reaching its normal period within two cycles as in Fig. 12.3. Very strong inhibition (higher conductance or longer pulse duration) causes a seizure-like rebound burst followed by normal walking activity. Phase of stimulation had little effect on transitions, except at very low values of $g_I$ (data not shown).

A brief inhibition of one neuron, as in Fig. 12.3, is the fastest way to produce a transition. It is possible to produce a switch with inhibitory pulses to both neurons. These transitions always take longer since both neurons will burst at the same time upon release of the inhibition. Persistent inhibition of sufficient strength will also cause a switch to walking (after several seconds) or actually make the paw-shake regime disappear entirely (data not shown).

## 12.3 Neuromechanical Simulations of Walking and Paw-Shake

The previous sections described a multifunctional CPG that can generate fast paw shake-like and slow walking-like rhythmic reciprocal activity depending on parameters of the HCO and input to the system in the form of a conductance pulse. In the following sections, we present preliminary results of neuromechanical simulations of cat walking and paw-shake developed using the AnimatLab software for neuromechanical simulations (Cofer et al. 2010). In these simulations, the musculoskeletal model of the hindlimb is controlled by a CPG that consists of a half-center oscillator (HCO) generating rhythmic activity with frequency of either 2 or 10 Hz. Motoneurons activating 12 flexor and extensor muscles in each hindlimb (Fig. 12.4) receive excitatory input from the corresponding half-center oscillator and also excitatory and inhibitory inputs from muscle length- and force-sensitive afferents (Fig. 12.5).

The goal of these neuromechanical simulations was to examine if a single multifunctional HCO could reproduce two very different rhythmic behaviors, walking and paw-shake. These behaviors differ not only in the cycle duration but also in movement mechanics and patterns of muscle activity (Smith et al. 1985; Smith and Zernicke 1987; Prilutsky et al. 2004, 2009; Klishko et al. 2012; Mehta and Prilutsky 2014). In particular, during walking, flexor and extensor muscles are activated reciprocally (Markin et al. 2012), whereas during paw-shake response, anterior and posterior muscles, rather than flexors and extensors, appear to demonstrate reciprocal activation (Smith et al. 1985; Pratt et al. 1991; Klishko et al. 2012).

### *12.3.1 Musculoskeletal Model of Cat Hindlimb*

A cat hindlimb model was assembled in the Body Plan Editor of AnimatLab as a system of 5 rigid segments representing the pelvis, thigh, shank, tarsals and toes. Inertial properties of each segment, i.e., segment mass, position of the center of segment mass, and moment of inertia with respect to the frontal axis through the center mass of the segment, were computed from the measured cat mass and segment

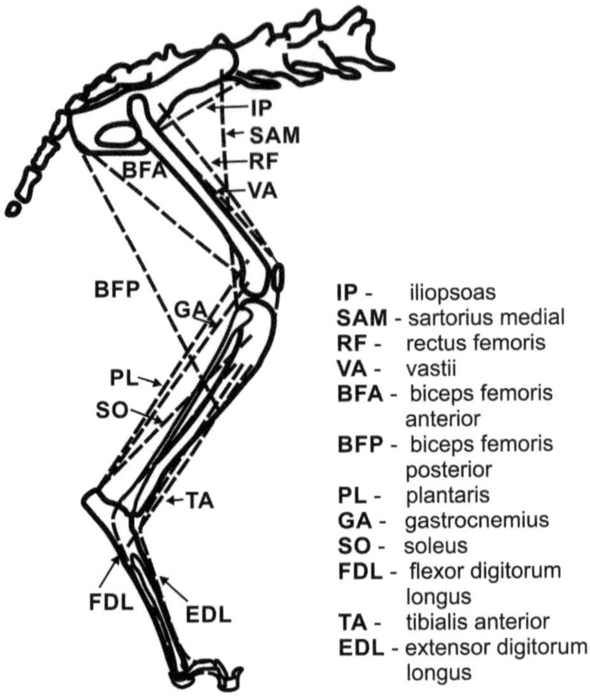


**Fig. 12.4** Musculoskeletal model of the cat hindlimb (model of the second hindlimb is identical). The hindlimb skeleton is represented by a system of 5 rigid segments (pelvis, thigh, shank, tarsals and toes) interconnected by frictionless hinge articulations (hip, knee, ankle and metatarsophalangeal, MTP, joints). The model is actuated by 12 Hill-type muscles that include iliopsoas (*IP*, hip flexor), sartorius medial (*SAM*, hip and knee flexor), rectus femoris (*RF*, hip flexor and knee extensor), BFA (hip extensor), BFP (hip extensor and knee flexor), vastii (knee extensor), tibialis anterior (*TA*, ankle flexor), extensor digitorum longus (*EDL*, ankle and MTP flexor), plantaris (*PL*, knee flexor and ankle extensor), gastrocnemius (*GA*, knee flexor and ankle extensor), soleus (*SO*, ankle extensor), flexor digitorum longus (*FDL*, ankle and MTP extensor)

length using the regression equations (Hoy and Zernicke 1985). The hindlimb segments were interconnected by frictionless hinge joints. The hindlimb skeleton was actuated by 12 muscles (Fig. 12.4). Points of each muscle origin and attachment were estimated based on the geometric model of a cat hindlimb (Goslow et al. 1973). Muscle paths were modeled as sets of straight lines connecting selected via points so that the moment arms of each muscle at joints it crosses corresponded roughly to the values reported in the literature (Boyd and Ronsky 1998; MacFadden and Brown 2007, 2010). Muscles and muscle spindles (see below) were represented by Hill-type models consisting of an elastic component (tendon) attached in series with the parallel combination of a parallel elastic component (passive connective tissues surrounding muscle fibers and compartments) and a contractile component (the cross-bridges). The physiological properties of the contractile component were described in the model by the muscle membrane voltage-contractile force curve, isometric force-length curve, force-velocity relationship, the optimum muscle length

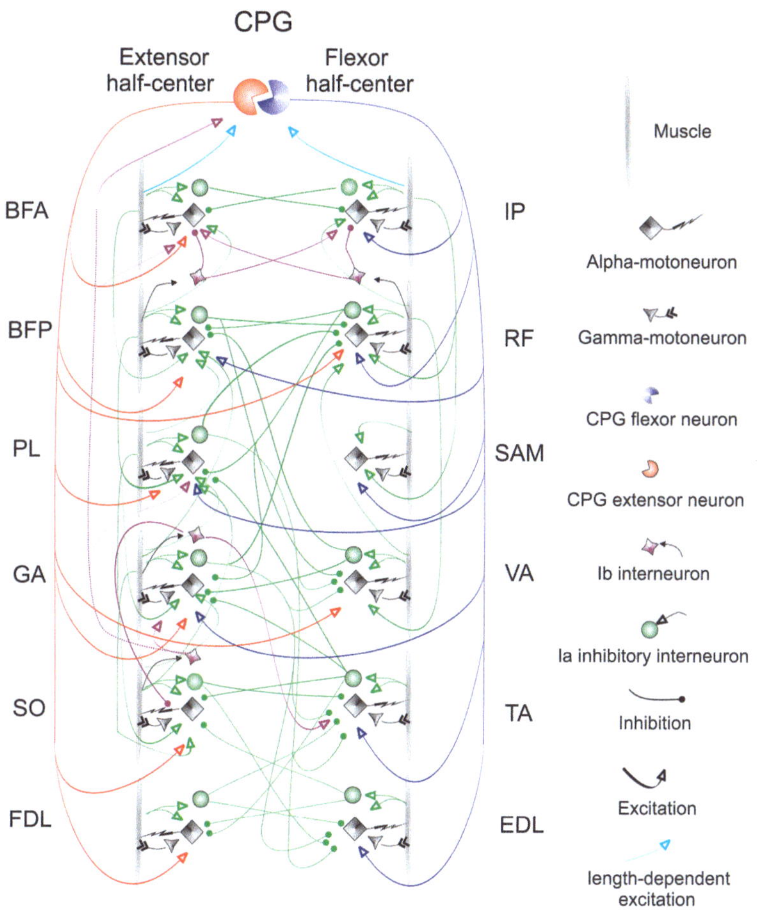


**Fig. 12.5** Neural control model of cat hindlimb. The model consists of a central pattern generator (*CPG*) and basic reflex pathways. CPG generates symmetric flexor and extensor rhythmic activity with a frequency of either 10 Hz (paw-shake rhythm) or 2 Hz (locomotion rhythm). The extensor and flexor half-centers provide excitatory input to motoneurons of one-joint extensors and flexors, respectively. Motoneurons of two-joint muscles may receive excitations from both half-centers in flexor and extensor phases. Reflex pathways include stretch reflex with reciprocal inhibition that is mediated by Ia spindle afferents and Ia-inhibitory interneurons, respectively; force-dependent inhibition and excitation from two-joint muscles to one-joint synergists and one-joint antagonists, respectively; force-dependent excitation of the CPG extensor half-center from SO motoneuron during the extensor phase of the CPG; and length-dependent excitation from hip flexor and extensor muscle to the CPG flexor- and extensor-half center, respectively (see text for more details). For muscle abbreviations see Fig. 12.4

and maximum force (Table 12.1; see also Cofer et al. 2010). Each muscle in the model had sensors of length (the spindle) and force (the Golgi tendon organ). The spindle force was generated in response to gamma motoneuron input and stretch of the muscle contractile component. Spindle force generated depolarizing membrane current in the spindle neuron that fired when the membrane potential exceeded the

**Table 12.1** Parameters of the muscle model

| Muscle | $B$ (Ns/m) | $K_{pe}$ (N/m) | $K_{se}$ (kN/m) | $F_{max}$ (N) | $PE_{len}$ (%) | $L_{rest}$ (cm) | $L_{width}$ (cm) |
|---|---|---|---|---|---|---|---|
| IP | 30.9 | 385 | 13.351 | 150 | 55 | 7.0 | 5.0 |
| BFA | 8.2 | 227 | 6.464 | 120 | 36 | 12.0 | 5.0 |
| BFP | 40.0 | 806 | 27.837 | 250 | 56 | 11.5 | 5.5 |
| SAM | 3.3 | 100 | 2.357 | 20 | 91 | 12.1 | 5.5 |
| RF | 25.0 | 642 | 15.844 | 122 | 20 | 11.1 | 3.0 |
| VA | 20.0 | 3125 | 34.014 | 250 | 27 | 11.1 | 8.0 |
| PL | 18.4 | 370 | 11.905 | 200 | 90 | 12.1 | 4.2 |
| GA | 18.4 | 370 | 11.905 | 200 | 23 | 12.5 | 4.2 |
| SO | 2.0 | 36 | 1.667 | 21 | 58 | 9.1 | 12.0 |
| TA | 10.1 | 191 | 4.643 | 65 | 57 | 10.1 | 2.8 |
| EDL | 10.0 | 105 | 2.484 | 40 | 35 | 11.7 | 1.7 |
| FDL | 20.0 | 433 | 7.738 | 130 | 17 | 12.1 | 2.0 |

Muscle abbreviations are the same as in Fig. 12.4

$B$ is viscosity of the linear force-velocity relationship of the muscle contractile component, $K_{pe}$ is stiffness of the parallel elastic component, $K_{se}$ is stiffness of the serial elastic component, $F_{max}$ is maximum muscle isometric force at optimal (resting) length of the contractile component, $PE_{len}$ is percentage of the parallel elastic component resting (optimal) length in the total muscle-tendon unit length, $L_{rest}$ is resting (optimal) length of the muscle, $L_{width}$ is the muscle length range for active force production in the isometric force ($F_L$) -length ($L$) relationship: $F_L = 1-(L-L_{rest}/L_{width})^2$. For details on the AnimatLab muscle model see (Cofer et al. 2010)

threshold. Tendon Golgi organ sensory signal was proportional to force produced by the muscle.

For simulations of walking, each hindlimb was connected to the trunk, which could move without friction along an elastic string (stiffness coefficient 5 MN/m) thus providing partial weight support to substitute support from the forelimbs that were not included in the model.

### *12.3.2 Neural Control Model*

A model of the neural control circuitry activating hindlimb muscles during walking and paw-shake was developed in the Behavior Editor of AnimatLab. The HCO was modelled by two spiking integrate-and-fire single compartment neurons mutually inhibiting each other via hyperpolarizing IPSP. Output of extensor and flexor half-centers activated motoneurons represented by non-spiking voltage- and time-dependent ionic conductance neurons (Vogels et al. 2005) via the same non-spiking interneurons that transformed spiking trains from the HCO into integrated membrane potential. The output of a single motoneuron represented an integrated output (membrane voltage) of a motoneuronal pool activating a single muscle. Motoneurons activated corresponding muscles that produced force in accordance with the stimulus (membrane voltage)-force curve (Cofer et al. 2010). Each motoneuron received excitatory inputs from the HCO and excitatory and inhibitory inputs from various reflex pathways (Table 12.2).

All motoneurons received length- and velocity-dependent monosynaptic excitatory inputs from the Ia spindle afferents of anatomical synergists (Fig. 12.5, open arrow heads; Table 12.2). Reciprocal inhibition among anatomical antagonists (e.g., soleus and tibialis anterior, vastii and gastrocnemius) was mediated via Ia inhibitory interneurons (Fig. 12.5, green circles; Table 12.2; Feldman and Orlovsky 1975; Geertsen et al. 2011). Excitation of one-joint antagonists and inhibition of one-joint agonists from two-joint muscles (e.g., excitation of tibialis anterior and inhibition of soleus from gastrocnemius) were mediated by Golgi tendon force-dependent pathways (Fig. 12.5; Table 12.2; Laporte and Lloyd 1952; Eccles et al. 1957; Nichols 1999, see also Chap. 3 by Nichols et al. 2015). The neural control model also incorporated afferent pathways that regulated the operation of the CPG during locomotion. Excitatory length-dependent afferent input from a hip flexor iliopsoas to the flexor half-center triggered transition from extensor to flexor phase of rhythmic activity, whereas excitatory length-dependent afferent signals from hip extensor biceps femoris anterior to the extensor half-center triggered transition from flexor to extensor phase (Fig. 12.5; Table 12.3; Kriellaars et al. 1994; Hiebert et al. 1996; McVea et al. 2005). A force-dependent afferent input during the extensor phase of the cycle from ankle extensor soleus to the extensor half-center and to extensor motoneurons enhanced activity of hindlimb extensors (Fig. 12.5; Table 12.3; Conway et al. 1987; McCrea 2001; Pearson 2008).

## *12.3.3 Neuromechanical Simulations*

### 12.3.3.1 Paw-Shake

In order to qualitatively reproduce patterns of muscle activity typical for paw-shake, i.e. alternating reciprocal activity of anterior and posterior hindlimb muscles (Smith et al. 1985; Pratt et al. 1991; Klishko et al. 2012), as opposed to alternating activity of flexor and extensor muscles observed during locomotion (Grillner 1981; Rossignol 2006; Markin et al. 2012), gains of afferent excitatory and inhibitory feedback actions on motoneurons were modified. Specifically, autogenic (from a muscle to its own motoneuron) and heterogenic (from a muscle to synergist motoneurons) Ia afferent feedback gains were increased (Table 12.2) given the very high firing rates of Ia afferents from various cat muscles during paw-shake (Prochazka et al. 1989). In addition, the excitatory force (Ib)-dependent input from ankle extensor soleus to the extensor half-center oscillator and extensor motoneurons during the extensor phase was removed to make the extensor and flexor phases equal in duration (Table 12.3).

Before each simulation, the initial hindlimb position approximately corresponding to that observed in experiments (see below) was set by tonic activity of IP and BFP motoneurons and maintained for 2 s. In this initial position hip, knee, ankle and

**Table 12.2** Weight coefficients of neural pathways to motoneurons during walking

| MN or muscle | $CPG_F$ (μS) | $CPG_E$ (μS) | $Ia_0$ (kHz/m) | $C_{Ia+}$ (nAs)/$D_{Ia+}$ (nA) | $C_{Ia-}$ (nAs)/$D_{Ia-}$ (nA) | $C_{Ib+}$ (nAs)/$D_{Ib+}$ (nA) | $C_{Ib-}$ (nAs)/$D_{Ib-}$ (nA) |
|---|---|---|---|---|---|---|---|
| IP | 0.1 | – | 100 | IP(0.1/0) RF(0.1/0) | BFA(0.3/0) | BFP(0.001/0) | RF(0.001/0) |
| BFA | – | 0.002 | 50 | BFA(0.1/0) BFP(0.08/0) | IP(0.1/0) | RF(0.001/0) | BFP(0.005/0) |
| BFP | 0.003 | 0.01 | 30 | BFP(0.05/0) GA(0.1/0) BFA(0.1/0) | VA(0.2/0)<br>RF(0.1/0) | | |
| SAM | 0.3 | 0 | 30 | SAM(0.2/0) IP(0.1/0) | | | |
| RF | 0.03 | 0.001 | 100 | RF(0.4/0) IP(0.1/0) VA(0.2/0) | BFP(0.5/0)<br>GA(0.4/−20)<br>PL(0.1/0) | | |
| VA | – | 0.003 | 40 | VA(0.2/0) RF(0.1/0) | GA(0.1/0)<br>PL(0.1/0)<br>BFP(0.05) | BFP(0.001/0) | RF(0.001/0) |
| PL | 0.001 | 0.05 | 30 | PL(0.1/−8) BFP(0.05/0) | RF(0.2/0)<br>TA(0.1/0) | | |
| GA | 0.001 | 0.05 | 30 | GA(0.2/0) BFP(0.05/0) | RF(0.2/0)<br>TA(0.1/0) | | |
| SO | – | 0.1 | 40 | SO(0.1/0) GA(0.01/0) PL(0.01/0) | EDL(0.1/0) | | GA(0.01/0) |
| TA | 0.1 | – | 50 | TA(0.2/−15) | GA(0.1/0)<br>PL(0.1/0)<br>SO(0.1/0)<br>FDL(0.05/0) | GA(0.002/0) | |

**Table 12.2** (continued)

| MN or muscle | $CPG_F$ (μS) | $CPG_E$ (μS) | $Ia_0$ (kHz/m) | $C_{Ia+}$ (nAs)/$D_{Ia+}$ (nA) | $C_{Ia-}$ (nAs)/$D_{Ia-}$ (nA) | $C_{Ib+}$ (nAs)/$D_{Ib+}$ (nA) | $C_{Ib-}$ (nAs)/$D_{Ib-}$ (nA) |
|---|---|---|---|---|---|---|---|
| EDL | 0.02 | – | 40 | EDL(0.1/−10) | GA(0.1/−15)<br>PL(0.1/−15)<br>SO(0.3/−10)<br>FDL(0.1/−10) | | |
| FDL | – | 0.05 | 40 | FDL(0.1/−10) | TA(0.5/−55)<br>EDL(0.1/−10) | | |

*MN* is motoneuron, *MN* abbreviations are the same as muscle abbreviations in Fig. 12.4. $CPG_F$ and $CPG_E$ are conductances of nicotinic acetylcholine (Ach) synapses mediating excitatory inputs to motoneurons from the flexor and extensor CPG half-centers, respectively; $Ia_0$ is the firing rate constant of the Ia stretch receptor; $C_{Ia+}$ and $D_{Ia+}$ are constants in the relationship describing the autogenic or heterogenic motoneuron excitatory input current ($I$, nA) as a function of the Ia afferent firing rate (R, Hz): $I = C_{Ia+}R + D_{Ia+}$; $C_{Ia-}$ and $D_{Ia-}$ are constants in the relationship describing the motoneuron inhibitory input current ($I$, nA) as a function of the Ia inhibitory interneuron firing rate (R, Hz): $I = C_{Ia-}R + D_{Ia-}$; $C_{Ib+}$ and $D_{Ib+}$ are constants in the relationship between the motoneuron excitatory input current ($I$, nA) as a function of the Ib afferent firing rate (R, Hz): $I = C_{Ib+}R + D_{Ib+}$; $C_{Ib-}$ and $D_{Ib-}$ are constants in the relationship between the motoneuron inhibitory input current ($I$, nA) as a function of the Ib afferent firing rate (R, Hz): $I = C_{Ib-}R + D_{Ib-}$. During walking, IP gamma-motoneuron was stimulated with 32 nA input current; all other motoneurons did not have gamma-motoneuron input

The following changes in the weight coefficients were made to simulate paw-shake. To set the initial hindlimb posture before paw-shake, IP and BFP motoneurons were stimulated with input current 17 and 10 nA, respectively. To increase sensitivity of spindle Ia afferents, the corresponding gamma-motoneurons received the following input currents: IP, 30 nA; BFA, 29.8 nA; BFP, 32 nA; RF 22 nA; SAM, 40 nA; GA, 32 nA; PL, 30 nA; VA, 24 nA; TA, 26 nA; SO, 20 nA; EDL, 24 nA; FDL, 20 nA. In addition, constants $C_{Ia+}/D_{Ia+}$ were changed for the following Ia autogenic and heterogenic inputs to motoneurons: BFA Ia to BFA MN, 0.3/0; RF Ia to RF MN, 0.4/−30; VA Ia to RF MN, 0.1/−4; GA Ia to GA MN, 0.2/−15.

AnimatLab models of neurons and synapses and their properties are described in (Cofer et al. 2010)

**Table 12.3** Weight coefficients of afferent pathways to CPG flexor and extensor half-centers during walking

| Afferent pathway | $C_{F+}$(nAs) | $C_{E+}$(nAs) |
|---|---|---|
| SO Ib | – | 0.3 |
| IP Ia | 0.01 | – |
| BFA Ia | – | 0.01 |
| IP II | 0.01 | – |
| BFA II | – | 0.01 |

Muscle abbreviations are the same as in Fig. 12.4. $C_{F+}$ and $D_{E+}$ are constants in the relationship describing the afferent excitatory input current ($I$, nA) to the CPG flexor and extensor half-center as a function of the afferent firing rate $(R, Hz)$: $I = C_{F+}R$ or $I = C_{E+}R$. Additional excitatory input from SO Ib afferents during the extensor phase of walking was also provided to hindlimb extensor motoneurons with the following weight constants: $C_{BFA+}$=0.15, $C_{RF+}$=0.08, $C_{VA+}$=0.1, $C_{PL+}$=0.03, $C_{GA+}$=0.1, and $C_{SO+}$=0.01

In paw-shake simulations, all weight coefficients of the Ib excitatory pathways operating during the extensor phase to the CPG extensor half-center and extensor motoneurons were set to zero

metatarsophalangeal joint angles were 105°, 49°, 93°, and 178°, respectively. Paw-shaking was generated by fast regime of the HCO (see Sect. 12.2.) that produced alternating bursting activity in the flexor and extensor half-centers with frequency of 10 Hz. Patterns of computed muscle forces were compared to low-pass filtered activity of cat hindlimb muscles that were recorded in cats with chronically implanted intramuscular EMG electrodes (Figs. 12.6 and 12.7).

Detailed description of surgical implantations and experiments can be found elsewhere (Gregor et al. 2006; Prilutsky et al. 2011; Mehta and Prilutsky 2014). Briefly, all surgical and experimental procedures were in agreement with US Public Health Service Policy on Humane Care and Use of Laboratory Animals and approved by the Institutional Animal Care and Use Committee of Georgia Institute of Technology. Surgery for implantation of EMG electrodes was performed under general anesthesia and aseptic conditions. Vital physiological parameters (temperature, respiration, heart rate and blood pressure) were monitored throughout the surgery. Pairs of teflon-insulated multi-stranded stainless steel fine wires (CW5402; Cooner Wire, Chatsworth, CA, USA) attached to a multi-pin Amphenol connector fixed on the scull were implanted in selected hindlimb muscles (majority of muscles shown in Fig. 12.4). After surgery, the animal received pain medication and antibiotics as needed and recovered for 2 weeks.

Prior to implantation surgery cats were trained to walk on a Plexiglas enclosed walkway with embedded small force plates (Bertec Corporation, Columbus OH, USA) using positive reinforcement with food. Reflective markers placed on major hindlimb joints with double sided adhesive tape and a 6-camera motion capture system Vicon (Oxford, UK) were used to record kinematics of walking and paw-shake. Paw-shake responses were evoked by attaching a piece of adhesive tape to the paw pad on the right hindlimb. EMG activity, kinematics and ground reaction forces during paw-shake and walking were recorded synchronously at a sampling rate of 3000, 120, and 360 Hz, respectively. Kinematic and ground reaction force measurements were used to identify stance and swing phases during walking and cycles during paw-shakes.

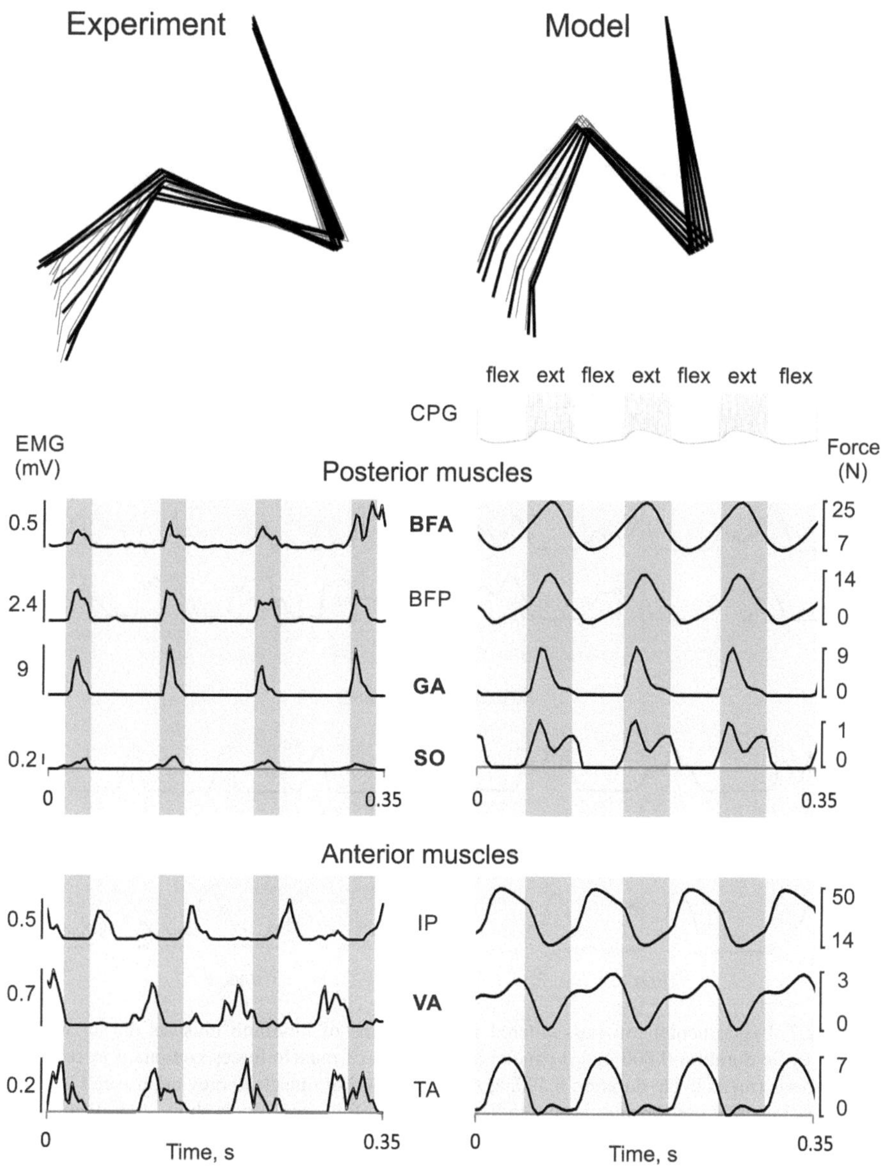


**Fig. 12.6** Experimental low-pass filtered EMG patterns of hindlimb muscles during paw-shake (mean cycle duration 0.100 s; *left* panels) and patterns of muscle forces computed in computer simulations (mean cycle duration 0.088 s; *right* panels). For muscle abbreviations see Fig. 12.4; abbreviations of extensor muscles are indicated by bold font. *Top* panels show stick-diagrams of the hindlimb during a recorded and simulated paw-shake cycle; *gray lines* indicate downward foot movement in one half-cycle, *black lines* correspond to upward foot movement in the next half-cycle. Second panel in the right column demonstrates simulated bursting activity of the flexor (*light lines*) and extensor (*darker lines*) half-centers of the CPG. Next panels show experimental low-pass filtered EMG (*left*) and simulated force patterns (*right*) of selected posterior muscles. Their activity and force production occur mostly during the extensor phase of the CPG activity (*dark gray rectangles*). The *bottom* panels demonstrate experimental low-pass filtered EMG (*left*) and simulated force patterns (*right*) of selected anterior muscles. Their activity and force production occur mostly during the flexor phase of the CPG activity (*light gray rectangles*). Experimental results are from one representative cat

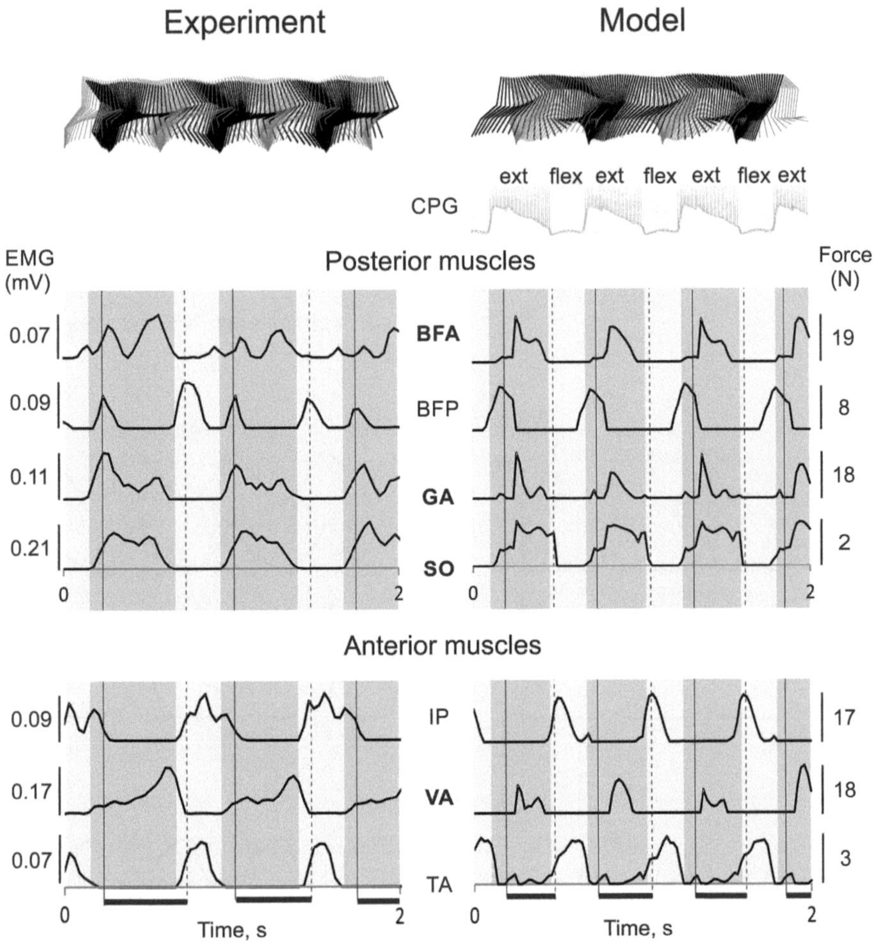


**Fig. 12.7** Experimental low-pass filtered EMG patterns of hindlimb muscles during walking (mean cycle duration 0.800 s; *left* panels) and patterns of muscle forces computed in computer simulations (mean cycle duration 0.500, s; *right* panels). For muscle abbreviations see Fig. 12.4; abbreviations of extensor muscles are indicated by bold font. *Top* panels show stick-diagrams of *left* (*gray lines*) and *right* (*black lines*) hindlimbs during recorded and simulated walking. Second panel demonstrates bursting activity of the flexor (*light lines*) and extensor (*darker lines*) half-centers of the CPG. Next panels show experimental low-pass filtered EMG (*left*) and simulated force patterns (*right*) of selected posterior muscles. The bottom panels demonstrate experimental low-pass filtered EMG (*left*) and simulated force patterns (*right*) of selected anterior muscles. One-joint hindlimb extensors (BFA, GA, SO, VA) are active during extensor phases of CPG activity (*dark gray rectangles*), while one-joint flexors (IP, TA) are active during flexor phases of CPG activity (*light gray rectangles*); two-joint BFP demonstrates two bursts of EMG activity in the cycle—at the swing-stance and stance-swing transitions, however simulations show only one period of BFP force production at the swing-stance transitions. *Horizontal black bars* at bottom indicate stance phases

Computer simulations of cat paw-shakes demonstrated that posterior muscles (BFA, BFP, GA and SO, Fig. 12.4) produced peak muscle forces in phase with the extensor half-center of the CPG, whereas the anterior muscles (IP, VA and TA) produced peak forces during the flexor phase (Fig. 12.6, right panel). Note that VA, one-joint knee extensor, produced force when it did not receive excitatory input from the CPG, while it did not produce force during most of the extensor phase in which it received excitation from the extensor half-center. The low-pass filtered EMG patterns were qualitatively similar to the simulated muscle force patterns except TA appears to generate EMG activity bursts in both phases (Fig. 12.6, left panel).

#### 12.3.3.2 Walking

In order for the model to generate walking-like patterns of muscle activity and hindlimb kinematics, relatively few changes in the reflex pathways were made apart from decreasing rhythm of the HCO from 10 to 2 Hz. The changes included the decrease in gain of Ia afferent feedback from all muscles (Table 12.2) and the introduction of a force-dependent excitation from SO to the CPG extensor half-center and extensor motoneurons during the extensor phase (Table 12.3). The latter change made the extensor phase duration longer than the flexor phase even when the hindlimbs were elevated above the ground and there was no contact of the paw with the ground (air stepping).

Neuromechanical simulations of walking produced hindlimb kinematics and muscle force patterns that were qualitatively similar to those recorded in a representative cat (Fig. 12.7). Specifically, extensor muscles BFA, VA, GA and SO produced force during the extensor (stance) phase of walking, whereas flexors IP and TA produced force during the flexor (swing) phase. Simulations generated force of a two-joint BFP muscle during the swing-stance transitions, although this muscle normally produces two activity bursts at the swing-stance and stance-swing phase transitions (see Chap. 5 by Shevtsova et al. 2015 and Fig. 12.7).

Although the mechanisms underlying the formation of different muscle synergies during paw-shake and walking are not clear and require additional studies, the preliminary computer simulations of paw-shake responses suggested that the knee extensor VA activity burst during the "wrong" flexor phase of the HCO may be caused by Ia length and stretch velocity sensitive afferents from fast elongating VA during this phase (Klishko et al. 2011, 2012). A similar mechanism might explain the absence of EMG burst of VA during the extensor phase of the HCO; although VA motoneuron receives excitation from the extensor half-center during this phase, the reciprocal inhibition via the Ia-inhibitory interneurons (Fig. 12.5) might overcome the CPG excitatory input due to high velocity of stretch of knee flexors, i.e. VA antagonists (BFP, GA and PL) and the corresponding high activity of their Ia afferents (Prochazka et al. 1989).

## 12.4 Possible Mechanisms of Bistability in the Multifunctional Half-Center Generator

To survive in nature animals should be able to exhibit a large and diverse repertoire of basic behaviors. The rhythmic behaviors like walking, swimming, breathing, etc. are controlled by specialized neuronal circuits, central pattern generators (CPGs). Combining the circuits for different behaviors could make the nervous system more energetically efficient, thus increasing survival chances. There is a great deal of evidence that groups of interneurons can participate in the control of multiple behaviors (Berkowitz et al. 2010). This consideration leads to the concept of multifunctional CPGs. The behavior produced by a multifunctional CPG would be dependent on context and sensory information. Sensory neurons can affect rhythmic activity by directly exciting/inhibiting the CPG neurons. They could also recruit additional neurons to participate in the rhythm suppression of a subset of neurons in such a way as to bias the circuit towards one pattern. The presence of neuromodulators, which is often context dependent, can reconfigure the output of the CPG by changing the electrical properties of neurons and synapses (Hooper and Moulins 1989; Marder et al. 1997; Doi and Ramirez 2008). It is also possible that multiple behaviors could be controlled by a single, multistable CPG. This CPG would be capable of producing multiple behaviors only requiring a single transient input (central or from sensory feedback) to switch between patterns. Our model is a multistable CPG that can produce two different patterns of activity (Fig. 12.2).

There is a large amount of evidence for the existence of multifunctional neurons that participate in more than one motor rhythm (Soffe 1993; Jing and Weiss 2001; Berkowitz 2010). Experiments in cats where a scratching rhythm slowly changed into a walking rhythm provided strong early evidence for overlap in neurons controlling multiple behaviors (Berkinblit et al. 1978), however see (Frigon and Gossard 2010). Much of the best evidence for multifunctional neurons in vertebrates comes from the study of turtles. All spinal interneurons active during one type of scratching in the turtle are also active during other types of scratching, although firing rates vary. There is less overlap for other behaviors that involve the legs, such as swimming and leg withdrawal, but overlap still exists (Berkowitz 2002, 2010). Simultaneous stimulations evoking different motor behaviors and modeling have also demonstrated that there must be either overlap or strong interactions between scratching and swimming CPGs (Hao et al. 2011). In these examples, the exact mechanism determining which pattern the interneurons will produce is not known, although it is likely that sensory information/receptive fields play a role.

*Xenopus* tadpoles also display two rhythmic behaviors (fast and slow) that appear to originate from a multifunctional CPG: struggling and swimming (2–10 Hz bursting and 10–25 Hz spiking, respectively; speed of oscillation varies across length of the body). These rhythms, and the transition between them, share many features with the behaviors we have modeled. The struggling rhythm occurs when pressure is applied to a tadpole's body (such as being grasped by a predator). This rhythm can also be evoked with repetitive electrical stimulation of the skin sensory pathway. The swimming rhythm can be evoked with a single stimulation of the same sensory neurons and can persist for a long time. Swimming will occur immediately upon release of pressure or stimulation that causes struggling, and usually begins at a high speed and slows, either to a stop or to a steady velocity. 76 % of the motoneurons and 68 % of the premotor interneurons involved in struggling are also active during swimming. The additional neurons active during struggling are all of the same classes as those active during swimming (Soffe 1993). The input from the sensory neurons (transient vs. continual) appears to determine which rhythm will be evoked (Soffe 1997). Studies of the transition from struggling to escape swimming revealed that the transition could be gradual under certain conditions (Green and Soffe 1996). Larval zebrafish display similar struggling and swimming behaviors with overlap in neuronal activation. However, in zebrafish, separate classes of inhibitory interneurons are recruited as well (Liao and Fetcho 2008).

Similar to tadpoles, leeches also display two patterns, swimming and crawling, generated by a multifunctional CPG. 93 % of the neurons active during swimming are active during crawling, which represent about half of the total neurons involved in crawling. Neurons have been identified that can perturb one or both rhythms if stimulated (Briggman and Kristan 2006). Unlike tadpole struggling and swimming, the presence of neuromodulation plays a role in determining which rhythm these cells produce; application of serotonin will elicit fictive swimming behavior, whereas dopamine will stop swimming and produce crawling (Crisp and Mesce 2004). It is known that serotonin modulates several voltage-dependent currents, including the persistent sodium current (Angstadt and Friesen 1993). In this case, multifunctionality requires the presence of a neuromodulator, at least one of which, dopamine, has been shown to come from higher level inputs (Crisp and Mesce 2004).

Unlike the previous examples, multistability as a mechanism for multifunctional CPGs means that sustained input is not necessary to produce multiple behaviors. Transient sensory input can produce stable changes in network activity. Multistability of circuits and single neurons present a basic feature for mechanisms underlying multifunctional CPGs. Multistability in single neurons has been described in a variety of neurons. Bistability between spiking and silence has been seen in neurons and in models (Jalife and Antzelevitch 1979; Hahn and Durand 2001; Paydarfar et al. 2006; Malashchenko et al. 2011), as well as bistability between spiking and silence (Fuentealba et al. 2004), bursting and spiking (Cymbalyuk et al. 2002; Shilnikov et al. 2005; Fröhlich and Bazhenov 2006), multiple bursting regimes (Butera 1998) and multiple spiking regimes (Egorov et al. 2002; Cymbalyuk and Shilnikov 2005). Switching between silence and spiking in motoneurons has been ascribed to the activation of an L-type calcium channel in the dendrites (Carlin et al. 2000).

Networks of neurons, like CPGs, can display multistability as well, either through multiple modes of firing (silent, spiking, bursting), or through different phase relationships. Network can have multistability without its component neurons being multistable (Bose et al. 2001). In our model we consider a classical paradigm of half-center oscillator (Rybak et al. 2006). This is a common motif in CPGs for producing alternating activity. Multistability in networks can arise through the dynamics of intrinsic or synaptic currents as opposed to the presence or absence of external input. Our HCO model is multifunctional due to the multistability arising from the dynamics of the two neurons' slow intrinsic currents. With the exact same parameters, multiple rhythms coexist, all of which are stable and will persist indefinitely if left undisturbed. Similar models have been made that display multistability through different mechanisms. Synaptic dynamics in a network can result in multistability. The presence of short term synaptic depression can lead to bistability between two modes: one with a weak synapse and one with a strong synapse. Synaptic depression is both voltage dependent and slow. Synaptic conductance will decrease over the course of consecutive action potentials and recover during silence. Thus, changes in synaptic conductance are similar in mechanism and effect to the changes in $I_{CaS}$ conductance in our model. A two neuron model with depressing synapses can also produce a fast and slow rhythm with the ability to switch back and forth between rhythms with pulses of injected current (Manor and Nadim 2001). Changes in synaptic current are part of the multistability of our model, however, these changes are a result of changes in intrinsic currents ($I_{CaS}$ and $I_{NaS}$), not their cause. Multistability arising from synaptic dynamics does not depend on intrinsic currents (Bose et al. 2001). The time constant of inhibition can also impart bistability onto a network (Bose and Kunec 2001). Network multistability in terms of the burst phase between four neurons has also been modeled (Chuan et al. 2004).

## 12.5 Future Studies

### *12.5.1 Predictions*

Based on our findings, it is possible to make several predictions about paw-shaking and walking in cats. In the future, these predictions of the neuromechanical model can be tested. A rapid transition to paw-shaking requires an optimally timed pulse of conductance. This is supported by observations that paw-shaking is always initiated during swing phase of walking (Carter and Smith 1986). It is also necessary for the stimulus to last on the order of the time constant of inactivation for $I_{CaS}$, which translates to close to the duration of one burst. In the model, at the end of the stimulus pulse, paw-shaking will be slower than normal, then speed up as it approaches its steady state period (usually the difference in period of about 40–30 ms). This is in conflict with data seen in actual cats, which show that paw-shaking generally increases in period from the first to last burst (Smith et al. 1985; Koshland and Smith 1989). This discrepancy could be accounted for by the presence of additional currents or neurons in the cat or by differences in the voltage dependence of inactivation for $I_{CaS}$.

In terms of the transition back to walking from paw-shaking, the model predicts there will be an asymmetry to the bursts between the two half-centers of the HCO for the first 2–3 cycles post inhibition, which could manifest as a slight discoordination between muscles/legs immediately after paw-shaking. Although, as with the initial slowness of the paw-shaking rhythm, it is possible that the presence of additional neurons would help stabilize the rhythm faster than two neurons alone could.

### 12.5.2 *Pharmacological Manipulations*

Paw-shaking and walking could be considered a more exited state and more inhibited state, respectively. Any change in excitation or inhibition will affect both rhythms. We could predict that a brief blockade of the inhibitory synaptic interaction would cause a switch to paw-shaking at about the same speed as excitatory stimulation, and weakening of the synapse would promote paw-shaking. In contrast, strengthening the synapse could abolish the paw-shaking rhythm or make it much more difficult to elicit. Paw-shaking in adult chronic spinalized cats can be eliminated by application of noradrenergic agonist clonidine, which will also increase the period of walking (Barbeau et al. 1987). Experiments in cats show norepinephrine inhibits 43 % of interneurons in the ventral horn of the spinal cord, excites 6 % and has no effect on the rest (Jordan et al. 1977). Application of NMDA increases expression of paw-shaking, likely by exciting sensory neurons (Chau et al. 2002). Both of these results are in line with our model.

The above predictions could be investigated in details by evoking paw-shake responses in different phases of the walking cycle with different duration of the stimulus.

**Acknowledgements** This work was supported in part by NSF grant PHY-0750456 to GC, NSF grant IOS 1120291 to DHE, and NIH grants EB012855 and HD032571 and the Center for Human Movement Studies to BIP. The authors would like to thank Dr. David Cofer for his help with developing an earlier version of the cat hindlimb model in AnimatLab.

## References

Angstadt JD, Friesen WO (1993) Modulation of swimming behavior in the medicinal leech. J Comp Physiol A 172:235–248

Barbeau H, Julien C, Rossignol S (1987) The effects of clonidine and yohimbine on locomotion and cutaneous reflexes in the adult chronic spinal cat. Brain Res 437:83–96

Berkinblit MB, Deliagina TG, Feldman AG, Gelfand IM, Orlovsky GN (1978) Generation of scratching. II. Nonregular regimes of generation. J Neurophysiol 41:1058–1069

Berkowitz A (2002) Both shared and specialized spinal circuitry for scratching and swimming in turtles. J Comp Physiol A 188:225–234

Berkowitz A (2010) Multifunctional and specialized spinal interneurons for turtle limb movements. Ann N Y Acad Sci 1198:119–132

Berkowitz A, Roberts A, Soffe SR (2010) Roles for multifunctional and specialized spinal interneurons during motor pattern generation in tadpoles, zebrafish larvae, and turtles. Front Behav Neurosci 4:1–18

Bose A, Kunec S (2001) Synchrony and frequency regulation by synaptic delay in networks of self-inhibiting neurons. Neurocomputing 38–40:505–513
Bose A, Manor Y, Nadim F (2001) Bistable oscillations arising from synaptic depression. SIAM J Appl Math 62:706–727
Boyd SK, Ronsky JL (1998) Instantaneous moment arm determination of the cat knee. J Biomech 31:279–283
Briggman KL, Kristan WB (2006) Imaging dedicated and multifunctional neural circuits generating distinct behaviors. J Neurosci 26:10925–10933
Briggman KL, Kristan WB (2008) Multifunctional pattern-generating circuits. Annu Rev Neurosci 31:271–294
Brown TG (1911) The intrinsic factors in the act of progression in the mammal. Proc R Soc Lond Series B, Contain Pap Biol Character 84:308–319
Butera RJ (1998) Multirhythmic bursting. Chaos 8:274–284
Canavier CC, Baxter DA, Clark JW, Byrne JH (1993) Nonlinear dynamics in a model neuron provide a novel mechanism for transient synaptic inputs to produce long-term alterations of postsynaptic activity. J Neurophysiol 69:2252–2257
Carlin KP, Jones KE, Jiang Z, Jordan LM, Brownstone RM (2000) Dendritic L-type calcium currents in mouse spinal motoneurons: implications for bistability. Eur J Neurosci 12:1635–1646
Carter MC, Smith JL (1986) Simultaneous control of two rhythmical behaviors. II. Hindlimb walking with paw-shake response in spinal cat. J Neurophysiol 56:184–195
Chau C, Giroux N, Barbeau H, Jordan L, Rossignol S (2002) Effects of intrathecal glutamatergic drugs on locomotion I. NMDA in short-term spinal cats. J Neurophysiol 88:3032–3045
Chuan L, Clark JW, Canavier CC, Baxter DA, Byrne JH (2004) Multimodal behavior in a four neuron ring circuit: mode switching. IEEE Trans Biomed Eng 51:205–218
Cofer D, Cymbalyuk G, Reid J, Zhu Y, Heitler WJ, Edwards DH (2010) AnimatLab: a 3D graphics environment for neuromechanical simulations. J Neurosci Methods 187:280–288
Conway BA, Hultborn H, Kiehn O (1987) Proprioceptive input resets central locomotor rhythm in the spinal cat. Exp Brain Res 68:643–656
Crisp KM, Mesce KA (2004) A cephalic projection neuron involved in locomotion is dye coupled to the dopaminergic neural network in the medicinal leech. J Exp Biol 207:4535–4542
Cymbalyuk G, Shilnikov A (2005) Coexistence of tonic spiking oscillations in a leech neuron model. J Comput Neurosci 18:255–263
Cymbalyuk GS, Gaudry Q, Masino MA, Calabrese RL (2002) Bursting in leech heart interneurons: cell-autonomous and network-based mechanisms. J Neurosci 22:10580–10592
Doi A, Ramirez J-M (2008) Neuromodulation and the orchestration of the respiratory rhythm. Respir Physiol Neurobiol 164:96–104
Eccles JC, Eccles RM, Lundberg A (1957) Synaptic actions on motoneurones caused by impulses in golgi tendon organ afferents. J Physiol 138:227–252
Egorov AV, Hamam BN, Fransén E, Hasselmo ME, Alonso AA (2002) Graded persistent activity in entorhinal cortex neurons. Nature 420:173–178
Feldman AG, Orlovsky GN (1975) Activity of interneurons mediating reciprocal 1a inhibition during locomotion. Brain Res 84:181–194
Friesen OW (1994) Reciprocal inhibition: a mechanism underlying oscillatory animal movements. Neurosci Biobehav Rev 18:547–553
Frigon A, Gossard JP (2010) Evidence for specialized rhythm-generating mechanisms in the adult mammalian spinal cord. J Neurosci 30:7061–7071
Fröhlich F, Bazhenov M (2006) Coexistence of tonic firing and bursting in cortical neurons. Phys Rev E 74:031922
Fuentealba P, Timofeev I, Bazhenov M, Sejnowski TJ, Steriade M (2004) Membrane bistability in thalamic reticular neurons during spindle oscillations. J Neurophysiol 93:294–304
Geertsen SS, Stecina K, Meehan CF, Nielsen JB, Hultborn H (2011) Reciprocal Ia inhibition contributes to motoneuronal hyperpolarisation during the inactive phase of locomotion and scratching in the cat. J Physiol 589:119–134

Getting PA (1989) Emerging principles governing the operation of neural networks. Annu Rev Neurosci 12:185–204
Goslow GE Jr, Reinking RM, Stuart DG (1973) The cat step cycle: hind limb joint angles and muscle lengths during unrestrained locomotion. J Morphol 141:1–41
Green CS, Soffe SR (1996) Transitions between two different motor patterns in *Xenopus* embryos. J Comp Physiol A 178:279–291
Gregor RJ, Smith DW, Prilutsky BI (2006) Mechanics of slope walking in the cat: quantification of muscle load, length change, and ankle extensor EMG patterns. J Neurophysiol 95:1397–1409
Grillner S (1981) Control of locomotion in bipeds, tetrapods, and fish. In: Brooks V (ed) Handbook of physiology. Section I. The nervous system, vol II, pp 1179–1236. American Physiological Society, Bethesda
Grillner S (2006) Biological pattern generation: the cellular and computational logic of networks in motion. Neuron 52:751–766
Guttman R, Lewis S, Rinzel J (1980) Control of repetitive firing in squid axon membrane as a model for a neuroneoscillator. J Physiol 305:377–395
Hahn P, Durand D (2001) Bistability dynamics in simulations of neural activity in high-extracellular-potassium conditions. J Comput Neurosc 11:5–18
Hao Z-Z, Spardy LE, Nguyen EBL, Rubin JE, Berkowitz A (2011) Strong interactions between spinal cord networks for locomotion and scratching. J Neurophysiol 106:1766–1781
Hiebert GW, Whelan PJ, Prochazka A, Pearson KG (1996) Contribution of hind limb flexor muscle afferents to the timing of phase transitions in the cat step cycle. J Neurophysiol 75:1126–1137
Hodgkin AL, Huxley AF (1952) A quantitative description of membrane current and its application to conduction and excitation in nerve. J Physiol 117:500–544
Hooper S, Moulins M (1989) Switching of a neuron from one network to another by sensory-induced changes in membrane properties. Science 244:1587–1589
Hounsgaard J, Kiehn O (1989) Serotonin-induced bistability of turtle motoneurones caused by a nifedipine-sensitive calcium plateau potential. J Physiol 414:265–282
Hounsgaard J, Hultborn H, Jespersen B, Kiehn O (1984) Intrinsic membrane properties causing a bistable behaviour of α-motoneurones. Exp Brain Res 55:391–394
Hoy MG, Zernicke RF (1985) Modulation of limb dynamics in the swing phase of locomotion. J Biomech 18:49–60
Hughes SW, Cope DW, Tóth TI, Williams SR, Crunelli V (1999) All thalamocortical neurones possess a T-type $Ca^{2+}$ 'window' current that enables the expression of bistability-mediated activities. J Physiol 517:805–815
Jalife J, Antzelevitch C (1979) Phase resetting and annihilation of pacemaker activity in cardiac tissue. Science 206:695–697
Jing J, Weiss KR (2001) Neural mechanisms of motor program switching in Aplysia. J Neurosci 21:7349–7362
Jordan LM, McCrea DA, Steeves JD, Menzies JE (1977) Noradrenergic synapses and effects of noradrenaline on interneurons in the ventral horn of the cat spinal cord. Can J Physiol Pharmacol 55:399–412
Klishko AN, Cofer D, Cymbalyuk G, Gregor RJ, Edwards DH, Prilutsky BI (2011) Contributions of proprioceptive feedback and CPG to coordination of two-joint muscles during a paw shake response: a computer simulation study. Program No. 920.10. Society for Neuroscience. Society for Neuroscience, 2011. Online, Washington, DC
Klishko AN, Cofer D, Cymbalyuk G, Edwards DH, Prilutsky BI (2012) Paw-shake response and locomotion: can one CPG generate two different rhythmic behaviors? Twenty first annual computational neuroscience meeting: CNS 2012, vol 13(Suppl 1), p 70, BioMed Central Ltd., Atlanta/Decatur, GA, USA
Koshland GF, Smith JL (1989) Mutable and immutable features of paw-shake responses after hindlimb deafferentation in the cat. J Neurophysiol 62:162–173
Kriellaars DJ, Brownstone RM, Noga BR, Jordan LM (1994) Mechanical entrainment of fictive locomotion in the decerebrate cat. J Neurophysiol 71:2074–2086

Laporte Y, Lloyd DP (1952) Nature and significance of the reflex connections established by large afferent fibers of muscular origin. Am J Physiol 169:609–621
Lechner HA, Baxter DA, Clark JW, Byrne JH (1996) Bistability and its regulation by serotonin in the endogenously bursting neuron R15 in Aplysia. J Neurophysiol 75:957–962
Liao JC, Fetcho JR (2008) Shared versus specialized glycinergic spinal interneurons in axial motor circuits of larval zebrafish. J Neurosci 28:12982–12992
Lieske SP, Thoby-Brisson M, Telgkamp P, Ramirez JM (2000) Reconfiguration of the neural network controlling multiple breathing patterns: eupnea, sighs and gasps. Nat Neurosci 3:600
Lockery S, Kristan W (1990) Distributed processing of sensory information in the leech. II. Identification of interneurons contributing to the local bending reflex. J Neurosci 10:1816–1829
MacFadden LN, Brown NA (2007) Biarticular hip extensor and knee flexor muscle moment arms of the feline hindlimb. J Biomech 40:3448–3457
MacFadden LN, Brown NA (2010) The influence of modeling separate neuromuscular compartments on the force and moment generating capacities of muscles of the feline hindlimb. J Biomech Eng 132:081003
Malashchenko T, Shilnikov A, Cymbalyuk G (2011) Bistability of bursting and silence regimes in a model of a leech heart interneuron. Phys Rev E 84:041910
Manor Y, Nadim F (2001) Synaptic depression mediates bistability in neuronal networks with recurrent inhibitory connectivity. J Neurosci 21:9460–9470
Marder E, Bucher D (2001) Central pattern generators and the control of rhythmic movements. Curr Biol 11:R986–R996
Marder E, Calabrese RL (1996) Principles of rhythmic motor pattern generation. Physiol Rev 76:687–717
Marder E, Jorge-Rivera JC, Kilman V, Weimann JM (1997) Peptidergic modulation of synaptic transmission in a rhythmic motor system. Adv Organ Biol 2:213–233 (Elsevier, Bittar EE (ed))
Markin SN, Lemay MA, Prilutsky BI, Rybak IA (2012) Motoneuronal and muscle synergies involved in cat hindlimb control during fictive and real locomotion: a comparison study. J Neurophysiol 107:2057–2071
McCrea DA (2001) Spinal circuitry of sensorimotor control of locomotion. J Physiol 533:41–50
McCrea DA, Rybak IA (2008) Organization of mammalian locomotor rhythm and pattern generation. Brain Res Rev 57:134–146
McVea DA, Donelan JM, Tachibana A, Pearson KG (2005) A role for hip position in initiating the swing-to-stance transition in walking cats. J Neurophysiol 94:3497–3508
Mehta R, Prilutsky BI (2014) Task-dependent inhibition of slow-twitch soleus and excitation of fast-twitch gastrocnemius do not require high movement speed and velocity-dependent sensory feedback. Front Physiol 5:410
Newman JP, Butera RJ (2010) Mechanism, dynamics, and biological existence of multistability in a large class of bursting neurons. Chaos 20:023118
Nichols TR (1999) Receptor mechanisms underlying heterogenic reflexes among the triceps surae muscles of the cat. J Neurophysiol 81:467–478
Nichols TR, Bunderson N, Lyle M (2015) Neural regulation of limb mechanics: insights from the organization of proprioceptive circuits. In: Prilutsky BI, Edwards DH Jr (eds) Neuromechanical modeling of posture and locomotion. Springer, New York (in press)
Paydarfar D, Forger DB, Clay JR (2006) Noisy inputs and the induction of on–off switching behavior in a neuronal pacemaker. J Neurophysiol 96:3338–3348
Pearson KG (2008) Role of sensory feedback in the control of stance duration in walking cats. Brain Res Rev 57:222–227
Pearson KG, Rossignol S (1991) Fictive motor patterns in chronic spinal cats. J Neurophysiol 66:1874–1887
Popescu IR, Frost WN (2002) Highly dissimilar behaviors mediated by a multifunctional network in the marine mollusk *Tritonia diomedea*. J Neurosci 22:1985–1993
Pratt CA, Chanaud CM, Loeb GE (1991) Functionally complex muscles of the cat hindlimb. IV. Intramuscular distribution of movement command signals and cutaneous reflexes in broad, bifunctional thigh muscles. Exp Brain Res 85:281–299

Prilutsky BI, Gregor RJ, Nichols TR (2004) Coordination of cat ankle extensors during the paw-shake before and after self-reinnervation of medial and lateral gastrocnemius muscles. Program No. 69.12. In 2004 neuroscience meeting planner (ed), Society for Neuroscience. Online, San Diego, CA

Prilutsky B, Klishko AN, Farrell B, Harley L, Philips G, Bottasso CL (2009) Movement coordination in skilled tasks: insights from optimization. In Shinohara M (ed) Advances in neuromuscular physiology of motor skills and muscle fatigue, pp 139–171. Research Signpost, Kerala

Prilutsky BI, Maas H, Bulgakova M, Hodson-Tole EF, Gregor RJ (2011) Short-term motor compensations to denervation of feline soleus and lateral gastrocnemius result in preservation of ankle mechanical output during locomotion. Cells Tissues Organs 193:310–324

Prochazka A, Hulliger M, Trend P, Llewellyn M, Durmuller N (1989) Muscle afferent contribution to control of paw shakes in normal cats. J Neurophysiol 61:550–562

Rinzel J (1978) On repetitive activity in nerve. Fed Proc 37:2793–2802

Ritzmann RE, Tobias ML, Fourtner CR (1980) Flight activity initiated via giant interneurons of the cockroach: evidence for bifunctional trigger interneurons. Science 210:443–445

Rossignol S (2006) Plasticity of connections underlying locomotor recovery after central and/or peripheral lesions in the adult mammals. Philos Trans R Soc Lond B Biol Sci 361:1647–1671

Rybak IA, Shevtsova NA, Lafreniere-Roula M, McCrea DA (2006) Modelling spinal circuitry involved in locomotor pattern generation: insights from deletions during fictive locomotion. J Physiol 577:617–639

Shevtsova NA, Hamade K, Chakrabarty S, Markin SN, Prilutsky BI, Rybak IA (2015) Modeling the organization of spinal neural circuits controlling two-joint muscles. In: Prilutsky BI, Edwards DH Jr (eds) Neuromechanical modeling of posture and locomotion, vol 12. Springer, New York (in press)

Shilnikov A, Calabrese RL, Cymbalyuk G (2005) Mechanism of bistability: tonic spiking and bursting in a neuron model. Phys Rev E 71:056214

Smith JL, Zernicke RF (1987) Predictions for neural control based on limb dynamics. Trends Neurosci 10:123–128

Smith JL, Hoy MG, Koshland GF, Phillips DM, Zernicke RF (1985) Intralimb coordination of the paw-shake response: a novel mixed synergy. J Neurophysiol 54:1271–1281

Soffe S (1993) Two distinct rhythmic motor patterns are driven by common premotor and motor neurons in a simple vertebrate spinal cord. J Neurosci 13:4456–4469

Soffe SR (1997) The pattern of sensory discharge can determine the motor response in young *Xenopus* tadpoles. J Comp Physiol A 180:711–715

Syed NI, Winlow W (1991) Respiratory behavior in the pond snail *Lynmaea stagnalis*. J Comp Physiol A 169:557–568

Turrigiano GG, Marder E, Abbott LF (1996) Cellular short-term memory from a slow potassium conductance. J Neurophysiol 75:963–966

Vogels TP, Rajan K, Abbott LF (2005) Neural network dynamics. Annu Rev Neurosci 28:357–376

Wang X-J, Rinzel J (1992) Alternating and synchronous rhythms in reciprocally inhibitory model neurons. Neural Comput 4:84–97

# Index


**A**
Abelew, F.J., 50
Abelew, T.A., 50, 74, 92
Abraham, L.D., 94, 95
Ache, J.M., 184
Ackermann, M., 201
Ackland, D.C., 106, 111
Afferent activity, 23, 303
  force-dependent, 28
  patterns of, 276, 299
  velocity-dependent, 43
Afferent feedback
  control of, 22
  CPG, control of, 28, 29, 34
  effects of, 275
  force-dependent, 43
  length-dependent, 22
  motion-dependent, computation of, 295, 297
  role of, 43
  weights of, 41
Ahn, A.N., 180, 188, 189
Ahn, J., 247
Akay, T., 274
Alexander, R.M., 110, 117
Alexandrov, A.V., 206, 246
Allen, J.L., 217
Alvarez F.J, 36, 50, 52
Anderson, F.C., 106, 154, 299
Andersson, O., 22
Angelaki, D.E., 255
Aoi, S., 202, 226, 236, 275
Arendt, D., 164
Ariano, M.A., 285
Atwood, H.L., 177
Audu, M.L., 299
Auyang, A.G., 104

**B**
Bai, Y.F., 5
Balance, 4, 91, 96, 198, 202
  characteristics of, 261
  control during gait, 263, 264
  control of, 246
  dynamics of, 262
  frontal plane model of, 209
  mechanisms, 212, 247
  multisegmental models of, 205, 206
  simple models of, 203
Baratta, R.V., 285
Barrett, D., 322
Basmajian, J.V., 211
Bauby, C.E., 247
Beer, R., 164
Benolken, M.S., 249
Berger, D.J., 216
Bernstein, N., 211
Besier, T.F., 211
Bey, M., 180
Biewener, A.A., 186
Bingham, J., 75, 82, 107, 218
Bingham, J.T., 5, 14, 106, 209, 276
Biomechanics, 17, 23, 115, 198
Bizzi, E., 215, 229, 239
Blair, J., 316, 324, 326
Blickhan, R., 200
Blümel, M., 165, 167, 169–173, 175, 176, 180, 181
Bonasera, S.J., 70, 73, 74, 81, 84, 86–88, 94
Bondy, B., 275
Boonstra, T.A., 246, 263
Booth, V., 155, 156, 157
Borzelli, D., 211
Bosboom, E.M.H., 180
Bosco, G., 115, 252




B. I. Prilutsky, D. H. Edwards (eds.), *Neuromechanical Modeling of Posture and Locomotion,* Springer Series in Computational Neuroscience,
DOI 10.1007/978-1-4939-3267-2

Bottinelli, R., 181
Brand, R.A., 105, 106, 112, 201, 215
Brenière, Y., 264
Brink, E.E., 74
Brock, O., 4
Brown, I.E., 171, 173, 180, 190, 191, 282, 284, 289, 292
Brown, T.G., 122, 202, 274
Buchanan, T.S., 112, 211
Bullinger, K.L., 79
Bunderson, N., 75, 82, 218
Bunderson, N.E., 5, 7, 8, 75, 82, 91, 106, 107, 112, 114, 116, 218, 276
Burke, R.E., 228
Burkholder, T.J., 13, 75, 105, 107, 108
Burrows, M., 108
Butera, R.J., Jr., 157
Butt, S.J., 39

C
Cabelguen, J.M., 123, 131
Cappellini, G., 229, 239
Carlson-Kuhta, P., 74, 123, 154
Casabona, A., 252
Cat, 22, 36, 75, 89, 240, 280
  hind-limb, 13
  level walking, 41
  model of, 23
Cavagna, G.A., 186, 187, 199, 310
Celichowski, J., 180
Cenciarini, M., 249, 257, 262, 263
Central pattern generator (CPG), 22
Chakrabarty, S., 123
Chang, Y.H., 104
Cheng, E.J., 175
Cheng, G.C., 200
Chen, J., 314
Chiel, H., 164
Chow, C.K., 110
Chvatal, S.A., 215, 216
Clark, A.E., 207
Clark, B.D., 50
Clark, D.J., 216
Cofer, D., 5, 218
Collins, D.F., 22, 23, 274
Collins, J.J., 201, 203
Collins, S.H., 117, 200
Computational modeling, 4, 94
Control, 4, 6, 8, 34, 70, 76, 117, 198, 202, 226, 236
  of body, 264
  of locomotor, 152
Conway, B.A., 231, 240, 274
Cope, T.C., 50, 52
Corana, A., 290
Coros, S., 4
Correa, T.A., 105, 106
Costello, W.J., 177
Crago, P.E., 295
Crowninshield, R.D., 106, 112, 201, 215
Cusumano, J.P., 247

D
Damsgaard, M., 211
Data averaging, 280
d'Avella, A., 215
d'Avella, A., 229, 239
Davies, W.D.T., 247
Davy, D.T., 299
Dean, J.C., 247
De Groote, F., 289
Delp, S.L., 5, 9, 12, 105, 108
De Luca, C.J., 203, 211
de Rugy, A., 208, 216
Dickinson, M.H., 199
Diener, H.C., 204, 205, 212
Dietz, V., 209
Dingwell, J.B., 247
Dominici, N., 229, 239
Drew, T., 229, 239
Duarte, M., 203
Duysens, J., 230, 231, 240, 258, 274, 276, 299, 300
Dynamic model, 117

E
Edman, K.A.P., 173, 286, 287, 288
Ekeberg, O., 23, 226, 227, 231, 235, 275, 276, 301
Energy efficiency, 187
Eng, C.M., 105
Enochson, L., 248
Erdemir, A., 201, 215
Erez, T., 4
Evans, N., 154

F
Farahat, W.A., 188
Ferrell, W.R., 303
Fictive locomotion, 23, 30, 35, 36, 52, 75, 123–126, 152, 155
  analysis of, 135
Fish, F., 310
Fitts, P.M., 4
Flapping propulsion, 310, 312, 322, 327
Force feedback, 70, 73, 74, 91, 93
  heterogenic, 87, 88

Forssberg, H., 240
Forward dynamics, 303
Fox, R., 321
Fujisawa, N., 246
Full, R.J., 105, 180, 188–190, 200
Fyffe, R.E., 36

**G**
Gait, 13, 122, 154, 199–201, 240, 246, 310, 324
  power-optimal, 327
Gareis, H., 292, 293
Gasser, H.S., 117
Gawthrop, P., 203
Geijtenbeek, T., 5
Geyer, H., 5, 117, 200
Gilliver, S.F., 181
Gomi, H., 104
Goodworth, A.D., 247, 249, 263
Gorassini, M., 25, 276, 295, 299, 300, 303
Gordon, A.M., 4, 15, 17, 286, 287
Gordon, M.E., 116
Goslow, G.E., Jr., 187
Gottlieb, G.L., 104
Govind, C.K., 177
Gregor, R.J., 154, 277, 298
Gribble, P.L., 104
Grillner, S., 22, 23, 123, 152, 226, 228, 231, 240, 274
Grottel, K., 180
Guertin, P., 22, 123, 125, 274
Guertin, P.A., 23
Guschlbauer, C., 165, 167, 173, 175

**H**
Haeufle, D.F.B., 173
Halbertsma, J.M., 123
Hall, A.L., 201
Hamade, K., 123
Hasan, Z., 74
Hase, K., 110
Hausdorff, J.M., 264
Hawkins, D., 180
Heckman, C.J., 105, 282, 291, 292
Heglund, N.C., 187
He, J., 275, 276, 279, 282, 284, 292
Henry, S.M., 110, 209
Herr, H., 5
Herr, H.M., 188
Herzog, W., 180, 299
Hettich, G., 246
Hiebert, G.W., 22, 231
Higginson, J.S., 201
Hill, A.V., 117, 164, 171, 287
Hof, A.L., 264
Hoffer, J.A., 275, 299, 303
Hogan, N., 247
Hollister, A.M., 105
Hooper, S.L., 167, 176, 180, 183–185
Horak, F.B., 246, 250
Houk, J.C., 295, 303
Hoyle, G., 177
Hoy, M.G., 24, 200, 281
Hsu, W.L., 104
Huijing, P.A., 105
Hultborn, H., 152
Human, 4, 94, 180, 184, 186, 198, 231
  models of, 198, 227
Huxley, A.F., 4, 177

**I**
Ijspeert, A.J., 5
Inman, V.T., 232
Insperger, T., 205
Interlimb coordination, 226, 238
  control of, 231, 238
  sensory regulation based, 241
Ivanenko, Y.P., 229, 230, 239, 240
Ivashko, D.G., 23, 226, 275, 276, 301, 303
Iwasaki, T., 310, 316, 324, 326

**J**
Jacobson, D.H., 110
Jahromi, S.S., 177
Jain, S., 4
Jami, L., 38, 39
Jankowska, E., 36–39, 122
Jewell, B.R., 164, 171
Johansson, R., 250
Jordan, L.M., 34, 36, 274
Jordan, M.I., 211, 229, 239
Jo, S., 5, 202, 207, 226, 229, 230
Josephson, R.K., 105, 117

**K**
Kawato, M., 104
Kay, B.A., 267
Khatib, O., 4
Kiehn, O., 23, 39, 274
Kiemel, T., 246, 263
Kim, S., 246
Klishko, A.N., 276
Koditschek, D.E., 105, 190, 200
Komuro, T., 177
Konow, N., 81, 95
Kriellaars, D.J., 22

Krouchev, N., 297
Kuo, A.D., 205, 206, 247, 264
Kurtzer, I., 215
Kutch, J.J., 212

**L**
Lafreniere-Roula, M., 125, 128, 130, 228, 231
Lam, T., 22, 30, 45, 104
Lang, F., 177
Lanuza, G.M., 39
Latash, M.L., 104, 229, 239
Lavoie, S., 298
Length feedback, 73, 80, 81, 84, 87, 88, 90, 91, 93, 95, 107
  effects of, 90
Liddell, E.G.T., 4
Lieber, R.L., 108
Limb mechanics, 70, 71, 73, 75, 81, 93, 277
Limb scaling, 11
Limb stiffness, 70, 74, 75, 89, 93, 95, 96
Lin, D.C., 117
Liu, M.M., 299
Liu, M.Q., 201
Li, Y., 246
Lloyd, D.G., 211
Lloyd, D.P., 37
Lobe, G.E., 171, 180, 190, 191
Lockhart, D.B., 112, 204, 212
Locomotion, 4, 22, 23, 41, 42, 81, 84, 91, 93, 95, 123, 199, 229, 276
  phase of, 34, 48
  steady-state, 43, 48
Loeb, G.E., 5, 94, 95, 208, 274, 276, 292, 299, 300, 303
Loram, I.D., 203, 263
Loren, G.J., 105
Low-pass filtering, 265, 279
  muscle, 181
Lundberg, A., 34, 37, 122
Lutz, G.J., 105

**M**
Maas, H., 50, 52, 105, 275, 277, 286, 290, 299, 303
MacGregor, R.I., 141
Macpherson, J.M., 110, 215, 229, 239, 246, 250
Mansouri, M., 207
Manter, J.T., 298
Markert, G., 255
Markin, S.N., 5, 27–29, 41, 42, 53, 123, 128, 218, 226, 275, 276, 287, 295, 297, 301, 303
Martelli, S., 201
Massaquoi, S.G., 5, 202, 207, 226, 229, 230
Matheson, R., 184
Maufroy, C., 23, 264
Maughan, R.J., 181
Maurer, C., 203
McCrea, D.A., 22, 23, 25, 27, 30, 34–36, 123, 125, 128, 130, 152, 153, 228, 231, 274
McGeer, T., 117, 200
McKay, J.L., 106, 110, 112, 208, 211, 215, 216, 217
McLean, S.G., 201
McMahon, T.A., 200, 259
McVea, D.A., 30, 45
Menard, A., 22
Merfeld, D.M., 255
Mergner, T., 246, 263
Mileusnic, M.P., 303
Milner-Brown, H., 211
Mochon, S., 200
Modeling, 13
  activation, 176
  batoid swimming
    equations of motion, 315, 316
    Euler-Lagrange equation, 313
    generalized forces, 314, 315
    propulsive model and motion variables, 311, 312
  computational, 4
  descriptive, 4
  experimental data, 247, 249
  long-term goal of, 275
  mathematical, 103, 105
  musculoskeletal, 200
  PBSt and RF behavior
    non-resetting deletions, 150, 152
    resetting deletions, 145, 147
  PE and damper elements, 167, 169
  predictive neuromechanical, 5
    package, 6
    tools facilitate, 5
    vertical leap, 14, 17
  SE element, 165, 167
  single neurons and neural populations, 139, 141
Moored, III.K., 310, 321
Morasso, P.G., 203
Morgan, D.L., 76, 117
Morris, G., 108
Morris, L.G., 183, 184
Motor control, 70, 72, 73, 76, 78, 94, 103, 250
Movement generation, 180, 186, 191, 199, 200, 226

Muller, H., 208
Mummolo, C., 247
Muscle, 8, 9, 24, 33, 73, 86, 89, 107, 112, 113, 115–117, 164, 189, 211, 213, 260, 261, 274, 275, 277, 282, 291, 294, 295, 300, 315
  complex models, 207
  dynamics, 282
  EMG activity, 41, 276
  force generating capacity of, 10, 11
  gastrocnemius, 84
  Hill-type, 13, 15
  hindlimb, 70
  intrinsic properties and muscle diversity
    energy-absorbing muscle, 188
    extreme low pass filters, 181
    Fmax forces, 186, 187
    mean data, 177, 179, 180
    neuromuscular systems, energy storage in, 187
    passive force scales, 183, 185
    preflexes, 190
    sarcomere FL and FV curves, 176, 177
    spike-number vs. spike-frequency contraction amplitude dependence, 183
  mechanical properties of, 70
  model, 165
    activation, 176
    components, 173, 175, 176
    LCE, FCE, on, 169, 171
    PE and damper elements, 167, 169
    SE element, 165, 167
    VCE, FCE, on, 171, 173
  moment arms of, 11, 12
  MTU lengths for, 299
  PD, 183, 184
  perturbations of, 116, 118
  synergy, 215, 232
    analysis, 229
    concept of, 229
    leg movement control based, 239
  tendon model, 289
  two-joint BFP and RF, 30, 32
  vastus, 86
Muscle intrinsic properties, 176
Musculoskeletal modeling, 295

**N**
Nashner, L.M., 246
Nassour, J., 23
Nataraj, R., 207
Neptune, R.R., 201, 217, 294
Neuromechanical integration, 13, 208, 218
Neuromechanical modeling, 5
  implication for, 6
Neuromechanics, 4
Neuromusculoskeletal model, 17, 226, 227, 232, 236, 275
Nichols, T.R., 13, 22, 34, 39, 70, 72–76, 78, 81–84, 86–88, 90–94, 105, 107
Niedergerke, R., 177
Nishikawa, K., 76
Nishimaru, H., 34, 36
Nomura, T., 226, 240
Norman, R., 299
Nozaki, D., 115

**O**
Obstacle avoidance, 226, 231, 236, 238, 240
  muscle activities for, 229
  performance of, 241
O'Donovan, M.J., 50, 81, 87
Ogihara, N., 202, 226
Ollivier-Lanvin, K., 41
O'Neill, M.C., 111
Optimal gaits, 311, 324, 328, 330
Optimization, 4, 212, 215, 293, 295, 298, 325
  cost function, 8
  dynamic, 290
  techniques, 15, 17, 201
  tools and control strategies, 6, 208
Orlovsky, G.N., 22, 34, 39, 226, 228, 240, 274
Orsal, D., 123, 131
Otnes, R.K., 248

**P**
Pandy, M.G., 106, 154, 299
Park, S., 206, 246
Parnas, I., 177
Parson, J., 310
Pasma, J.H., 246, 263
Passive muscle force, 185
Paul, C., 202
Pearson, K., 226, 227, 231, 235, 275, 276, 301
Pearson, K.G., 22, 23, 30, 35, 39, 45, 73, 74, 104, 231, 274
Perreault, E.J., 75, 95
Perreault, M.C., 22, 30, 45
Perret, C., 123, 131
Perry, J., 200
Peterka, R.J., 203–205, 247, 249, 250, 255–257, 262, 263
Peterson, C.L., 201
Phase resetting, 226, 231, 240
  application of, 235
  use of, 233, 235, 238
Pintelon, R., 248, 249

Poppele, R.E., 115
Postural control, 104, 203, 207, 226
  models of, 203, 213
Posture, 11, 13–15, 17, 82, 83
  advantages of, 90
  control, 231, 233, 236
  effect of changing, 109
  gravity-independent, 184, 186
  limb, 116, 183
  neutral, 108
  predictive neuromechanical models of, 4
Pratt, C.A., 34, 36, 123
Preflex, 190, 191
Prilutsky, B.I., 23–25, 39, 41, 50, 52, 81, 95, 106, 154, 276, 277, 289, 291, 298
Prinz, A.A., 215
Prochazka, A., 25, 94, 95, 226, 227, 258, 276, 295, 297, 299, 300, 303
Proske, U., 76

**Q**
Quevedo, J., 22

**R**
Raasch, C.C., 216
Rack, P.M., 76
Rat, 187, 238
  model, 227, 236, 239, 240
    locomotion in, 236
    posture control, 236
  skeletal model, 227
Rigid body model, 8, 9, 11, 12, 105
Roberts, T.J., 81, 95
Rome, L.C., 105
Ronsky, J.L., 106, 289
Rosenberger, L., 310, 322
Rossignol, S., 22, 34, 36, 41, 83, 94, 231, 274
Ross, K.T., 22, 70, 73, 74, 79, 80, 84, 87, 88, 91, 93, 94
Roth, E., 208
Roy, R.R., 283, 285, 289, 290, 292, 295
Royuela, M., 177
Rubenson, J., 175, 181
Rudomin, P., 303
Ruina, A., 200, 201
Russo, R., 311, 317, 322
Ryall, R.W., 36
Rybak, I.A., 23, 27, 34, 36, 39, 50, 52, 104, 123, 125, 128, 130, 131, 137, 139, 141, 146, 150, 152, 153, 155, 156, 159, 228, 231, 240, 274, 275, 301, 303
Rymer, W.Z., 74, 117

**S**
Sacks, R.D., 25, 283, 285, 289, 290, 292, 295
Safavynia, S.A., 204, 205, 211
Safronov, B.V., 156
Sainburg, R.L., 72
Saito, M., 314
Sandercock, T.G., 105, 282, 291, 292
Santello, M., 94
Sarcomere, 11, 164, 176–178, 287
Sartori, M., 211, 217
Sarvestani, I.K., 5
Schaefer, J., 310, 311
Scheys, L., 105
Schieppati, M., 203
Scholz, J.P., 104, 211
Schomburg, E.D., 230, 231, 240
Schoner, G., 104, 211
Schoukens, J., 248, 249
Scott, S.H., 173, 292
Scovil, C.Y., 106, 289
Scrivens, J.E., 209
Selverston, A.I., 181
Sensitivity, 15, 17, 106, 257, 266
  analysis, 5
  muscle, 105
  quantifying, 13
Sensorimotor control, 202
  balance reveal principles, 203
  mechanisms of, 211
  problem of neural, 202
Sensorimotor feedback, 212
Sensorimotor integration, 205, 241, 274
Sensory integration, 254, 255, 257, 260
  mechanism, 261
  principles, 263, 264
Sensory reweighting, 256, 264, 267
Shao, Q., 211
Sherrington, C.S., 4
Shevtsova, N.A., 23, 25, 27, 123
Shik, M.L., 226, 228, 274
Shoham, S., 280
Siebert, T., 180, 286, 287, 290
Silva, P.L., 81
Silverman, A.K., 201
Simulation, 9, 50, 106, 133, 150–152, 167, 240, 296, 301, 323
  computational tools, 218
  computer, 46, 142, 241, 274
  dynamics, 303
  neuromechanical, 41
  numerical, 312
  predictive, 6
  properties, 232
Sinkjaer, T., 209

Smith, J.L., 74, 75, 123, 154, 297
Smutz, W.P., 105
Sohn, M.H., 201, 212, 215
Southgate, D.F., 106
Spector, S.A., 25, 283, 285, 290
Spinal cord, 22, 39, 71, 74, 94
  different states of, 77
  injury, 88
  levels, 228, 231
  mammalian, 23, 274
Sponberg, S., 76
Srinivasan, M., 200, 201
Stability, 8, 13, 48, 104, 106, 112, 176
  analysis, 9, 209
  dynamic, 264, 267
  level of, 14, 209
  posture, 191, 209
  static, 267
Stahl, V.A., 72, 81, 86, 94
Stance, 81, 209, 246, 263, 264, 266
  configuration, 109
  control model, 249
  experiments, 266
  muscle and tendon stretch at, 187
  muscle passive force, 186
  phase, 25, 27, 28, 48, 267, 279, 281
    acceptance, 93
    duration of, 29, 44, 48, 50
    enhances, 274
    locomotion, 35
    prolongation of, 30, 46
  terminal, 299, 301
Stapley, P.J., 212
Stauffer, E.K., 75
Stecina, K., 30, 45
Steele, K.M., 201
Stein, P.S.G., 123
Stein, R., 76, 211
Stein, R.B., 47, 50, 110
Stepping, 22, 226, 264, 310
  frequency, 265, 267
  motor control of, 240
  on level treadmill, 94
  processes of, 240
  rectangular pulse for, 231
Sternad, D., 208
Stevenson, A.J., 91, 92
Stuart, D.G., 75, 152
Summers, A., 310
Sutherland, D.H., 200
Swimming, 310
  batoid, 310, 321, 322
    dynamical models, 330
    energetic cost of, 324
    gaits of, 326
    modeling, 311, 313
  extreme modes of, 310
  statistical, 323
  steady, 310, 316, 324, 325, 330
Sypert, G.W., 75

**T**
Taga, G., 23, 202, 226
Takeoka, A., 274
Terry, K., 247
Thelen, D.G., 15, 17, 211, 215
Thuma, J.B., 167, 176, 177, 178
Ting, L.H., 5, 14, 91, 92, 106, 110, 112, 201, 204, 205, 207–209, 211, 212, 215–217, 229, 239
Todorov, E., 5, 108, 211, 229, 239
Tomita, N., 5
Torres-Oviedo, G., 91, 209, 211, 216
Toth, T.I., 23
Tuthill, C., 88, 91
Two-joint muscles, 23, 27, 30, 154
  activity pattern of, 152
  locomotor activities of, 125
  motoneurons controlling, 123

**V**
Valero-Cuevas, F.J., 106, 212
van den Bogert, A.J., 201, 280
van der Kooij, H., 203, 205, 206, 246, 254, 257, 262
van Ingen Schenau, G.J., 86
Vogel, S., 322
Vogel, W., 156

**W**
Wadden, T., 23
Walking, 25, 39, 41, 50, 52, 154, 187, 226, 267, 274
  basic patterns for, 229, 230, 240
  behaviors, 227, 240
  biomechanical subtasks of, 201
  cat, 30, 52, 275, 276, 279, 289, 290, 299, 301
    cycle, 286
    neuromechanical simulations of, 41
  cycle of, 25, 45, 47, 48, 280, 286, 290, 293, 300
  function of muscles, 201
  generation of, 232, 233, 236
  kinematics of, 23, 45, 298
  kinetics of, 298
  mechanics of, 199, 277, 292

motions, 200, 231
simulated mechanics, 42
steady, 236
swing phase of, 200
Walter, J., 211, 217
Warren, W.H., 267
Weber, D.J., 276, 278, 280, 299
Welch, T.D.J., 5, 204, 205, 208, 211, 212
Wells, R., 154
Westbury, D.R., 76
Whelan, P.J., 22, 39, 231, 274
Whittington, B., 117
Wickiewicz, T.L., 181
Wilkie, D.R., 164, 171
Wilmink, R.J., 70, 74, 84, 86, 87, 90
Wilson, E., 180
Windhorst, U., 36
Winter, D.A., 110, 203, 211, 232, 246
Wisse, M., 200

**Y**
Yakovenko, S., 23, 47, 226, 227, 275, 276
Yamasaki, T., 226, 240
Yamazaki, N., 202, 226
Yano, M., 5
Young, R.P., 81, 112

**Z**
Zajac, F.E., 10, 15, 72, 95, 108, 116, 205, 211, 216, 282, 284, 289, 292
Zangger, P., 123
Zarrugh, M., 325
Zatsiorsky, V.M., 74, 95, 154, 203
Zehr, E.P., 91, 92
Zernicke, R.F., 24, 200
Zheng, M., 310
Zhong, G., 39
Zupan, L.H., 255

MIX
Papier aus verantwortungsvollen Quellen
Paper from responsible sources
FSC® C105338

If you have any concerns about our products,
you can contact us on
**ProductSafety@springernature.com**

In case Publisher is established outside the EU,
the EU authorized representative is:
**Springer Nature Customer Service Center GmbH**
**Europaplatz 3, 69115 Heidelberg, Germany**

Printed by Libri Plureos GmbH
in Hamburg, Germany